THE HADLEY CIRCULATION: PRESENT, PAST AND FUTURE

ADVANCES IN GLOBAL CHANGE RESEARCH

VOLUME 21

The titles published in this series are listed at the end of this volume.

THE HADLEY CIRCULATION: PRESENT, PAST AND FUTURE

Edited by

Henry F. Diaz

Climate Diagnostics Center,
Oceanic and Atmospheric Research, NOAA, Boulder, CO, U.S.A.

and

Raymond S. Bradley

Climate System Research Center,
Department of Geosciences, University of Massachusetts, Amherst, MA, U.S.A.

KLUWER ACADEMIC PUBLISHERS
DORDRECHT / BOSTON / LONDON

A C.I.P. Catalogue record for this book is available from the Library of Congress.

ISBN 1-4020-2943-8(HB)
ISBN 1-4020-2944-6 (e-book)
ISSN 1574-0919 Advances in Global Change Research

Published by Kluwer Academic Publishers,
P.O. Box 17, 3300 AA Dordrecht, The Netherlands.

Sold and distributed in North, Central and South America
by Kluwer Academic Publishers,
101 Philip Drive, Norwell, MA 02061, U.S.A.

In all other countries, sold and distributed
by Kluwer Academic Publishers,
P.O. Box 322, 3300 AH Dordrecht, The Netherlands.

Cover illustration: an historical account of the trade winds, and monsoons, observable in the seas between the Tropicks, with an attempt to assign the physical cause of said winds, Philosophical Transactions, 16: 153-168 © The Royal Society

Printed on acid-free paper

Printed in the Netherlands.

TABLE OF CONTENTS

Section B: Variability of the Hadley Circulation in the Past

Section C: Causes of Variability in the Hadley Circulation: Past and Future

ACKNOWLEDGMENTS

The editors wish to thank all of the contributing authors for their commitment and perseverance in seeing this effort through. We are very pleased with the results and hope that they all share in this sentiment. We also wish to thank the reviewers of the chapters for giving generously of their time, for their expert evaluation, and for their constructive suggestions. Jon Eischeid's assistance with graphics preparation is gratefully acknowledged. Diana Miller lent invaluable assistance throughout the editing process and in helping to develop the camera-ready manuscripts. The meeting was supported by the U.S. National Science Foundation (NSF), the U.S. National Oceanic and Atmospheric Administration (NOAA), and the International Geosphere-Biosphere Programme, Past Global Changes Project (IGBP-PAGES) This was one of a series of meetings held under the umbrella of the PAGES-PANASH initiative (Paleo-environments of the Northern and Southern Hemispheres), which is one of the major foci of PAGES (see: http://www.pages.unibe.ch/about/research/focus1.html). The illustration on the book cover is taken from Sir Edmund Halley's 1686 paper in *Philosophical Transactions of the Royal Society of London* (vol. 16: 153–168), reprinted with permission of the Royal Society of London. We also acknowledge our own support from the Office of Science (Biological and Environmental Research), U.S. Department of Energy, Grant No. DE-FG02-98ER62604.

PREFACE

Climate dynamicists generally characterize the Hadley circulation in terms of some derived meteorological parameters, such as the mass stream function (the nondivergent part of the flow) or the velocity potential (the divergent circulation), both of which are based on measurements of the three-dimensional wind field. Yet, we know very little about how such indices have varied in the past—beyond the most recent decades. Paleoclimatologists are unable to reconstruct such indices, so long-term reconstructions of the Hadley circulation must be based on indirect characteristics that can be in some way plausibly linked to the dynamics of the system. Reconstructed quantities, such as precipitation amount, position and strength of the trade winds, and the location of the Intertropical Convergence Zone (ITCZ), have all been derived from different types of paleoclimatic (proxy) data, and could be potentially useful in understanding key aspects of past variability in the Hadley system.

While these studies all provide an important perspective on changes that have taken place within the Hadley circulation, there has been little effort to tie individual studies together, to obtain a more comprehensive perspective on the overall variability of the system. With this in mind, a three-day meeting was held at the International Pacific Research Center, Honolulu, Hawaii, in November 2002. This was the first time that climatologists, paleoclimatologists, and modelers had met with the specific goal of examining this important part of the climate system. The goal was to provide a forum for discussion of modern system dynamics, paleoclimatic records that are either currently available or that should be obtained, and potentially important forcing factors. There were three main sessions, focused on current knowledge of the Hadley circulation, on paleoclimatic records that relate to past dynamics of the Hadley circulation, and on prospects for future changes in the context of increased greenhouse gas concentrations.

This book is an outgrowth of that meeting. We hope that it provides a useful overview of current research and a sense of where there is the need for further inquiry. In particular, bridging the disciplinary boundaries between those engaged in paleoclimate research, and those focused on modeling and empirical studies of modern observational records remains a challenge. We believe that further interactions between those with different temporal perspectives on the Hadley and Walker circulations will greatly advance our understanding of the full spectrum of variability of these systems.

Henry F. Diaz, Boulder, Colorado
Raymond S. Bradley, Amherst, Massachusetts

CONTRIBUTING AUTHORS

Keith Alverson
PAGES International Project Office
CH-3011, Bern, SWITZERLAND

Tercio Ambrizzi
Instituto de Astronomía, Geofísica e Ciencias Atmosféricas
Department of Atmospheric Sciences
University of São Paulo, Rua do Matao, 1226
São Paulo, SP, BRAZIL 05508-090

Julie M. Arblaster
National Center for Atmospheric Research
P.O. Box 3000
Boulder, CO 80307-3000, U.S.A.

Raymond S. Bradley
Climate System Research Center
Department of Geosciences
University of Massachusetts
Amherst, MA 01003-9297, U.S.A.

Stephen J. Burns
Department of Geosciences
University of Massachusetts
Amherst, MA 01003-9297, U.S.A.

John C.H. Chiang
Department of Geography and Berkeley Atmospheric Sciences Center
University of California
Berkeley, CA 94720-4740, U.S.A.

Amy Clement
Rosenstiel School of Marine and Atmospheric Sciences
University of Miami
4600 Rickenbacker Causeway
Miami, FL 33149, U.S.A.

Kerry H. Cook
Department of Earth and Atmospheric Sciences
Cornell University
Ithaca, NY 14853-1504, U.S.A.

Aiguo Dai
National Center for Atmospheric Research
P.O. Box 3000
Boulder, CO 80307-3000, U.S.A.

Henry F. Diaz
Climate Diagnostics Center
National Oceanic and Atmospheric Administration
325 Broadway
Boulder, CO 80305, U.S.A.

Michael N. Evans
University of Arizona
Laboratory of Tree-Ring Research
Tucson, AZ 85721, U.S.A.

Dominik Fleitmann
Department of Geological and Environmental Sciences
Stanford University
Stanford, CA 94305-2115, U.S.A.

Michael K. Gagan
Research School of Earth Sciences
The Australian National University
Canberra, ACT 0200, AUSTRALIA

Françoise Gasse
CEREGE, UMR 6635, BP 80, 13545
Aix-en-Provence Cedex 4, FRANCE

Martin P. Hoerling
Climate Diagnostics Center
National Oceanic and Atmospheric Administration
325 Broadway
Boulder, CO 80305, U.S.A.

Gerald Holdsworth
Arctic Institute of North America
University of Calgary
Calgary, Alberta T2N 1N4, CANADA

Alexey Kaplan
Lamont-Doherty Earth Observatory of Columbia University
Palisades, NY 10964, U.S.A.

Athanasios Koutavas
Department of Earth Atmospheric and Planetary Sciences
Massachusetts Institute of Technology
Cambridge, MA 02139, U.S.A.

Jan Kramers
Institute of Geological Sciences
University of Bern
CH - 3012 Bern, SWITZERLAND

Jean Lynch-Stieglitz
School of Earth and Atmospheric Sciences
Georgia Institute of Technology
Atlanta, Georgia 30332, U.S.A.

Augusto Mangini
Heidelberg Academy of Sciences
69117 Heidelberg, GERMANY

Albert Matter
Institute of Geological Sciences
University of Bern
CH - 3012 Bern, SWITZERLAND

Gerald A. Meehl
National Center for Atmospheric Research
P.O. Box 3000
Boulder, CO 80307-3000, U.S.A.

Shoshiro Minobe
Division of Earth and Planetary Sciences, Graduate School of Science
Hokkaido University
Sapporo, 060-0810, JAPAN

Kent Moore
Department of Physics
University of Toronto
Toronto, Ontario M5S 1A7, CANADA

Manfred Mudelsee
Department of Earth Sciences
Boston University
Boston, MA 02215 U.S.A.

Ulrich Neff
Heidelberg Academy of Sciences
69117 Heidelberg, GERMANY

Bette L. Otto-Bliesner
Climate and Global Dynamics Division
National Center for Atmospheric Research
Boulder, CO 80307, U.S.A.

Judith Perlwitz
NASA/GISS at Columbia University
New York, NY 10025, U.S.A.

Roger S. Pulwarty
NOAA-CIRES Climate Diagnostics Center
325 Broadway
Boulder, CO 80305, U.S.A.

Xiao-Wei Quan
NOAA-CIRES Climate Diagnostics Center
325 Broadway
Boulder, CO 80305, U.S.A.

David Rind
NASA/GISS at Columbia University
New York, NY 10025, U.S.A.

C. Neil Roberts
School of Geography
University of Plymouth, Drake Circus
Plymouth, PL4 8AA, UK

Everaldo B. Souza
Instituto de Astronomía, Geofísica e Ciencias Atmosféricas
Department of Atmospheric Sciences
University of São Paulo, Rua do Matao, 1226
São Paulo, SP, BRAZIL 05508-090

Lonnie G. Thompson
Department of Geological Sciences and Byrd Polar Research Center
The Ohio State University, 108 Scott Hall
1090 Carmack Road
Columbus, OH 43210-1002, U.S.A.

Chunzai Wang
NOAA Atlantic Oceanographic and Meteorological Laboratory
4301 Rickenbacker Causeway
Miami, FL 33149, U.S.A.

Warren M. Washington
National Center for Atmospheric Research
P.O. Box 3000
Boulder, CO 80307-3000, U.S.A.

Peter J. Webster
School of Earth and Atmospheric Sciences and School of Civil and Environmental Engineering
Georgia Institute of Technology
Atlanta, Georgia 30332-0340, U.S.A.

T.M.L. Wigley
National Center for Atmospheric Research
P.O. Box 3000
Boulder, CO 80307-3000, U.S.A.

Shang-Ping Xie
International Pacific Research Center and Department of Meteorology
University of Hawaii
Honolulu, HI 96822, U.S.A.

THE HADLEY CIRCULATION: PRESENT, PAST, AND FUTURE
An Introduction

Henry F. Diaz[1] and Raymond S. Bradley[2]
[1]*Climate Diagnostics Center, NOAA, 325 Broadway, Boulder, Colorado 80305, U.S.A.*
[2]*Climate System Research Center, Department of Geosciences, University of Massachusetts, Amherst, Massachusetts 01003-9297, U.S.A.*

Near the turn of the eighteenth century, two eminent British scientists—Sir Edmund Halley (of comet fame) and Sir George Hadley—put forth a conceptual framework aimed at explaining the nature of the observed wind patterns in the tropics. These early attempts at synthesizing what was known about the causes of the large-scale atmospheric circulation were amazingly prescient, providing the first scientific explanations of the phenomenon that now bears Hadley's name (see Chapter 1, by P. Webster). This large-scale meridional overturning circulation spans half the area of the globe, and variability within this system affects the lives of billions of people. Along with the large-scale zonal tropical cells named after Sir Gilbert Walker, the Hadley circulation comprises fundamental regulators of the earth's energy budget. Although the Hadley circulation is a well-known concept, surprisingly little attention has been paid to understanding the variability of the system on long time scales. This book is a step towards addressing the question of the nature and causes of changes in the Hadley circulation on multiple time scales.

Our detailed understanding of the Hadley and Walker circulations is limited to studies based on instrumental measurements within the last century (and in particular over the even shorter record of satellite-derived observations) and insights derived from coupled modeling studies aimed at simulating climate variability in the tropics and over the globe. A longer-term perspective is provided by a variety of proxy records from tropical glaciers, corals, lake and marine sediments, and cave deposits that record century- and millennial-scale changes in climate patterns. Such proxy records of climate can record the long-term variability of the El Niño/Southern Oscillation (ENSO) phenomenon (Diaz and Markgraf 1992, 2000; Markgraf 2001),

H.F. Diaz and R.S. Bradley (eds.), The Hadley Circulation: Present, Past and Future, 1–5.

changes in the Indo-Pacific warm pool (IPWP), and changes in seasonality that tie in directly to the strength of the Hadley circulation.

The conventional view of the Hadley circulation, in terms of its mean meridional circulation, is of two symmetrical cells that migrate seasonally. In reality, the circulation is dominated most of the year by a single cell, with the classical two-cell system only identifiable in the April and October transition seasons. The latitudinal extent and intensity of the Hadley circulation are driven by convective processes in the rising branch of the system, and cooling at higher latitudes. The descending air interacts with transient baroclinic waves in mid-latitude westerlies. In this sense, the Hadley circulation is constrained by extratropical as well as tropical processes (see Chapters 1 and 2, by Webster and Cook, respectively). Analysis of both the zonal mean and regional Hadley cells shows an intensification of the overturning circulation in recent decades (see Chapter 3, by Quan, Diaz, and Hoerling). The changes are seasonally dependent, so it is important to analyze changes in the Hadley/Walker circulation throughout the year. By the same token, zonally averaged changes may reflect strong regional circulation changes, so it is important to understand the mechanisms that lead to these spatially heterogeneous changes—in particular, the spatial variations of sea surface temperature (SST) in the tropics (see chapters by Minobe [Ch. 5]; Wang [Ch. 6]; Webster [Ch. 1]; and Xie [Ch. 4]).

Regional changes in the Hadley circulation have significant impacts on the environment and human populations affected by those changes. Several chapters focus on regional changes in circulation associated with the seasonal insolation cycle on millennial time scales (Gasse and Roberts [Ch. 11], Fleitmann et al. [Ch. 9], Koutavas and Lynch-Stieglitz [Ch. 12], and Chiang [Ch. 16]), or on decadal time scales [Moore, Alverson, and Holdsworth (Ch. 13)], whereas Meehl et al. (Ch. 17), focus on regional aspects of the response of the Hadley circulation in the twentieth century to changes in insolation, greenhouse gases, ozone, and aerosol loading. Two chapters specifically focus on long-term circulation changes in the major ocean basins—the Pacific Ocean in the Moore et al. chapter (Ch. 13), and the Atlantic in Chiang's chapter (Ch. 16). A third chapter (Ch. 7, by Ambrizzi, Souza, and Pulwarty) focuses on decadal circulation changes over South America in the last 50 years, and their relationship to the ENSO phenomenon.

Paleoclimate research generally focuses on specific features related to the Hadley circulation—the Intertropical Convergence Zone (ITCZ), trade winds and subtropical high-pressure regions, or the continental monsoons. However, such studies have generally been unconnected to ongoing research involving climate diagnostic studies of how the climate system works, and modeling studies of tropical variability involving the Hadley and Walker circulation system as a whole. This book represents an effort to bring together in one volume, analyses that bridge these topics in both space and time. Our aim is to foster communication among researchers working on topics related to the Hadley and Walker circulations in a modern climate context, and those in the paleoclimate community working on topics relevant to understanding longer-term variations of these global-scale systems (see, also, Trenberth and Otto-Bliesner 2003). We hope that this interaction will help to develop insights into areas of mutual interest and collaboration among investigators from both of these climate communities.

In a longer-term context, the question of abrupt changes in climate driven from the tropics, and the concept of threshold instabilities, is an important open question for scientific inquiry. Changes in the Hadley/Walker circulation are driven largely by precipitation changes through changes in diabatic heating, forced by changes in sea surface temperature patterns. These interrelationships provide a nexus for paleoclimatic research aimed at reconstructing precipitation changes as well as changes in SST patterns, particularly over the western Pacific warm pool (see Ch. 10, by Gagan and Thompson). The chapter by Evans and Kaplan (Ch. 8) partly addresses the issue raised in the Preface; namely, the problem of developing appropriate paleoclimate indices that can provide a measure of the strength of the Hadley and Walker circulations.

A further area of paleoclimate research that is relevant to improved understanding of the long-term variability of the Hadley/Walker circulation deals with documenting changes in the mean meridional temperature gradient, as several of the authors show (for example, chapters by Rind and Perlwitz [Ch. 14], Koutavas and Lynch-Stieglitz [Ch. 12], and Otto-Bliesner and Clement [Ch. 15]).

The question of future changes in the Hadley and Walker circulation associated with greenhouse warming is a critical one, and is examined from different perspectives by several of the chapters; namely, by Meehl et al. (Ch. 17), Otto-Bliesner and Clement (Ch. 15), and Rind and Perlwitz (Ch. 14). In Chapter 14, by Rind and Perlwitz, and Chapter 15, by Otto-Bliesner and Clement, the results of a suite of climate model simulations under very different climate conditions are discussed, and these studies illustrate the complexity of the response of the climate system to different external forcings and boundary conditions. Some model simulations with increasing greenhouse gases show, on average, an El Niño–like response of sea surface temperatures in the Pacific, which would also intensify the Hadley circulation and, *inter alia*, increase the aridity in subtropical regions. However, others fail to show such changes, and instead simulate slow weakening of the Hadley cells over time, pointing to the need for reconciliation of these differences.

The Hadley circulation during northern winter and spring (December through May) appears to have intensified after 1950, particularly since the 1970s (Quan, Diaz, and Hoerling, Ch. 3; Evans and Kaplan, Ch. 8). This intensification may be a response to more frequent El Niño events during that time. Doubled-CO_2 coupled climate model simulations point towards a weakening of the Hadley circulation from twentieth-century mean values by the end of the twenty-first century. Hence, it is not clear whether the changes observed over the past few decades represent a transient response (or a natural fluctuation) or a response to global warming that is not being adequately represented in the general circulation models (GCMs). The future strength of the Hadley circulation in a changing climate remains uncertain.

Regardless of uncertainties concerning the detailed changes in the Hadley system to be expected under global warming, the critical components of the system that will influence the evolution of the Hadley circulation are clearer. These involve changes in the earth's overall radiative balance, in the monsoon systems, and in the ocean circulation. Changes in these fundamental climate system components, likely will lead to changes in the meridional temperature gradient due to differential warming of the planet, which in turn will modulate the Hadley circulation.

It is essential to develop a predictive capability regarding future changes in the strength and position of the major rainfall belts (the ascending limb of the Hadley cells) and the dry subtropical zones that make up the descending limb of the circulation (an area containing about 70% of the population of the globe). Such changes will in turn modify the extratropics through changes in the meridional fluxes of heat, momentum, and humidity. To accomplish these goals we will need comprehensive monitoring of the climate system, climate modeling, and judicious interpretations of paleoclimatic reconstructions of key components of the Hadley and Walker circulation systems to provide a long-term perspective.

The complexity of the coupled ocean-atmosphere-land system, which is a principal source of uncertainty in projecting the sort of changes we might expect to see in the Hadley circulation in the future, is highlighted in most of the book's chapters. By the same token, the unique contributions of paleoclimatology toward enhancing our understanding of long-term aspects of the Hadley system, are also amply illustrated in this volume. They provide a benchmark to guide further paleoclimate reconstructions of Hadley circulation–sensitive parts of the climate system that might contribute toward improved understanding of natural changes in climate and in key components, such as the Hadley and Walker circulations.

We hope that the talent and multidisciplinary expertise that has been brought to bear on this important scientific question will find a extensive audience, motivating others to delve into the problem of climate variability and its effect on the largest planetary heat engine—the Hadley and Walker circulation system.

REFERENCES

Diaz, H.F., and V. Markgraf (eds.). 1992. *El Niño: Historical and Paleoclimatic Aspects of the Southern Oscillation.* Cambridge, UK: Cambridge University Press, 476 pp.

Diaz, H.F., and V. Markgraf (eds.). 2000. *El Niño and the Southern Oscillation: Multiscale Variability and Global and Regional Impacts.* Cambridge, UK: Cambridge University Press, 496 pp.

Markgraf, V. (ed.). 2001. *Interhemispheric Climate Linkages.* San Diego: Academic Press, 454 pp.

Trenberth, K., and B. Otto-Bliesner. 2003. Toward integrated reconstruction of past climates. *Science* 300: 589–591.

SECTION A

The Role of the Hadley Cell in Atmospheric Circulation

Chapter 1

THE ELEMENTARY HADLEY CIRCULATION

Peter J. Webster
School of Earth and Atmospheric Sciences and School of Civil and Environmental Engineering, Georgia Institute of Technology, Atlanta, Georgia 30332-0340, U.S.A.

Abstract In the most basic terms, the Hadley circulation can be thought of as a large-scale overturning of the atmosphere driven by latitudinal heating gradients, extending roughly between the Tropics of Cancer and Capricorn covering roughly half the surface area of the planet. Rising air occurs near the equator with subsidence in the subtropics. The circulation has a strong seasonal variability. It is manifested during the equinoxes as a pair of relatively weak cells with a common rising zone near the equator termed the Intertropical Convergence Zone (ITCZ). A much stronger cross-equatorial cell marks the solstitial seasons with rising motion in the summer hemisphere and widespread descending air in the winter hemisphere. The meridional circulations are instrumental in determining where tropical rainfall occurs and where the great deserts are located. Variability of the location and intensity of the Hadley circulation (or its regional manifestation such as the monsoons), through the ages has helped shape the history of mankind, either spawning regions of civilization by providing an abundance of rainfall for agriculture or destroying them by periods of drought. The variability of the Hadley circulation is also manifested on interannual times scales as an important component of the waxing and waning of El Niño in the Pacific Ocean, perturbing seasonal climates worldwide.

The Hadley circulation was the first phenomenon to be described by using the physical insight of the natural system emerging out of the Renaissance. Both Halley (1686) and Hadley (1735) provided basic accounts of the physical processes that drive the meridional cells. However, a detailed examination of the phenomenon, using data sets that are now available, shows that many questions cannot be answered in the confines of the Halley-Hadley model. For example, what limits the latitudinal extent of the cells? What is the role of the Hadley system in balancing the planetary heat budget? What factors determine the vertical scale of the Hadley circulation? Why is there considerable longitudinal variability in the strength of the circulation? How does the ocean interact with the atmospheric Hadley circulation and is there an oceanic counterpart?

An attempt is made to answer these questions from a fundamental physical

H.F. Diaz and R.S. Bradley (eds.), The Hadley Circulation: Present, Past and Future, 9–60.

perspective. It is found, for example, that the vertical transport of heat and the heat balance of the tropics in the ascending branch of the Hadley circulation are difficult to understand without considering "undiluted hot convective towers," first considered by Riehl and Malkus (1958). An explanation of the depth of tropical convection follows by consideration of the magnitude of the sea surface temperature (SST) and the stability of the tropical atmosphere. Furthermore, both the atmosphere and the ocean meridional cells contribute to the poleward transport of heat. In the atmosphere, it is the instabilities of the Hadley cell (the middle latitude eddies or waves in the westerlies) that complete the transport of heat towards the poles. It is shown that the atmospheric Hadley circulation drives an oceanic circulation that acts as a negative climate feedback. Finally, a simple model of the combined ocean-atmosphere system is presented that underlines the importance of both the oceanic and the atmospheric Hadley circulations in balancing the heat budget of the planet.

1. INTRODUCTION

In both hemispheres, steady and brisk winds flow across the subtropical oceans in a general westward equatorial direction, converging gradually towards the equator. These fair weather winds, referred to as the trade winds,[1] merge together in a more stormy and rainy environment in a region of low pressure called the equatorial trough. This area was termed the doldrums by early mariners because of the generally sluggish light winds, which provide only slow and uncertain progress or "tread" in addition to frequent storms and squalls. In the vicinity of the equator, the converging warm tropical air is forced to rise, and on reaching the upper troposphere it flows poleward until it reaches the subtropics, where it descends and flows towards the equator in the surface layers. The high humidity of the air rising over the equator results in the largest rainfall rate on the planet. Furthermore, the descending branch of the tropical cell is responsible for the vast dry regions of the subtropical oceans and continental areas, forming great deserts over land and the subtropical high-pressure belt over the oceans. In addition, the strongly seasonal monsoon circulations, which bring copious amounts of summer rainfall to over 60% of the global population, are regional manifestations of these meridional circulations. Collectively, this

[1] It is generally thought that the name "trade winds" has a commercial origin because of their use by earlier traders crossing the Atlantic Ocean for the Americas. However, Philander (1998) suggests that the name has a nautical origin, reflecting the steadiness of the winds coming from the word "tread" which refers to the steady path of a ship's progress.

broad-scale circulation pattern, which dominates the general circulation and covers nearly half of the surface area of the globe, is called the Hadley circulation.

To a significant degree, the Hadley circulation has determined where human civilization has grown and flourished. At the same time, temporal variations in the location and strength of the Hadley circulation may have been responsible for the demise of early civilizations in North Africa, Mesopotamia, and North America (e.g., Weiss and Bradley 2001). On a much shorter time scale, the circulation is intimately linked with the aperiodic El Niño/Southern Oscillation (ENSO) phenomenon, which brings drought or flood to different parts of the tropics on interannual time scales (Bjerknes 1966, 1969). Thus, in terms of its vast scale and its influence on the habitability of planet, the Hadley circulation is arguably the most important of all climate features. Yet, despite the fact that most scientists are familiar with the phenomenon, and while it is the climate feature one first identified as an entity, the physical processes that drive the Hadley circulation and determine its variability are not well understood at all.

With the importance of the Hadley circulation in mind, it is clear that a physical understanding of the system is essential. An absence of this knowledge makes it very difficult both to advance knowledge of the present climatology, thus limiting our ability to predict the variability of the present climate and weather, and to interpret proxy data laid down by past climates. It will be equally difficult in the absence of a thorough physical understanding to determine future configurations of planetary climate in the presence of changing constituents of the atmosphere and through the influences of longer-term solar variability.

1.1 Climatology of the Hadley Circulation

Despite a lack of a thorough physical understanding of the system, the fundamental forces that drive the Hadley circulation are well known. In fact, they are the same factors that drive all motions on the planet: pressure gradient forces arising from differential radiative heating between the equator and the poles and the rotation rate of the planet. Figure 1-1a shows the zonally averaged solar radiation absorbed at the surface of the planet (S, top panel), the long-wave radiation emitted to space by the planet (I_E, middle panel), and the net columnar radiation (R_{TOT}, defined as $S - I_E$, bottom panel) plotted as a function of latitude. Three averages (annual, ANN; boreal summer, June through August, JJA; and boreal winter, December through February, DJF) are plotted. The absorbed solar radiation possesses an extremely strong seasonal variability, with maximum values in the summer subtropics.

The variability of S is in sharp contrast to the distributions of emitted long-wave radiation (I_E), which is relatively constant in the tropics and subtropics but shows substantial seasonal changes at high latitudes. Overall, at all times

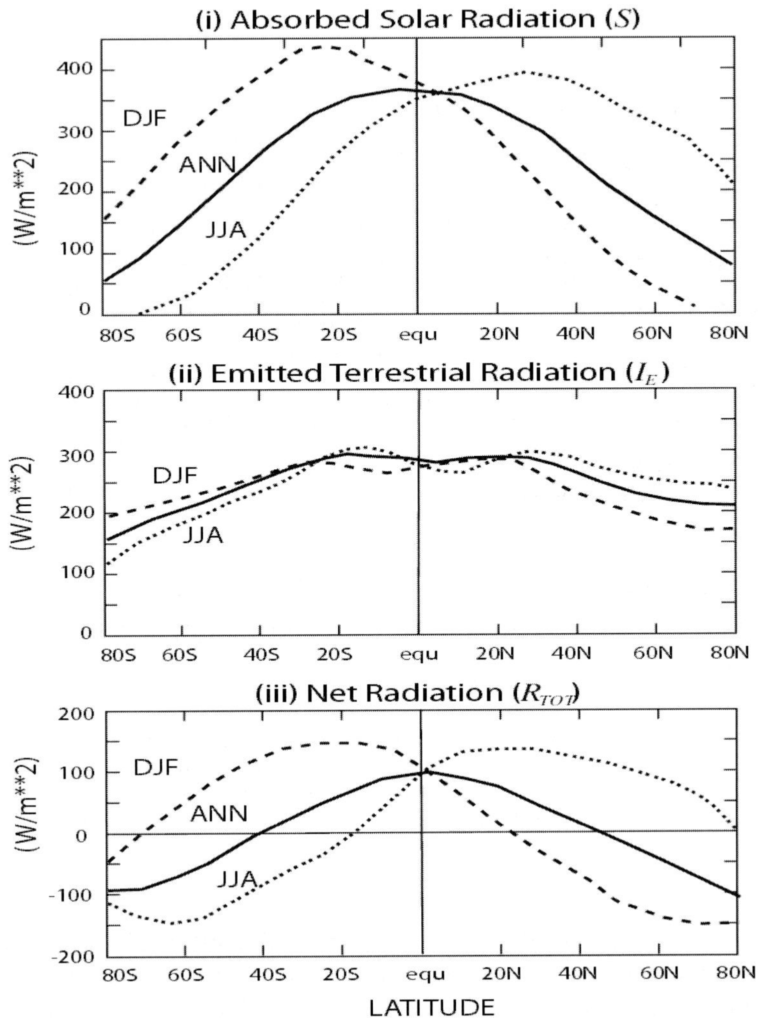

Figure 1-1a. Absorbed solar radiation (S, upper panel), emitted long-wave radiation at the top of the atmosphere (I_E, middle panel), and net radiation (R_{TOT}, bottom panel) as a function of latitude for the annual average (heavy solid line), boreal summer (June–August, JJA, dotted line) and boreal winter (December–February, DJF, dashed line). Note that in the equatorial regions the emitted terrestrial radiation stays essentially constant. The strong seasonal variability in the net radiation comes from solar radiation variability. Units W m^{-2}.

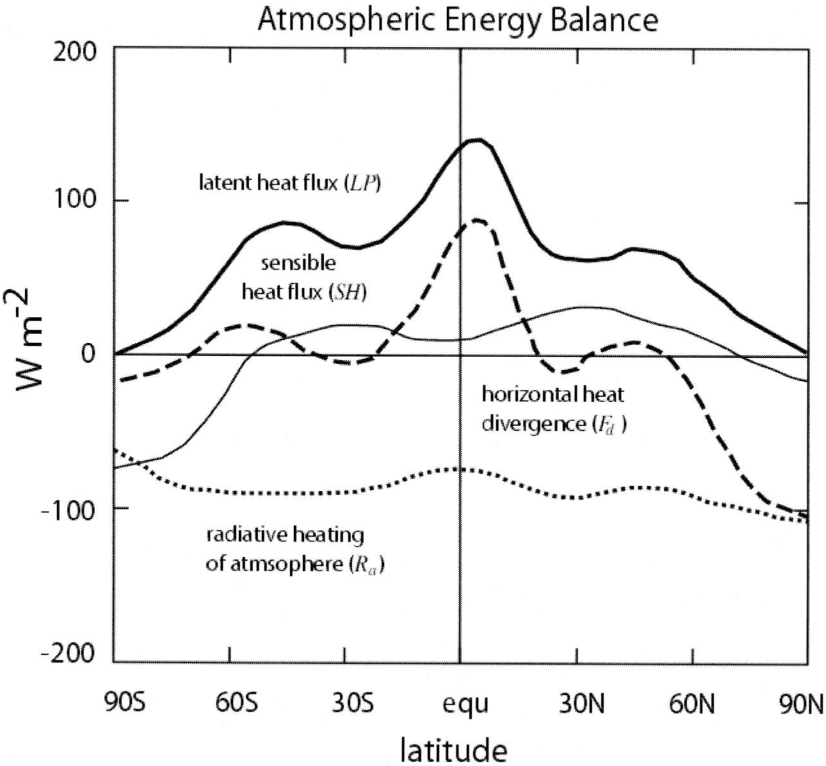

Figure 1-1b. Atmospheric energy budget showing heating due to condensation (*LP*, heavy solid line, where *L* is the latent heat of condensation and *P* is the precipitation rate), sensible heating (*SH*, light solid line), net radiative heating of the column (R_a, dotted line), and the advection of heat by atmospheric motions (F_d, dashed line). Units W m^{-2}.

of the year, there is net heating in the tropics and substantial cooling at higher latitudes, especially in the winter hemispheres. In general, the summer hemispheres show net heating between the equator and the poles and possess a much smaller equator-to-pole radiational heating gradient than in the winter hemispheres. However, it is important to note that while there exists net heating in the tropics and net cooling at higher latitudes, the long-term seasonal average temperature distribution remains relatively constant. This being the case, it is clear that there must be a net transport of heat between the tropics and higher latitudes and that this transport can only be accomplished by fluid motion in the atmosphere and the ocean forced by pressure gradient forces resulting from the radiational heating imbalances. Furthermore, as the winter and summer temperatures of both hemispheres remain much the same from year to year (ignoring for the moment possible global warming effects), the heat transports and radiational cooling to space that provide this long-term equilibrium must occur on subseasonal time scales.

To learn more about the processes that produce the long-term thermal equilibrium of the planet, it is useful to look closely at the columnar atmospheric heat budget. Figure 1-1b shows the latitudinal distributions of the components of the columnar atmospheric energy balance. The energy balance is given by $\Delta F = LP + SH + R_A$, where LP represents the release of latent heat through precipitation P, where L is the latent heating of condensation coefficient, S represents the sensible heating of the atmosphere, R_A is the net heating of the column (the difference between the net radiation at the top of the atmosphere and the surface), and ΔF is the net heating of the column. The units of all of the terms are W m^{-2}. The figure shows substantial heating from the condensation of evaporation at the surface and, to a lesser degree, from the heating of the atmosphere at the planetary surface. The heating occurring at any latitude is balanced by a combination of radiative cooling of the atmosphere to space and latitudinal transports of the residual energy by the fluid motions of the atmosphere and the oceans.

Figures 1-2a–d show latitude-height cross sections of the long-term average meridional stream function (ψ, lower panels) and the zonal wind component ($[\bar{u}]$, upper panels) for the boreal spring (March–May, MAM), JJA, the boreal fall (September–November, SON), and DJF, respectively. Following Peixoto and Oort (1992), the mass stream function ψ is defined by:

$$[\bar{v}] = \frac{g}{2\pi a \cos\varphi}\frac{\partial \psi}{\partial p} \quad \text{and} \quad [\bar{\omega}] = -\frac{g}{2\pi a^2 \cos\varphi}\frac{\partial \psi}{\partial \varphi} \tag{1}$$

where $[\bar{v}]$ and $[\bar{\omega}]$ are the climatological values of the meridional and vertical velocity components averaged around latitude circles. The bars in Equation (1) denote time averages and the square brackets zonal averages around latitude circles. The vertical velocity in Equation (1) is the time rate of change of pressure (*dp/dt*), and as pressure decreases with height a negative value of $[\bar{\omega}]$ denotes upward motion. The terms a and g represent the radius of the planet and acceleration due to gravity and φ is latitude, respectively.

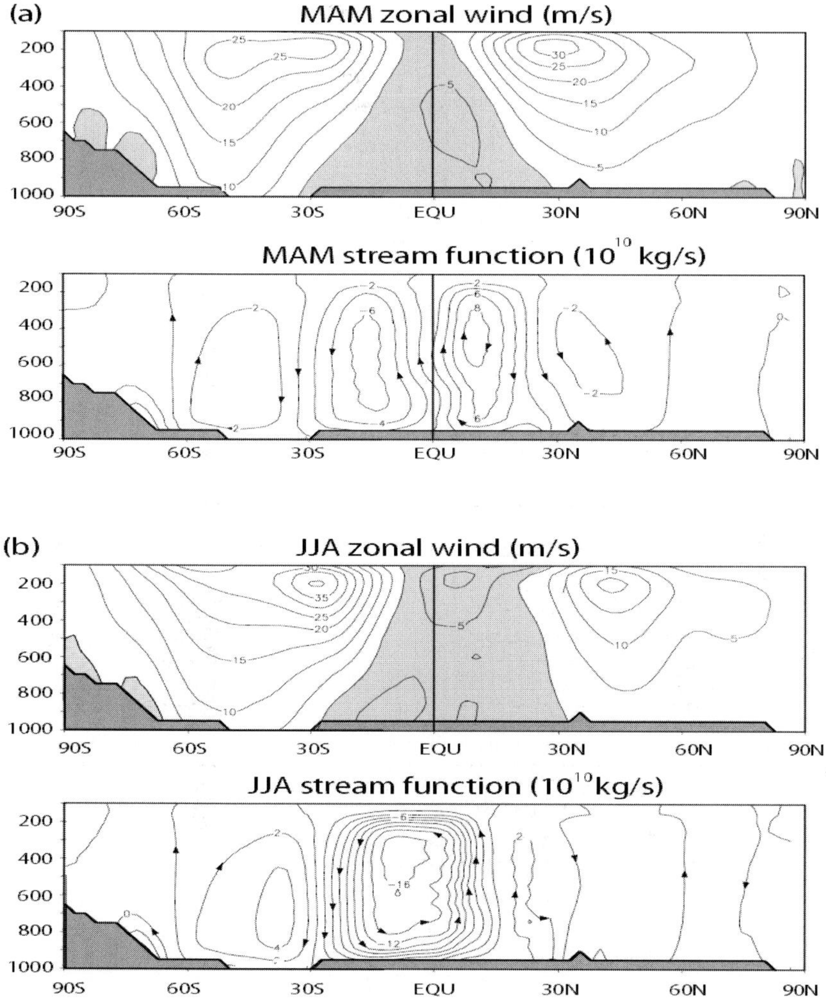

Figure 1-2. The zonally averaged circulation of the atmosphere showing zonally averaged zonal wind component ($[\overline{u}]$, upper panel, m s^{-1}) and mass stream function (ψ, lower panel, 10^{10} kg s^{-1}) for (a) March–May (MAM), (b) June–August (JJA), (c) September–November (SON), and (d) December–February (DJF), plotted against latitude and pressure (hPa). Zonally averaged elevation of the planet and negative (easterly) zonal winds are shaded. The stream function denotes the average flow in the latitude-height plane. Weaker Ferrel cells are evident in all seasons at higher latitudes. NCEP/NCAR reanalysis data were used to compute 50-year seasonal averages.

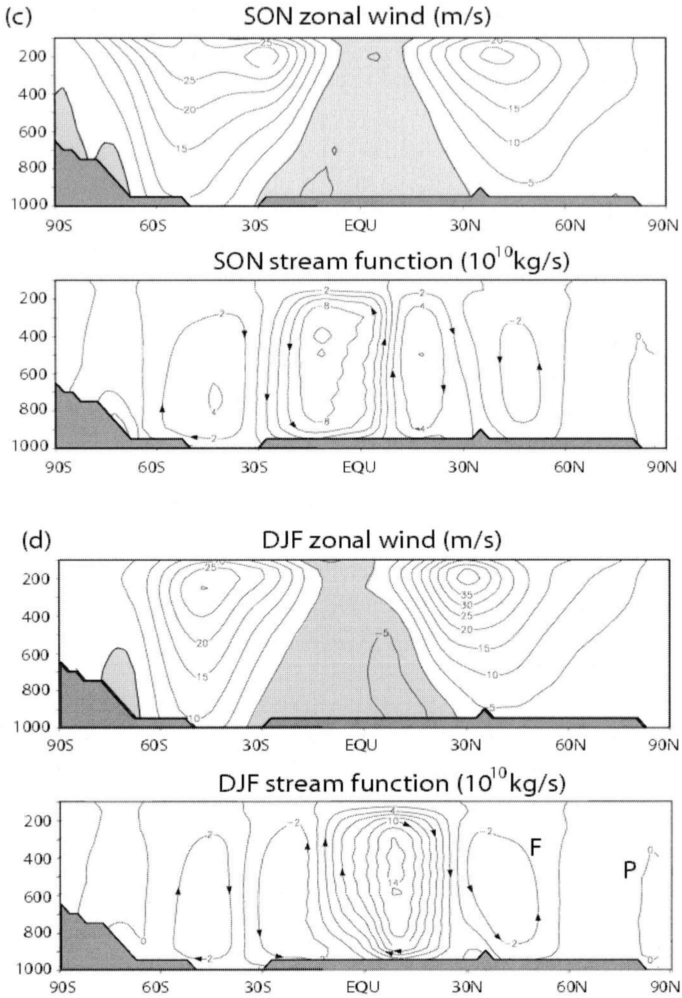

Figure 1-2. Continued.

The meridional mass stream function, as defined in Equation (1), shows the average trajectories of air parcels in the latitude-height plane. Note from Equation (1) that the vertical velocity is proportional to the latitudinal gradient of the stream function. Therefore, a tighter packing of the streamlines in latitude is indicative of a stronger vertical velocity. Similarly, a greater vertical stacking of the streamlines implies a stronger meridional wind component. In total, the zonal velocity component $[\bar{u}]$ and the fields of ψ completely represent the time- and zonally averaged flow in the latitude-height plane. The equinoctial meridional circulations (lower panels of Figs. 1-2a and c) each possess a pair of meridional overturnings between the equatorial regions and the subtropics on both sides of the equator that combine near the equator to produce a relatively narrow region of ascent defining the mean Intertropical Convergence Zone (ITCZ). At all times of the year the vertical extent of the ascent, and thus the vertical scale of the Hadley circulation, is between 12 and 15 km. In the upper troposphere, air diverges away from the ITCZ region towards the poles. The character of the solstitial circulations is very different (lower panels of Figs. 1-2b and d) to that found at the equinoxes. The relatively weak dual cells are replaced by much stronger single cells that straddle the equator, with ascent in the summer hemisphere and descent in the winter subtropics. The vertical velocities of these solstitial cells are roughly twice the magnitude of equinoctial values. In all seasons, weak reverse cells occur in both hemispheres at higher latitudes, but these are faint replicas of the tropical cells.

The zonal wind distributions (top panels, Figs. 1-2a–d) are divided into broad regimes: lower tropospheric easterly regimes in the tropics (the trades) and broad westerlies with maximum strength in the upper troposphere in the mid-latitudes near 300–200 hPa levels. In the solstitial seasons the stronger of the two westerly jet streams is located in the winter hemisphere poleward of the strong upper tropospheric cross-equatorial flow and strong winter hemisphere trades. In contrast, during the equinoctial seasons the mid-latitude westerly jets and the low-level trades are fairly symmetrical about the equator.

Figure 1-3 plots the meridional distribution of rainfall over the oceans and the continental regions separately for DJF and JJA. The two distributions are remarkably different, particularly in the relative positions of the maxima. Precipitation over land occurs much more poleward than over the ocean, suggesting the strong influence of the heated continents in the summer hemispheres. In contrast, the maximum precipitation over the oceans appears to follow the annual variation of the sea surface temperature (SST)[2] (bottom panel). Noting that the precipitation maxima are generally synonymous with the ascending part of the Hadley cell, the large differences in the locations of the land and ocean precipitation maxima suggest that there is substantial variation of the circulation in different longitude bands.

[2] There is an exception to this rule. Except in times of El Niño, the ITCZ, and the warmest SST in the East Pacific, resides in the Northern Hemisphere during both the boreal summer and winter.

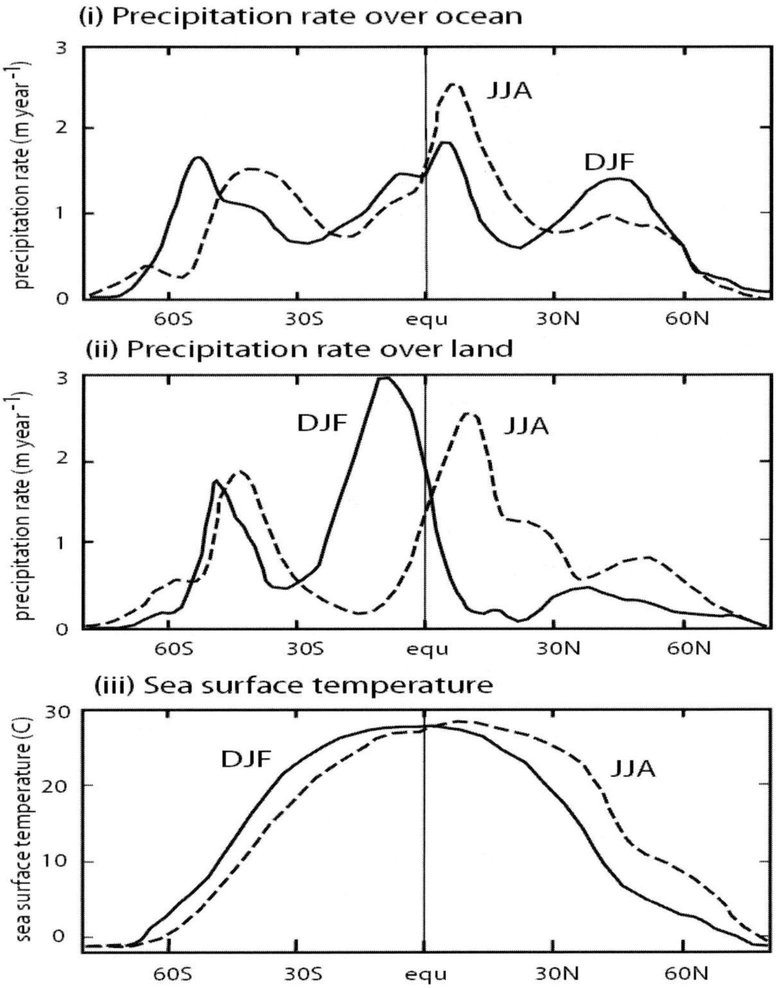

Figure 1-3. Precipitation rate (P, units m year^{-1}) over the global oceans (i) for JJA and DJF and land areas (ii). Ocean precipitation possesses a maximum in the Northern Hemisphere in both seasons due to the much colder water in the Southeast Pacific, which does not support strong convection in either season. Warmer SSTs (panel iii) in the Northern Hemisphere are consistent with increases in precipitation rate during JJA. Precipitation over land occurs at higher latitudes, reflecting the impact of heated continents and the monsoon circulations. Rainfall plots adapted from Jaeger (1976).

1.2. A List of Questions

The climatological features of the Hadley circulation described above are well known and appear in many textbooks (e.g., Peixoto and Oort 1992), with seasonal variability of the Hadley circulations being described in terms of radiative heating and angular momentum arguments. Yet, there are a number of characteristics of the Hadley circulation that are rarely addressed and which raise some interesting questions. For example:

(1) Why is the Hadley circulation limited in latitudinal extent? Why doesn't it extend all the way to the poles rather than being constrained in its location to the tropics and the subtropics?

(2) What determines the vertical scale of the Hadley circulation? Why isn't it 2 or 50 km rather than the observed 12–15 km?

(3) What factors determine the location of the near-equatorial precipitation or, more fundamentally, the region of ascending motion in the Hadley circulation?

(4) Why is the Hadley circulation so much stronger in the solstitial seasons compared to the equinoctial seasons?

(5) Why is there longitudinal variability in the strength of the Hadley circulation?

(6) Is there an oceanic counterpart of the Hadley regime? If so, is it a component of the overall system such that the Hadley cell may be thought of as a coupled ocean-atmosphere phenomenon?

These are essential questions that require consideration if we are to understand the nature of the Hadley circulation and, by extension, the vital components of the ENSO phenomenon and the monsoons.

We begin the investigation of these six questions by discussing the primary physical processes that create the Hadley cell. A historical perspective is developed starting with discovery of the circulation over 300 years ago. The successes and shortcomings of these early explanations are discussed. The dynamical constraints that determine the vertical and horizontal extent of the circulation are explored and the factors that determine where precipitation occurs are elucidated. It is argued that the monsoon circulation may be considered as a modified Hadley circulation. Finally, we develop the concept of an "oceanic Hadley-like circulation" and discuss its cooperative role, teamed with the atmospheric Hadley circulation, in balancing the heat budget of the planet.

2. PHYSICAL NATURE OF THE HADLEY CIRCULATION

2.1. Early Explanations

The first attempts to explain the physical nature of the tropical climate came from two scientists: Sir Edmund Halley (1656–1742) and Sir George Hadley (1685–1786), both of whom were intrigued with the weather observations accumulating from around the world. For over 200 years, mariners had kept careful records of their shipping and trading routes and recorded in some detail the wind, weather, and state of the ocean. These observations were of such widespread interest and economical value that compilations were published and made available to mariners setting out on commercial ventures around the world (Royal Society 1699; Kutzbach 1987). When these observations were viewed collectively, scientists such as Halley and Hadley found persistent features of regional weather that showed both strong seasonality and year-to-year repeatability. First, Halley (1686) and then Hadley (1735) attempted to find physical rules that would explain the patterns of wind and weather they had uncovered.

Utilization of the concept of seasonality of the global wind systems certainly predated the expansion of European commerce in the fifteenth and sixteenth centuries. For example, the peoples of the Pacific Ocean traded over vast distances for centuries prior to the arrival of the Europeans, making use of the trade winds (e.g., Kirch 2000). There is also much historical evidence for the use of the seasonally reversing monsoon winds as the centerpiece of flourishing trade between South Asia and East Africa and throughout most of the Indian Ocean basin as early as 4,000 BP (Warren 1987). These same trade routes, anchored in timing to the annual cycle, are still utilized to the present day. What made the efforts of Halley and Hadley so special, though, was that they were the first to apply the emerging understanding of physical concepts, developed during the Renaissance, to the problem of explaining the direction and seasonal variation of the tropical wind systems.

Sir Edmund Halley was a mathematician, physicist, and astronomer of considerable renown, remembered most for his work on the orbits of comets.[3] However, it is arguable that his astronomical work pales in comparison to some of his other scientific contributions. He provided an astronomical basis for the determination of longitude at sea, a problem of enormous consequence at the time, even though his method was difficult to implement. He solved the problem of representing three-dimensional quantities in two-dimensional form by introducing the concept of contours or isogenic lines. The first chart provided a global view of magnetic variation from true north, providing a method of correction for compass readings made by mariners. He used the same construct to produce the first surface wind climatology. Figure 1-4 shows climatologies of the Atlantic and the Indian Oceans from Halley (1686).

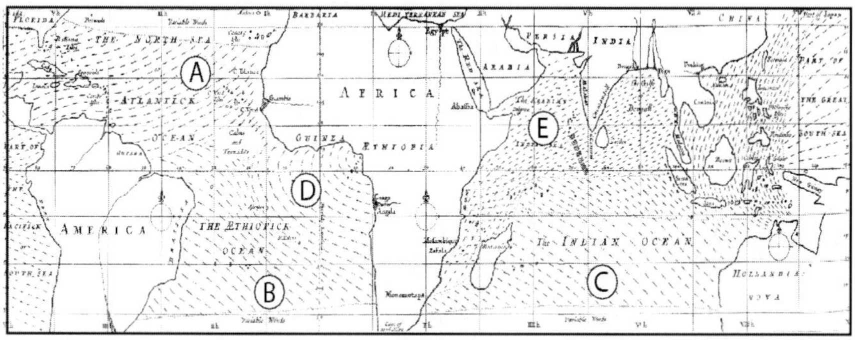

Figure 1-4. Halley's original map of the surface winds in the Indian and Atlantic Oceans. The map accurately represents the constancy of the trade winds throughout the year in the North and South Atlantic Ocean (A, northeast trades; B, southeast trades) and the South Indian Ocean (C, southeast trades). It also depicts where the winds reverse seasonally off the coast of West Africa (D, African monsoon) and in the North Indian Ocean (E, Indian monsoon). Adapted from Halley (1686).

From a meteorological point of view, Halley's most enduring work is an identification of the basic forcing mechanisms that drive the trade winds and the monsoon circulations shown in Figure 1-4. It was the ". . .

[3] Halley did forecast the return of the comet that bears his name. In fact, it has reappeared on time four times since his initial forecast, although the first appearance occurred after his death. In addition to the scientific works described briefly here, he was also a diplomat, a professor of geometry at Oxford University, and a captain in the Royal Navy. In addition to all of these accomplishments, he performed one task of enormous importance. He convinced Isaac Newton to consolidate his scientific research and publish *Philosophiae Naturalis Principia Mathematica.* And, in fact, he even paid for the publication costs (Bryson 2003).

action of the Suns Beams upon the air and water . . ." which produced a dynamic effect on the property of the overlying surface such that ". . . according to the Laws of Staticks, air which is less rarified or expanded by heat [i.e., over the colder regions] . . . must have a Motion towards those parts . . . which are more rarefied [i.e., air over warmer regions] . . . to bring it to an Equilibrium . . . ." (Halley 1686). Essentially, Halley had described the dynamic forcing of air by differential radiative heating. Applying this general principle to the observed surface winds, Halley went on to hypothesize that ". . . as the cold and dense Air, by reason of its greater Gravity, presses on the hot and rarified [i.e., low-level convergence of air resulting from differential heating] . . . [it] must ascend . . . and being ascended it must disperse to preserve the Equilibrium [i.e., upper air divergence to conserve mass] by a contrary current which must move from those parts where the greatest heat is: So that by a kind of Circulation, the North-East Trade below, will be attended by a South-Westerly above, and the South-Easterly with a North West Wind above . . . ." .Halley had managed to describe the manner in which differential heating would result in fluid motion. Although he appeared to omit an explanation of why there was an easterly component of the surface trade wind regime, the anticipation of an upper-level return flow was a remarkable achievement predating upper-air observations by almost 300 years.

A more complete explanation for the orientation of the tropical wind fields was left to Sir George Hadley, a lawyer and meteorologist, after whom the tropical meridional circulation was finally named. Noting that ". . . the causes of the general trade-winds have not been fully explained by any of those who wrote on that subject . . . " Hadley (1735) invoked the impact of the ". . . diurnal motion of the Earth . . ." as the factor that determined the directions of both the surface and upper tropospheric winds. He noted that ". . . setting aside the diurnal motion of the Earth, the tendency of the air would be from every side towards that part [of the planet] where the Sun's action is most intense at the time, and so a N.W. [northwesterly] wind would be produced in the morning and a N.E. [northeasterly] in the afternoon, by turns, on this side of the parallel of the Sun's declination, and a S.W [southwesterly] and S.E. [southeasterly] on the other . . . ." Hadley continued his deduction by supposing the atmosphere was at rest relative to the rotating planet and letting the air be forced to move from higher latitudes to the equator by the same forces proposed by Halley. This exercise allowed him to conclude that ". . . as the surface of the Earth at the equator moves so much faster [i.e., greater tangential velocity] than at the Tropicks [from which it follows that] . . . the air as it moves from the Tropicks towards the equator, having less velocity than parts of the Earth it arrives at, will have a relative motion contrary to the diurnal motion of the Earth, which combined

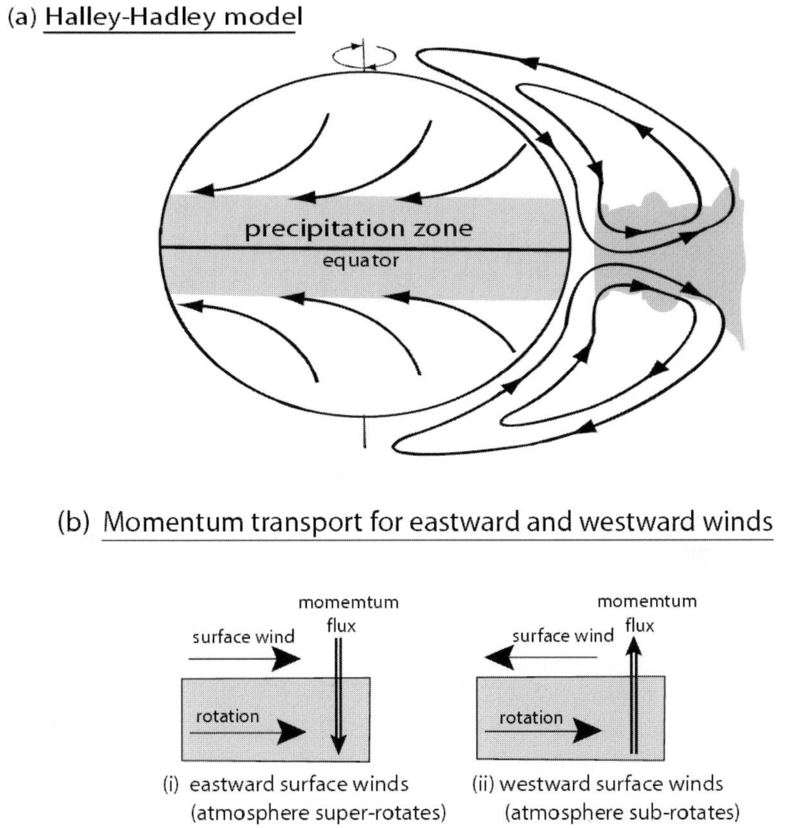

Figure 1-5. (a) The Halley-Hadley model showing one direct cell in each hemisphere. Only one precipitation zone would be produced by this system, located in the vicinity of the equator. The impact of rotation is to turn the equatorial surface flow towards the west over the entire globe. The resultant easterly winds would mean that the atmosphere was rotating at a rate less than the rotation of the planet (i.e., subrotating). (b) Schematic diagram of the transfer of momentum between the atmosphere and the solid earth for westerly (eastward) and easterly (westward) winds. If the winds over the entire planet were one-signed (as in panel a), the transfer of momentum would also be of one sign. Conservation of angular momentum would insist that the atmosphere would accelerate at the expense of the solid earth (or vice versa). To account for the observed quasi-steady state of the kinetic energy of the atmosphere, and to explain the existence of extratropical westerlies, a series of additional cells were proposed.

with the motion towards the equator, a N.E. wind will be produced on this side of the equator, and a S.E. on the other . . . becoming stronger and more and more easterly and be due east at the equator . . . ." Noting that the theoretical winds, so produced, were much stronger than observed, he invoked frictional dissipation at the surface to reduce its speed. Hadley also extended the argument to show the upper tropospheric return flow, noting that the raised air at the equator had a greater tangential speed than at higher latitudes and then, under the action of Halley's forces, the air would have an ever increasing westerly component as it moved towards the poles. Noting that the ascending air at the equator will move poleward, he used the same argument to explain the generation of upper atmospheric westerlies. A schematic of the Halley-Hadley model is shown in Figure 1-5a.

Together, Halley and Hadley described the physical essence of the mean meridional circulations of the tropics.[4] Halley (1686), by extension, also provided an explanation for the regional monsoons and the role of differential heating between the land and the ocean in producing seasonal reversibility of the winds. In effect, Halley's explanation also extends to zonally asymmetric circulations along the equator as being forced by longitudinal gradients of thermal forcing. An example of such a circulation is the surface flow along the equator, which converges in the western Pacific Ocean and rises in the vicinity of the Indonesian Archipelago, forced primarily by heating gradients associated with the ocean warm pool in the western Pacific Ocean and the colder upwelling waters of the eastern Pacific Ocean. These are the circulations responsible for the macro-scale correlations found in atmospheric variables by Walker (1924, 1928) such as the Southern Oscillation. The zonal circulations were later called the "Walker circulations" (Bjerknes 1969).

2.2. Higher-Latitude Issues of the Halley-Hadley Theory

While the Halley-Hadley theory explains a significant portion of the variance of the mean seasonal state of the tropical atmosphere, it does not explain the entire climate state of the higher latitudes. The theory calls for two cells, each occupying an entire hemisphere, with rising motion occurring near the equator and descent over the rest of the planet, in essence providing one maximum in precipitation near the equator. In contrast, Figure 1-2 shows that the Hadley cell is confined to the tropics. Figure 1-3 shows, in

[4] Perhaps a fairer name for the circulation would have been the Halley-Hadley circulation. After all, Hadley's application of the conservation of angular momentum rested completely on Halley's explanation of the thermal forces that produce an equatorward motion of surface air parcels and his understanding of the conservation of mass. It is not known to this author who was first to call the circulation the "Hadley circulation."

addition to precipitation aligned with the rising branches of the Hadley cell, maxima in the extratropics as well. Furthermore, if a Hadley cell is driven just by surface pressure gradient forces moving air towards the equator, then all surface winds would have an easterly component as they converge towards the equator, while the return flow would be completely westerly as air moves poleward. Three questions remain: (1) How can we explain the surface westerlies at higher latitudes within the confines of the Halley-Hadley theory? (2) Why don't the observed cells extend all the way to the poles? and (3) What causes the secondary precipitation maxima in the extratropics? The first question was answered to a large degree by Hadley (1735). Answers to the second and third questions depend on the instability of the Hadley circulation itself.

The law of conservation of angular momentum requires that the total angular momentum of the earth system remains constant. That is, the combined angular momentum of the atmosphere, solid earth, and oceans is fixed[5]. Because of frictional effects it is possible for momentum to be transferred between the atmosphere and the ocean/land interface. If the surface winds were to have an easterly component everywhere, the atmosphere would gain westerly momentum from the solid earth/ocean at all points on the globe. That is, the atmosphere would gain angular momentum while the solid earth/ocean would lose angular momentum as shown in Figure 1-5b. However, as the length of day (a measure of the angular momentum of the solid component of the system) remains constant within narrow bounds, as does the integrated kinetic energy of the winds, then the time-average net transfer of angular momentum between the atmosphere-ocean-land and the atmosphere must equate to zero. This balance can occur only if there are surface westerlies as well as easterlies.

The necessity for the existence of both easterlies and westerlies at the surface of the planet can be proved by using a very simple argument. Consider a featureless planet without topography. Assume that the average eastward force per unit area by the atmosphere on the surface of earth at some latitude ϕ is given by $F(\varphi)$ that may be written as $F(\varphi)=-\alpha[\overline{u}]$ where α is some frictional coefficient. Then the total torque on the atmosphere about earth's axis is:

$$aF(\varphi)\cos\varphi = -a\alpha\left[\overline{u}\right]\cos\varphi \tag{2}$$

5 Strictly, the moon is also part of the earth system and should be included in our argument. However, for simplicity, we ignore the moon, noting only that through tidal friction effects that the radius of the orbit of the moon and rotation rate of the planet are slowly changing but on time scales that are not of concern to us here.

Noting that the area of a zonal strip between latitudes φ and $\varphi + d\varphi$ is $2\pi a^2 \cos\varphi \, d\varphi$, we can write an expression for the total torque in a strip of width $d\varphi$ as:

$$2\pi a^3 F(\varphi)\cos^2\varphi \, d\varphi \tag{3}$$

The total torque of the surface winds on the solid earth can be calculated as the integral of expression (3) across all latitudes. For long-term equilibrium the total torque must equal zero—i.e.,

$$\int_{-\pi/2}^{\pi/2} F(\varphi)\cos^2\varphi \, d\varphi = -\int_{-\pi/2}^{\pi/2} \alpha\left[\overline{u}\right]\cos^2\varphi \, d\varphi = 0 \tag{4}$$

Between Equations (2) and (4) and noting that $\cos^2\varphi$ and α are positive, it is clear that for the net torque to be zero, not only must $F(\varphi)$ change sign but the net torque exerted by westerly winds must equal the net torque by the easterly winds. So there must be both surface easterlies and westerlies. Therefore, a pair of Hadley cells with surface easterlies everywhere and westerlies aloft (Fig. 1-5a) cannot produce a wind system that remains constant in time.[6]

It is clear from Figure 1-2 that the Hadley circulation does not extend beyond the subtropics, in contrast to the Halley-Hadley model shown in Figure 1-5a. To explain the surface westerlies, it was argued that a second cell was required of the opposite sign. This is the mid-latitude Ferrel cell, marked as "F" in Figure 1-2d. The Ferrel cell has a region of common descent with the Hadley cell in the subtropics and rising motion at higher latitudes close to the extratropical precipitation maxima shown in Figure 1-3. It was conjectured that the turning of the poleward lower tropospheric wind was the reason for the prevalent surface westerlies. Weak polar cells ("P" in Fig. 1-2d) can also be identified. These polar cells were used to explain the weak polar easterlies. In total, six cells (two Hadley, two Ferrel, and two polar cells) were needed to describe the mean structure of the atmosphere.

Hadley (1735) had a simpler idea. He noted that, "The same principle as necessarily extends to the production of West Trade-Winds without

---

[6] In this argument we have assumed a featureless planet. If there are mountains, then the torque that the mountains impart on the atmosphere must be taken into account. Thus, in Equation (4), there would be an extra term on the left-hand side of the equation. In the Northern Hemisphere the frictional torque imparted by mountains is about 25% of the total frictional torque, but comparatively much smaller in the Southern Hemisphere (Peixoto and Oort 1992).

the Tropicks. . . ." [i.e., the extratropical westerlies]. Noting that the heated air at the equator will rise ". . . to make room for the air from cooler parts . . ." it will ". . . spread itself abroad over the other air and so its motion must in the upper regions must be to the north and south of the equator. Being got up at a distance from the surface of the Earth, it will soon lose great part of its heat and therefore acquire density and gravity to make it approach the surface again . . . ." As was explained earlier, this subsiding air will have a strong westerly component so that the surface ". . . thereby become a westerly wind . . . ." Hadley's arguments are seductive. They are consistent with the need for there to be surface easterlies and westerlies on the surface of the globe in order to balance the eastward and westward torques. Furthermore, they remove the necessity of having to explain the existence of surface westerlies by the Coriolis turning of the weak Ferrel cell poleward flow in the mid-latitudes.

However, neither the Ferrel cell nor Hadley's concept of descending westerly momentum can explain the latitudinal limitation of the Hadley cell or the existence of eastward propagating transient systems that dominate the weather of higher latitudes. Clearly, other arguments are necessary to explain the observed nature of the general circulation.

2.3. The Hadley Circulation and the Poleward Transport of Heat

During the last few decades it has become clear that the indirect Ferrel cell is merely a by-product of the very strong poleward transport of energy by the waves in the westerlies. These waves result from the process of baroclinic instability that is, in essence, the instability of the Hadley circulation itself. Charney (1947) and Eady (1949) independently developed the concept of baroclinic instability. Baroclinic instability arises from the rapid growth of perturbations in an environment where the basic flow varies in a particular manner in the environment. Of particular importance is the variation of the zonal flow $[\overline{u}]$. It is found that if positive shear exists (e.g., surface westerlies or easterlies surmounted by stronger westerlies) and if it is located sufficiently far from the equator, wave-like disturbances will grow rapidly that receive their energy from the basic flow itself. Using basic flows similar to what is observed in the subtropics, Eady and Charney showed that the fastest growing modes were the same scale as those observed in the extratropics: the familiar transient waves associated with extratropical weather. If the shear is strong enough, small motions within the flow will grow rapidly, transferring potential energy to kinetic energy. As they grow at the expense of the background flow, they reduce the temperature gradient and, as all instabilities do, tend to stabilize the system. During this process of stabilization, the waves transfer heat and momentum polewards, allowing the energy balance of the planet to be achieved.

Considered in tandem, the Hadley circulation and the extratropical baroclinic waves are synergetic. The Hadley circulation arises as a means of transferring excess tropical heat towards the poles in an attempt to achieve planetary thermal equilibrium, as was discussed earlier. As the planet is rotating, the upper tropospheric poleward flow becomes increasingly westerly as latitude increases. For the Hadley circulation to accomplish the necessary poleward heat transfer, the meridional temperature gradient must increase to values much larger than are observed. The well-known thermal wind equation:

$$\frac{\partial[\bar{u}]}{\partial z} \approx -\frac{g}{f[\bar{T}]a}\frac{\partial[\bar{T}]}{\partial \varphi} \tag{5}$$

shows the consequences of an increasing latitudinal temperature gradient. Here $[\bar{u}(z,\varphi)]$ is the zonally averaged zonal wind component; $[\bar{T}(z,\varphi)]$ is the temperature of the atmosphere; and g, f, and a are the gravitational constant, the Coriolis parameter ($2\Omega\sin\varphi$ where φ is latitude), and planetary radius, respectively. The equation shows that the vertical shear ($\partial[\bar{u}]/\partial z$) will increase as the latitudinal temperature gradient (i.e., as $\left|\partial[\bar{T}]/\partial\varphi\right|$ increases). This is the vertical shear in the westerly winds to which Eady and Charney referred and which is unstable to small perturbations. Viewed broadly, one can think of the baroclinic waves as more efficient transporters of heat than the Hadley circulation, tending to stabilize the flow and reduce the meridional temperature gradient and the vertical shear. Also, rising motion in the low-pressure sector of the baroclinic waves produces the mid-latitude precipitation maxima seen in Figure 1-3.

Two factors could alter the shear and possibly the location of baroclinic waves: the rotation rate (Ω s^{-1} in the thermal wind equation 5) and magnitude and form of the factors that might alter the latitudinal temperature gradient. Hunt (1979) used a general circulation model (GCM) to test the sensitivity of Hadley circulation to rotation rate, varying it upwards and downwards by a factor of 5. The Hadley circulations produced in these experiments were found to be tightly bound about the equator for higher rotation rates, but more expansive in latitude than they are at present for slower rotation rates. In addition, the band of extratropical waves also moved their location correspondingly equatorward. These relocations are consistent with larger values of vertical shear being closer to the equator, which is evident from Equation (5). Although changes in rotation rate on the scale considered by Hunt have not occurred in the more recent history of the planet, the experiments illustrate the role rotation plays in determining the structure of the tropical climate.

The second possible factor, the alteration of the latitudinal temperature gradient, allows for more plausible scenarios. Changes in the latitudinal distribution of albedo were probably associated with the oscillation of high-latitude ice margins and variations of vegetation between the ice ages and the interglacials. Such variations may influence the magnitude and location of $\left|\partial[\bar{T}]/\partial\varphi\right|$ and with it the location of the zone of extratropical disturbances. For example, Shin et al. (2003) used a coupled general circulation model to study the climate at the last glacial maximum (LGM) and found a stronger Hadley cell, which was attributed to the stronger latitudinal SST gradient, and enhanced precipitation in the extratropical storm tracks. However, the stronger Hadley circulation appears at odds with interpretations of proxy data for the last glacial maximum that suggested much drier tropics, at least in the western Pacific Ocean/Indonesian region (e.g., Webster and Streten 1978). Probably more thought and more numerical experiments are warranted to explore this important topic.

2.4. Disposition of Energy in the Hadley Circulation

In addition to the important role the Hadley circulation plays in transporting heat poleward, it is a critical player in the vertical transport of heat in the tropics. The heat balances displayed in Figure 1-1 might suggest a rather simple energy balance picture. An excess of heating occurs in the tropics (because of greater solar heating compared to long-wave cooling) and a deficit occurs at higher latitudes (greater long-wave cooling than solar heating). Balance is accomplished by oceanic and atmospheric meridional heat transports. Figure 1-6a shows a schematic of the processes listed above, including estimates of their respective magnitudes. Balancing the heat budget of the tropics is complicated because heating and cooling of the tropical atmosphere take place at different vertical levels. The three main heating and cooling processes are:

(1) *Radiative heating*: As the atmosphere is largely transparent to incoming solar radiation, most of the energy that arrives at the top of the atmosphere is reflected by clouds or the planetary surface, absorbed in the upper layers of the ocean or close to the surface of the

land areas, and, at the long-wave end of the solar spectrum, absorbed by atmospheric water in ice, liquid, or vapor form. About 50% of the incident radiation at the top of the atmosphere is absorbed at the planetary surface. Solar heating of the atmosphere occurs indirectly by the transfer of heat at the surface through the turbulent transfer of sensible heat (generally large over land and smaller over the oceans) and latent heat through evaporative processes (generally smaller over land than the oceans). The atmospheric column is also heated by long-wave radiation from the surface that is offset to a large degree by the heating of the surface by downwelling atmospheric long-wave radiation. The latent heating is not realized immediately except as an increase in the specific humidity of the boundary layer.

(2) *Condensational heating:* As the moist air converges towards the equator under the action of pressure gradient forces, it is forced to rise and as it does so cools, condensing water vapor and producing deep convective clouds. Overall, the trade wind regime acts as a vast solar collector, with most of the energy being released in the ITCZ. In this manner, the ITCZ acts as the boiler box of the tropical heat engine. Through these processes, solar heating of the surface has relocated, in effect, to the upper troposphere where it can be radiated efficiently to space, be advected to higher latitudes, or some combination of both processes.

(3) *Radiative cooling:* With little water vapor or other greenhouse gases in the atmosphere above the tropopause, the planetary system can cool rather effectively to outer space. From Figure 1-1, we have noted that the cooling to space is fairly constant seasonally and decreases from the equator to the poles by about a factor of 2. It is interesting to observe that the heat loss to space by radiative processes in the upper troposphere is about two and a half times the net heat transferred poleward by atmospheric motions. The factors that determine the ratio of the two forms of heat loss by the tropics turn out to be determined by how quickly the atmosphere of a planet cools compared to how quickly heat can be moved from one location to another by atmospheric and oceanic motions.

Consider the situation where the tropical atmosphere transfers heat upwards in clouds as shown in Figure 1-6a but where this heat is lost at a rapid rate to space through radiative cooling. In this situation, none is available for transfer to higher latitudes. On the other hand, consider the same situation but now assume that the dynamical transports are so strong that an atmospheric parcel does not have time to cool radiatively. In the first instance, the atmospheric column would be said to be in radiative equilibrium with outer space. In the second instance, dynamics completely overshadow radiative effects. These effects may be termed radiative and dynamic limits, respectively. One might ask, Why aren't all of the atmospheric columns between the atmosphere and the pole in radiative balance, thus negating the need for the horizontal transfer of heat? Clearly on earth this radiative "solution" does not occur, although there are examples in the solar system where the conditions are nearly met. To determine the situation on earth, we need expressions for the radiative and dynamical time scales.

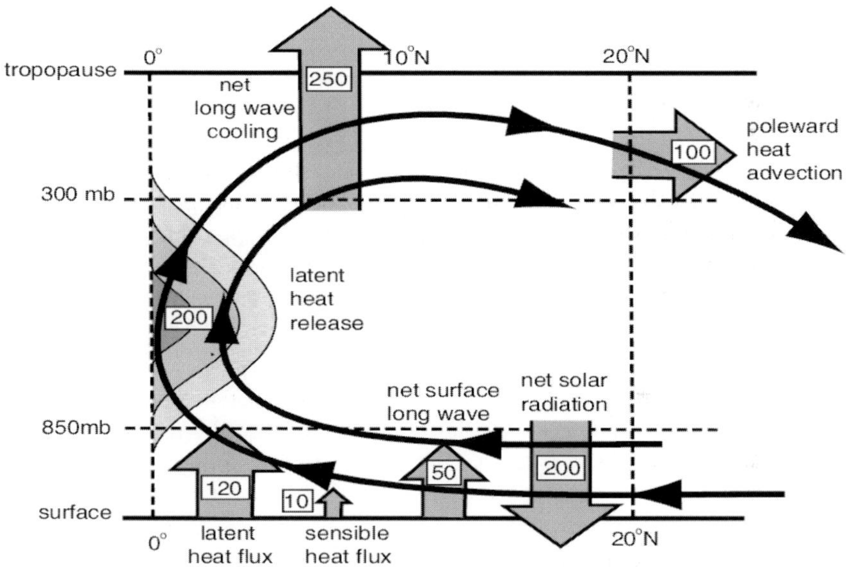

Figure 1-6a. Heat balance in the near-equatorial section of the Hadley circulation (solid dark arrows). Heating of the surface occurs from either direct solar heating (200 W m^{-2}) or by down-welling atmospheric long-wave radiation. Heating of the atmosphere occurs at the surface by direct sensible heating (10 W m^{-2} over oceans, higher over land), by a net long-wave radiative flux (50 W m^{-2}), and in the mid-troposphere by the condensation of water vapor that initially evaporated at the surface (200 W m^{-2}). At the top of the atmosphere, radiative cooling to space with a magnitude of about 250 W m^{-2} takes place. The residual (100 W m^{-2}), is transported poleward initially by the Hadley circulation. It is important to note that the regions of heating and cooling in the tropical atmosphere are separated distinctly in the vertical. That is, the tropics are heated at the surface and middle atmosphere and are cooled in the upper troposphere.

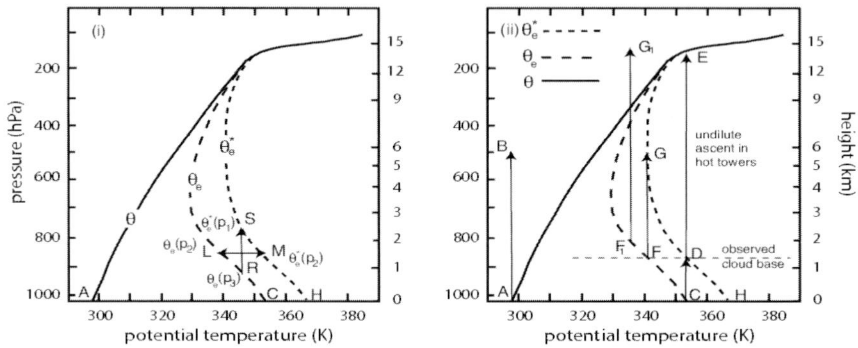

Figure 1-6b. The profiles of potential temperature θ, equivalent potential temperature θ_e, and equivalent potential temperature saturated at the local vertical temperature θ_e^* for the climatological tropical atmosphere. Panel (i) shows the three profiles where it can be seen that from about 600 mb and above, the atmosphere is stable ($\partial\theta_e^*/\partial z \geq 0$). Panel (ii) shows the consequences of ascent of moist parcels vertically lifted from different locations in the tropical column. It is clear that a parcel lifted from the surface (point C) requires the least work to lift it to its condensation point (D). Furthermore, this parcel, having the highest equivalent potential temperature in the lower troposphere, will release the greatest amount of latent heat if it is lifted through the atmosphere. The trajectory D–E shows the path of vertical ascent through the undilute cores of Riehl and Malkus's (1958) hot towers.

The temperature change of an atmosphere is given by (e.g., Curry and Webster 1998):

$$\frac{dT}{dt} = \frac{g}{C_p}\frac{\partial F}{\partial p} \tag{6}$$

where C_p is the specific heat of the atmosphere and F is the net radiative flux divergence between the atmosphere and outer space where:

$$F = 0 - \sigma T_e^4 \tag{7}$$

and it is assumed that there is no flux of long-wave radiation from outer space and the planet radiates to outer space at the equivalent temperature of the planet defined as $T_e = [S(1-\alpha)/4\sigma]^{1/4}$ where S is the incident solar radiation, α the planetary albedo, and σ the Stefan-Boltzman constant. Between Equations (6) and (7), we can find an expression for the cooling rate of the planet:

$$\frac{dT}{dt} = -\frac{g}{C_p}\frac{\sigma T_e^4}{np_0} \tag{8}$$

The factor n has been introduced to provide a measure of the long-wave optical depth. For a strongly absorbing atmosphere, we set $n = 1$ (see Curry and Webster 1998, chapter 14). For a less-absorbing atmosphere such as earth or Mars, which have optically thinner atmospheres, $n < 1$. From Equation (8) it is clear that the cooling rate of the planet is determined by a combination of the mass of the planet (given by surface pressure p_0), as well as its heat capacity (C_p), radiative equilibrium temperature (T_e), and its long-wave optical properties (n). Thus, a hot atmosphere (large T_e) with little mass (small p_0) will cool quicker than a cooler and more massive atmosphere.

We define a radiative time scale (τ_{rad}) as the time it takes for the planet to cool by a certain fraction from its initial temperature.[7] This is referred to as the *e*-folding rate and provides a time scale for the cooling of the planet by 0.37 of its initial value. τ_{rad} can be calculated by integrating Equation (6) in between the times when the temperature is T_e and when it cools to T_e/e. That is:

$$\tau_{rad} = \int_{t(T_e/e)}^{t(T_e)} dt = \int_{T/e_e}^{T_e} \frac{dt}{dT} dT \tag{9}$$

and using Equation (8) gives:

$$\tau_{rad} = \frac{np_0C_p}{g\sigma}\int_{T_e/e}^{T_e}\frac{1}{T^4}dT = \frac{np_0C_p}{4g\sigma(T_e^3 - T_e^3/e^3)} \approx \frac{np_0C_p}{4g\sigma T_e^3} \tag{10}$$

---

[7] It might seem that a more logical time scale might be the halving time of temperature; that is, the time it would take to halve the initial temperature. However, it is mathematically convenient to choose the base of the natural logarithm, *e*, and consider when the temperature drops to 1/*e* of its initial value. "*e*" has a value of approximately 2.71828. Such a time scale is referred to as the *e*-folding time scale.

Finally, we require a dynamical time scale. This is defined simply as the time for the wind of characteristic speed U to move across over some characteristic planetary distance taken here as the radius of the planet, a. Then:

$$\tau_{dyn} = \frac{a}{U} \tag{11}$$

The ratio of the dynamical and radiative time scales determines whether radiative or dynamical effects will dominate a climate. Between Equations (10) and (11) we find:

$$\varepsilon = \frac{\tau_{dyn}}{\tau_{rad}} = \frac{4ag\sigma T_e^3}{Unp_0C_p} \tag{12}$$

We can understand the relative allocation of energy between radiative cooling at the top of the atmosphere and lateral poleward advection by considering the extremes of the ε:

(1) $\varepsilon >> 1$: For such situations, the dynamical time scale must be very much larger than the radiative time scale, meaning radiative effects will dominate. On Mars, for example, $\varepsilon \approx 200$ because of the smallness of the atmospheric mass (6 hPa surface pressure compared to 1,000 hPa on earth) and a very weak greenhouse effect. With such a large value of ε, the horizontal dynamical fluxes of heat are small compared to the local radiative cooling to space, meaning that a parcel of air advected along by the winds will cool rapidly to space so that the initial temperature signature of the parcel would be lost relatively quickly. The dominance of radiative effects illustrates why the observed equator-to-pole temperature gradient and the dayside-nightside temperature gradient along a line of latitude on Mars is so much larger than that observed on earth.

(2) $\varepsilon << 1$: In this case, the dynamic transports of heat dominate over radiative effects. In such a system, a parcel advected by the motion of the atmosphere will cool very slowly and maintain its initial signature as it moves poleward. Venus has a value of ε of about 0.003 because of the massiveness of the atmosphere (the surface pressure on Venus is about 10,000 times and 100 times that of Mars and earth, respectively). Thus, during the time it takes a parcel to be ad-

vected from the equator to the pole, the atmospheric parcel would have cooled imperceptibly. Consequently, the equator-to-pole temperature gradient is almost zero and there is little difference between daytime and nighttime temperatures.

(3) $\varepsilon \approx 1$: A value of unity suggests parity between dynamical and radiative effects on an atmospheric parcel. Earth, for example, has a value of ε of about 0.4. On its poleward motion, an advected parcel will cool radiatively but not sufficiently rapidly that its initial thermodynamic signature will be lost by the time the parcel reaches its destination. The comparatively similar values of the two time scales is the reason why in the upper troposphere, there are both an advection of heat towards the poles by the Hadley circulation and significant local cooling to space of roughly equal magnitudes.

To a large degree, we have accounted for the disposition of heat once it reaches the upper troposphere. However, the manner in which the vertical transport of heat takes place in the upward branch of the Hadley cell is more difficult to explain for two reasons: (1) Air is not transported upward by large-scale ascent as is suggested by Figure 1-6b, and (2) there exists a distinct stable layer in the middle troposphere that makes vertical transport difficult. These two issues are strongly linked.

Satellite pictures suggest a very different view of the tropics to that suggested by both Figures 1-2 and 1-6a. By and large, the tropics are relatively cloud free and the predominant feature is weak subsiding air over broad cloudless regions. Interspersed within this subsidence are deep cloudy regions where very strong ascent is associated with convection that links the boundary layer with the upper troposphere. The net ascent shown in Figure 1-6a is really a composite of the regions of vigorous ascent and slow descending air outside the disturbed regions. We can understand the problem by considering some simple energy concepts. Figure 1-6b shows profiles of potential temperature θ, equivalent potential temperature θ_e, and equivalent potential temperature saturated at the local vertical temperature θ_e^*, which are defined in a convenient energy form as, respectively:

$$\left.\begin{aligned} C_p\theta &= C_pT + gz && \text{(a)} \\ C_p\theta_e &= C_pT + gz + Lq && \text{(b)} \\ C_p\theta_e^* &= C_pT + gz + Lq^* && \text{(c)} \end{aligned}\right\} \qquad (13)$$

These three equations represent the dry static energy, the moist static energy, and the saturated moist static energy, respectively. Here, L is the latent heating coefficient of condensation, q is the observed or ambient vapor pressure, $q^*(T)$ is the saturated vapor pressure at temperature T of the height of the parcel, and C_p is the specific heat of air at constant pressure. The θ and θ_e profiles shown in Figure 1-6b represent average conditions in the tropical atmosphere. The θ_e^* profile shows the equivalent potential temperature profile if the air were saturated at the ambient temperature at some height z. The letters A, C, and H represent the three temperatures at the surface, values of which can be found by setting $z = 0$ in (Eqs. 13a–c). From Equations (13) it can be seen that the temperature would rise by 55 K if all of the water vapor in the surface parcel were condensed (cf. Eqs. 13a and b) or by 65 K if the parcel were saturated (cf. Eqs. 13a and c). This 10 K difference at the surface between the ambient equivalent potential temperature and the saturated value changes with height, as the temperature of the atmosphere (and hence the saturated vapor pressure) is a function of height. The greatest difference between θ_e and θ_e^* occurs in the mid-troposphere. Saturation of a parcel and the release of latent heat occur if the parcel is lifted and cooled adiabatically (constant θ) and if the parcel encounters the θ_e^* curve, as occurs in panel (i) of Figure 1-6b, when a parcel is lifted and cooled along the line R–S. Finally, for the atmosphere to be unstable at some height, there is the condition that $\partial\theta_e / \partial z < 0$ (e.g., Curry and Webster 1998). Clearly, this is not the case above 600 mb (4 km), suggesting that the middle and upper troposphere are very stable.

Panel (ii) of Figure 1-6b shows the outcome of lifting air parcels in the tropical atmosphere. If the air is dry ($q = 0$), then a parcel will follow a trajectory A–B. As the potential temperature θ is constant for adiabatic ascent and as the potential energy gz increases as the parcel is lifted, then T must decrease with ascent. Thus the parcel will always be cooler than its environment and can never generate buoyancy, indicating an extremely stable state. Consider now a moist parcel located at the surface (point C). If the parcel is lifted to D, it becomes saturated and latent heat is released, which increases the buoyancy of the parcel. As it turns out, this is roughly the observed cloud base of deep convective clouds in the tropics. Consider a parcel located at higher elevation (F) in the atmosphere. Upon lifting, the parcel will eventually reach saturation but not until point G near 500 mb. Nearly 5 times the amount of work[8] is necessary to raise a parcel from F to G than to lift a parcel from C to D in order to achieve saturation and the release of latent heat. Thus, there is great efficiency in raising lower tropospheric parcels of moist air compared to moist parcels located higher in the troposphere. It should also be noted that the fact that the parcel is moist does not ensure that latent heat will be released if it is lifted and cooled. Consider a parcel at height F_1. No matter how high the parcel of moist air is lifted (e.g., to G_1), it never encounters the saturated equivalent potential temperature (θ_e^*) curve and it will always be cooler than its environment. Such a parcel will never generate buoyancy, and by the time that it reaches the upper troposphere (if somehow it were lifted to that height), it will be cooler than the environment.

The difficulties in raising parcels through the tropical mid-tropospheric stable layer were first identified in the landmark study by Riehl and Markus (1958). Consider once again the lifting of a parcel from level C to level D where saturation occurs (panel ii, Figure 1-6b). The parcel will continue to rise using the buoyancy generated by the release of latent heat of condensation. However, dry, low θ_e air from air surrounding the cloud will be entrained into the cloud, potentially reducing the buoyancy and curtailing the vertical ascent of the parcel and the height of the cloud. Riehl and Malkus proposed the concept of the "hot tower," which is a cumulonimbus cloud that possesses an "undilute" saturated core that would funnel moist and warm boundary layer air through the stable mid-troposphere, protecting it from dilution with the low θ_e air. To provide such protection, the cloud must be of significant horizontal dimensions. In addition, the ascending air must have the highest specific humidity, originating near the surface (position C). Also, the amount of work needed to raise a surface parcel is less than anywhere in the tropical atmospheric column. Large convective elements are common in the tropics, especially in the region of the ITCZ. Riehl and Malkus calculated that about 1,500–5,000 hot towers would be needed each day to perform the required vertical heat transport. Viewed from a tropics-wide perspective, this is not a large number.

Whereas the hot tower hypothesis answers many of the problems that we have encountered in transporting surface heat into the upper troposphere, one problem remains. Some form of mechanical lifting is necessary to saturate the parcel as it is raised from C to D. A number of mechanisms have been proposed, all of which depend on the generation of a low-level

[8] Work is defined as the product of the force needed to lift the parcel ($m\ g$ where m is the mass of the parcel) times the distance moved in the vertical Δz; that is, the work required to lift a parcel from point C to D is $m\ g(z_D - z_C)$.

cyclonic circulation and frictional convergence. In the conditional instability of the second kind (CISK) mechanisms, the converging moist air rises and the release of latent heat lowers the surface pressure, which increases the inflow and so on. This mechanism was introduced by Ooyama (1964) and Charney and Eliassen (1964) to explain tropical cyclone development, and later by Bates (1970) to explain instabilities of the ITCZ. Unfortunately, there is little observational evidence for CISK in nature. We will propose a second mechanism for lifting parcels to saturation in Section 4.4.

2.5. Driving Forces of the Hadley Circulation

Condensational heating of the equatorial atmosphere does two things: It raises the temperature of the atmospheric column, and it decreases the pressure at the surface relative to that in the subtropics. These two impacts are used to support the common argument that the Hadley circulation is driven by an equator-directed surface pressure gradient between the subtropical high-pressure zone and the equatorial trough. While this argument is not incorrect, it is incomplete. The Hadley circulation is also driven by a reversed pressure gradient in the upper troposphere. In fact, if there were no surface pressure gradients at all, there would still be meridional cells, as long as the equatorial regions were warmer than the subtropics.

Consider the following argument. If the equation of state ($p = \rho RT$) is combined with the hydrostatic equation ($\partial p/\partial z = -\rho g$), and it is assumed that the temperature of an atmospheric column is constant (i.e., T = constant), we obtain:

$$\partial \ln p/\partial z = -g/RT\,, \tag{14}$$

where p, ρ, and R are the atmospheric pressure, density, and gas constant, respectively. Equation (14) shows that that the warmer the atmospheric column, the slower the pressure decreases with height. Now consider two columns with mean temperatures T_{EQ} and T_{ST} representing an equatorial and a subtropical column, respectively, where $T_{EQ} > T_{ST}$. From Equation (14) it is clear that the pressure decreases less rapidly with height in the warm column compared to that in the cold column. Thus, if the equatorial surface pressures in the tropics and subtropics were the same, then at some height above the surface the pressure in the warm column will be greater than in the cold column, producing a pressure gradient aloft that would drive air poleward. We may see this formally by integrating Equation (14) in the vertical between the surface ($z = 0$) and some height in the column ($z = z_1$).

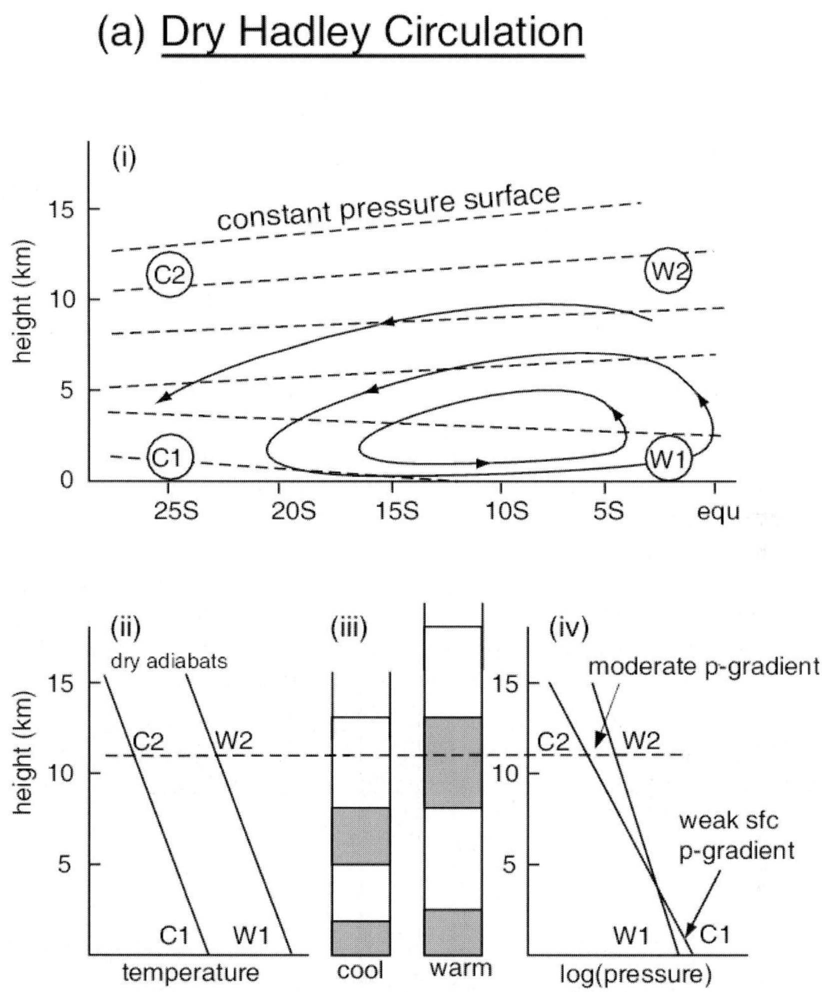

Figure 1-7a. Schematic of the Hadley circulation between the subtropics and the equator in an environment where there is no latent heat of condensation. The dashed lines in panel (i) show the slope of the pressure surfaces as indicated by Equations (7) and (8). Panel (ii) shows the distribution of temperature with height for the warm column over the heated continent (W) and the cold column in the subtropics (C) at two heights (1 and 2). In the dry atmosphere, the temperature decreases upward dry adiabatically. Panel (iii) shows the expansion of layers of different temperatures. Panel (iv) shows the variation in pressure with height. At the surface there is a weak pressure gradient where (p(C1) > p(W1)) while at higher levels the pressure gradient is reversed (p(W2) > p(C2)). The pressure gradient reversal occurs in the lower troposphere and a weak shallow meridional cell ensues.

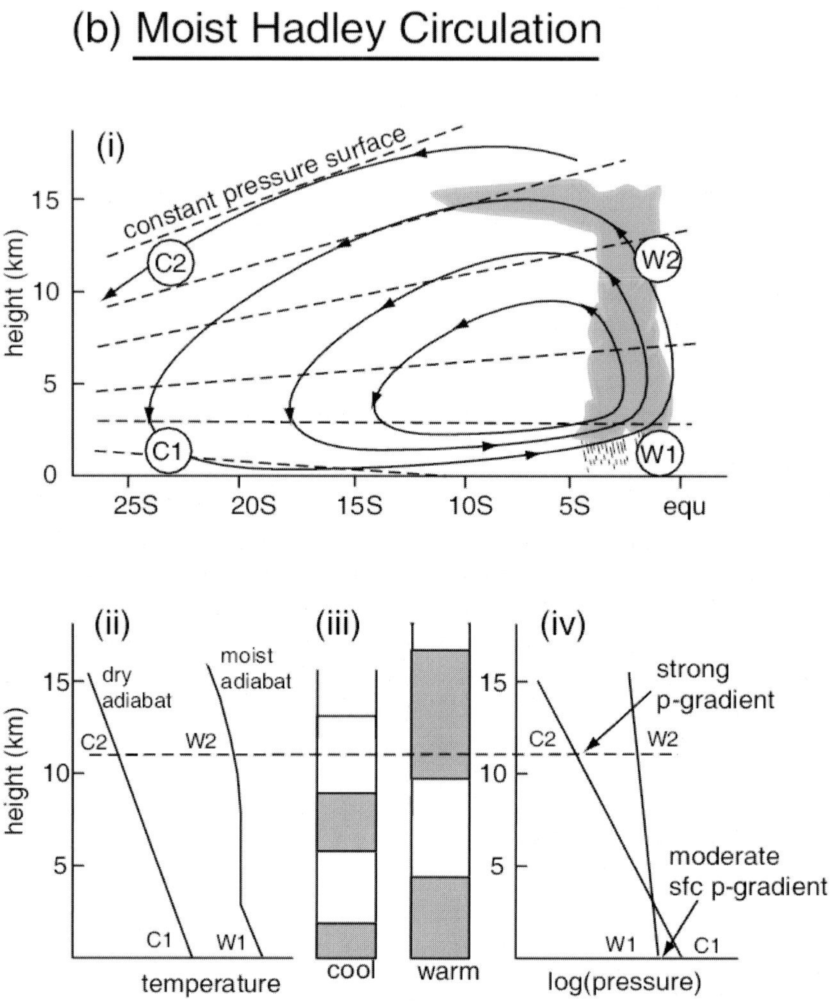

Figure 1-7b. As in Fig. 1-7a except where latent heating occurs in the ascending region of the Hadley cell. In the region of ascent, the vertical temperature lapse rate is close to moist adiabatic (panel ii), and expansion of the layers is more extreme in the heated region so that the upper-level pressure gradient between the warm and cold column is extremely large (i.e., (p(W2) >> p(C2)). The result is a system with more tilted isobars (dashed lines, panel i) and a deeper and more vigorous Hadley circulation (p(W2) > p(C2)). Like the monsoons and sea breezes, the Hadley circulation is driven largely by a reversed upper-level pressure gradient.

Performing this operation for the equatorial and subtropical columns, we obtain the following vertically dependent difference in pressure between the columns:

$$\Delta(\ln p_{z=z_1}) - \Delta(\ln p_{z=0}) = -\frac{gz_1}{R}\left(\frac{1}{T_{EQ}} - \frac{1}{T_{ST}}\right) \tag{15}$$

where Δ refers to the difference between the equatorial column and the subtropical column. If the differences in the surface pressure were zero, then:

$$\Delta(\ln p_{z=z_1}) = -\frac{gz_1}{R}\left(\frac{1}{T_{EQ}} - \frac{1}{T_{ST}}\right) \tag{16}$$

where $\Delta(\ln p_{z=z_1}) > 0$ if $T_{EQ} > T_{ST}$. Also, it should be noted that the pressure difference increases linearly with height. Furthermore, from observations, we know that $\Delta(\ln p_{z=z_1}) > \Delta(\ln p_{z=0})$ so that the Hadley circulation is driven by two opposing pressure gradients, one located at the surface and the second, the stronger, at higher levels in the atmosphere. The same basic physics described above drive thermal circulation ranging from small-scale sea breezes to very large-scale monsoons. These factors are shown schematically in Figure 1-7a for a dry Hadley circulation (i.e., assuming there are no moist processes) and in Figure 1-7b for a Hadley circulation with moist processes.

3. THE VERTICAL SCALE OF THE HADLEY CIRCULATION

The Hadley circulation occupies the entire depth of the troposphere extending up to heights greater than 15 km, consistent with the vertical extent of deep convective clouds. However, this is not the same as heights expected from simple fluid dynamical arguments. A series of studies (e.g., Matsuno 1966; Webster 1972; Gill 1980) examined the fundamental structure of waves in the tropical atmosphere that were, in essence, a subset of global modes identified and catalogued by Longuet-Higgins (1968). These studies showed that modes in a fully stratified atmosphere could be represented by modes in a shallow fluid of a particular depth. The depth of the fluid is referred to as the "equivalent depth" of the mode.[9] For slow large-scale modes, the vertical scale is between 2 and 3 km and much smaller than the observed vertical scale of the Hadley circulation.

The difference between the predicted and observed scales of tropical motion can be reconciled by noting that the theoretical models mentioned above did not consider moist processes or, more particularly, dissipative processes associated with convection. Chang (1977) found that the observed scales of the modes would be the same as those predicted with both heating and dissipation in the formulations. To a large degree, the vertical scale of the circulation is set by the heating profile associated with the latent heat release. As the maximum vertical velocity occurs where the maximum heating occurs, we can see from Figure 1-2 that it should be located in the middle troposphere centered near 7–8 km as is shown in Figure 1-6a. But why should tropical heating be located in the mid-troposphere?

Webster (1994) and Emanuel et al. (1994) argue that the magnitude of the tropical SST determines the vertical structure of the tropical atmosphere. From Clausius-Clapeyron considerations (e.g., Curry and Webster 1998), we know that the saturated vapor pressure of water increases exponentially with temperature so that an air parcel over the warm equatorial ocean will have a larger water vapor pressure than a parcel over a cooler ocean. Thus, surface air flowing from higher latitudes towards the equator in the trade winds (forced by a pressure gradient formed in response to the SST gradient) will possess increasingly saturated vapor pressure as the parcel warms. The warmer the SST at the end of the parcel's journey, the greater the moisture content of the air parcel. As the air masses converge, ascent of the moist air occurs, producing a nearly moist adiabatic temperature profile. In essence, the warmer the SST, the greater the release of latent heat in the atmospheric column and the higher a parcel will ascend. A schematic of the relationship between the magnitude of the SST, the vertical temperature profile, and the latent heat profile if ascent were to occur is shown in Figure 1-8. This figure is supported by observational evidence that the SST and the tropopause height appear to be closely related.

[9] Formally, the equivalent depth is contained within the separation coefficient when the vertical and horizontal parts of the equations are formally separated. Separation requires that only relatively simple thermal structures can be considered. But even with this simplicity, considerable insight can be gained about the fundamental nature of real atmospheric waves.

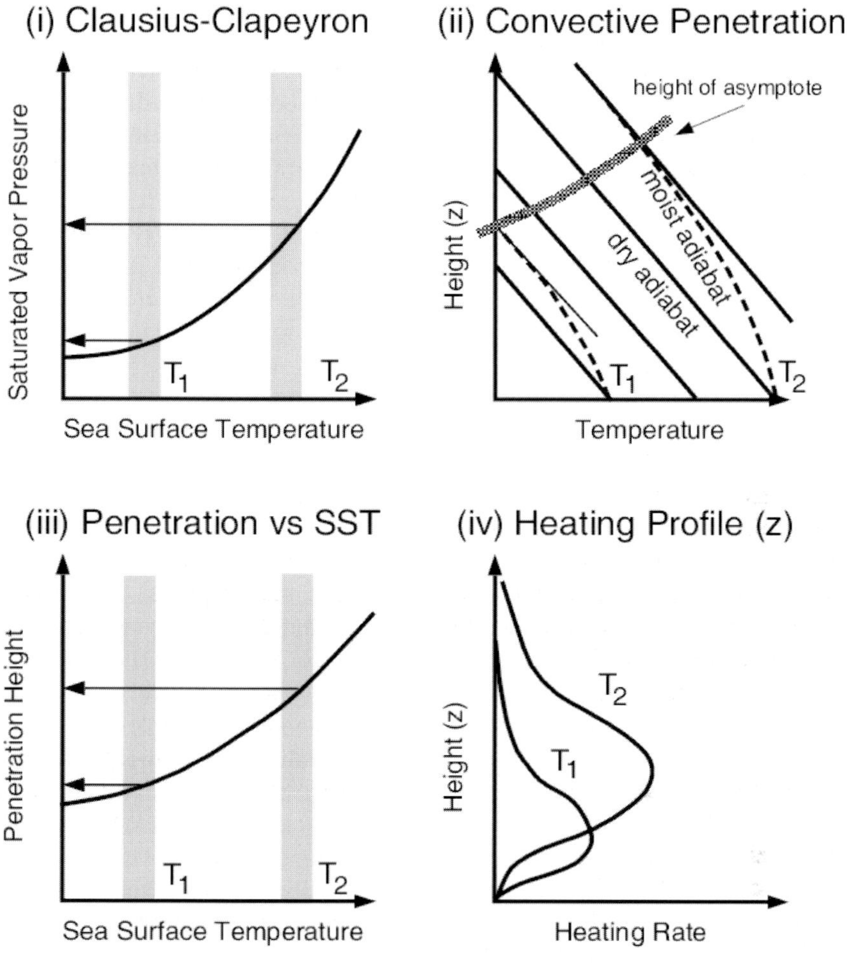

Figure 1-8. Schematic diagram relating the distribution of organized convection over the oceans with SST. Two columns with temperatures T_1 and T_2 (where $T_2 > T_1$) represent, for example, the subtropics and the equatorial trough. Panel (i) shows the Clausius–Clapeyron relationship (i.e., $e_s = e_s(T)$) and its relationship to the convective penetration height (panels ii and iii) over the ocean. Assuming the same relative humidity distribution in each column and that convection has occurred, the magnitude of the heating in the warm column will be greater than that in the cold column and will occur exponentially higher in the atmosphere (panel iv). Adapted from Webster (1994).

4. LOCATION OF THE ITCZ

Observationally, the location of the ITCZ can be identified by satellite as a minimum in outgoing long-wave radiation (OLR)[10]. Seasonal means of OLR show the ITCZ as a meandering narrow band of cloudiness close to the equator. In general, the ITCZ resides near the equator or in the summer hemisphere. An exception is in the eastern Pacific Ocean, where it resides to the north of the equator for the entire year, except occasionally in spring, when it forms as a double ITCZ on either side of the equator. Two factors appear to be responsible for its location: the magnitude of the equatorial SST and strength of the cross-equatorial pressure gradient. We will briefly consider each of these factors.

4.1. Influence of SST on the Location of the ITCZ

From the discussion in Section 3, it would seem that the deepest and strongest equatorial convection (and hence the location of the ITCZ) should coincide with the locations of the warmest SSTs. However, this is a common misconception. Figure 1-9 shows OLR plotted a function of SST for regions in the Indian Ocean and the equatorial Pacific. If SST were the sole determining factor of the strength of convection, then the OLR-SST relationship would be a straight line. However, in the Pacific Ocean we note that as SST increases, the OLR-SST relationship becomes increasingly less linear. In fact, the OLR takes on a broad range of values indicating both cloudy and cloudless regions in the regions of highest SST. In the Indian Ocean box, where the SST is generally warmer than in the Pacific Ocean, there is little relationship between SST and OLR. Thus, although there is a general increase in convection as SST increases, especially in the Pacific Ocean, it is clear that where the SST is warmest, other factors must come into play in determining the location of convection.

The latitudinal distributions of SST, the mean atmospheric sea level pressure (MSLP), and convection (inverse of OLR) in the Indian Ocean, the eastern Pacific Ocean, and the central Atlantic Ocean are shown in Figure 1-10. In the boreal winter, the SST maximum is poleward of the convection by about 5° of latitude. The minimum MSLP coincides with the SST maximum near 15°S. Similar displacements occur in the Atlantic and the eastern Pacific during summer. In the Atlantic Ocean during the boreal winter,

---

[10] OLR is a satellite product and is the integral of radiation emitted from the surface of the planet, the atmosphere, and clouds. The tops of deep convective clouds are cold and thus minima in the OLR represent deep clouds.

maxima in SST and convection occur in the same locations as the MSLP minima. An exception occurs in the summer Indian Ocean. We will return to this case presently. There, the mean monthly convection occurs over a broad latitudinal range.

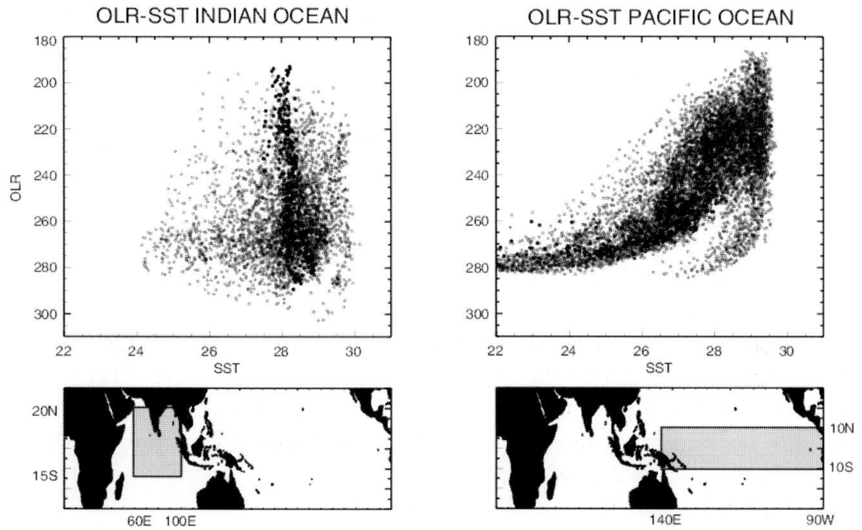

Figure 1-9. OLR (W m^{-2}, inverted scale) as a function of SST (°C) for the Indian Ocean (shaded region, left map, 20°N–15°S, 60°E–100°E) and the equatorial Pacific Ocean (shaded region, right map, 10°N–10°S, 140°E–90°W). OLR-SST pairs were computed from mean monthly data over the 11 y period 1980–1990 in 2° x 2° longitude-latitude squares within the shaded regions. In general, small values of OLR indicate deep clouds and high values indicate shallow convection or its absence. In the Pacific region, there is a general increase in the height of convection with increasing SST as suggested by the heating-SST relationships shown in Fig. 1-8. However, at the warmest temperatures (>28°C) there is a breakdown of this relationship. In the Indian Ocean, there is an absence of a relationship between OLR and SST altogether, although it is noted that the SST is in general >26°C. The lack of correlation at high SSTs and in the Indian Ocean suggests that other processes besides SST, such as atmospheric dynamics, are important in determining where convection occurs and also preclude the appearance of deep convection in the warmest regions of the tropical oceans.

In summary, Figures 1-9 and 1-10 suggest that maximum values in SST do not necessarily set the location of the ITCZ. We note from Figure 1-

10 that the greatest deviations between maximum SST and minimum OLR occur where there is a strong cross-equatorial pressure gradient. However, when there is a co-location of maximum SST and minimum OLR, there is a minimal cross-equatorial pressure gradient.

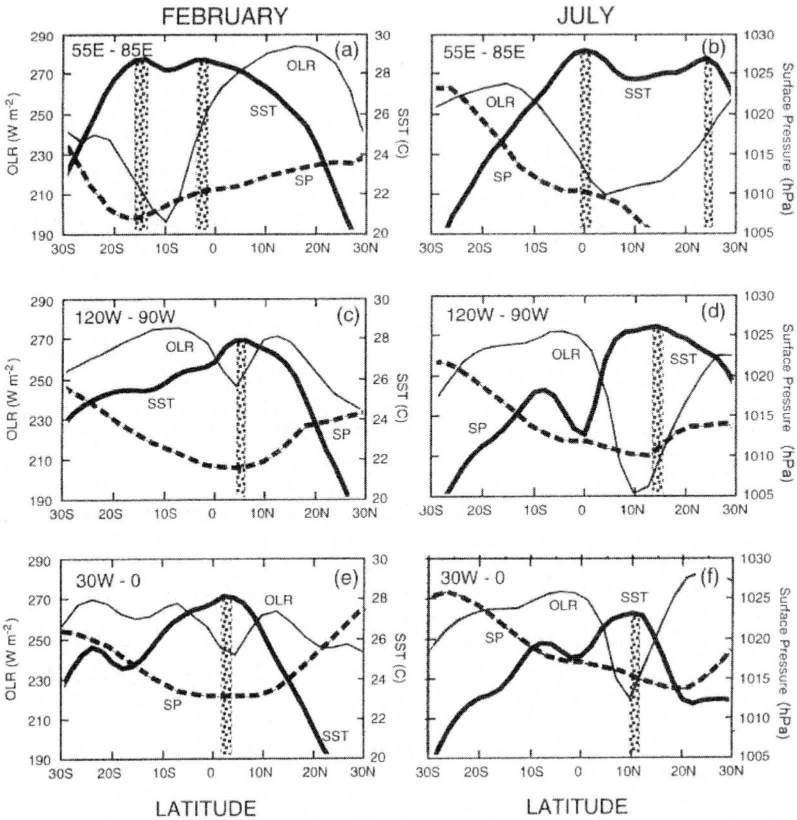

Figure 1-10. Distributions of OLR (thin solid lines), SST (heavy solid lines), and mean sea level pressure (SP: dashed lines, hPa) for three longitudinal bands: (a) and (b) Indian Ocean, 55°E to 85°E; (c) and (d) eastern Pacific Ocean, 120°W to 90°W; (e) and (f) eastern Atlantic Ocean, 30°W to 0°. (a), (c), and (e) are for February 1992, and (b), (d), and (f) are for July 1992. Stippled areas denote areas of maximum SST. Adapted from Tomas and Webster (1997).

4.2. Influence of Cross-Equatorial Pressure Gradients

Tomas and Webster (1997) noted that, in addition to the SST maximum occurring on the poleward side of strong convection, the location of the zero absolute vorticity line[11] ($\eta = 0$) lay just equatorward of the maximum convection. They also found that where the $\eta = 0$ contour was located was directly related to the magnitude of the cross-equatorial pressure gradient. In Figure 1-11a, two situations are shown that depict, respectively, weak and strong cross-equatorial pressure gradients. Figure 1-11b plots the location of the zero absolute vorticity gradients as a function of cross-equatorial pressure gradient. Irrespective of the domain over which the pressure gradient is calculated, the correlations are very high, indicating an extremely strong relationship between the location of convection and the pressure gradient. Noting that the regression curves in Figure 1-11b pass though zero, it is clear that if there is no pressure gradient that the zero absolute vorticity line (and convection) resides at the equator. When the cross-equatorial pressure gradient was positive (high pressure north of the equator, low pressure to the south), convection lay south of the equator, and vice versa. In general, the stronger the pressure gradient, the more the $\eta = 0$ contour was displaced away from the equator.

It is clear that the location of convection is associated with the cross-equatorial pressure gradient. But what produces this gradient? A first thought might be that the convection itself may be responsible. But if this were the case, convective maxima and MSPL minima would coincide, and this does not occur unless there is no cross-equatorial pressure gradient (see Figure 1-10).

The question thus becomes: Why is a cross-equatorial pressure gradient important in determining the location of convection? Without explanation, Lindzen and Hou (1988) made the interesting observation from a modeling study that a heating function of a given magnitude produced a much larger response if it were placed off the equator rather than on the equator. It turns out that the cross-equatorial pressure gradient effect and the sensitivity to where heating is placed in the Lindzen and Hou model are related. The relationship resides in the manner in which the dynamics of the tropics produce circulations that may enhance tropical convection under certain circumstances.

---

[11] Absolute vorticity, η, is given as the sum of the earth's vorticity plus the relative vorticity, i.e., $\eta = f + \zeta = 2\Omega \sin\varphi + (u_y - v_x)$.

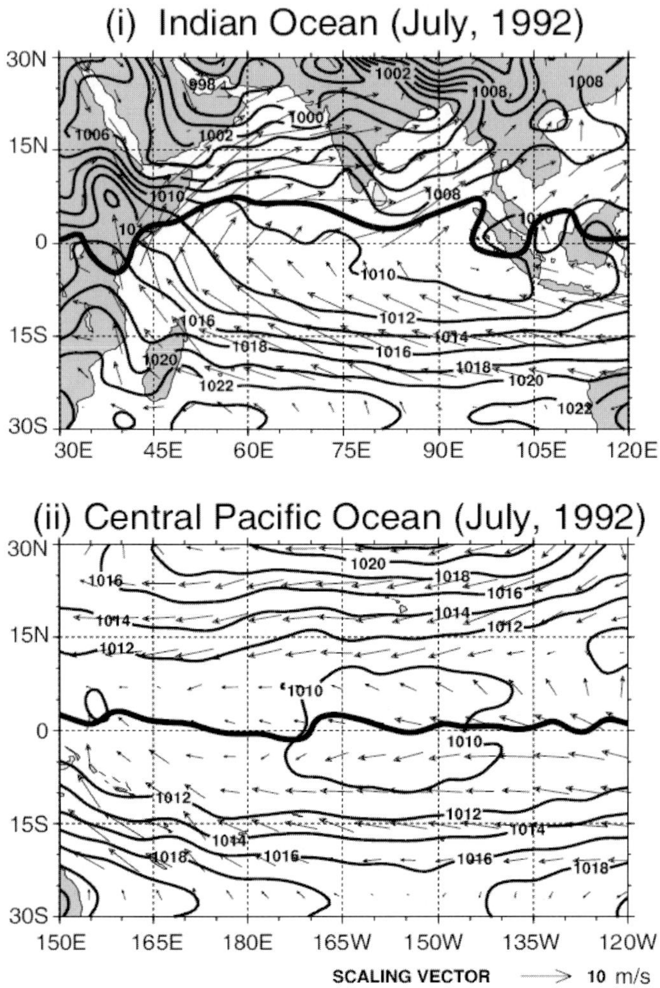

Figure 1-11a. Distributions of sea level pressure (hPa) and horizontal wind (m/s) at 925 hPa commonly encountered about the equator. (i) Indian Ocean, 30°E to 120°E, and (ii) Pacific Ocean, 150°E to 120°W. Bold lines denotes the zero absolute vorticity contours. In the western Pacific warm pool (ii), where there is an absence of a cross-equatorial pressure gradient, the zero absolute vorticity gradient lies close to the equator. In the monsoon region (i), where there is a strong cross-equatorial pressure gradient, there is a substantial displacement of the zero absolute vorticity line from the equator.

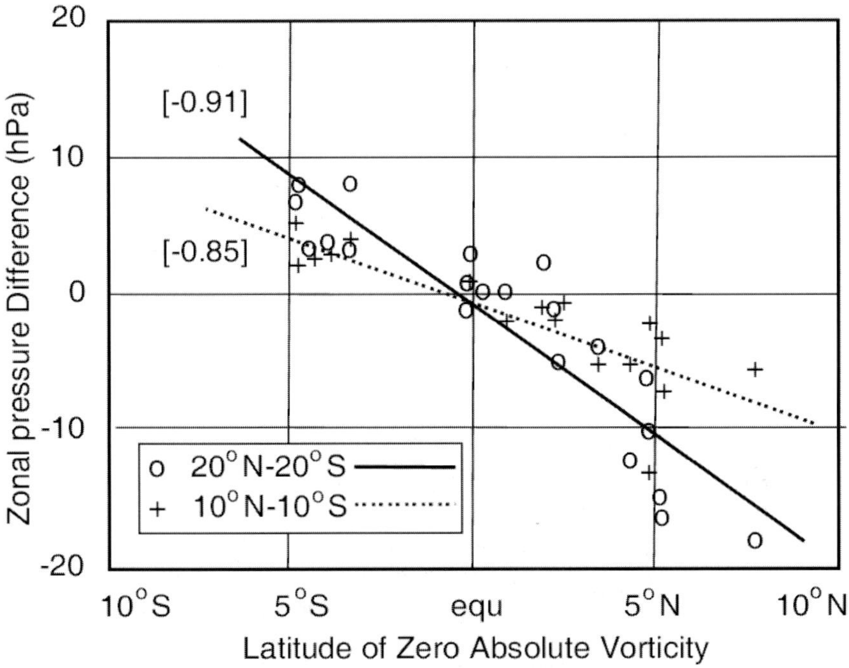

Figure 1-11b. Latitude of zero absolute vorticity at 850 hPa and the magnitude of the cross-equatorial pressure gradient calculated between 10°N and 10°S and between 20°N and 20°S. The dotted and solid lines are regression lines. Regression line and correlation are shown for calculations across two latitude zones. Correlations are exceptionally high (–0.85 and –0.91, respectively). European Center for Medium Range Weather Forecasts data were used. Adapted from Tomas and Webster (1997).

In the tropics, absolute vorticity is dominated by earth's vorticity (f) so that η is negative south of the equator and positive to the north. However, if there is advection from one hemisphere to the other under the action of a cross-equatorial pressure gradient, then it is possible for the $\eta = 0$ contour to reside away from the equator so that regions of negative η can reside north of the equator and vice versa. In this situation, the cross-equatorial pressure gradient advects absolute vorticity from one hemisphere to the other. Such a situation is dynamically unstable (Tomas and Webster 1997). The form of the instability is called "inertial instability," which is a subset of a larger class of instabilities called “symmetric instability” (Emanuel 1979). In the case of a high pressure in the Southern Hemisphere and a low pressure to the north of the equator, the advection of absolute vorticity will cause an “invasion” of anticyclonic vorticity into the Northern Hemisphere, resulting in an acceleration of the parcel away from its original position. The only way of stabilizing the system is to produce cyclonic vorticity to neutralize the anticyclonic vorticity. Creation of cyclonic vorticity is accomplished by “vortex tube stretching” through the establishment of a vertical circulation to the north of the equator (Tomas and Webster 1997). As the atmosphere is moist, the forced vertical ascent will produce a strong release of latent heat and an enhanced circulation. In fact, a close examination of the Lindzen-Hou results shows the existence of the $\eta = 0$ contour just equatorward of the enhanced convection.

4.3 Sequence of Processes for the Formation of Organized Off-Equator Convection

The simple instability theory, and some corroboration from observations, suggests a sequence of processes leading to convection away from the equator but not necessarily over the warmest SSTs. The sequence is as follows:

(1) *A cross-equatorial advection of absolute vorticity:* Organized, zonally orientated convection occurs off the equator when there is a north-south cross-equatorial pressure gradient. The strongest gradients occur over the Indian Ocean in both the boreal summer and winter, and over the Atlantic and eastern Pacific Oceans during the boreal summer. These latitudinal SST distributions produce large-scale pressure gradients spanning 15°–20° on either side of the equator. The macro-scale cross-equatorial pressure gradients, produced by the background SST distribution, must be sufficiently strong to produce a divergent wind field that advects absolute vorticity across the equator. The advection of absolute vorticity across the equator renders air parcels in the near-equatorial summer hemisphere to be inertially unstable.

(2) *The forcing of a stabilizing secondary circulation:* In order to relax the instability, a secondary circulation forms, which produces cyclonic vorticity to compensate for the importation of anticyclonic vorticity from the winter hemisphere. This secondary circulation oc-

curs as a convergence-divergence doublet about the zero absolute vorticity line. Together they form a vertical circulation.

(3) *Convective conditional instability:* The existence of a cross-equatorial pressure gradient is necessary but not sufficient for the formation of off-equator convection. In addition, the SST must be sufficiently high for there to be a conditionally unstable atmosphere and for latent heat release to accompany the secondary circulation. Without the release of latent heat, the ensuing circulation will remain shallow and be governed by the equivalent depth of the mode.

4.4. Vertical Transport of Heat at the Equator

In Section 2.4, we discussed the problem of moving heat from the boundary layer to the upper troposphere where the system can cool to space or advect heat polewards. Riehl and Malkus (1958) noted that broad-scale ascent could not accomplish the transport and that heat could only be transported vertically by very deep cumulus clouds within their "protected" cores. The problem is describing a mechanism that allows convection to be initiated. Specifically, what mechanism takes care of the work necessary to lift a parcel from the surface (point C) to saturation height (point D) in Figure 1-6b?

We suggest that the mechanisms responsible for the lifting are the secondary circulations resulting from the inertial instability associated with the advection of absolute vorticity across the equator under the action of cross-equatorial zonal pressure gradients. The upward vertical motion in the ascending part of the stabilizing cell essentially reduces the stability of the atmospheric column, allowing deep penetrative convection to develop that would help accomplish the vertical transport of heat. Tomas and Webster (1997) describe the generation of deep convection as spasmodic with roughly the inertial frequency of the latitude of the zero absolute vorticity line (roughly 5° from the equator and thus 4–5 days). They speculate that the aperiodic convection may be the source of easterly waves that propagate westward with deep convective elements. In this manner, the generation of deep convection and the consequent balancing of the equatorial heat budget is accomplished by instabilities of the ITCZ or, in other words, the instabilities of the ascending branch of the Hadley circulation.

5. MONSOONS AND THE HADLEY CIRCULATION

The largest cross-equatorial pressure gradients exist in the vicinity of the great continents of Asia, Africa, and Australia. During the summer, these continental regions receive copious precipitation at the same latitudes where other locations receive very little precipitation. In addition, the strength of these regional monsoon circulations has a very strong influence on the zonally averaged circulation distributions seen in Figure 1-2. For example, if the South Asian monsoon region were removed from the JJA mean stream function, the zonally averaged distribution would be much weaker and the region of ascent would be found to be much closer to the equator. Consideration of Figure 1-3 supports this contention. Figure 1-12 shows the zonal velocity component and the stream function averaged zonally across the narrow band 70°E–90°E for JJA and DJF. This particular band was chosen to give a cross section through the region of strongest precipitation in the South Asian monsoon.[12] The most apparent difference between the monsoon cross sections (Fig. 1-12) and the global charts is that monsoon fields are very much stronger. For comparison, consider the region of descent in the Southern Hemisphere in Figures 1-2b and 1-12a. Noting that the vertical velocity is proportional to the latitudinal gradient of the stream function, we have gradients between the equator and 30°S of 16 x 10^{10} kg s^{-1}/30°latitude for the zonally averaged case and 68 x 10^{10} kg s^{-1}/30°latitude for the summer monsoon. That is, vertical velocities associated with the monsoons are about a factor of 4 larger. Similar differences can be seen in the boreal winter (cf. Figs. 1-2d and 1-12b).

The zonal velocity structures for the summer and winter monsoons are very different. During winter, an extremely strong westerly jet stream resides equatorward of the Himalayas. During summer, this jet weakens and moves poleward of the mountains and is replaced by an upper tropospheric easterly jet stream. The easterly jet results from the strong heating of the atmosphere over South Asia, which reverses the temperature gradient between the equator and 30°N. Equation (5) shows that under these circumstances, the winds must become increasingly westward with height (i.e., if $\partial T/\partial\varphi > 0$ then $\partial u/\partial z < 0$). The resultant easterly jet extends westward across North Africa and determines to a large degree the demarcation of wet

[12] In Figure 1-2, the stream function was computed by using zonally averaged values of the vertical and meridional components with the constraint that at any latitude the vertically averaged meridional velocity component must average to zero in the long-term mean because of mass continuity. However, averages over finite meridional bands, like those shown in Figure 1-12, are not constrained, simply because flow can return at other longitudes. That is, within a narrow longitude band, conservation of mass would also need to take into account the zonal velocity.

and dry regions in the Sahel (Nicholson and Flohn 1980; Webster et al. 1998). The reversal of the temperature gradient is unique to all of the monsoon regions and probably is associated with the elevated sensible and latent heating associated with the Himalayas (Webster et al. 1998).

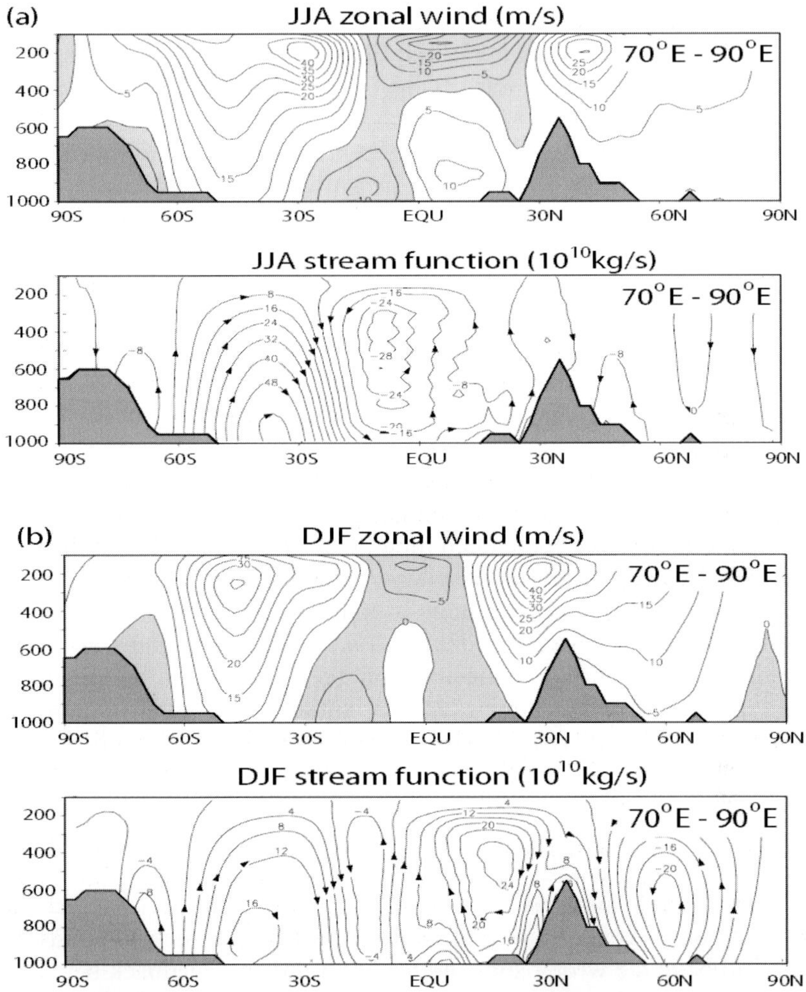

Figure 1-12. Same as Fig. 1-2 but for the monsoon regions. Zonally averaged zonal winds (upper panel, m s^{-1}) and mass stream function (lower panel, 10^{10} kg s^{-1}) for (a) June–August (JJA) and (b) December–February (DJF), plotted against latitude and pressure (hPa). Contours show averages between 70°E and 90°E. Zonally averaged elevation of the planet is shaded, with the Himalayas and the Tibetan Plateau predominating near 30°N and the Antarctic continent in the south. It should be noted that, unlike the zonally averaged mass stream function shown in Fig. 1-2, the stream function might not necessarily conserve mass, as zonal variability needs to be taken into account.

In essence, the same physics that control the Hadley circulation control the dynamics of the monsoon circulation, with a large number of other factors coming into play. First, in the Indian Ocean sector during the boreal summer, the cross-equatorial pressure gradient is stronger than elsewhere on the globe (Tomas and Webster 1997). Second, during the boreal spring the SST of the Indian Ocean is warmer than in any other tropical region. Third, the land area is elevated, so the impact of heating is emphasized because it occurs higher in the troposphere (Webster et al. 1998). Air forced across the equator by the strong cross-equatorial pressure gradient acquires a very high specific humidity. During the boreal winter, the cross-equatorial pressure gradient reverses due principally to the intensity of the Siberian High. However, Australia is a relatively flat continent in comparison to Asia and so there is no counterpart to the Northern Hemisphere elevated heat source and maximum heating lies close to the equator. In combination, these factors lead to a weaker winter monsoon than its summer counterpart.

6. OCEAN HADLEY CIRCULATION

So far, we have considered the impact of solar heating of the surface of the planet from a purely atmospheric perspective, noting that the atmosphere is heated from below through a combination of turbulent latent and sensible heat, and long-wave radiative fluxes. Over land areas, most of the surface heating is returned to the atmosphere. However, in the ocean a large amount of solar energy penetrates to substantial depths with *e*-folding penetrations of about 15 m (e.g., Webster 1994). As turbulent wind mixing allows heat to be stored within the ocean mixed layer, not all of the radiative energy arriving at the ocean surface is returned immediately to the atmosphere. Thus, the equatorial oceans have the potential of accumulating heat and increasing substantially in temperature.

Does the ocean have a role in maintaining the heat balance of the tropical regions? At first glance, it might appear that the oceanic transport of heat would have the wrong sign and lead to a positive feedback between the strength of the Hadley circulation and the monsoons. Winds flow towards the ITCZ in both the Hadley and monsoon regimes. If the heat transport were to be in the same direction as the winds, then the oceans would accumulate heat near regions of convection. This increased SST gradient would drive an even stronger Hadley circulation or monsoon and produce even stronger convection. That is, there would be a strong positive feedback.

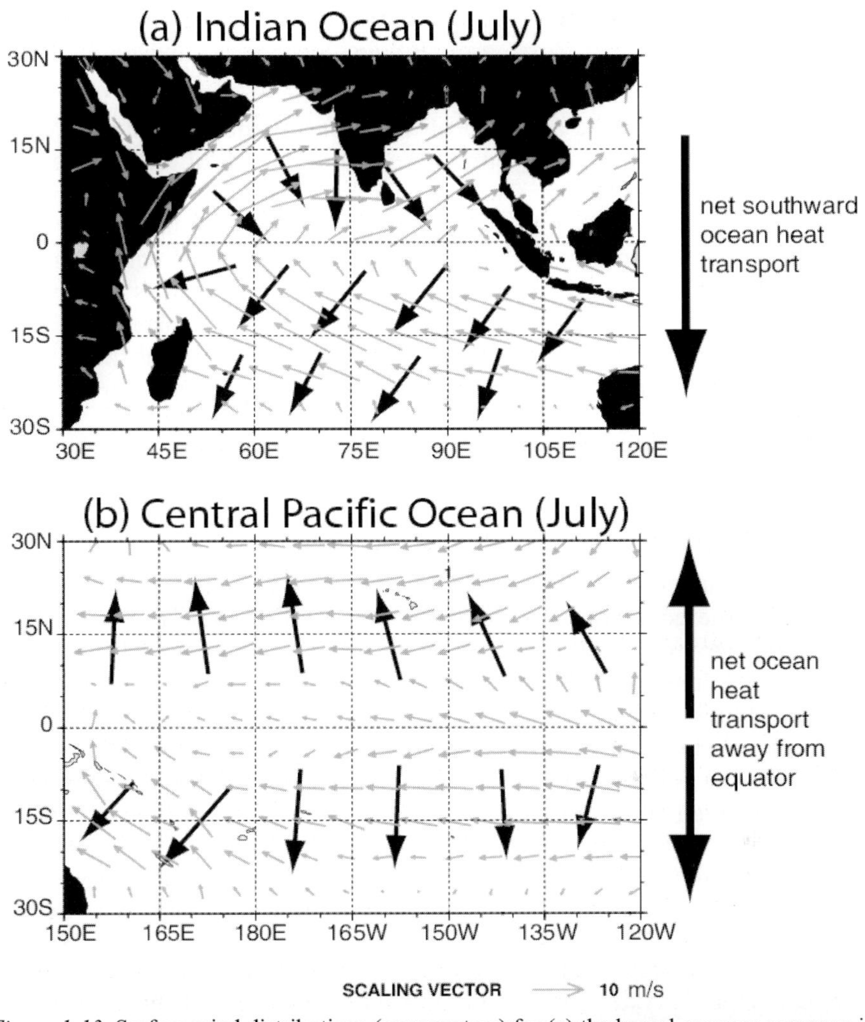

Figure 1-13. Surface wind distributions (gray vectors) for (a) the boreal summer monsoon in the Indian Ocean, and (b) the trade wind regime of the Pacific, and Ekman transport by the ocean (black arrows). The total Ekman transport mass transport is to the right of the surface wind in the Northern Hemisphere and to the left in the Southern Hemisphere. The overall effect of this wind-driven oceanic transport is to advect heat in the opposite direction to that advected by the divergent part of the wind field.

However, in reality, the ocean response to the Hadley and monsoon lower tropospheric winds act in the opposite sense and provide an exceedingly efficient negative feedback that regulates the strength of the circulation. This is because the effect of the rotation of the earth manifests itself by producing mass transports (totaled through the upper layers of the ocean) that are orthogonal to the direction of the surface wind—to the right of the surface wind in the Northern Hemisphere and to the left of the surface wind in the Southern Hemisphere. This effect is referred to as Ekman transport (see e.g., Loschnigg and Webster 2000). Thus, on a rotating planet, ocean mass and heat transport by the trade winds is away from the equator in both hemispheres. In the monsoon regions, where the flow is across the equator, the mass and heat transport are towards the winter hemisphere at all latitudes. Thus, although the atmospheric heat transport is towards the ITCZ or monsoon continental regions, the ocean transport is in the opposite direction (Fig. 1-13).

Noting that the Ekman transport is a wind-driven circulation and that it produces a heat transport that is opposite to that occurring in the atmosphere, it is easy to see why there is a negative feedback between the atmosphere and the ocean. Consider the situation where, for some reason, there is a stronger than average Hadley or monsoon cell, perhaps because the SST in the equatorial regions is warmer than normal. The lower-level winds will be stronger and, as a result, the ocean heat transport will be increased but in the opposite direction. The Hadley or monsoon cell will then be reduced in intensity because the ocean heat content near the ITCZ will be less, reducing the SST, and consequently, the intensity of convection. The opposite effect is also true. If the Hadley or monsoon is weak for some reason, wind-driven ocean transport will be also weak, heat will accumulate in the equatorial near-surface layers, and the atmospheric cell will return to average strength. Webster et al. (2002) describe this regulatory process in great detail for the monsoon circulation and use it as an argument to explain why the monsoon precipitation over South Asia does not deviate greatly from year to year and that there are rarely successive years of above-average or below-average rainfall.

Finally, it is interesting to contrast the manner in which the ocean and the atmosphere are heated and cooled. Figure 1-14 shows a comparison of the two systems. As was discussed at length in Sections 2.5 and 2.6, the atmosphere is heated at the planetary surface and also in the mid-troposphere, the latter through latent heat release in the ITCZ. Furthermore, it is cooled at the top of the atmosphere and by upper-tropospheric poleward transports of heat by the winds. The ocean, on the other hand, is heated and cooled at its surface through a net positive flux of heat at low latitudes and a net negative flux at high latitudes, respectively. Near the equator, the atmosphere is conditionally unstable, at least in the lower troposphere. The upper ocean immediately below, however, is statically stable with warm, fresh water overlying colder, more saline, deeper water. The surface cooling at higher latitudes, on the other hand, renders the upper ocean convectively unstable. Deep plumes of subsiding water result from this cooling, especially in the northern Atlantic Ocean and to some extent in regions around Antarctica. The heating and cooling of the atmospheric and oceanic Hadley cells and the regions where each is either convectively stable or unstable appear as reversed mirror images.

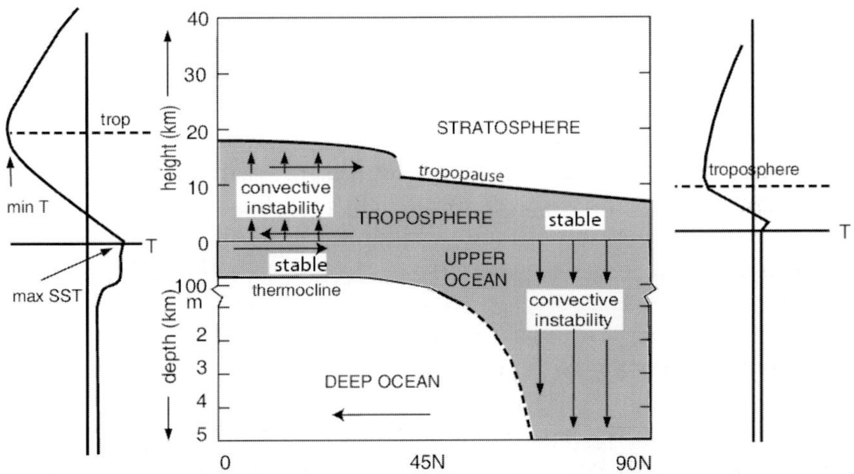

Figure 1-14. Schematic cross section of the atmosphere and the ocean showing the manner in which the ocean and the atmosphere interact. Arrows indicate the direction of heat flux in the atmosphere and the ocean. Vertical temperature sections through the atmosphere and the ocean are shown, left (tropics) and right (polar regions). In the tropics, the warm SST renders the troposphere convectively unstable and heat is transferred to the upper troposphere, where it is either radiated away or advected to higher latitudes. The ocean below, heated by the sun and freshened by rainfall, is extremely stable, on the other hand. The stability of the upper ocean minimizes vertical mixing of heat downwards, thus maintaining the high SSTs. At higher latitudes, on the other hand, cooling at the surface renders the ocean convectively unstable, in sharp contrast to the atmosphere above. Overall, the ocean and the atmosphere are reversed mirror images of each other. The shaded area may be referred to as the joint interactive zone of the ocean and the atmosphere, where interactions can take place on relatively short time scales. Adapted from Webster (1994).

7. SOME CONCLUDING REMARKS

The Hadley circulation and its local manifestations, such as the Asian-Australian monsoon system, impact most of the world's population both in the long term, by defining regions of habitability, and also in the short term by being involved in events such as ENSO. These circulations result from radiative imbalances between the equator and higher latitudes. However, because of the role of rotation, the importance of moist thermodynamics, and the involvement of the oceans, physical explanations of the Hadley circulation are rather complicated. Viewed collectively, the ocean and the atmosphere appear as a self-regulating system that is probably of importance in stabilizing the climate of the planet. It is important that the physics of the coupled ocean-atmosphere Hadley circulation be understood. For example, it is essential that data laid down by past climates be interpreted in a proper physical context. It is clear that future climate states will depend a great deal on how the Hadley circulation and the monsoons react to increases in greenhouse gases. Only with well-configured models will it be possible to determine the degree to which the negative feedbacks between the atmosphere and the oceans might mitigate changes imposed by anthropogenic effects or whether or not the feedbacks will be positive.

8. ACKNOWLEDGMENTS

This chapter resulted from a research grant from the program in Climate Dynamics of the National Science Foundation (NSF Grant: ATM 0328842). Thanks to the Royal Society and JSTOR at http://www.jstor.edu for permission to reproduce Figure 1-4 from Halley (1687) and for the availability of the papers by Edmund Halley and George Hadley. We are grateful for the use of the NCEP/NCAR reanalysis data set.

9. REFERENCES

Bates, J.R. 1970. Dynamics of disturbances on the intertropical convergence zone. *Quarterly Journal of the Royal Meteorological Society* 96: 677–700.

Bjerknes, J. 1966. A possible response of the atmospheric Hadley Circulation to equatorial anomalies of ocean temperature. *Tellus* 18: 820–829.

Bjerknes, J. 1969. Atmospheric teleconnections from the equatorial Pacific. *Monthly Weather Review* 97: 163–172.

Bryson, B. 2003. *A Short History of Nearly Everything*. New York, New York: Broadway, 560 pp.

Chang, C.P. 1977. Viscous internal gravity-waves and low-frequency oscillations in the tropics. *Journal of the Atmospheric Sciences* 34: 901–910.

Charney, J.C. 1947. The dynamics of long waves in a baroclinic westerly current. *Journal of Meteorology* 4: 135–163.

Charney, J.C., and A. Elaissen. 1964. On the growth of the hurricane depression. *Journal of the Atmospheric Sciences* 21: 68–75.

Curry, J.A., and P.J. Webster. 1998. *Thermodynamics of Atmospheres and Oceans*. International Geophysics Series, Vol. 65, San Diego, California, Academic Press, 471 pp.

Eady, E.T. 1949. Long waves and cyclone waves. *Tellus* 1(3): 33–52.

Emanuel, K.A. 1979. Inertial instability and mesoscale convective systems. I. Linear theory of inertial instability in rotating viscous systems. *Journal of the Atmospheric Sciences* 36: 2425–2449.

Emanuel, K.A., J.D. Neelin, and C.S. Bretherton. 1994. On large-scale circulations in convecting atmospheres. *Quarterly Journal of the Royal Meteorological Society* 120: 1111–1143.

Gill, A.E. 1980. Some simple solutions for heat-induced tropical circulations. *Quarterly Journal of the Royal Meteorological Society* 106: 447–462.

Hadley, G. 1735. Concerning the cause of the general trade-winds. *Philosophical Transactions of the Royal Society of London* 39: 58–62.

Halley, E. 1686. An historical account of the Trade Winds, and Monsoons, observable in the seas between the Tropicks, with an attempt to assign the physical cause of the said Winds. *Philosophical Transactions of the Royal Society of London*, 16: 153–168.

Hunt, B.G. 1979. Influence of the earth's rotation on the general circulation of the atmosphere. *Journal of the Atmospheric Sciences* 36: 1392–1408.

Kirch, P.V. 2000. *On the Road of the Winds: An Archaeological History of the Pacific Islands before European Conquest*. Berkeley, California: University of California Press, 424 pp.

Kutzbach, G. 1987. Concepts of monsoon physics in historical perspective: The Indian monsoon (seventeenth to early twentieth century). In, Fein, J.S., and P.L. Stephens (eds.). *Monsoons*. Chichester, U.K.: Wiley Interscience., pp. 159–210.

Lindzen, R.S., and A.Y. Hou. 1988. Hadley circulations for zonally symmetric heating centered off the equator. *Journal of the Atmospheric Sciences* 45: 2416–2427.

Longuet-Higgins, M.S. 1968. The eigenfunctions of Laplace's tidal equations over a sphere. *Philosophical Transactions of the Royal Society of London A*. 262: 511–607.

Loschnigg, J., and P.J. Webster. 2000. A coupled ocean-atmosphere system of SST regulation for the Indian Ocean. *Journal of Climate* 13: 3342–3360.

Matsuno, T. 1966. Quasi-geostrophic motions in the equatorial area. *Journal of the Meteorological Society of Japan*, Series II, 44: 25–43.

Nicholson, S.E., and H. Flohn. 1980. African environment and climate changes and the general atmospheric circulation in late Pleistocene and Holocene. *Climate Change* 2: 313–348.

Ooyama, K. 1964. A dynamical model for the study of tropical cyclone development. *Geofisica International* 4: 187–198.

Peixoto, J.P., and A.H. Oort. 1992. *Physics of Climate*. Melville, New York: American Institute of Physics, 520 pp.

Philander, S.G. 1998. Is the temperature rising? The uncertain science of global warming. Princeton, New Jersey: Princeton University Press, 262 pp.

Riehl, H., and J.S. Malkus. 1958. On the heat balance in the equatorial trough zone. *Geophysica* 6: 503–538.

Royal Society. 1699. *Directions for Sea-men, Bound for Far Voyages. Philosophical Transactions*, Royal Society (London).

Shin S.I., Z. Liu, B. Otto-Bliesner, E.C. Brady, J.E. Kutzbach, and S.P. Harrison. 2003. A simulation of the last glacial maximum climate using the NCAR-CCSM. *Climate Dynamics* 20(2–3): 127–151.

Tomas, R., and P.J. Webster. 1997. On the location of the intertropical convergence zone and near-equatorial convection: The role of inertial instability. *Quarterly Journal of the Royal Meteorological Society* 123: 1445–1482.

Walker, G.T. 1924. World Weather I . *Memoirs of the India Meteorological Department* 24: 75–131.

Walker, G.T. 1928. World Weather III. *Memoirs of the India Meteorological Department* 2: 97–106.

Warren, B.A. 1987. Ancient and medieval records of the monsoon winds and currents of the Indian Ocean. In, Fein, J.S., and P.L. Stephens (eds.). *Monsoons*. Chichester, U.K.: Wiley Interscience, pp. 137–158.

Webster, P.J. 1972. Response of the tropical atmosphere to local steady forcing. *Monthly Weather Review* 100: 518–541.

Webster, P.J. 1994. The role of hydrological processes in ocean-atmosphere interaction. *Reviews of Geophysics* 32: 427–476.

Webster, P.J., and N.A. Streten. 1978. Late Quaternary ice-age climates of tropical Australasia: Interpretations and reconstructions. *Quaternary Research* 10: 279–309.

Webster, P.J., C. Clark, G. Chirikova, J. Fasullo, W. Han, J. Loschnigg, and K. Sahami. 2002. The monsoon as a self-regulating coupled ocean-atmosphere system. *Meteorology at the Millennium*. San Diego, California: Academic Press, pp. 198–219.

Webster, P.J., T. Palmer, M. Yanai, R. Tomas, V. Magana, J. Shukla, and A. Yasunari. 1998. Monsoons: Processes, predictability and the prospects for prediction. *Journal of Geophysical Research* 103 (TOGA special issue): 14451–14510.

Weiss, H., and R.S. Bradley. 2001. What drives societal collapse? *Science* 291(5504): 609–610.

Chapter 2

HADLEY CIRCULATION DYNAMICS

Seasonality and the Role of Continents

Kerry H. Cook

Department of Earth and Atmospheric Sciences, Cornell University, Ithaca, New York14853-1504, U.S.A.

Abstract The equations that govern the Hadley circulation are reviewed, and the observed circulation is described. Atmospheric general circulation model (AGCM) simulations are used to evaluate the dominant zonally averaged momentum and thermodynamic balances within the Hadley regime.

A diagnostic application of the governing equations is used to identify the mechanisms of the Hadley circulation's seasonal evolution between equinox and solstice states. A "vertical driving" mechanism acts through the thermodynamic balance, and is important for regulating the circulation's strength when heating differences between seasons are close (within ~5°) to the equator. A "horizontal driving" mechanism acts through the horizontal momentum equations and is more effective off the equator. Unlike the results from axisymmetric models in which the prescribed heating is always close to the equator, the horizontal forcing mechanism is responsible for most of the Hadley circulation seasonality in the reanalysis and GCM simulations.

The presence of continental surfaces introduces longitudinal structure into tropical diabatic heating fields, and pulls them farther from the equator. The winter Hadley cells in a simulation with continents are much stronger than in a simulation with no continents, and the summer cell is half the intensity of that when continents are included. The strengthening of the winter cell occurs through an increase in low-level wind speeds, which enhances the zonal momentum flux from the surface into the atmosphere. The development of strong monsoon circulations in the Northern Hemisphere summer and the convergence zones of the Southern Hemisphere (South Pacific [SPCZ], South Atlantic [SACZ], and South Indian Ocean [SICZ] convergence zones) shifts mass out of the subtropics, lowers the zonal mean subtropical highs, and weakens the summer cell.

H.F. Diaz and R.S. Bradley (eds.), The Hadley Circulation: Present, Past and Future, 61–83.

1. INTRODUCTION

This chapter provides an introduction to Hadley cell dynamics, including a discussion of the processes that determine the circulation's climatology. The physics of the seasonal oscillation of the Hadley circulation is emphasized, since this intra-annual variability provides insight into possible changes in the circulation on other, e.g., paleoclimate, time scales. The role of the continents in driving the Hadley circulation is also discussed. Much of the heating that ultimately drives the circulation is delivered to the atmosphere over continental surfaces through latent and sensible heat fluxes, and vertical momentum transports are also enhanced over the continents, so changes in these surfaces can modify the circulation.

2. DEFINITION AND OBSERVATIONS OF THE HADLEY CIRCULATION

A Hadley circulation is a large-scale meridional overturning of a rotating atmosphere that has a heating maximum at the surface near or on the equator. The strength and geometry of the Hadley circulation can be quantified by using a stream function. The stream function expresses the fact that, for a two-dimensional flow, the conservation of mass equation couples motion in one direction with motion in the other direction, so one variable (the stream function) can fully describe the flow.

Using pressure as the vertical coordinate, conservation of mass requires

$$\frac{1}{a\cos\phi}\frac{\partial u}{\partial\lambda}+\frac{1}{a\cos\phi}\frac{\partial(v\cos\phi)}{\partial\phi}+\frac{\partial\omega}{\partial p}=0, \tag{1}$$

where u is the east/west (or zonal) velocity, v the north/south (meridional) velocity, ω the vertical p-velocity (dp/dt), a the earth's radius, λ longitude, and ϕ latitude. If Equation 1 is averaged over longitude, around the entire globe, then the first term on the left-hand side (LHS) of Equation 1 is zero and a two-dimensional flow is defined. Using square brackets to denote this longitudinal (zonal) average, the continuity equation is

$$\frac{1}{a}\frac{\partial[v]}{\partial\phi}+\frac{\partial[\omega]}{\partial p}=0 \tag{2}$$

Equation 2 states that if $[v]$ is known then $[\omega]$ is known, and vice versa. In other words, one variable can be used to fully define the two-dimensional flow. One could use either $[v]$ or $[\omega]$ as this single variable, but a more physical representation of the full flow field can be generated by using a stream function. The Stokes stream function, ψ, which is typically used to characterize the Hadley circulation, is defined by

$$[v] = \frac{g}{2\pi a \cos\phi} \frac{\partial \psi}{\partial p} \quad \text{and} \quad [\omega] = -\frac{g}{2\pi a^2 \cos\phi} \frac{\partial \psi}{\partial \phi}, \tag{3}$$

where g is the acceleration due to gravity. Note that Equations 3 satisfy Equation 2. Theoretically, stream function values can be calculated from observations of either $[v]$ or $[\omega]$, but $[v]$ is used for practical reasons because meridional velocities are more frequently and accurately observed. Solving for ψ and integrating from the top of the atmosphere, where it is assumed that $\psi = 0$ and $p = 0$, yields

$$\psi(\phi, p) = \frac{2\pi a \cos\phi}{g} \int_0^p [v(\phi, p)] dp. \tag{4}$$

According to Equation 4, the value of the Stokes stream function at a given latitude and pressure level is equal to the rate at which mass is being transported meridionally (with positive values indicating northward transport) between that pressure level and the top of the atmosphere. Note that the Hadley circulation, also know as the mean meridional circulation (MMC), is a zonal-mean quantity by definition.

Figure 2-1 shows the Stokes stream function for each month from the NCAR/NCEP reanalysis. (Reanalyses are blended observational and model output products that provide the best estimate of the climatology of many atmospheric variables, including the circulation. See, for example, Kalnay et al. [1996], for a discussion of the NCAR/NCEP product.) The MMC is dominated by a strong winter hemisphere cell and a very weak (or nonexistent) summer hemisphere cell during solstice months. Near the equinoxes, the cells are of comparable magnitude.

3. A SIMPLIFIED SET OF GOVERNING EQUATIONS FOR THE HADLEY CIRCULATION

To understand why such a circulation occurs, consider the equations that govern large-scale atmospheric flow. These equations are reviewed briefly here, and simplified for treating the tropical MMC. This simplification is based on an examination of the output from a climate model (described in Section 4) and a blended observational/modeling product (the NCEP/NCAR reanalysis), which provide a consistent picture of the dominant terms for maintaining and changing the Hadley cells.

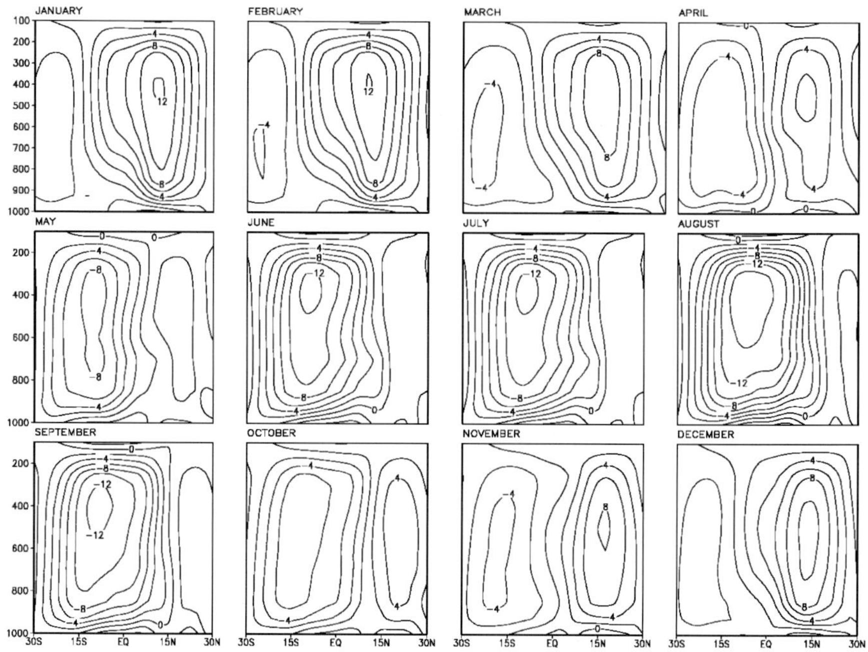

Figure 2-1. Stokes stream function from the NCEP/NCAR Reanalysis climatology for each month. Positive (negative) contours indicate clockwise (counterclockwise) circulation. Contour intervals are 2 x 10^{10} kg/s.

Newton's second law of motion ($F = ma$), the governing equation for motion (wind) in the atmosphere, can be written

$$\vec{a} = \frac{d\vec{v}}{dt} = \frac{\sum \vec{F}}{m}, \tag{5}$$

i.e., changes in velocity ($\vec{v}$) with time (accelerations) are calculated as the sum of the forces, $\vec{F}$, per unit mass, *m*. This so-called Lagrangian framework moves with a parcel of air as it moves around in the atmosphere, and is analogous to following a block of wood sliding down an incline plane in the classic physics problem. To consider any variable, β, (e.g., the wind velocity vector, temperature, or pressure) on a grid that is fixed in space, such as latitude and longitude, the Lagrangian derivative, $d\beta/dt$, is converted into the Eulerian partial derivative, $\partial\beta/\partial t$, by taking advection (transport of β by the wind) into account –

$$d\beta/dt = \partial\beta/\partial t + \vec{v}\cdot\nabla\beta. \tag{6}$$

The vector momentum equation (Eq. 6) can be written in terms of scalar components if a coordinate system is chosen. For simplicity, we choose local Cartesian coordinates with pressure as the vertical coordinate. The east/west wind, *u*, blows along the x axis with unit vector $\hat{i}$ pointing eastward, and the north/south wind, *v*, blows along the y axis with unit vector $\hat{j}$ pointing to the north. Equation 5 can then be written in component form as

$$\frac{\partial u}{\partial t} = -\vec{v}\cdot\nabla u + \sum\frac{F_x}{m} \tag{7}$$

$$\frac{\partial v}{\partial t} = -\vec{v}\cdot\nabla v + \sum\frac{F_y}{m}, \tag{8}$$

where Equation 6 has been used. Equations 7 and 8 are called the horizontal momentum equations because they indicate how momentum per unit mass (i.e., velocity) changes locally in time. (To first order, the vertical equation of motion reduces to a statement of hydrostatic balance, which expresses the idea that vertical velocity is not generated by imbalances between gravitation and vertical pressure gradient forces.)

For large-scale motion, the important forces to consider in the momentum equations are Coriolis, pressure gradient, and frictional forces (dissipation). The first simplification adopted here is to write an approximate form of the Coriolis force (per unit mass), keeping only the dominant terms; this is an excellent approximation for large-scale (thousands of kilometers) atmospheric motion, and has been verified here by an examination of the model output and the reanalysis. Equations 7 and 8 become

$$\frac{\partial u}{\partial t} = -\vec{v} \cdot \nabla_p u - \frac{\partial \Phi}{\partial x} - fv + D_x \tag{9}$$

and

$$\frac{\partial v}{\partial t} = -\vec{v} \cdot \nabla_p v - \frac{\partial \Phi}{\partial y} + fu + D_y, \tag{10}$$

where $f = 2\Omega\sin\phi$, where Ω is the rotation rate of the earth. Note that since pressure is used as the vertical (independent) coordinate, geopotential height, Φ, is the dependent variable that expresses the atmosphere's mass distribution; i.e., the locations of highs and lows.

A commonly used parameterization for the dissipation components, D_x and D_y, is based on the vertical wind shear

$$D_x = -\frac{g}{p^*}\frac{\partial}{\partial \sigma}\left[-\frac{\rho^2 g}{p^*}K_v\frac{\partial u}{\partial \sigma}\right]$$

and

$$D_y = -\frac{g}{p^*}\frac{\partial}{\partial \sigma}\left[-\frac{\rho^2 g}{p^*}K_v\frac{\partial v}{\partial \sigma}\right] \tag{11}$$

Here, p^* is surface pressure, ρ is density, and K_v is a momentum transfer coefficient, σ is a normalized pressure (vertical) coordinate commonly used in models, $\sigma \equiv {}^{p}\!/_{p^*}$. The fluxes of horizontal momentum from the ground to the lowest atmospheric level are typically expressed by the bulk aerodynamic formulation, with the wind stress components given by

$$\tau_x = -\rho C_D V u \text{ and } \tau_y = -\rho C_D V v. \tag{12}$$

V is the total wind speed at the lowest model level. The aerodynamic drag coefficient, C_D is set to 0.001 over ocean and 0.003 over land to represent enhanced momentum fluxes that occur in the more well-developed boundary layers over land.

The momentum equations are further simplified for a first-order analysis of the MMC by averaging over time and longitude. The time mean, denoted below by overbars, should be thought of as an average over many years so time derivatives are negligible. The geopotential height gradient term in the zonal momentum equation is eliminated when the zonal average is taken, and Equations 9 and 10 become

$$0 = f[\overline{v}] - [\overline{\vec{v} \cdot \nabla_p u}] + [\overline{D}_x] \tag{13}$$

and

$$0 = -f[\overline{u}] - \frac{[\partial \overline{\phi}]}{\partial y} - [\overline{\vec{v} \cdot \nabla_p v}] + [\overline{D}_y]. \tag{14}$$

Each term of the simplified *u*-momentum equation (Eq. 13) at 935 hPa is displayed in Figure 2-2a from a July model climatology. Climate model output is used for this evaluation because observations are not sufficiently complete to provide, for example, a climatology of the dissipation terms in Equations 12 and 14. The advection term is calculated as a residual, so any numerical area associated with, for example, calculating derivatives in the other terms is gathered here. The primary *u*-momentum balance is between the Coriolis force and frictional dissipation. Very close to the equator, where the Coriolis force vanishes, and in the summer hemisphere tropics, where the low-level meridional circulation is weak, advection of *u*-momentum balances friction. In the upper troposphere, represented by the 250 hPa level in Figure 2-2b, friction is negligible and the primary balance is between the Coriolis force and nonlinear advection. This balance suggests the relevance of transient and stationary eddies in maintaining the Hadley circulation (see Pfeffer 1980; Held and Phillips 1990; Kim and Lee 2001; Becker and Schmitz 2001; and others).

The *v*-momentum balance, shown at 935 hPa in Figure 2-2c and 250 hPa in Figure 2-2d, is primarily between Coriolis and meridional pressure gradient forces at all levels; i.e., the geostrophic balance. The friction and advection *v*-momentum tendencies are similar in magnitude to those in the *u*-momentum balance, but they are much smaller than the meridional geopotential height gradient and Coriolis forces.

The first law of thermodynamics provides an equation governing atmospheric temperature. The full equation is

$$c_v \frac{dT}{dt} + p \frac{d\alpha}{dt} = J \tag{15}$$

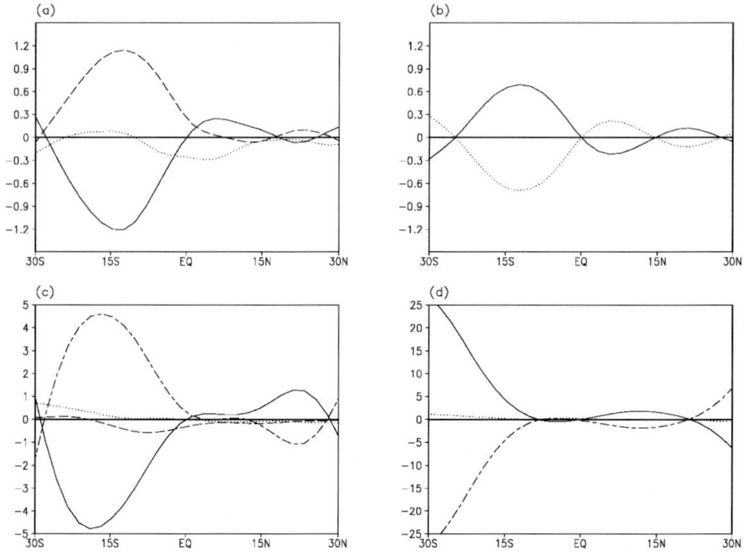

Figure 2-2. Components of the u-momentum balance (Eq. 13) from a model climatology for July at (a) 935 hPa and (b) 250 hPa, and components of the v-momentum balance (Eq. 14) at (c) 935 hPa and (d) 250 hPa. Coriolis forces per unit mass are indicated by solid lines, dissipation (friction) by dashed lines, advection by the dotted lines, and the geopotential height gradient force by the dot-dash line. Units on the vertical axis are 10^{-4} m s^{-2}.

where α is the specific volume (volume occupied by 1 kg of air, or inverse density). Equation 15 states that an air parcel can have two responses to the application of diabatic heating, J. (Diabatic heating of the atmosphere is due to radiation, latent heat release, and sensible heating.) One is a change in temperature (first term, LHS) and the other is adiabatic expansion or compression. Using the perfect gas law, one of Poisson's equations, and Equation 6, Equation 15 is rewritten

$$\left(\frac{\partial T}{\partial t} + \vec{v}_p \cdot \nabla T\right) - S_p \omega = \frac{J}{c_p}, \tag{16}$$

where $\vec{v}_p$ is the horizontal wind vector and S_p is the static stability parameter, defined in terms of potential temperature, Θ, as

$$S_p \equiv -\frac{T}{\Theta}\frac{\partial \Theta}{\partial p}. \tag{17}$$

Then, the climatological, zonally averaged thermodynamic equation is

$$\left[\vec{v}\cdot\nabla_p T\right]-\left[S_p\omega\right]=\left[\frac{\bar{J}}{c_p}\right] \tag{18}$$

Equation 18 states that an applied zonally averaged heating, $\left[\bar{J}/c_p\right]>0$, is balanced either by the advection of cooler air, $\left[\vec{v}\cdot\nabla_p T\right]>0$, or by adiabatic cooling (rising air), $\left[S_p\omega\right]<0$. In the deep tropics, on large space scales, atmospheric heating is primarily balanced by rising motion, because horizontal temperature gradients are weak. A longitude-height cross section of the adiabatic and diabatic heating terms in Equation 18 at 5°S in July is shown in Figure 2-3. (Again, model output is used to examine the full thermodynamic balance, because the heating, *J*, is not well known from observations.) It is clear that, to first order, they balance, and that the heating and vertical motion are concentrated over the continents and the western warm pool of the Pacific. Clearly, the heating field that drives the Hadley circulation is not zonally uniform, even though the circulation itself is, by definition, zonally uniform.

A zonally averaged view of the thermodynamic balance is provided in Figure 2-4, which shows diabatic (solid line) and adiabatic (dashed line) heating rates along with the temperature advection term (dotted line) for July at 568 hPa. From about 5° latitude in the winter (Southern) hemisphere to 23° latitude in the summer (Northern) hemisphere, strong diabatic heating is balanced by adiabatic cooling, with a little help from temperature advection. In the winter hemisphere subtropics, large-scale sinking (adiabatic warming) in the trade wind regime balances diabatic cooling due to the long-wave radiative flux through the low-moisture air of the world's desert regions.

The continuity equation (Eq. 2) completes the set of governing equations. In local Cartesian coordinates,

$$\frac{\partial[\bar{v}]}{\partial y}+\frac{\partial[\bar{\omega}]}{\partial p}=0. \tag{19}$$

Equations 13, 14, 18, and 19 constitute a simplified set of equations governing the MMC, and can be used to discuss how and why the Hadley circulation occurs and varies.

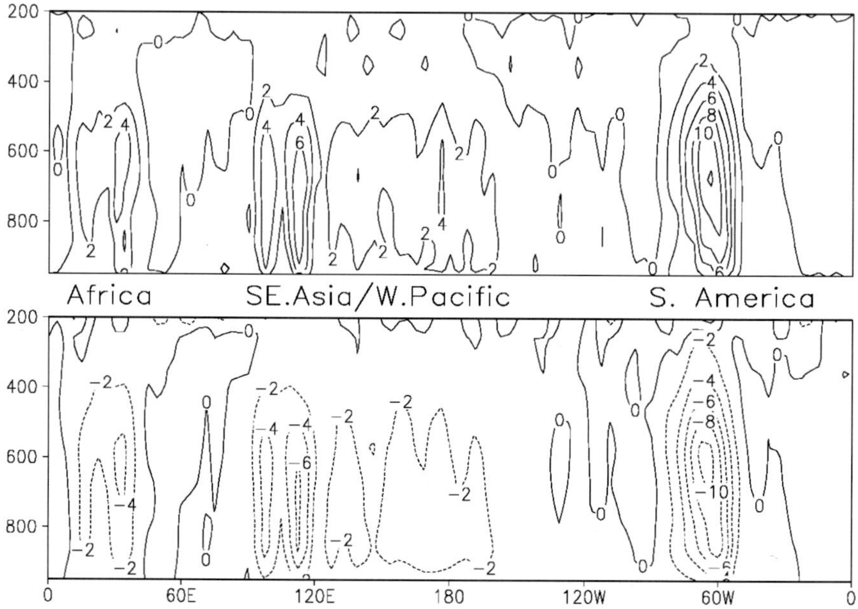

Figure 2-3. Dominant terms in the thermodynamic balance in the tropics (Eq. 18) in July at 3.35°N from a model climatology; (a) diabatic heating, J/c_p and (b) adiabatic cooling $-S_p\omega$. Units are 10^{-5} K s^{-1}.

Solar heating is delivered to the earth's atmosphere from below. The atmosphere is, to first order, transparent to incoming solar radiation, so much of this radiative energy reaches the surface and heats it; the resulting emission of long-wave radiation from the surface is the largest direct source of heating the atmosphere. Also, because of the shape of the earth, more solar energy is delivered at low latitudes than at high latitudes. Two driving mechanisms for the Hadley circulation derive from this structure in atmospheric heating.

First consider the effects of heating the atmosphere from below. Solar heating of the surface is translated into diabatic heating of the atmosphere through surface fluxes of sensible and latent heat (evaporation). The former is deposited into the lower troposphere, and the latter primarily into the middle troposphere; both cool the surface. The tropical air responds by rising to balance the diabatic heating by adiabatic cooling due to uplift (Eq.

18 and Fig. 3-3), and the upward branch of the Hadley circulation forms. The zonal mean meridional velocity responds to conserve mass (Eq. 19), and a Hadley circulation is generated.

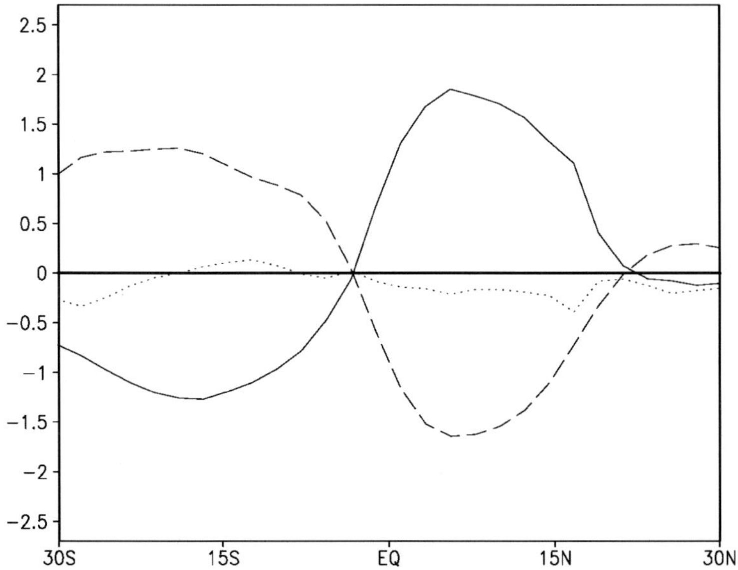

Figure 2-4. Thermodynamic budget (Eq. 18) at 568 hPa in July from a model simulation. Solid line is the diabatic heating term, dashed line is the adiabatic term, and the dotted line is temperature advection. Units are 10^{-5} K_s^{-1}.

Now consider the effects of having warmer surface temperatures at low latitudes. By its definition, the meridional geopotential height gradient at a level *p* is related to the average meridional temperature gradient in the atmosphere below level p; i.e.,

$$\phi(p) \equiv -R \int_{p_s}^{p} T d\ln p, \tag{20}$$

where *R* is the gas constant. Stronger solar heating at the subsolar latitude causes warmer surface temperatures and lower surface pressures. If the heating maximum is on the equator, for example, $\partial\phi/\partial y > 0$ in the Northern Hemisphere and $\partial\phi/\partial y < 0$ in the Southern Hemisphere. According to the primary (geostrophic) balance of meridional momentum (Eq. 14),

$[\bar{u}] \approx -\frac{1}{f}\frac{\partial \phi}{\partial y}$, so easterly flow ($[\bar{u}] < 0$) is generated in the subtropics of both hemispheres (the trades) since $f > 0$ in the Northern Hemisphere and $f < 0$ in the Southern Hemisphere. Easterly flow generates westerly acceleration due to friction and, according to Equation 13, equatorward meridional velocity. By mass conservation (Eq. 19), this equatorward flow must be balanced by upward velocity at the surface, and meridional divergence at the tropopause (where the vertical stability of the lower stratosphere tends to cap vertical motion).

These two processes for driving the Hadley circulation are interdependent and inseparable. For example, one can see the easterly flow of the trade wind regime as a consequence of the Coriolis force acting on the meridional return flow generated through the thermodynamics equation, although it is not clear that the trade wind regime would have its large horizontal extent in the absence of meridional geopotential height (temperature and surface pressure) gradients. But thinking of them as distinct is useful for organizing one's thoughts about how the Hadley circulation is generated, and why it varies on seasonal to paleoclimatic time scales.

4. MODEL SIMULATIONS

Simulations with a three-dimensional climate model are used to investigate the seasonality of the Hadley circulation and the role of continents in determining climatology. The type of model used is a general circulation model (GCM; see Washington and Parkinson [1986] for a more complete description of these models than is possible here.) As in all GCMs, the governing equations are the complete, nonlinear, and time-dependent primitive equations (which were simplified in Section 3). This class of models is capable of producing a realistic representation of the Hadley circulation and its seasonal changes, and provides information about relevant variables for which observed climatologies are not available (e.g., dissipation and diabatic heating rates).

Several model simulations with different prescribed surface boundary conditions are presented. The integration lengths are also different, since boundary conditions with more structure require longer integrations to form a steady climatology. Each run is initialized from a dry isothermal atmosphere at rest, and output from January through June of the first year of the integration is discarded as a spin-up period.

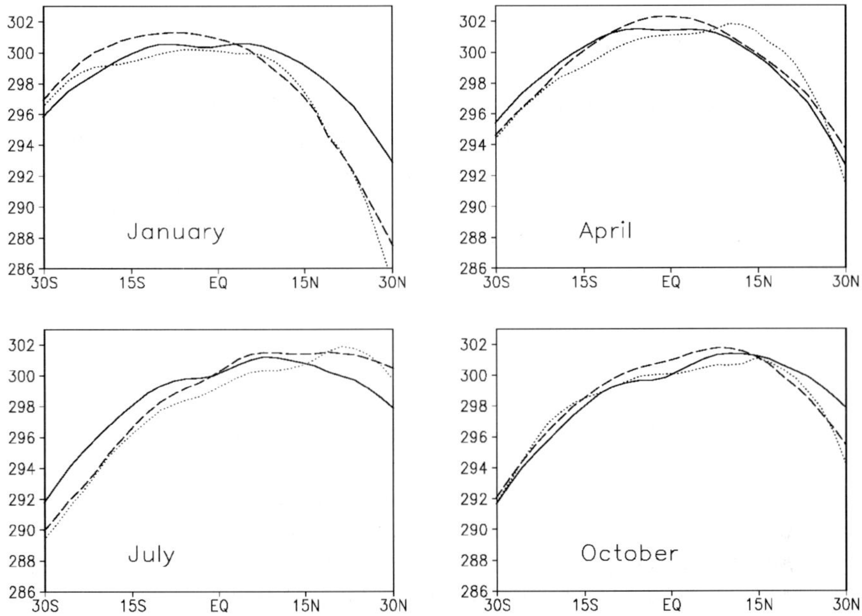

Figure 2-5. Surface temperatures (K) in the GCM simulations. Solid lines indicate zonally averaged observed SSTs (as used in the no-continent simulation); dashed lines are idealized SSTs as imposed in simple models of the Hadley circulation, and dotted lines are from a GCM simulation with idealized continents and observed zonally uniform SSTs.

One simulation has an all-ocean surface, with observed zonally uniform sea surface temperatures (SSTs) from Shea et al. (1990), denoted by the solid lines in Figure 2-5. Note that the observed SSTs are not simple cosine functions of latitude, as they would be if they closely reflected the solar forcing. The SST distribution is nearly flat across the equator in January through May, with slight off-equatorial maxima. During Northern Hemisphere summer and fall—e.g., July and October in Figure 2-5—the SST distribution is less symmetric about the equator, with a single maximum of about 301 K well off the equator. The observed zonal mean SST is influenced by ocean boundary currents and upwelling/downwelling processes. Thus, although continents are not explicitly included in the GCM boundary conditions, the imposed zonally averaged SSTs reflect their influence.

The other two GCM simulations discussed here include continents. Unlike the ocean surface, which has fixed temperatures, land surface temperature is calculated in the model as the result of a surface heat budget. One simulation has flat, featureless continents and observed zonally uniform

SSTs. The resulting zonally averaged surface temperature from this simulation is shown by the dashed lines in Figure 2-5. The summer hemisphere temperatures are warmer, and the winter hemisphere cooler, than in the all-ocean simulation, reflecting the ability of land to heat and cool faster than the ocean. During the equinox seasons, the simulation with idealized land surfaces tends to be warmer than the all-ocean case in the tropics, and the asymmetry of the July (Northern Hemisphere monsoon season) surface temperature distribution is maintained through October.

A simulation with realistic surface features, including topography, realistic soil moisture and surface albedo distributions, and realistic SSTs with longitudinal structure, was also performed. Surface temperatures from this run's climatology are indicated by the dotted lines in Figure 2-5. They are significantly different from the surface temperature distribution in the featureless continent case. These surface features, however, do not introduce significant differences in the MMC compared with the idealized continent simulation, primarily because the differences is temperature are largely associated with different elevations of the surface. For this reason, the analysis below is focused on the simpler case (featureless continents) to address a first-order understanding of the circulation.

5. SEASONALITY OF THE HADLEY CIRCULATION

The simplified set of governing equations written above can be used to provide insight into how and why the Hadley circulation changes seasonally. Examining how the terms in each equation change during the transition from equinox to solstice circulations in the GCM simulation with idealized continents (described above) explains why the summer cell weakens and the winter cell intensifies during this period. The April to July time period is chosen (Fig. 2-6), since the April circulation is neatly symmetric and the strongest winter cell occurs in July (Southern Hemisphere).

Figure 2-7 displays terms from the thermodynamic equation (Eq. 18) for April from the GCM simulation. Compared with July, the diabatic heating and vertical velocity are much closer to and more symmetric about the equator. The heating maximum is stronger in April than in July, but heating amounts are not well correlated with the circulation strength (integrated over the entire Hadley regime) in any of the GCM simulations or in the NCEP/NCAR reanalysis (Cook et al. 2004).

Differences in the 935 hPa zonal momentum balance between April (Fig. 2-8a) and July (Fig. 2-2a) indicate that a stronger winter cell involves increases in the dominant terms; i.e., the westerly acceleration of the trade wind (easterly) flow by friction and its deceleration by the Coriolis force.

Recall that dissipation depends on vertical structure in the zonal wind (Eq. 11). Latent and sensible heating of the atmosphere diminish as winter advances. This change increases the vertical stability of the atmosphere, so the zonal wind shear becomes larger, enhancing the injection of u-momentum into the lower atmosphere and generating a larger meridional velocity (Eq. 13).

In contrast to the zonal momentum balance, the low-level meridional momentum balance does not change very much between the equinox and winter. The winter (Southern) hemisphere geopotential height gradient and Coriolis terms (Fig. 2-2c) are only slightly larger than in the autumn case (Fig. 2-8b). The most notable difference is the equatorward shift of the maxima in both terms. Since a larger zonal velocity is required to balance a given meridional geopotential height gradient closer to the equator (where the Coriolis parameter, f, is smaller), this shift is consistent with the enhancement of the circulation as winter develops.

To understand the weakening of the Hadley circulation in the spring to summer transition, consider the Northern Hemisphere momentum balances in Figure 2-8. In contrast to the winter hemisphere, large changes in the magnitude of the v-momentum balance terms accompany the weakening of the Hadley cell (compare Northern Hemispheres in Figs. 2-2c and 2-8b). The meridional geopotential height gradient weakens by more than a factor of 4 when the continental surfaces in the subtropics warm, and the Coriolis force weakens by a similar amount. The deceleration of the low-level easterlies (i.e., v-momentum Coriolis force) is reflected in a weaker frictional acceleration in the u-momentum balance (compare Figs. 2-2a and 2-8a), weaker meridional flow, and a weaker Hadley circulation.

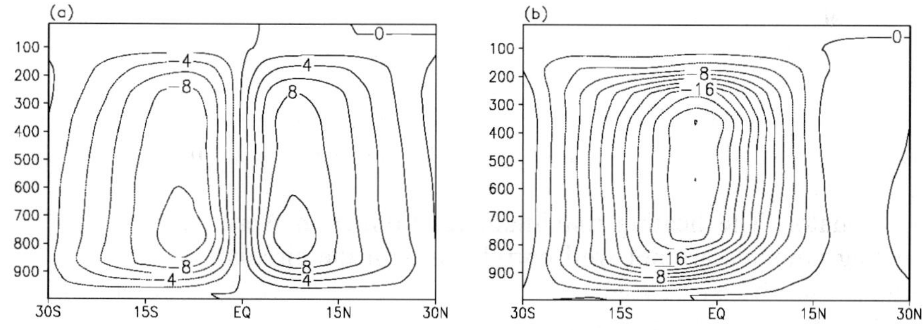

Figure 2-6. Stokes' stream function for (a) April and (b) July from the idealized continent GCM simulation. Contour intervals are 2 x 10^{10} kg/s.

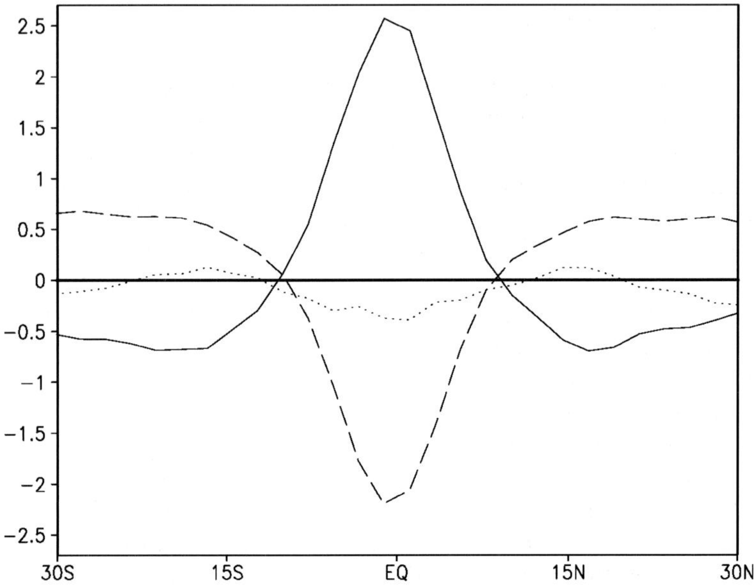

Figure 2-7. Thermodynamic budget (Eq. 18) at 568 hPa in April from a GCM simulation with idealized continents and zonally uniform observed SSTs. Solid line is the diabatic heating term, dashed line is the adiabatic term, and the dotted line is temperature advection (calculated as a residual). Units are 10^{-5} K s^{-1}.

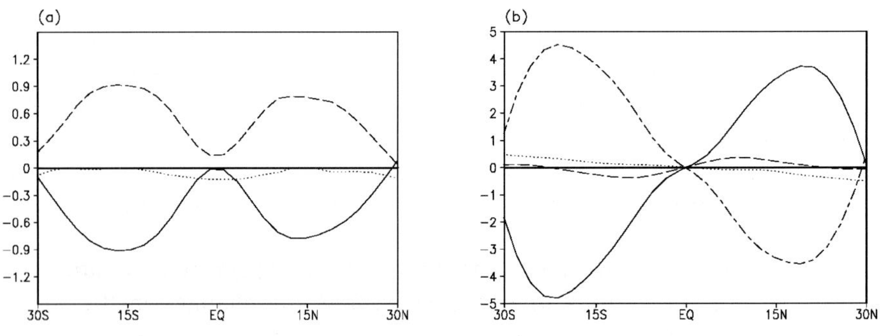

Figure 2-8. Terms of the (a) *u*-momentum (Eq. 13) and (b) *v*-momentum (Eq. 14) balances for April at 925 hPa in the idealized-continent GCM simulation. Solid lines in both (a) and (b) are Coriolis terms, dashed lines represent friction, and dotted lines are the advection terms. In (b), the meridional geopotential height gradient term is denoted by the dot-dashed line. (Units as in Fig. 2-2.)

6. CONTINENTAL HEATING AND THE HADLEY CIRCULATION

As was discussed in Section 1, the Hadley circulation is a zonally averaged quantity by definition, but it is not driven by zonally uniform heating. The ultimate driving force of the Hadley circulation is, of course, the solar energy flux into the climate system, and this energy is delivered into the top of the atmosphere without longitudinal structure. However, most of the solar energy that fuels the troposphere is first absorbed by the surface and converted to long-wave radiation and sensible heating that is deposited in the lower atmosphere from the surface, or converted into latent heat by evaporating water and deposited into the middle troposphere when that water condenses. After this pass through the surface, the energy distribution is no longer zonally uniform.

Figure 2-9 illustrates this point. Surface temperature, which is closely related to sensible heat fluxes and evaporation rates, in July differs by up to 10 K at a given latitude, with significantly higher values in the western ocean basins and over land in the summer hemisphere. Precipitation is also organized by the land/sea distribution, and varies by almost 1 order of magnitude across the tropics even in this coarse-resolution view.

A comparison between the all-ocean GCM simulation with observed zonally uniform SSTs and the simulation with featureless continents and the same SSTs is used to explore the role of continents. Figures 2-10a and b show the Stokes stream function in January and July, respectively, from these two simulations. Without continents, the MMC is stronger in the winter hemisphere than in the summer hemisphere, with the up branch centered near the equator. When featureless continents are introduced at the surface, the winter cell becomes even stronger, and the summer cell weaker, and the center of the up branch moves farther off the equator. As can be seen in Figures 2-10c and d, the presence of continents is associated with a halving of the strength of the Southern Hemisphere summer cell, and the Northern Hemisphere summer cell essentially disappears. Meridional mass transport by the both winter cells approximately doubles when continents are present.

A comparison of the momentum and thermodynamic equations between the two simulations in July reveals how the changes in the surface boundary conditions bring about the differences in the Hadley cells (Cook 2003). Recall that adding continents introduces two differences in the surface boundary conditions; namely, it changes the surface temperature distribution and introduces a rougher surface (more vigorous boundary layer).

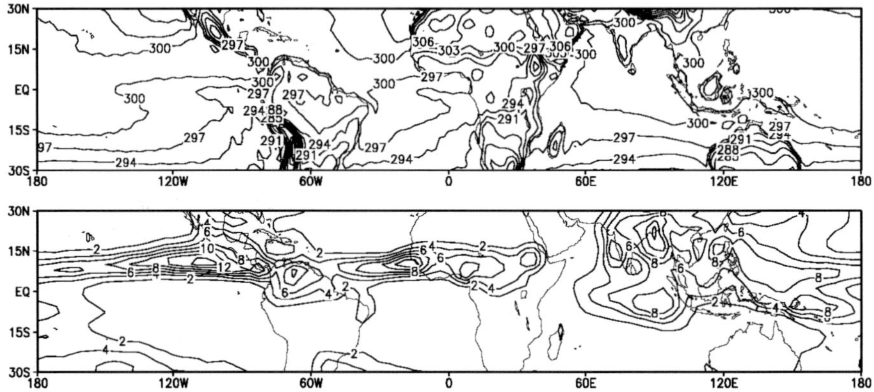

Figure 2-9. July distributions of surface temperature from the NCEP/NCAR reanalysis (top) and precipitation from satellite/gauge blended observations (bottom). Temperature contours are 3 K, and precipitation contours are 2 mm/d.

Figure 2-11 shows the thermodynamic balance in the all-ocean case for July. Compared with the simulation with continents, shown in Figure 2-4, both the diabatic heating and vertical velocity are located closer to the equator and are more concentrated. The maximum values are larger than in the continents case, despite the fact that the winter circulation is weaker in the absence of continents.

The July u- and v-momentum balances for the simulation with no continents are presented in Figure 2-12. Despite the striking intensification of the winter cell due to continents, the v-momentum balance is not very different between the two simulations in the Southern Hemisphere (compare Figs. 2-12b and 2-2c). Surface temperatures in the winter hemisphere are colder over land surfaces, and the surface meridional temperature gradient is stronger as a result, but the cooling is confined to the surface in the vertically stable winter hemisphere, and even at 935 hPa the meridional temperature gradient is very similar in the two simulations.

The u-momentum balance in the winter (Southern) hemisphere, however, is significantly altered by the presence of continents (compare Figs. 2-2a and 2-12a). The increased roughness of the surface (see Eq. 12) enhances the upward flux of u-momentum from the surface and the friction term in Equation 13 increases. This is balanced by an increase in meridional velocity, and the Hadley circulation intensifies.

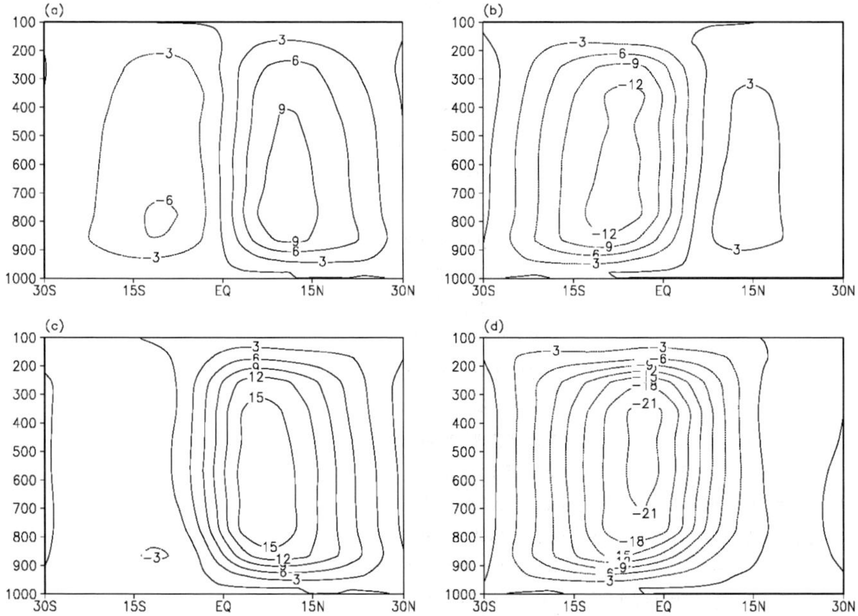

Figure 2-10. Stokes stream function for (a) January and (b) July from a GCM simulation with no continents. Stokes stream function for (c) January and (d) July from a GCM simulation with idealized continents. Contour intervals are 2 x 10^{10} kg/s.

The role of continents in flattening the meridional temperature gradient in the summer hemisphere is clearly seen in the low-level *v*-momentum balance. While the simulation with continents present had essentially constant zonal-mean surface temperature in the Northern Hemisphere tropics (Fig. 2-2c), the meridional temperature gradient in the all-ocean case is appreciable, being about half the magnitude of the winter hemisphere gradient. Since the strong vertical mixing (convection) of the summer atmosphere communicates the surface temperature structure into the low and middle troposphere, the circulation can respond and the result is a stronger summer cell in the simulation with no continents.

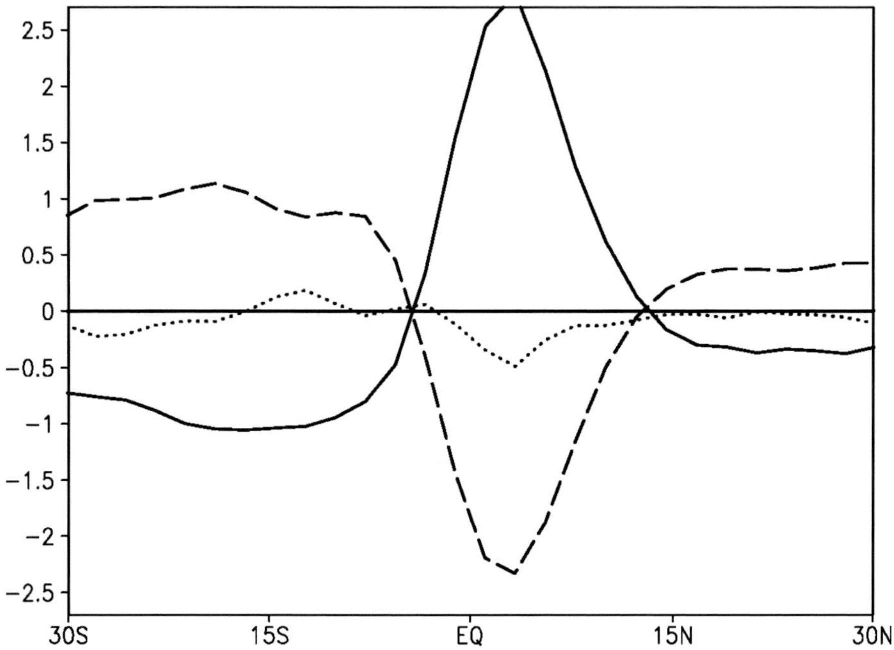

Figure 2-11. Thermodynamic budget (Eq. 18) at 568 hPa in July from a GCM simulation with no continents and zonally uniform observed SSTs. Solid line is the diabatic heating term, dashed line is the adiabatic term, and the dotted line is temperature advection (calculated as a residual). Units are 10^{-5} K s^{-1}.

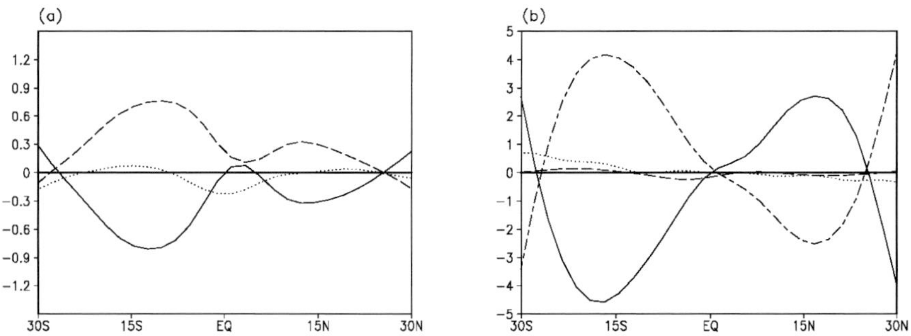

Figure 2-12. Terms of the (a) u-momentum (Eq. 13) and (b) v-momentum (Eq. 14) balances for July in the no-continent GCM simulation. Solid lines in both (a) and (b) are Coriolis terms, dashed lines represent friction, and dotted lines are the advection terms. In (b), the meridional geopotential height gradient term is denoted by the dot-dashed line.

7. SUMMARY

The Hadley circulation is defined in terms of a mass stream function, usually the Stokes stream function. It quantifies air mass transport in the tropics and subtropics and is, by definition, a two-dimensional (zonally averaged) quantity. In the annual mean, the Hadley circulation consists of two equally strong cells, with rising air in the tropics and sinking in the subtropics. But an examination of the monthly mean climatology of the MMC indicates that the winter hemisphere cell is much stronger than the summer hemisphere cell, and this asymmetric circulation dominates for much of the year.

A set of equations, simplified from the full primitive equations, captures the first-order physical processes of the Hadley circulation dynamics. Consideration of the zonally averaged, climatological thermodynamic balance shows that vertical motion results from heating the troposphere in the tropics, in contrast to the mid-latitude response, which tends to balance heating with the horizontal transport (advection) of cooler air. Constraints of mass conservation in the zonally averaged framework require low-level meridional flow into regions of upward motion, and outflow aloft at the base of the vertically stable stratosphere (i.e., near the tropopause). The circulation is further intensified by the resulting release of latent heat.

The zonally averaged horizontal momentum equations express the role of meridional temperature and pressure gradients imposed by the shape of the solar forcing in driving the Hadley circulation. Higher temperature and lower surface pressure at the latitude of maximum heat flux from the surface impose meridional geopotential height gradients that are associated with zonal velocities through the meridional momentum balance, which is essentially geostrophic even within 5° latitude of the equator. According to the zonal momentum balance, zonal frictional acceleration is primarily balanced by meridional flow. Again, the continuity equation connects meridional convergence with vertical motion, and a Hadley circulation results.

The seasonality of the Hadley circulation is not completely understood by thinking only about low-level meridional convergence driven by a zonally uniform diabatic heating maximum near the equator. For most of the year, the heating is well off the equator, especially locally, and momentum balances and zonal velocities are important elements for explaining the features and variations of the circulation. The intensification of the winter cell comes about through the u-momentum balance, when enhanced vertical wind shear and frictional dissipation are balanced by meridional flow. The

v-momentum balance, which is largely a reflection of the role of meridional temperature gradients on the circulation, is not a driving factor for the winter cell intensification because of the high vertical stability of the atmosphere. The weakening of the cell in the spring-to-summer transition, however, is closely related to the flattening of the meridional temperature gradients as the surface responds to heating excursions off the equator. Convection communicates the weakening meridional temperature and geopotential height gradients through the depth of the troposphere, and the zonal circulation weakens as well according to the geostrophic balance. The tight coupling between the two horizontal directions of motion in the rotating atmosphere, in this case via the dissipation term in the u-momentum equation, means that the meridional velocity must weaken as well.

The role of the continents in determining the Hadley circulation climatology was investigated because land/sea contrasts at the earth's surface are responsible for the marked longitudinal structure in the heating that drives the circulation. In addition to being associated with enhanced fluxes of momentum from the surface, the lower heat capacity of the continents, as compared with the ocean surface, decreases the summer hemisphere meridional temperature gradients and strengthens the winter hemisphere gradient. This modification of the surface temperature distribution is responsible for weakening the summer hemisphere cells. However, the increased meridional temperature gradients associated with land in the winter hemisphere are not mapped very effectively into the troposphere because the atmosphere is vertically stable in winter. Instead, the increase in surface roughness over the continents is responsible for the enhancement of the winter cell compared to the case with no continents.

The Hadley circulation is the largest circulation system on the planet, directly influencing half the surface area of the earth. Understanding how it may have been different in the past, and how it may change in the future, is essential for improving our understanding of long-period climate variability. Geological evidence of past climate is often a measurement at a point, and the challenge of deriving information about the Hadley circulation from this evidence is aided by caution and a consideration of the physics of the circulation.

8. REFERENCES

Becker, E., and G. Schmitz. 2001. Interaction between extratropical stationary waves and the zonal mean circulation. *Journal of the Atmospheric Sciences* 58: 462–480.

Cook, K.H. 2003. Role of continents in driving the Hadley cells. *Journal of the Atmospheric Sciences* 60: 957–976.

Cook, K.H., L.L. Greene, and B.N. Belcher. 2004. Seasonal forcing of the Hadley circulation. *Quarterly Journal of the Royal Meteorological Society* (submitted).

Held, I.M., and P.J. Phillips. 1990. A barotropic model of the interaction between the Hadley cell and a Rossby wave. *Journal of the Atmospheric Sciences* 47: 856–869.

Kalnay, E., M. Kanamitsu, R. Kistler, W. Collins, D. Deaven, L. Gandin, M. Iredell, S. Saha, G. White, J. Woollen, Y. Zhu, M. Chelliah, W. Ebisuzaki, W. Higgins, J. .Janowiak, K.C. Mo, C. Ropelewski, J. Wang, A. Leetma, R. Reynolds, R. Jenne, and D. Joseph. 1996. The NCEP/NCAR 40-year reanalysis project. *Bulletin of the American Meteorological Society* 77: 437–471 (see also NCEP/NCAR Reanalysis Electronic Atlas, webmaster@cdc.noaa.gov).

Kim, H.K., and S. Lee. 2001. Hadley cell dynamics in a primitive equation model. Part II: Nonaxisymmetric flow. *Journal of the Atmospheric Sciences* 58: 2859–2871.

Pfeffer, R.L. 1980. Wave-mean flow interactions in the atmosphere. *Journal of the Atmospheric Sciences* 38: 1340–1359.

Shea, D.J., K.E. Trenberth, and R.W. Reynolds. 1990. A global monthly sea surface temperature climatology. NCAR Tech. Note, NCAR/TN-345+STR. (Available from National Center for Atmospheric Research, P.O. Box 3000, Boulder, CO 80307-3000.)

Washington, W.M., and C.L. Parkinson. 1986. *An Introduction to Three-Dimensional Climate Modeling*. Mill Valley, California: University Science Books, 422 pp.

Chapter 3

CHANGE IN THE TROPICAL HADLEY CELL SINCE 1950

Xiao-Wei Quan, Henry F. Diaz, and Martin P. Hoerling
NOAA-CIRES Climate Diagnostic Center, 325 Broadway, Boulder, Colorado 80305 U.S.A.

Abstract The change in the tropical Hadley cell since 1950 is examined within the context of the long-term warming in global surface temperatures. The study involves analyses of observations, including various metrics of the Hadley cell, and ensemble 50-year simulations by an atmospheric general circulation model (AGCM) forced with the observed evolution of global sea surface temperature since 1950. Consistent evidence is found for an intensification of the Northern Hemisphere winter Hadley cell since 1950. This is shown to be an atmospheric response to the observed tropical ocean warming trend, together with an intensification in El Niño's interannual fluctuations, including larger amplitude and increased frequency after 1976. The intensification of the winter Hadley cell is shown to be associated with an intensified hydrological cycle consisting of increased equatorial oceanic rainfall, and a general drying of tropical/subtropical landmasses. This Hadley cell change is consistent with previously documented dynamic changes in the extratropics, including a strengthening of westerly atmospheric flow and an intensification of mid-latitude cyclones.

1. INTRODUCTION

The tropical Hadley cell, by definition, is the zonal mean meridional mass circulation in the atmosphere bounded roughly by 30°S and 30°N. It is characterized by equatorward mass transport by the prevailing trade wind flow in the lower troposphere, and poleward mass transport in the upper tro-

H.F. Diaz and R.S. Bradley (eds.), The Hadley Circulation: Present, Past and Future, 85–120.

posphere. This lateral mass circulation links the mean ascending motion in the equatorial zone with subsidence in the subtropics, and represents a major part of the large-scale meridional overturning between the tropics and subtropics. In this chapter we examine the long-term change of the tropical Hadley cell as an expression of climate change associated with a warming trend in global mean surface temperature during the last half of the 20th century (IPCC 1996, 2001).

Previous observational studies have confirmed that the global annual mean surface temperature has increased about 0.6ºC during the past century (e.g., Jones et al. 1999). The warming has occurred in a steplike manner having two phases, one from about 1910 to 1945, and a second after the mid-1970s. Recent observational studies of the global pattern of temperature changes have revealed that, since the late 1970s, the warming trend in global land surface air temperature (LSAT) is larger than the warming trend in sea surface temperature (SST) (IPCC 2001). The recent temperature changes over the land-masses of the northern mid-latitudes during winter appears related to changes in atmospheric circulation (e.g., Hurrell 1996; Gaffen et al. 2000; Parker 2000; Santer et al. 2000). Pronounced changes in the wintertime atmospheric circulation have occurred since the mid-1970s over the Northern Hemisphere. The variations over the North Atlantic are related to changes in the North Atlantic Oscillation (NAO), and the changes over the North Pacific involve variations in the Aleutian Low with teleconnections downstream over North America.

To what extent are these changes in subtropical and mid-latitude circulation systems linked to a tropical source, rather than being a mere expression of intrinsic extratropical climate noise? And furthermore, are these circulation changes an indication of the atmospheric response to an intensified tropical Hadley cell? There is evidence for an intensification in marine surface wind since 1950, as recorded by observations from ships sailing over the global oceans (Diaz et al. 1992, 1994). The intensification in the observed ship-based marine surface wind is most significant in their zonal (east/west) component during the northern winter since the late 1970s. The intensification of the northern winter zonal wind partially reflects the intensification of the Aleutian Low and prevailing westerlies over the mid-latitude central and eastern Pacific (e.g., Trenberth and Hurrell 1994; Graham 1994), and the strengthening of mid-latitude westerly winds over the North Atlantic Ocean associated with a trend of the NAO toward its positive phase (e.g., Hurrell 1995; Thompson et al. 2000). Observational studies also have found that since 1948, the frequency and intensity of extreme cyclones have increased markedly over the North Pacific Ocean during the northern winter (Graham and Diaz 2001).

It has also been argued that the intensification of the extratropical northern winter circulation is a result of increasing ocean surface temperature, particularly in the tropics (e.g., Hoerling et al. 2001a for the tropical origins of the NAO change; and Graham et al. 1994; Lau and Nath 1994, 1996 for the tropical forcing of changes over the North Pacific; and Trenberth et al. 1998). Model simulations indicate that the atmospheric angular momentum increases in response to a warming of tropical SSTs, and there is an indication that the observed atmospheric angular momentum has itself increased since 1950 (Huang et al. 2001, 2003). Increasing angular momentum (i.e., increasing westerly flow) can be due to a strengthened tropical Hadley cell and/or increased eddy forcing from the mid-latitudes. That an intensified Hadley cell due to diabatic forcing may be particularly relevant is implied by model evidence for a substantial increase in zonal mean equatorial rainfall since 1950 (Kumar et al. 2004), which would lead a stronger mean ascending branch of the Hadley cell. Such changes in diabatic heating are consistent with the overall tropical SST warming (Hurrell et al. 2004).

A significant component of the warming trend in the global ocean (Levitus et al. 2000) has been the strong warming in the tropical ocean during recent decades (e.g., Lau and Weng 1999). As is shown in Figure 3-1, since 1950, strong warming in the tropics has occurred in the Indian Ocean, the western Pacific, along the coasts of Southeast Asia, off equatorial Indonesia, and in the Atlantic Ocean. The origins for these warming trends are still under investigation, though the Indo-West Pacific warming has been argued to be inconsistent with intrinsic unforced coupled ocean-atmosphere interaction alone (Knutson et al. 1999).

The tropical Indian and western Pacific Oceans possess the warmest water in the global ocean, with SSTs often higher than 28°C, and this is usually referred to as the "warm-pool" region. Because of the already warm surface, atmospheric convection is sensitive to small temperature changes in the warm-pool region. It is interesting to note that the warming in the warm-pool region has occurred in a manner that is somewhat different from the warming in the eastern tropical Pacific. The SSTs in the tropical Indian and West Pacific region have been increasing since 1950, with a steplike increase in the late 1970s (lower panel in Fig. 3-2). This change is large compared to the small natural variability (Latif et al. 1997), suggesting it is consistent with the notion that the warming has been externally forced (Hoerling et al. 2004).

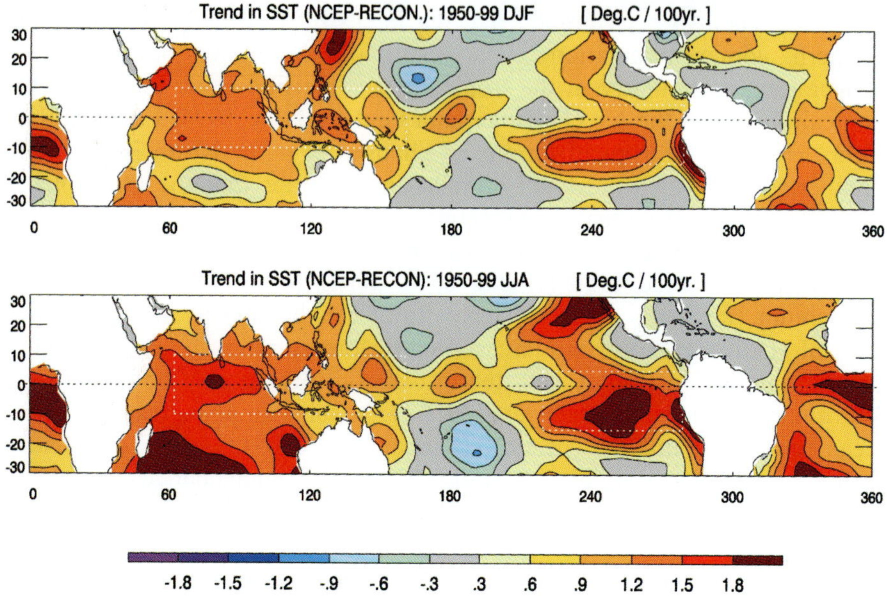

Figure 3-1. Spatial distribution of the linear trend of tropical SSTs. The data used are from Smith et al. (1996).

The manner in which the atmospheric circulation has adjusted to such changes in the tropical oceans is still an open question. Recent satellite observations indicate that the tropical Hadley cell has been strengthening (e.g. Chen et al. 2002), but others argue that the signal of the change in the Hadley cell is not large enough to exceed the range of uncertainty in current observational systems (e.g., Trenberth 2002). In this chapter, we analyze the temporal evolution of the tropical Hadley cell since 1950. An analysis on longer-term change in the tropical Hadley cell based on marine surface wind observations can be found in Evans and Kaplan (2004, this volume). Data used for the analyses in this chapter are described in section 2. The long-term changes in the Hadley cell indicated in observational data are examined in section 3, followed by analyses on the change of the Hadley cell in atmospheric model simulations in section 4. Conclusions and further discussions about the nature of the long-term changes in the Hadley cell are presented in section 5.

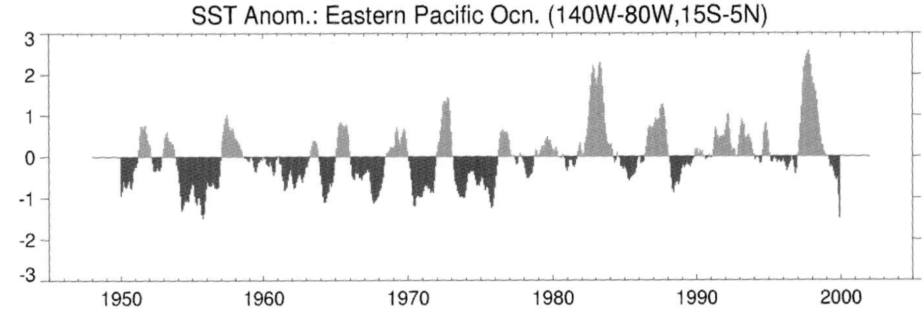

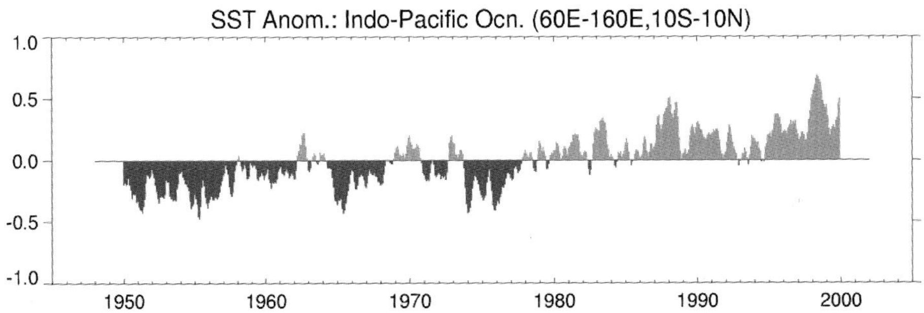

Figure 3-2. Time series of spatial mean SST anomalies with respect to 1950–99 climatology.

2. Data

Three independent data sets are used in this study. The monthly mean data of the NCEP/NCAR reanalysis (Kistler et al. 2001) covers the period from January 1948 to the present. Values of monthly mean zonal and meridional (north/south) wind components are available at 17 pressure levels at 2.5° longitude by 2.5° latitude grid points. The NCEP/NCAR reanalysis data are not a purely observed data set. They are a mix of real observations with model simulations using the method of temporal and spatial assimilation in an atmospheric general circulation model (AGCM). Insofar as different data platforms have been used in constructing the reanalysis, long-term trends calculated from it may be nonphysical.

In order to clarify whether the change of the tropical Hadley cell diagnosed from the reanalysis data is a physical signal of true climate change, we also analyze an ensemble of 50-year AGCM simulations. The ensemble includes 10 simulations of an AGCM developed at the European Center for Medium Range Weather Forecast (ECMWF), and the Max-Plank Institute at Hamburg, ECHAM-3 (Roeckner et al. 1992). All 10 simulations were identically forced with the observed monthly global SST evolution from 1950 through 1999, and the 10 simulations differ from each other by starting from different initial conditions. Simulations of the monthly mean precipitation, and 850 and 200 hPa wind fields, are analyzed. The spatial resolution of the model experiments is about 2.8° longitude by 2.8° latitude.

The third data set is the satellite rain gauge combined precipitation data set produced by the Global Precipitation Climatology Project (GPCP, see, Huffman et al. 1997). The GPCP precipitation data are available in the form of monthly means on 2.5° by 2.5° grid points, and cover the period from 1980 to the present.

3. CHANGE OF THE TROPICAL HADLEY CELL IN THE NCEP REANALYSIS

3.1. Climatology of the Hadley Cell

A conventional way to depict the tropical Hadley cell is to use the stream function of zonal mean meridional and vertical velocity in the meridional vertical plane (e.g., Oort and Yienger 1996). The 1950–99 long-term average of the annual mean and also the seasonal cycle of the stream function in the NCEP/NCAR reanalysis are shown in Figure 3-3. As represented by the contour-lines of the annual mean stream function (top panel in Fig. 3-3), the tropical Hadley cell is a major component of the global mass circulation, which consists also of the Ferrel cell in mid-latitudes, and the polar cell in high latitudes. In the annual mean climatology, there are two Hadley cells in the tropics, one on each side of the equator, and both are much more intense than the Ferrel and polar cells in the extratropics. The annual mean tropical Hadley cell in the Southern Hemisphere (SH) is stronger than its counterpart in the Northern Hemisphere (NH). The dividing latitude of the two Hadley cells, roughly corresponding to the latitude of mean ascent, is located north of the equator, reflecting the fact that the Intertropical Convergence Zone (ITCZ) remains in the NH throughout the year. A comprehensive review of the issues concerning the position of the

ITCZ is given by Xie (2004) in this volume. An analysis of the biases in previous calculations of the Hadley circulation using *in situ* rawinsonde data can be found in the paper by Waliser et al. (1999).

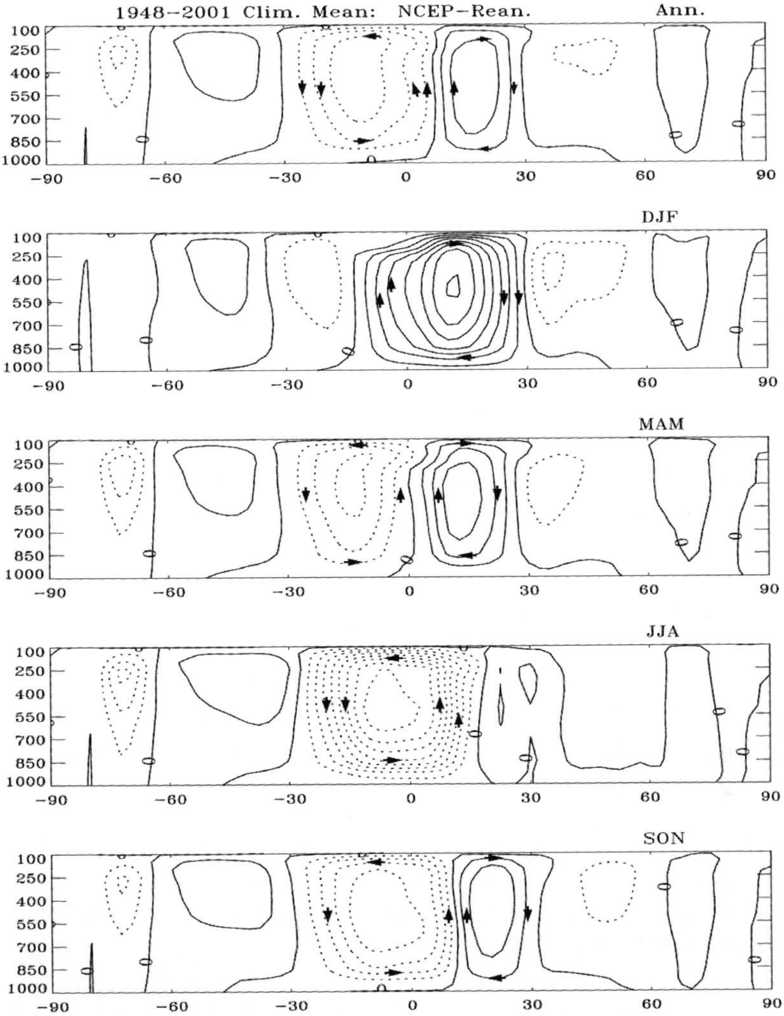

Figure 3-3. Stream function of the zonal mean meridional wind circulation based on the NCEP/NCAR reanalysis. Units are in 10^{10} kg s $^{-1}$ and the contour interval is 2 x 10^{10} kg s $^{-1}$.

The tropical Hadley cells vary strongly with the changes in seasons. For example, the ascending branch of the Hadley cell migrates seasonally across the equator in response to the solar annual cycle, and the structure (for instance, the polarity, and single versus double cell) of the tropical Hadley cell adjusts accordingly. A study by Dima and Wallace (2003) shows that the seasonal cycle of the climatological mean tropical Hadley cells is dominated by two components of roughly comparable amplitude: a seasonally invariant pair of Hadley cells with rising motion centered near and just to the north of the equator and subsidence in the subtropics (e.g., the Hadley cells of annual mean, March–May (MAM, and September–October (SON) in Fig. 3-3), and a seasonally reversing, solstitial cell with ascent in the outer tropics of the summer hemisphere and subsidence in the outer tropics of the winter hemisphere (e.g., the Hadley cells of December–February (DJF) and June–August (JJA in Fig. 3-3). The seasonality of the tropical Hadley cell represents only part of the much stronger seasonality and regional migration of the monsoon systems over Asia and the Americas (e.g., Webster 1987; Philander 1990; Trenberth et al. 2000; Webster, this volume).

3.2. Interannual Variability in the Hadley Cell

The interannual variability of the tropical Hadley cell is dominated by the ocean-atmosphere variability associated with the El Niño/Southern Oscillation (ENSO). This connection has previously been illustrated by composite methods that difference the mean state of the Hadley cell during the warm phase of ENSO from its cold phase counterpart. Such composite differencing extracts the linear component of the Hadley cell's response. Oort and Yienger (1996) applied this method to radiosonde observations and found that the Hadley cell's linear response to ENSO consists of a pair of anomalous direct meridional cells symmetric about the equator. Their analysis indicates that the two anomalous cells are strengthened (weakened) during El Niño (La Niña) events.

Using the same methodology, the composite linear response of the Hadley cell to ENSO in the NCEP/NCAR reanalysis data is shown in the top panel in Figure 3-4. The years included in the composite are listed in Table 3-1. Waliser et al. (1999) indicated that differences in structure exist between the radiosonde-based and the reanalysis-based composites, and they also noted that the composite linear response in the reanalysis is weaker than that in the radiosonde data. Their analysis also showed that the differences between the two composites are largely attributed to the sparse spatial coverage of the in situ data. Nevertheless, our results are in qualitative agreement with those of Oort and Yienger (1996) insofar as the linear El Niño signal during northern winter consists of anomalous ascent located south of the equator, with subtropical sinking in both hemispheres. In contrast to the results of Oort and Yienger (1996), our analysis also shows a secondary anomalous southward overturning meridional cell extending from about 15°N to 30°N, and shows a stronger anomalous circulation in the upper troposphere.

Table 1. List of El Niño and La Niña event years (1950–2002) used in the composite analysis. The year is considered El Niño (La Niña) when the DJF Niño-3.4 SST index exceeds ±1 standard deviation.

El Niño	La Niña
1958	1951
1966	1955
1969	1956
1973	1965
1983	1971
1987	1974
1992	1976
1995	1985
1998	1989
	1999
	2000

We present new evidence that the nonlinear component of the Hadley cell's response to ENSO is comparable to the linear signal. The nonlinear component is defined here as the sum of Hadley cell anomalies for El Niño and La Niña. In the case that these anomalies are of exactly equal amplitude but opposite polarity, the nonlinear component would be judged to be zero. Figure 3-4 compares the composite anomalous Hadley cell pattern for the El Niño events (Fig. 3-4, 2nd panel from top) with its counterpart for the La Niña events (Fig. 3-4, 3rd panel from top). The two composites are different in two aspects: (1) the composite of the anomalous Hadley cell shows a stronger response to the El Niño than to the La Niña events, and (2) the anomalous Hadley cell for the El Niño events is characterized by a single strong anomalous cell over the equator, while a near-equatorial anomalous Hadley cell during La Niña events is virtually absent. This nonlinearity is further illustrated in the bottom panel in Figure 3-4, which shows the sum of the composites for El Niño and La Niña events. It is clear that the one-cell structure in the nonlinear response of the Hadley cell to the ENSO extreme phase is comparable to its linear signal (Fig. 3-4, top panel).

We further examine the composites for the subperiods of 1950–75 and 1976–2002 in Figures 3-5 and 3-6, respectively, to assess the robustness of the ENSO signal in the Hadley cell. The composite patterns are obtained by using the same method as that used for Figure 3-4, except that all the anomalies in Figures 3-5 and 3-6 are based on the climatological mean for the 1950–75 and 1976–2002 periods, respectively. The purpose of using different climatologies for the two shorter periods is to reduce the effect of any possible artificial change in the climate mean state in the reanalysis data as discussed in section 2. The linear signals are qualitatively similar for the two periods, though with stronger amplitudes in recent decades (top panels in Figs. 3-5 and 3-6). The nonlinear component in the Hadley cell's responses to ENSO is virtually absent in the composite of 1950–75, and is apparent in the 1976–2002 period (bottom panel in Fig. 3-6). This difference is due mainly to an interdecadal change in the tropical Hadley cell's response to the warm events. The Hadley cell had responded to the El Niño events with a largely intensified southward overturning anomalous cell and a slightly increased northward overturning anomalous cell before 1976 (2nd panel from top in Fig.3-5), compared to a much stronger northward overturning anomalous cell with a slightly increased southward overturning cell for the El Niño events after 1976 (2nd panel from top in Fig. 3-6). There is also a stronger northward overturning anomalous cell over the Northern Hemispheric subtropics for the La Niña events after 1976 (3rd panel from top in Fig. 3-6).

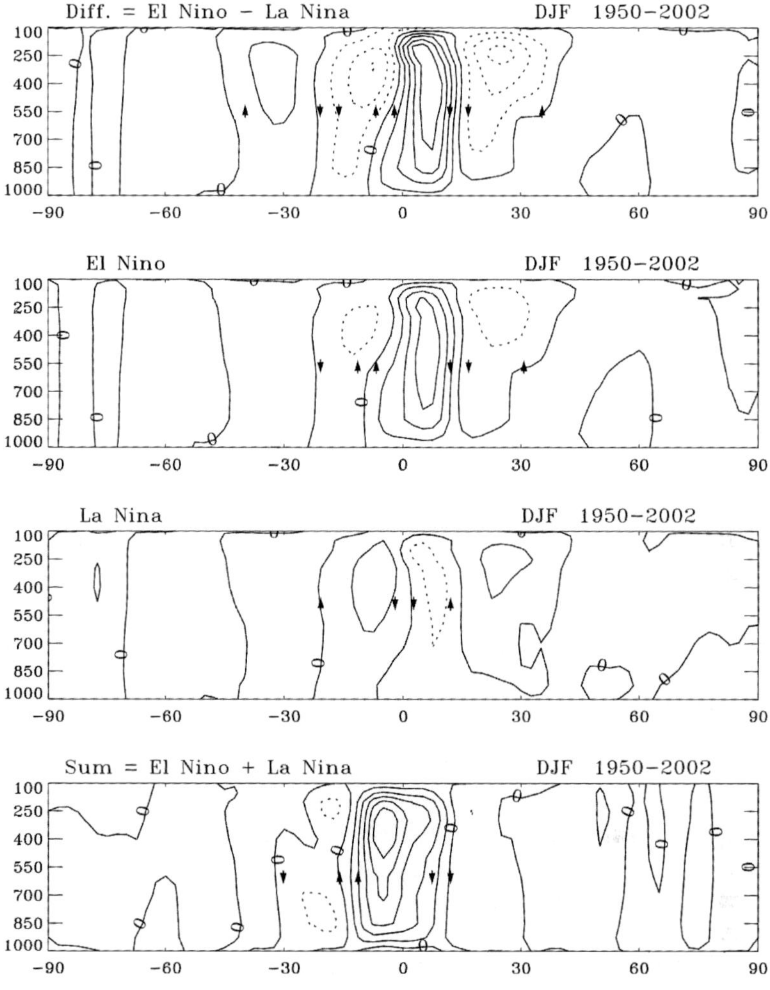

Figure 3-4. Difference in the anomalous zonal mean meridional stream function between El Niño and La Niña periods (top panel). The composite anomalous pattern of the zonal mean meridional stream function for El Niño (the 2nd panel from top) and La Niña (3rd from top) periods. And the sum of the two composites (bottom panel). Contour intervals are 4 x 10^9 kg s^{-1} in the top three panels, and 2 x 10^9 kg s^{-1} in the bottom panel.

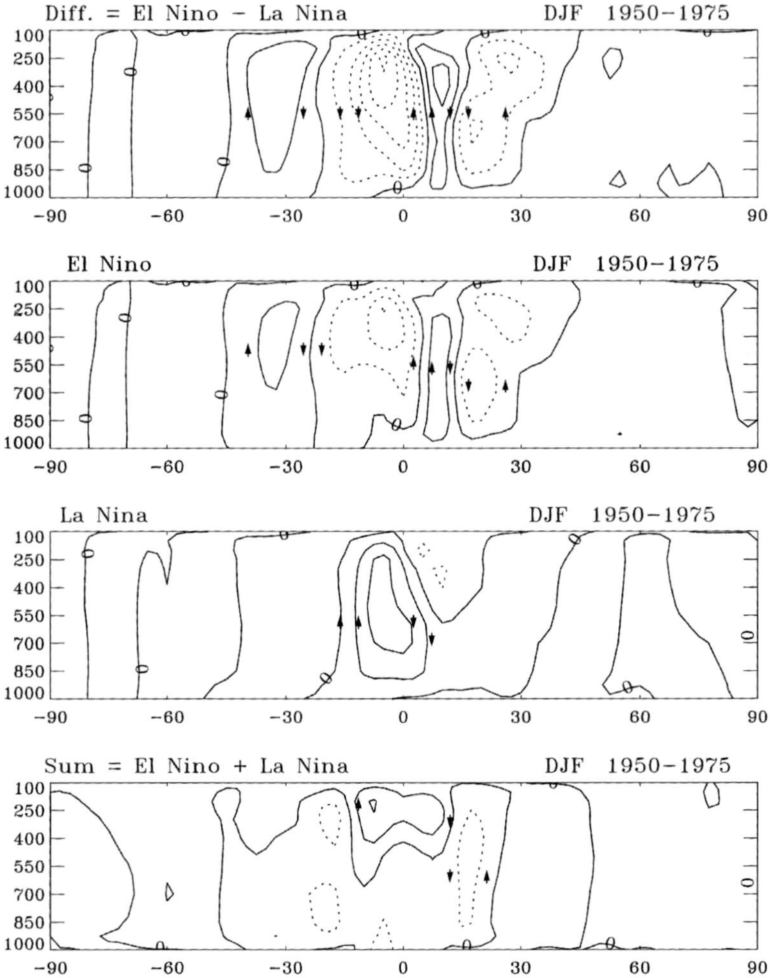

Figure 3-5. Difference in the anomalous zonal mean meridional stream function between El Niño and La Niña events (top panel) during the period 1950–75. The composite anomalous pattern of the zonal mean meridional stream function for El Niño (2nd panel from top) and La Niña (3rd from top) periods, and the sum of the two composites (bottom panel). Contour intervals are 4 x 10^9 kg s^{-1}.

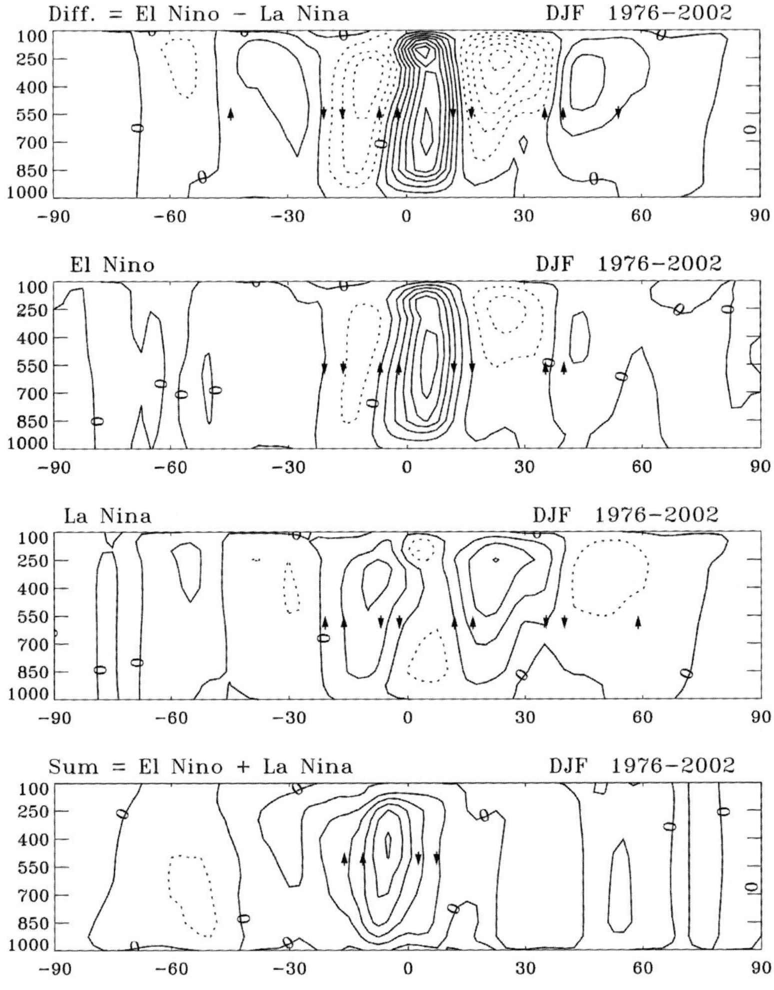

Figure 3-6. Same as Fig. 3-5, but for the period 1976–2002.

It should be emphasized that these epoch differences are being estimated from small sample sizes, and thus are subject to appreciable sampling error. Nevertheless, it will be shown in section 4 that similar epoch differences in the Hadley cell's responses to ENSO occur in a large ensemble of climate simulations.

3.3. Interdecadal Change in the Hadley Cell

In order to describe the temporal variation of the Hadley cell's intensity, we use the maximum value of the zonal mean stream function occurring within the latitudinal zone of 0°–30°N. This index measures the strength of the northward overturning Hadley cell, and the minimum value of the zonal mean stream function within 0°–30°S measures the strength of the southward overturning Hadley cell (Oort and Yienger 1996). Time series for the monthly mean strength of the northward and southward overturning Hadley cells are shown in Figure 3-7. It is interesting to note that the northward overturning Hadley cell shows a trend toward intensification for the past 50 years. But the southward overturning Hadley cell does not show a similar long-term trend. The southward overturning Hadley cell shows an interdecadal swing with the index being low (i.e., intensified southward overturning) during the decades of the 1950s and 1960s, high (or weakened southward overturning) during the 1970s and 1980s, and low again during the 1990s. Since the northward overturning Hadley cell dominates during the northern winter, and the southward overturning Hadley cell prevails during the southern winter (Fig. 3-3), the difference between the long-term changes in the northward and southward overturning Hadley cells implies that the tropical Hadley cell has intensified in the northern winter but not in the southern winter. This seasonal difference in the tropical Hadley cells' long-term change is further illustrated in Figure 3-8, which shows the difference between the temporal average of the zonal mean meridional stream function for two periods: 1948–75 and 1976 to the present for the annual mean, and four cardinal seasons. We have selected these two periods in light of the different Hadley cell behaviors seen in both the reanalysis and the ensemble AGCM simulations (section 4) before and after 1975. The intensification in the annual mean Hadley cell is due mainly to the intensification of the northward overturning Hadley cell during the northern winter (DJF) and spring (MAM). The interdecadal difference in the zonal mean stream function for the two periods appears to be very small during the northern summer (JJA) and autumn (SON).

To further diagnose the structure, in addition to the amplitude, of the variation in the tropical Hadley cells since 1950, we calculated the vertical shear of the zonally averaged meridional velocity between 200 and 850 hPa. This depiction of the tropical Hadley cells is illustrated in Figure 3-9, in which the close correspondence between the seasonal reversal of the solstice Hadley cell and the change of the vertical shear in the zonal mean meridional wind can be clearly seen.

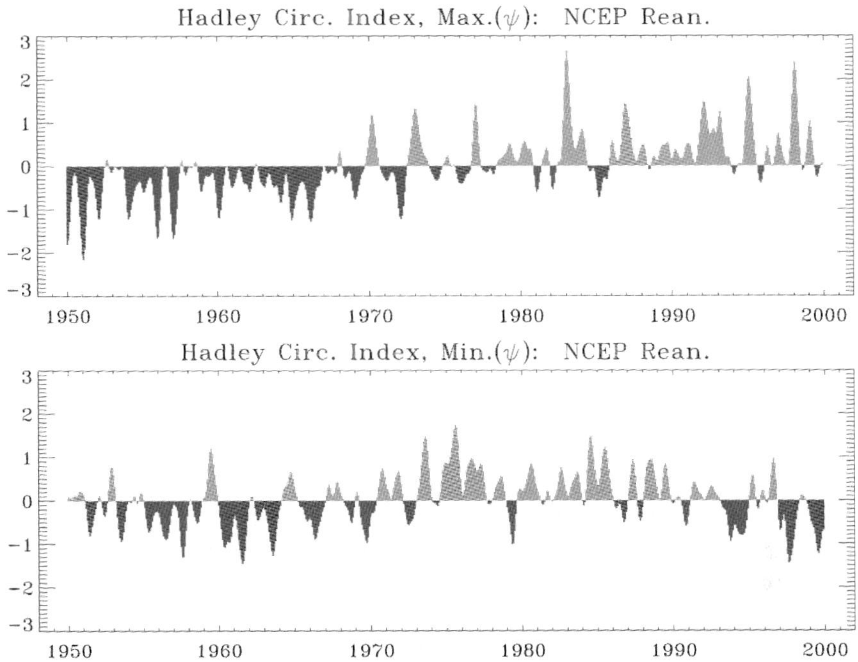

Figure 3-7 Time series of index for the tropical Hadley cells. Temporal variation of the northward/southward overturning cell is represented by the maximum/minimum of the zonal mean meridional stream function (upper/lower panel). Positive/negative values indicate stronger northward/southward overturning. The time series are smoothed with an 11-month weighted filter. Units are in 10^{10} kg s^{-1}

Since the long-term changes in the tropical Hadley cells during the northern winter and spring are quite different from their counterparts during the northern summer and fall, it is necessary to separately examine the long-term Hadley cell change for winter and summer. The temporal variation in the zonal distribution of the upper-minus-lower anomalous meridional wind is examined for JJA and DJF, respectively, in Figure 3-10. This diagnosis of the long-term change of the tropical Hadley cell is consistent with the results from the analysis based on the zonal mean meridional stream function.

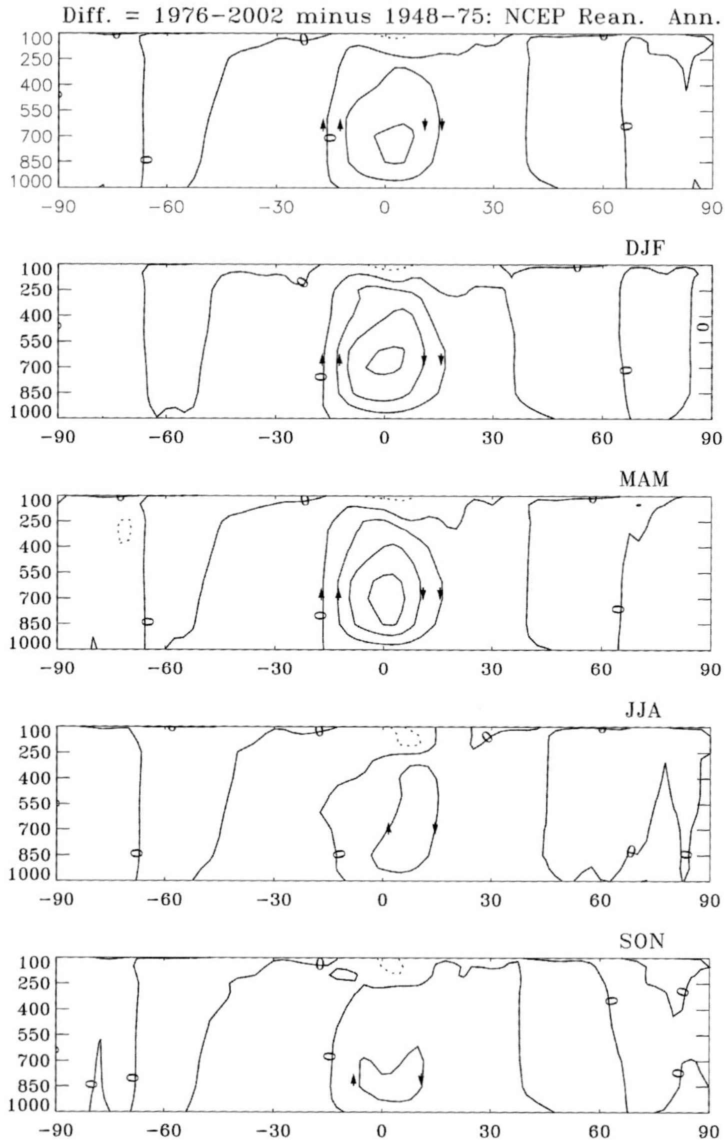

Figure 3-8. Difference in the time averages of the zonal mean meridional stream function in the NCEP/NCAR reanalysis: the average for 1976–2002 minus the average for 1950–75. Contour intervals are 2 x 10^{10} kg s^{-1}

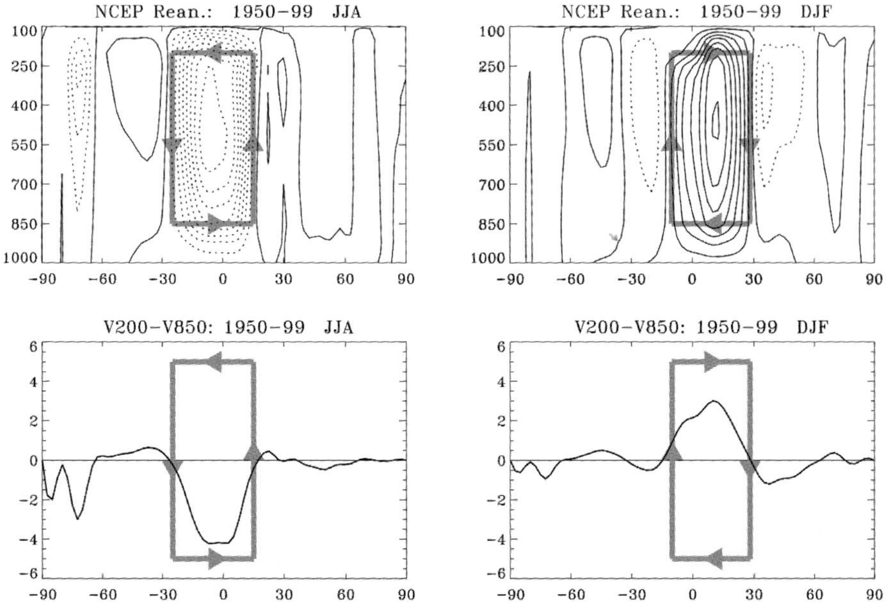

Figure 3-9. Climatological mean of the zonal mean meridional stream function (upper two panels), and the difference between the zonal mean of the meridional wind at 200 and 850 hPa (lower two panels) for JJA and DJF. The difference is obtained by subtracting the 850 hPa meridional wind value from the 200 hPa wind value at each latitudinal grid point.

Both the long-term trend toward intensification in the winter/northward overturning Hadley cell, and the interdecadal swing in the change of the summer/southward overturning Hadley cell, as depicted in Figures 3-7 and 3-8, are also seen clearly in Figure 3-10.

A "regime change" of the northern winter Hadley cell apparently occurred in the mid-1970s, which had three elements. First, the Hadley cell was generally weaker in the early decades. Second, the interannual pulses of the Hadley cell before the mid-1970s were weaker compared to the pulses of the Hadley cell after then. And third, the Hadley cells have been responding to the warm ENSO phase with an intensified northward overturning cell during the later period.

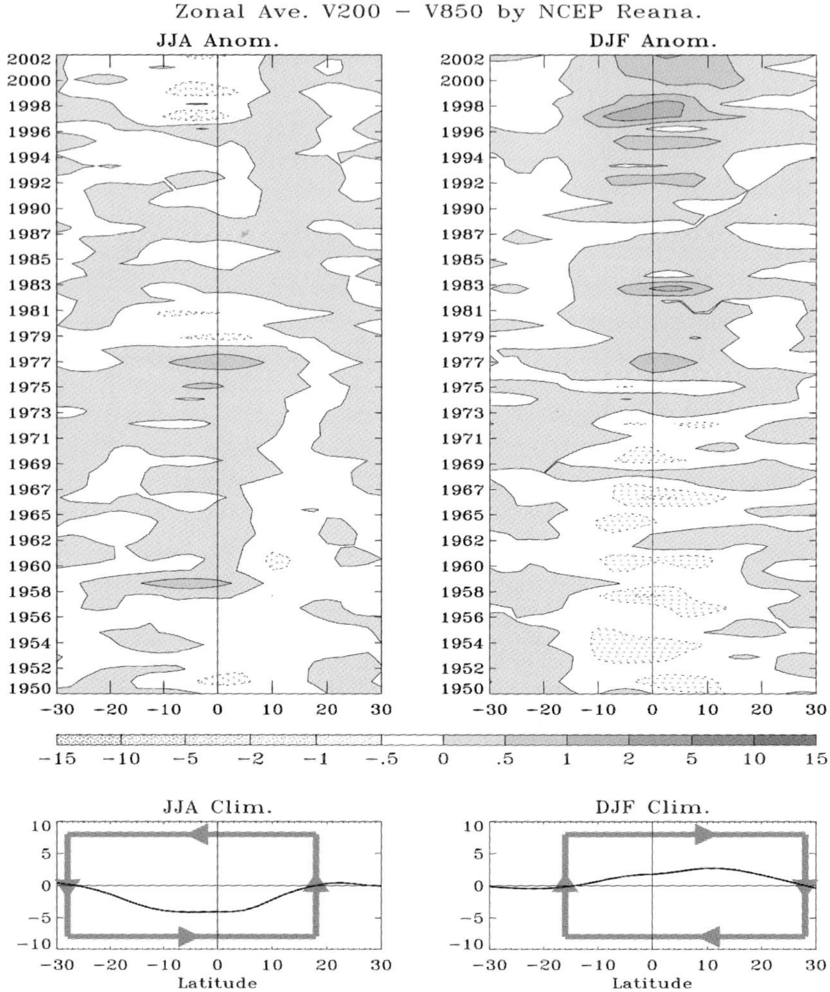

Figure 3-10. Time-latitude cross section of the anomalous zonal mean vertical shear between the meridional wind at 200 and 850 hPa (V200 minus V850) in the NCEP/NCAR reanalysis for JJA and DJF, respectively. The climatological means of the zonal mean vertical wind-shear for JJA and DJF are shown in the bottom two panels with the schematic arrows interpreting the meaning of the positive/negative signs of the values.

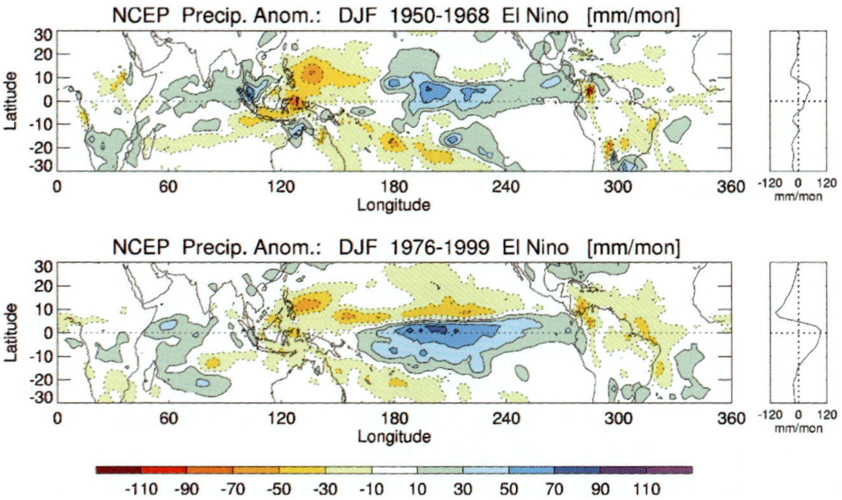

Figure 3-11. Composites of the anomalous precipitation in the NCEP/ NCAR reanalysis during the winter (DJF) of El Niño events in the 1950–68 (top panel) and 1976–1999 (bottom panel) periods. Zonal means of the anomalous precipitation are shown in the panels on the right-hand side.

To further clarify the apparent regime change in the atmospheric circulation over the tropical Pacific, we have also examined the spatial distribution of the tropical rainfall anomalies during the warm ENSO events for the two subperiods. These have been extracted from the reanalysis data, and should be viewed with caution since those data did not assimilate observed gauge rainfall estimates. Figure 3-11 shows the comparison of the composite anomalous rainfall pattern for the winter (DJF) of El Niño events during the 1950–68 period, against its counterpart during the 1976–1999 period. During the earlier period, the anomalous rainfall pattern during El Niño events was characterized by a belt of increased rainfall over the tropical Pacific located mostly on the north side of the equator. In contrast, the belt of increased rainfall over the tropical Pacific has shifted largely to the south side of the equator during the warm ENSO events since 1976. There is thus some indication that the interdecadal change in the anomalous rainfall belt over the tropical Pacific is consistent with the interdecadal change in the tropical Hadley cells' response to the warm ENSO events. Further discussion is found in section 4.3.

4. CHANGE OF THE TROPICAL HADLEY CELL IN AGCM SIMULATIONS

4.1. Validation and Attribution of the Observed Hadley Cell Change

An important question is whether the temporal evolution of the tropical Hadley cell since 1950 shown in the NCEP/NCAR reanalysis data is physical, and consistent with an atmospheric response to the forcing by changes in SST. To address this question, we apply the same analysis to ensemble AGCM simulations and compare 1950–99 temporal evolution of the model simulations with those in the reanalysis data. A similarity between the change in the ensemble model simulations and that found in the reanalysis data would provide some justification for claiming the existence of a forced signal, and that the effect of artificial climate shifts in the reanalysis data due to nonphysical processes of data input changes is minimal.

Ensemble averages of the model simulated temporal variation of the vertical shear between 200 and 850 hPa zonal mean meridional wind for the past five decades are shown for DJF and JJA in Figure 3-12. Three major features of the simulated variation of the tropical Hadley cell are similar to those seen in the NCEP/NCAR reanalysis (Fig. 3-10): (1) An intensification of the northward overturning Hadley cell during the Northern Hemisphere winter (DJF) in recent decades, (2) a strong ENSO signature in the interannual variations of the simulated Hadley cell, and (3) little long-term trend in the southward overturning Hadley cell in the Northern Hemisphere summer (JJA). As in the observations, there is evidence for a regime change in the behavior of the northern winter Hadley cell, though occurring later in the AGCM. Thus, the simulated Hadley cell was weaker in the 1950s, 60s, and 70s compared to later decades, and more intense interannual variations are seen after 1976.

The regime change related to ENSO-induced rainfall patterns is shown in Figure 3-13, which compares composites of the tropical rainfall anomalies for the El Niño events during the 1950–68 and 1976–99 periods, respectively. Similar to the reanalysis data (Fig. 3-11), increased rainfall over the tropical Pacific Ocean occurs mostly north (south) of the equator during the ENSO events of earlier (later) decades. Particularly evident is the increased spatial coverage of equatorial enhanced rainfall during recent El Niño events.

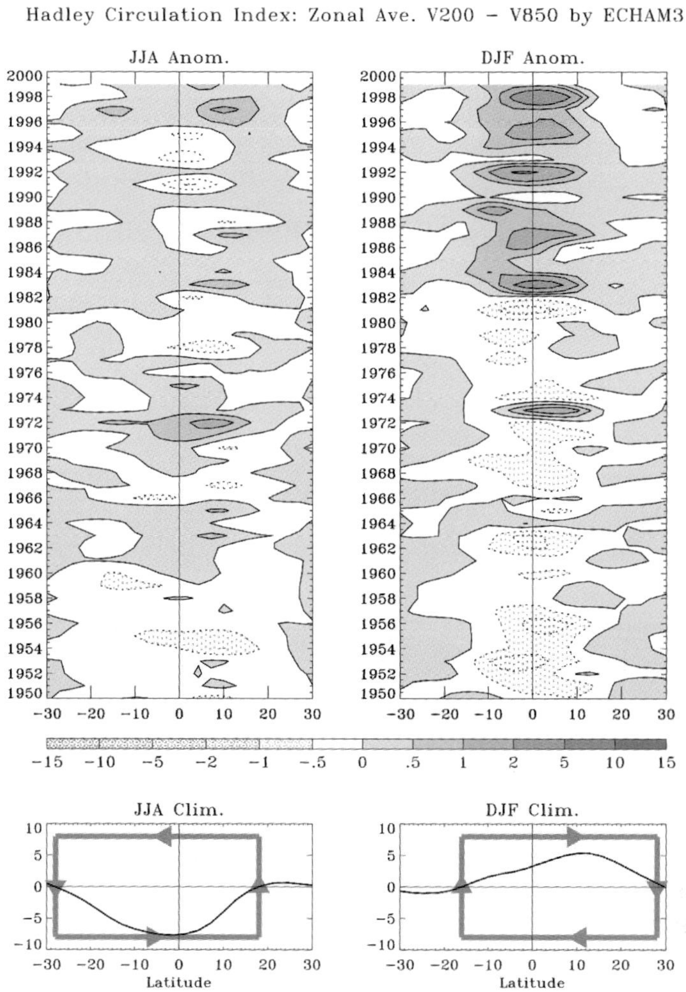

Figure 3-12. Same as Fig. 3-10, but for the ensemble mean of the ECHAM3 simulations.

Based on the qualitative agreement between the time history of simulated and the reanalysis Hadley cell intensities, we propose that the change in Hadley cell strength during the northern winter is consistent with an oceanic change. To the extent that the relevant oceanic changes are not themselves forced by the Hadley cell, then the trend in the Hadley cell is judged to have been forced by the oceans as revealed in these AGCM simulations.

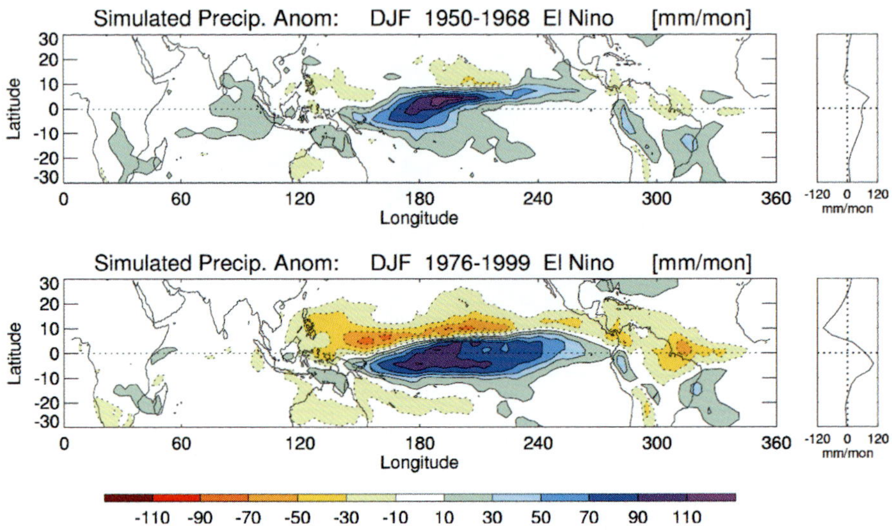

Figure 3-13. Same as Fig. 3-11, but for the ensemble mean of the ECHAM3 simulations.

4.2. The Seasonal Dependence of the Interdecadal Change in the Tropical Hadley Cell

The seasonal dependence of the interdecadal change in the tropical Hadley cell is further examined in Figure 3-14, which shows the time series of the vertical shear of the 200 hPa minus 850 hPa zonal mean meridional wind at the equator for DJF and JJA, respectively. The DJF time series shows an increasing trend of about 0.02 m s^{-1} yr^{-1}. The differences among the 10 simulations are relatively small (indicated by the vertical distance between the dashed lines and the solid lines in their outer side in Fig. 3-14). On the other hand, the JJA time series does not show a significant trend during the 50-year period, and has larger sample-to-sample differences among individual simulations.

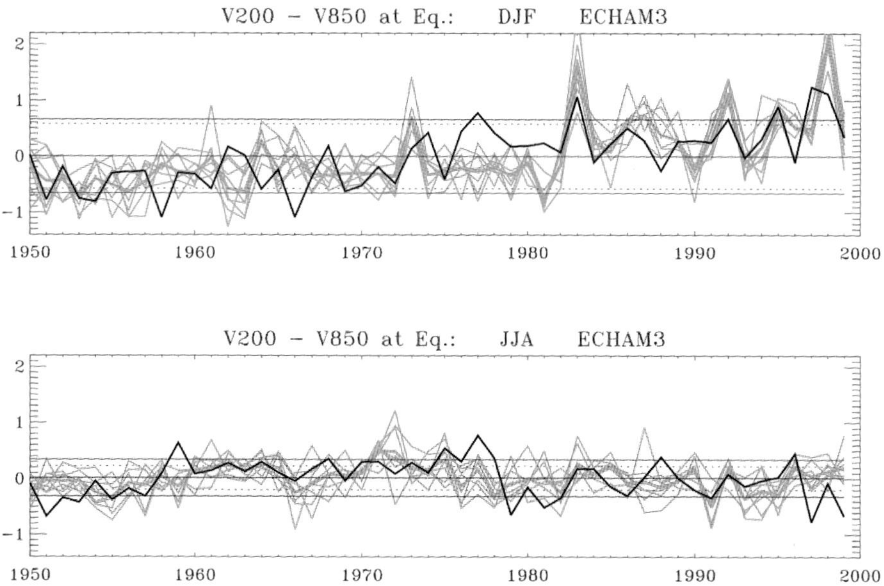

Figure 3-14. Time series of the vertical shear of the 200 and 850 hpa zonal mean meridional wind: DJF (top panel) and JJA (bottom panel) over the equator. The gray thick solid curves represent the ensemble average, and gray thin dotted curves for each individual simulation. The standard deviation due to the temporal variation in the ensemble average is indicated by the distance between the dashed line and the average (central line). The standard deviation that includes inter-sample differences is represented by the solid lines shown outside of the dashed lines. The black solid curves show the vertical wind shear in NCEP reanalysis.

The seasonal dependence of the simulated change in the Hadley circulation since 1950 appears to reflect different impacts of the SST warming on the oceanic-dominated monsoons during boreal winter versus the continental-dominated monsoons during boreal summer. During winter, a warming of the oceans throughout the deep tropics yields an increase in zonally averaged rainfall, and hence an intensification of the zonally symmetric meridional overturning (Fig. 15, top panel). The simulated summertime rainfall trend is more regional, and lacks a zonally symmetric component, especially in the southern tropics (Fig. 3-15, bottom panel).

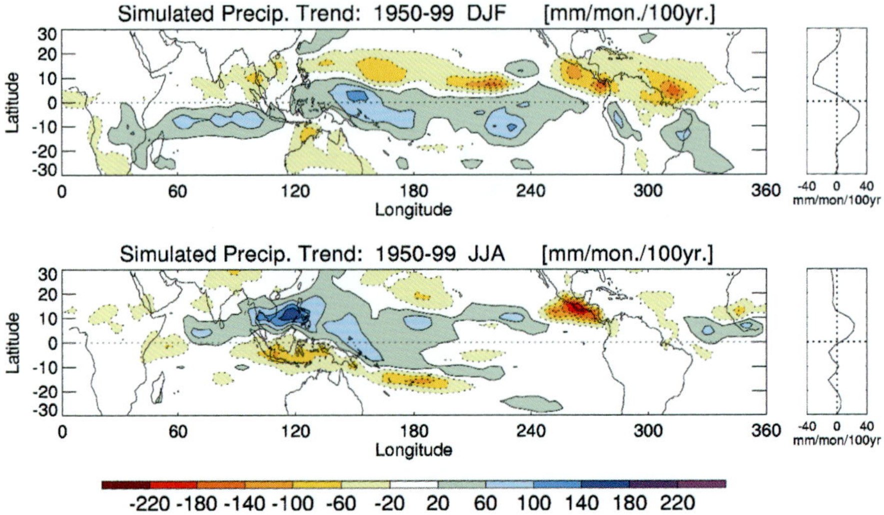

Figure 3-15. Linear trend in the ensemble average of the ECHAM3 simulated precipitation for the period of 1950–99 during DJF (top panel) and JJA (bottom panel). Zonal means of the trend are shown in the panels on the right-hand side.

The regional patterns of the simulated epochal changes in winter and summer 200 hPa divergent mass circulations are shown in Figure 3-16. The oceanic warming, through its forcing of increased oceanic rainfall, intensifies winter regional monsoon circulations over the Indian Ocean and western Pacific region. We speculate that the intensified divergent mass circulation over the equatorial eastern Pacific in recent decades is related to the larger, more frequent El Niños of recent years (Fig. 3-16, top panel). Further evidence for this hypothesis will be given in the next section. Note the strong zonal symmetry of the increased poleward mass transport at 200 hPa during winter, a structure projecting on the zonally symmetric Hadley cell.

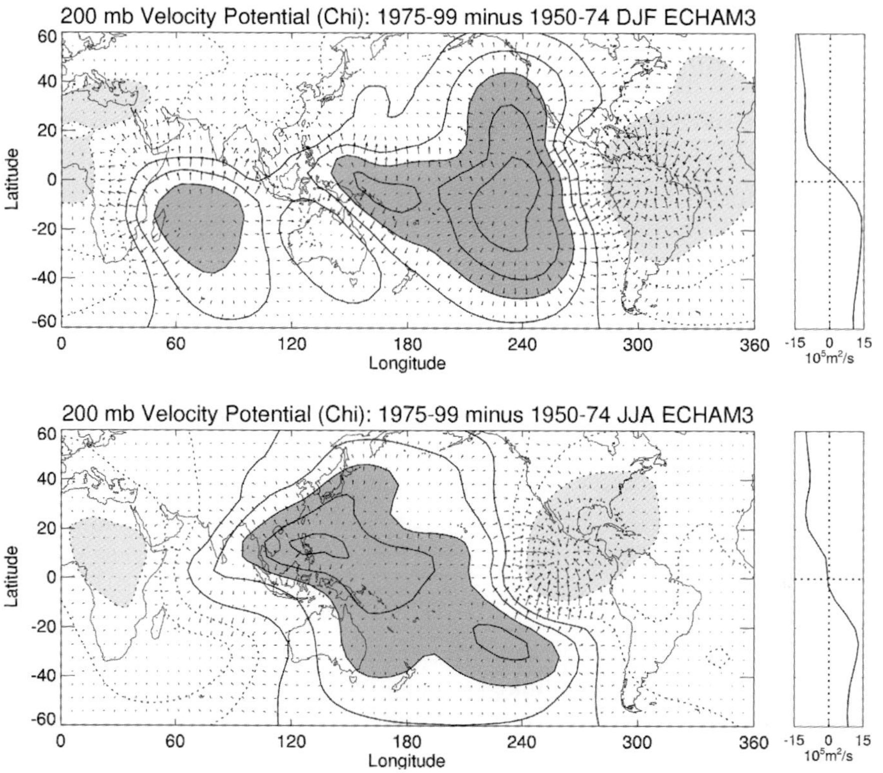

Figure 3-16. Change in the ensemble mean of the ECHAM3 200 hPa divergence wind field for DJF (top panel) and JJA (bottom panel). Contours for the velocity potential fields start from $\pm 5 \times 10^5$ m^2 s^{-1} with a contour interval of 10^5 m^2 s^{-1}. Dark/light shades indicate areas where the velocity potentials are higher/lower than +/– 25 x 10^5 m^2 s^{-1}. Zonal means of the velocity potential differences are shown in the panels on the right-hand side.

During the northern summer, the oceanic warming forces intensification of convection over mainly the western Pacific Ocean, which appears to be coupled with the intensification of descending motion in the tropics; for example, over North Africa and over the northern tropical American monsoon area. It is also noticeable that the upper level divergent flow over the Southern Hemisphere is stronger than over the Northern Hemisphere during the summer, which may be attributed to weaker SST cooling in the southern extratropics compared to the strong SST cooling in the northern extratropics (e.g., Parker et al 1994; also see Fig. 8 in Quan et al. 2003).

4.3. The Role of the Warming Trend of Tropical "Warm-Pool" SSTs

The mean warming of the tropical oceans also appears to be an important factor that contributes to the intensification of the tropical Hadley cell during the Northern Hemisphere winter. To demonstrate this impact, the DJF intensification of the Hadley cell simulated by the ECHAM3 model can be partitioned into ENSO-related and non-ENSO-related portions. Figure 3-17 (left panel) shows the time series of the linear regression between the Hadley cell index and the spatial average of SST anomalies for the region (15°S–15°N, 160°W–80°W, see top panel in Fig. 3-2). The residual from this linear regression (i.e., total value minus regressed value) is shown in the right panel in Figure 3-17. About half of the DJF intensification of the Hadley cell can be explained by the linear response to the increased amplitude of El Niño in the central and eastern tropical Pacific Ocean. The remaining part is non-ENSO in origin, of which we believe the most relevant oceanic change to be the mean state, especially the warming over the tropical Indian and western Pacific Oceans as indicated further below.

As has already been shown, the temporal change in Hadley cell intensity since 1950 has been linked with an intensification of the tropical hydrological cycle. Manifestations of the enhanced hydrological cycle include, during northern winter, the increased southern tropical rainfall together with northern tropical drying, and a reversed rainfall pattern during summer (lower panel, Figs. 3-11 and 3-13). One contribution to the Hadley cell change that we have emphasized is the change in statistical properties of El Niño. This is further shown in Figure 3-18 (top panel), which shows the substantial increase in equatorial Pacific rainfall response to the El Niños occurring after 1970, as simulated by the ECHAM3 model. The effect on the Hadley circulation is clearly seen in the time history of the ENSO contribution to the Hadley cell (left panel in Fig. 3-17). On the other hand, the right-side panel in Figure 3-17 appears to be more driven by the increase in warm-pool precipitation that is especially evident after 1980 (Fig. 3-18, bottom panel). This trend toward increased rainfall—evident in both models and observations—had previously been documented, and is consistent with the underlying SST warming (e.g., Hoerling et al. 2001a, 2004; Hurrell et al. 2004).

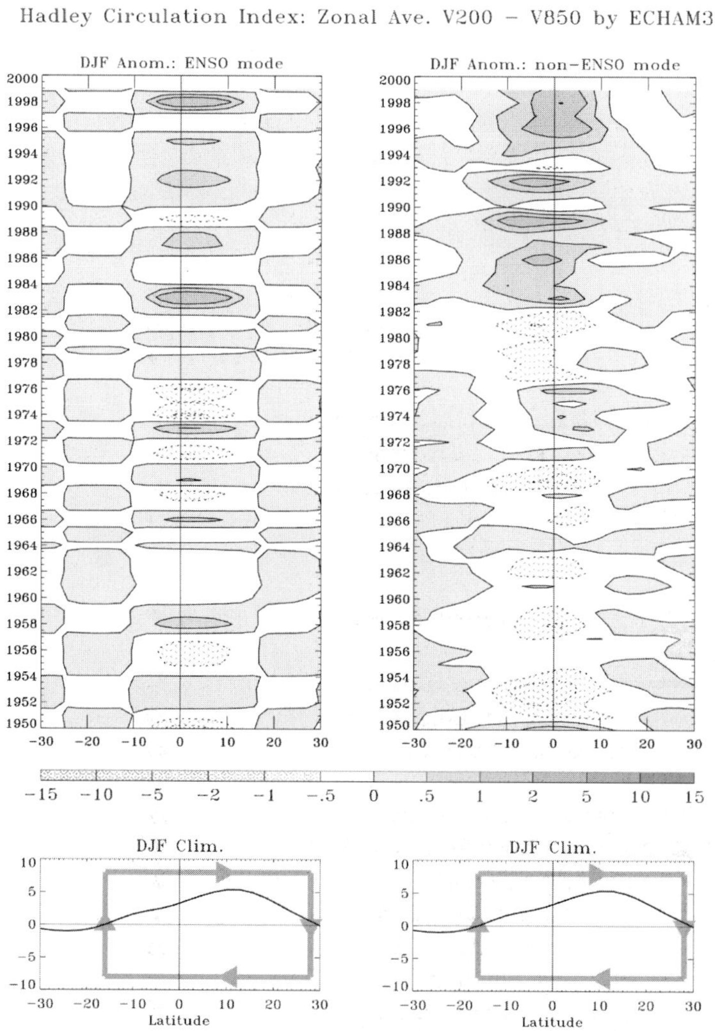

Figure 3-17. Time-latitude cross section of the DJF anomalous zonal mean vertical shear between the meridional wind at 200 and 850 hPa (V200 minus V850) in the ensemble mean of the ECHAM3 simulations for the ENSO component (left panel) and the residual (right panel), respectively. The climatological means of the zonal mean vertical wind shear for DJF are shown in the bottom two panels with the schematic arrows interpreting the meaning of the positive/negative signs of the values.

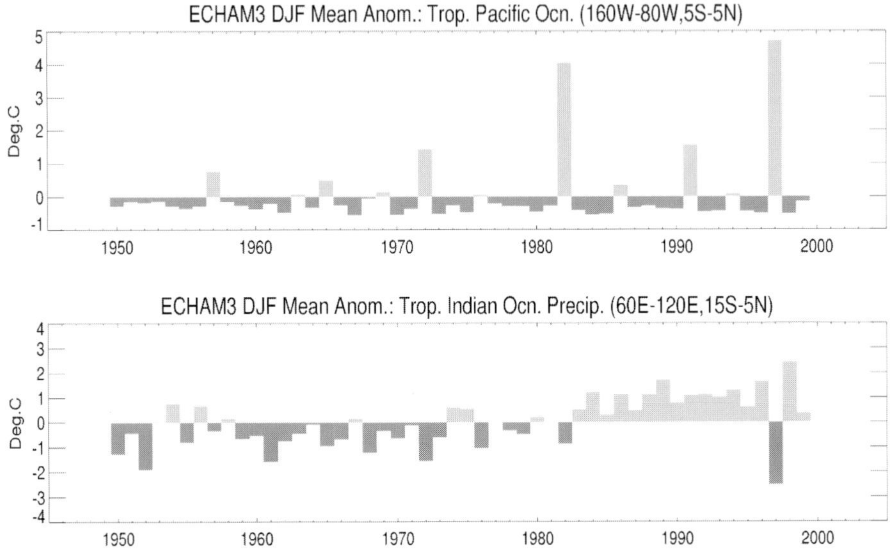

Figure 3-18. Time series of DJF mean anomalous precipitation in the ensemble mean of the ECHAM3 simulations.

For the recent decades, for which independent observational data are available, we have compared the simulated precipitation with satellite precipitation estimates over the open oceans. Figure 3-19 shows a comparison between the model-simulated precipitation and the observed precipitation from GPCP. The top panel shows the time series of the spatial average of 3-month mean precipitation in the tropical zone (10ºS–10ºN) for the Northern Hemisphere winter. The model simulation shows a reasonable agreement with the observations for interannual variations during the 1980–99 period. To verify the model's amplitude of rainfall variations in response to tropical SSTs, we compare the composite amplitude of anomalous precipitation of the GPCP data with that of the model simulation in the lower panel of Figure 3-19. The composite anomalies are made from the DJF of 1982/83, 1986/87, and 1997/98 for the warm events, and January–February 1980, 1988/89 DJF, and December of 1999 for the cold events (cf. the SST time series in the middle panel of Fig. 3-19). The amplitude of the model simulated rainfall anomalies are comparable to the observations. We note that the increase of rainfall over the tropical Pacific Ocean is consistent with a study by Morrisey and Graham (1996).

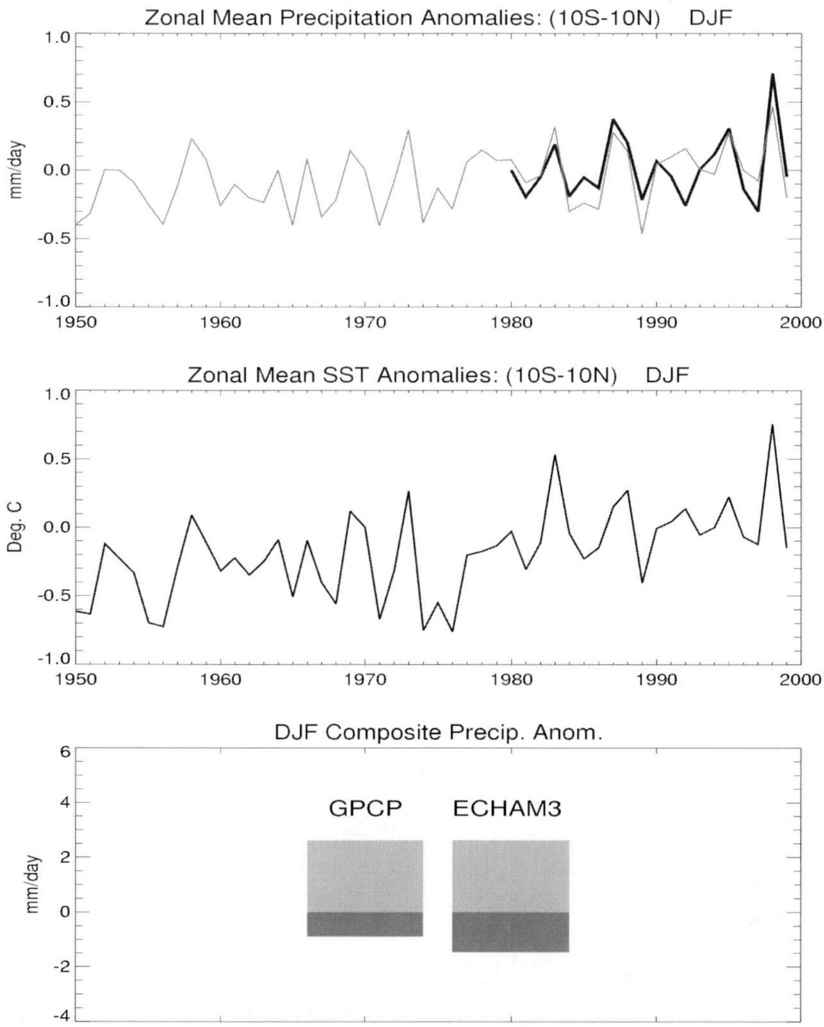

Figure 3-19. Time series of the spatial mean of the DJF anomalous precipitation over the global tropical zone (10°S–10°N) in the ensemble mean of the ECHAM3 simulation (thin line, top panel), the GPCP observations (dark line, top panel), and the DJF anomalous SST (middle panel). The values of composite anomalous rainfall are compared in the bottom panel for El Niño (gray)/ La Niña (black) for the GPCP observation (left side) and the ECHAM3 ensemble mean.

5. SUMMARY AND DISCUSSION

The consistent results emerging from our analyses of the reanalysis and model simulations suggest the following conclusions:

- The winter (DJF) Hadley cell has increased in intensity since 1950.
- The temporal variation of the tropical Hadley cell during 1950–99 is closely related to the variation of the sea surface temperatures in the tropical oceans. At interannual timescales, the tropical Hadley cell is linked to the interannual variation of the El Niño/Southern Oscillation. The Hadley cell responds particularly strongly to the El Niño events, but more weakly to the La Niña events. Nonlinear differences also exist between the spatial structures in the Hadley cells' responses to the opposite phases of ENSO. The change in the statistical properties of ENSO since 1950—in particular the increased frequency and amplitude of El Niño events since 1976—have contributed to a strengthening trend of the NH winter Hadley cell.
- The warming in the tropical Indo-West Pacific warm pool is an equally important forcing factor that has been driving an acceleration of the boreal winter Hadley cell.
- The time history of the southward overturning Hadley cell during the Southern Hemisphere winter lacks a trend, though it does exhibit strong decadal variations.
- A strong seasonal dependence of the 50-year trend in the tropical Hadley cell reflects different impacts of tropical SST warming trends on the ocean-dominated monsoons during the northern winter versus the continent-dominated monsoons during the southern winter.

The analysis revealed a substantial nonlinearity in the Hadley cell/ENSO connection. A possible cause is the nonlinearity in thermodynamic processes that determines the strength of the rainfall response to interannual variations in tropical Pacific SSTs. In particular, a warm SST anomaly may cause larger rainfall and diabatic heating changes than an equal cold SST anomaly (e.g., Zhang 1993). To the extent that there exists a zonal mean component to this rainfall response, thermodynamic nonlinearity offers one possible explanation for why the Hadley cells' response to the cold ENSO events is weaker than its counterpart to the warm ENSO events. Global circulation impacts from such a thermodynamic nonlinearity have been found in the ENSO-PNA teleconnection (Hoerling et al. 2001b), and in the seasonal cycle of atmospheric climate predictability (Quan 1998; Quan et al. 2004).

Another possible explanation for the nonlinearity in the Hadley cell/ENSO connection is the difference in the geographical location of the maximum rainfall anomalies associated with warm versus cold ENSO states. A study by Hoerling et al. (1997) showed that the maximum rainfall anomalies along the equator are located east of the date line during warm ENSO events, but west of the date line during the cold events, implying large differences in diabatic heating patterns associated with the two extreme ENSO phases (e.g., DeWeaver and Nigam 2002). In the above context, the spatial structure of the atmospheric response to anomalous rainfall/diabatic heating in the ascending branch of the Walker circulation (west of the date line) is not a linear inverse to its anomalous counterpart in response to anomalous forcing in the descending branch (east of the date line); neither is the spatial pattern of the Hadley cell change in cold ENSO events the inverse opposite to the pattern of the Hadley cell change during warm ENSO events.

As is indicated in the above analyses, the tropical Hadley cells have responded to the interannual variations of ENSO quite differently before and after the mid-1970s. This interdecadal change in the tropical Hadley cells' response to ENSO describes one aspect of a regime change in the global atmospheric circulation after the mid-1970s. Other aspects of the regime change in the global atmospheric circulation have been found by other authors, and include (1) the intensification of cyclones over the North Pacific Ocean during the Northern Hemisphere winter (e.g., Graham and Diaz 2001); (2) a phase change in the preferred interannual occurrence of the North Atlantic Oscillation (e.g., Hurrell 1995; Thompson et al. 2000); (3) a weakened correlation between the Indian Monsoon and ENSO during the Northern Hemisphere summer (e.g., Kumar et al 1999; Kinter III et al. 2002); (4) a change in the correlation between ENSO and interannual swings of Australian monsoon rainfall (Power et al. 1999); and (5) change in the onset process of ENSO over the tropical Pacific Ocean (Wang 1995).

The change of the tropical Hadley cell is dynamically consistent with the intensification of cyclones over the North Pacific, and each is further consistent with the change in ENSO statistics. The enhanced northward overturning cell during El Niño events after 1976 contributes to an intensification of the westerlies over the subtropical North Pacific, which in turn creates a more favorable vertical shear profile for cyclone intensification.

On the other hand, the enhanced northward overturning Hadley cell leads to a stronger easterly (trade) wind flow in the tropics, consistent with the results by Wang (1995). Wang found that the trade winds over the tropical southeast Pacific intensified during the onset phase of El Niño events occurring after 1976 compared to the earlier El Niños.

What is a possible cause for such regime change in atmospheric circulation? Wang (1995) suggested that the regime change in the ENSO evolution is linked to the background warming of the eastern tropical Pacific Ocean. Our analyses of the regime change of the tropical Hadley cell offers another plausible hypothesis for how the warming in the eastern tropical Pacific has led to the regime change in the atmospheric circulation. The warming in the tropical eastern Pacific has occurred more rapidly in the area south side of the equator, especially during the Northern Hemisphere winter (Fig. 3-1). In response to the stronger warming in the southern tropical Pacific, the maximum rainfall anomalies shifted from the north side of the equator during the El Niño events before the 1970s, to the south side of the equator after the 1970s (see Figs. 3-11 and 3-13). Because of the southward shift of the maximum anomalous rainfall, the northward overturning Hadley cell became more dominant during the El Niño events after the 1970s, which is further manifested by the intensification of the cyclones over the North Pacific and the trades over the tropical southeastern Pacific.

Another factor that may contribute to the regime change of the tropical Hadley cell is the change in meridional SST gradient in the Pacific Ocean. The meridional SST gradient has been largely increased in the North Pacific Ocean by the warming in the tropical Pacific and cooling in the northern extratropical Pacific (e.g., Graham 1994; Trenberth and Hurrell 1994; Zhang et al. 1997). In contrast, change in the meridional SST gradient in the South Pacific Ocean is only moderate because the SST cooling in the southern extratropical Pacific has been much weaker than its counterpart in the northern extratropical Pacific (Parker et al. 1994). The increased tropical-to-extratropical SST gradient in the North Pacific provides a favorable condition for stronger southerly winds in the lower troposphere over the subtropical North Pacific and an intensified northward overturning Hadley cell during the northern winter. These conditions in turn help the persistence/development of the SST cooling in the northern extratropical Pacific (Lau and Nath 1996), forming a positive feedback in the ocean-atmosphere-ocean coupled variation. The absence of a similar trend of an intensified southward overturning Hadley cell during the Southern Hemisphere winter indicates that such an ocean-atmosphere feedback process has been much weaker or has not even been in existence over the South Pacific Ocean during the southern winter since 1950.

Some uncertainties still exist in the reanalysis-model comparison. A major uncertainty for the Northern Hemisphere winter season is the difference in the timing of the regime change in the tropical Hadley cell. The change occurred around the late 1960s to early 1970s (cf. the right panel in Fig. 3-10) in the reanalysis, but the late 1970s to early 1980s in the model simulation (the right panel in Fig. 3-12). The change in tropical Indo-Pacific SSTs in the late 1970s seems to be a dominant factor in the model simulation. Some other process must have also been influential in the reanalysis in causing the change to occur in the late 1960s. One possible concern is the change in the quantity and spatial coverage of the observed data included in the reanalysis. A rapid increase in the total number of observations in the Southern Hemisphere took place in the mid- to late 1960s (Kistler et al. 2001), and the change in the spatial coverage might have caused a shift of the climate mean state of the Southern Hemispheric circulation in the reanalysis data around that time. However, some significant climate change did occur during the late 1960s. For example, a persistent drought in tropical North Africa started to develop in the late 1960s, accompanied by a change in the north-south interhemispheric gradient of sea surface temperature in the Atlantic Ocean (e.g., Ward 1998). Additionally, the change in the Atlantic Ocean may also affect the circulation regime over the Indo-Pacific Oceans (e.g., Chang et al. 2001). However, the uncertainty is larger for the boreal summer. The interdecadal change in the model-simulated 200 hPa divergence wind field during the Northern Hemisphere summer (Fig. 3-16, bottom panel) is quite different compared to its counterpart in the reanalysis (e.g., Kumar et al. 1999; Krishnamurthy and Goswami 2000; Chang et al. 2001; Quan et al. 2003).

6. REFERENCES

Chang, C.P., P. Harr, and J. Ju. 2001. Possible roles of Atlantic circulations on the weakening Indian Monsoon rainfall-ENSO relationship. *Journal of Climate* 14: 2376–2380.

Chen, J., B.E. Carlson, and A.D. Genio. 2002. Evidence for strengthening of the tropical general circulation in the 1990s. *Science* 295: 838-841.

DeWeaver, E., and S. Nigam. 2002. Linearity in ENSO's atmospheric response. *Journal of Climate* 15: 2446–2461.

Diaz, H.F., C.B. Fu, and X.W. Quan. 1992. A comparison of surface geostrophic winds with COADS ship wind observations. *Proceedings of the International COADS Workshop,* Boulder, Colorado, January 13–15, 1992, pp. 131–141.

Diaz, H.F., X.W. Quan, and C.B. Fu. 1994. Marine surface wind changes during 1978–1992: An estimation based on COADS. *Proceedings of the International COADS Winds Workshop,* Kiel, Germany, May 31–June 2, 1994, pp. 48–67.

Dima, I., and J.M. Wallace. 2003. On the seasonality of the Hadley Cell. *Journal of Atmospheric Sciences* 60: 1522–1527.

Gaffen, D.J., B.D. Santer, J.S. Boyle, J.R. Christy, N.E. Graham, and R.J. Ross. 2000. Multidecadal changes in the vertical temperature structure of the tropical troposphere. *Science* 287: 1242–1245.

Graham, N.E. 1994. Decadal-scale climate variability in the tropical and north Pacific during the 1970s and 1980s: Observations and model results. *Climate Dynamics* 10: 135–162.

Graham, N.E., and H.F. Diaz. 2001. Evidence for intensification of North Pacific winter cyclones since 1948. *Bulletin of the American Meteorological Society* 82: 1869–1893.

Graham, N.E., T.P. Barnett, R. Wilde, M. Ponater, and S. Schubert. 1994. On the role of tropical and midlatitudes SSTs in forcing interannual to interdecadal variability in the winter northern hemisphere circulation. *Journal of Climate* 9: 1416–1441.

Hoerling, M.P., A. Kumar, and M. Zhong. 1997. El Niño, La Niña, and the nonlinearity of their teleconnections. *Journal of Climate* 10: 1769–1786.

Hoerling, M.P., J.W. Hurrell, and T. Xu. 2001a. Tropical origins for recent north Atlantic climate change. *Science* 292: 90–92.

Hoerling, M.P., A. Kumar, and T. Xu. 2001b. Robustness of the nonlinear climate response to ENSO's extreme phases. *Journal of Climate* 14: 1277–1293.

Hoerling M.P., J.W. Hurrell , T. Xu , G.T. Bates, and A. Phillips. 2004. Twentieth Century North Atlantic climate change. Part II: Understanding the effect of Indian Ocean warming. *Climate Dynamics* (in press).

Huang, H.P., K.M. Weickmann, and C.J. Hsu. 2001. Trend in atmospheric angular momentum in a transient climate change simulation with greenhouse gas and aerosol forcing. *Journal of Climate* 14: 1525–1534.

Huang, H.P., K.M. Weickmann, and R.D. Rosen. 2003. Unusual behavior in atmospheric angular momentum during the 1965 and 1972 El Niños. *Journal of Climate* 16: 2526–2539.

Huffman, G.J., R.F. Adler, P. Arkin, A. Chang, R. Ferraro, A. Gruber, J. Janowiak, A. McNab, B. Rudolf, and U. Schneider, 1997. The Global Precipitation Climatology Project (GPCP) combined precipitation dataset. *Bulletin of the American Meteorological Society*, 78, 5–20

Hurrell, J. 1995. Decadal trends in the North Atlantic Oscillation: Regional temperature and precipitation. *Science* 269: 676–679.

Hurrell, J. 1996. Influence of variations in extratropical wintertime teleconnections on Northern Hemisphere temperature. *Geophysical Research Letters* 23: 665–668.

Hurrell J.W., M.P. Hoerling, A. Phillips, and T. Xu. 2004. Twentieth Century North Atlantic climate change. Part I: Assessing determinism. *Climate Dynamics* (in press).

IPCC (Intergovernmental Panel on Climate Change) 1996. *Climate Change 1995: The Science of Climate Change*, J.T. Houghton et al. (eds.). Cambridge: Cambridge University Press, 572 pp.

IPCC. 2001. *Climate Change 2001: The Science Basis*, J.T. Houghton et al. (eds.). Cambridge: Cambridge University Press, 881 pp.

Jones, P.D., M. New, D.E. Parker, S. Martin, and I.G. Rigor. 1999. Surface air temperature and its changes over the past 150 years. *Reviews of Geophysics* 37: 173–199.

Kinter III, J.L., K. Miyakoda, and S. Yang. 2002. Recent change in the connection from the Asian Monsoon to ENSO. *Journal of Climate* 15: 1203–1215.

Kistler, R., E. Kalnay, W. Collins, S. Saha, G. White, J. Woollen, M. Chelliah, W. Ebisuzaki, M. Kanamitsu, H. van den Dool, R. Jenne, and M. Fiorino. 2001. The NCEP/NCAR 50-year reanalysis: Monthly means, CD-ROM and documentation. *Bulletin of the American Meteorological Society* 82: 152–168.

Knutson, T.R., T.L. Delworth, K.W. Dixon, and R.J. Stouffer. 1999. Model assessment of regional temperature trends. *Journal of Geophysical Research* 104: 30981–30996.

Krishnamurthy, V., and B.N. Goswami. 2000. Indian monsoon-ENSO relationship on inter-decadal timescale. *Journal of Climate* 13: 579–595.

Kumar, K.K., B. Rajagopalan, and M.A. Cane. 1999. On the weakening relationship between the Indian monsoon and ENSO. *Science* 284: 2156–2159.

Kumar, A., F. Yang, L. Goddard, and S. Schubert. 2004. Differing trends in the tropical surface temperatures and precipitation over land and oceans. *Journal of Climate* 17: 653–664.

Latif, M., R. Kleeman, and C. Eckert. 1997. Greenhouse warming, decadal variability, or El Niño? An attempt to understand the anomalous 1990s. *Journal of Climate* 10: 2221–2239.

Lau, K.-M., and H. Weng. 1999. Interannual, decadal-interdecadal, and global warming signals in sea surface temperature during 1955–1997. *Journal of Climate* 12: 1257–1267.

Lau, N.C., and M.J. Nath. 1994. A modeling study of the relative roles of tropical and extratropical SST anomalies in the variability of the global atmosphere-ocean system. *Journal of Climate* 7: 1184–1207.

Lau, N.C., and M.J. Nath.1996. The role of the "atmospheric bridge" in linking tropical Pacific ENSO events to extratropical SST anomalies. *Journal of Climate* 9: 2036–2057.

Levitus, S., J. Antonov, T.P. Boyer, and C. Stephens. 2000. Warming of the world ocean. *Science* 287: 2225–2229.

Morrisey, M.L., and N.E. Graham, 1996. Recent trends in rain gauge precipitation measurements from the tropical Pacific: evidence for an enhanced hydrological cycle. *Bulletin of the American Meteorological Society*, 77, 1207-1219

Oort, A.H., and J.J. Yienger. 1996. Observed interannual variability in the Hadley circulation and its connection to ENSO. *Journal of Climate* 9: 2751–2767.

Parker, D.E., P.D. Jones, C.K. Folland, and A. Bevan. 1994. Interdecadal changes of surface temperature since the late nineteenth century. *Journal of Geophysical Research* 99: 14373–14399

Parker, D.E. 2000 Temperatures high and low. *Science* 287: 1216.

Philander, S.G. 1990. *El Niño, La Niña, and the Southern Oscillation*. San Diego: Academic Press, 293 pp.

Power, S., T. Casey, C. Folland, A. Colman, and V. Mehta. 1999. Inter-decadal modulation of the impact of ENSO on Australia. *Climate Dynamics* 15: 319–324.

Quan, X.W. 1998. Interannual variability associated with ENSO: Seasonal dependence and interdecadal change. Ph.D. Thesis, University of Colorado at Boulder.

Quan, X.W., H.F. Diaz, and C.B. Fu. 2003. Interdecadal change in the Asian-Africa summer monsoon and its associated changes in global atmospheric circulation. *Global and Planetary Change* 37: 171–188.

Quan, X.W., P.J. Webster, A.M. Moore, and H.R. Chang. 2004. Causes of the seasonality in SST forced atmospheric short-term climate predictability. *Journal of Climate* (in press).

Roeckner, E., and coauthors. 1992. Simulation of the present-day climate with the ECHAM model: Impact of model physics and resolution. *Max-Plank-Institute für Meteorologie, Report* 93, 171 pp. (Available from MPI für Meteorologie, Bundesstr. 55, D-20146 Hamburg, Germany.)

Santer, B.D., T.M. Wigley, D.J. Gaffen, L. Bengtsson, C. Doutriaux, J.S. Boyle, M. Esch, J.J. Hnilo, P.D. Jones, G.A. Neehl, E. Roeckner, K.E. Taylor, and M.F. Wehner. 2000.

Interpreting differential temperature trends at the surface and the lower troposphere. *Science* 287: 1227–1231.

Smith, T.M., R.W. Reynolds, R.E. Livezey, and D.C. Stokes. 1996. Reconstruction of historical sea surface temperature using empirical orthogonal functions. *Journal of Climate* 9: 1402–1420.

Thompson, D.W.J., J.M. Wallace, and G.C. Hegerl. 2000. Annular modes in the extratropical circulation. Part II: Trends. *Journal of Climate* 13: 1018–1036.

Trenberth, K.E. 2002. Changes in tropical clouds and radiation. *Science*. 296: 2095.

Trenberth, K.E., and J.W. Hurrell. 1994. Decadal atmosphere-ocean variations in the Pacific. *Climate Dynamics* 9: 303–319.

Trenberth, K.E., G.W. Branstator, D. Karoly, A. Kumar, N.-C. Lau, and C. Ropelewski. 1998. Progress during TOGA in understanding and modeling global teleconnections associated with tropical sea surface temperatures. *Journal of Geophysical Research* 103: 14291–14324.

Trenberth, K.E., D.E. Stepaniak, and J.M. Caron. 2000. The global monsoon as seen through the divergent atmospheric circulation. *Journal of Climate* 13: 3969-3993.

Waliser, D.E., Z. Shi, J.R. Lanzante, and A.H. Oort. 1999. The Hadley circulation: Assessing NCEP/NCAR reanalysis and sparse in situ estimates. *Climate Dynamics* 15: 719–735.

Wang, B. 1995.Interdecadal change in El Niño onset in the last four decades. *Journal of Climate* 8: 267–285.

Ward, M.N. 1998. Diagnosis and short-lead time prediction of summer rainfall in tropical North Africa at interannual and multidecadal timescales. *Journal of Climate* 11: 3167–3191.

Webster, P.J. 1987. The elementary monsoon. In, Fein, J.S., and P.L. Stephens (eds.). *Monsoons*. New York: John Wiley & Sons, pp. 3–32.

Zhang, C.D. 1993. Large-scale variability of atmospheric deep convection in relation to sea surface temperature in the tropics. *Journal of Climate* 6: 1898–1913.

Zhang, Y., J.M. Wallace, and D.S. Battisti. 1997. ENSO-like interdecadal variability: 1900–93. *Journal of Climate* 10: 1004–1020.

Chapter 4

THE SHAPE OF CONTINENTS, AIR-SEA INTERACTION, AND THE RISING BRANCH OF THE HADLEY CIRCULATION

Shang-Ping Xie
International Pacific Research Center and Department of Meteorology, University of Hawaii, Honolulu, Hawaii 96822, U.S.A.
xie@hawaii.edu

Abstract This chapter begins with a brief history of Intertropical Convergence Zone (ITCZ) research. It then goes on to summarize recent progress in understanding why the ITCZ is locked in the Northern Hemisphere in the eastern Pacific and Atlantic Oceans, and how this northward-displaced ITCZ affects the space-time structure of tropical climate variability.

1. INTRODUCTION

The differential solar radiation in the meridional direction is the ultimate drive for the global Hadley circulation, dictating that its rising branch and heavy rainfall should be located near the equator. This solar forcing of the atmosphere is indirect, however, since most absorption of solar radiation takes place at the surface of earth. Over the tropical oceans, most of the absorbed solar energy is used for surface evaporation and the resultant water vapor is gathered by winds to fuel deep convection that is organized into zonally oriented rain bands. The ocean's effect on tropical convection and hence the rising branch of the Hadley circulation is obvious; tropical rain belts are anchored on the warmest waters, with spatial patterns that can markedly deviate from the distribution of insolation. In particular, the rain band over the eastern Pacific and Atlantic Oceans, called the Intertropical

H.F. Diaz and R.S. Bradley (eds.), The Hadley Circulation: Present, Past and Future, 121–152.

Convergence Zone (ITCZ), is mysteriously displaced to the north of the equator in the annual-mean climatology, a distribution inexplicable from solar forcing alone[1].

This chapter reviews the progress made in the past decade in understanding the coupled ocean-atmospheric dynamics that govern the rising branch of the Hadley circulation and places this progress in a historical perspective. This chapter focuses on the ITCZ over the eastern Pacific and Atlantic, while Webster (Chapter 1, "The Elementary Hadley Circulation," this volume) discusses convection in the Indo-western Pacific sector. Wang et al. (in press) is a global survey of air-sea interaction and its role in climate variability, including a comparative view for the three tropical oceans.

The rest of the chapter is organized as follows. Sections 2 and 3 give historical and observational background, respectively. Section 4 investigates ocean-atmosphere interactions that maintain the climatic asymmetry of the northward-displaced ITCZ, and Section 5 considers the effect of land-sea distribution. Section 6 discusses the climatic consequence of the northward-displaced ITCZ. Following a discussion of some remaining issues in Section 7, Section 8 summarizes the main results.

2. HISTORY OF THE STUDY OF TROPICAL WINDS AND RAINS

> "It is not the work of one, nor of few, but of a multitude of Observers, to bring together the experience required to compose a perfect and complete History of these Winds."
>
> Edmond Halley (1686)

Before the invention of steam engines, knowledge of the direction, speed and steadiness of sea surface winds was of vital importance for the navigation of sailing boats. By the late seventeenth century, the traffic between Europe and the New World had grown to such a level that Halley (1686) was able to compile a quite accurate map of surface-wind streamlines for the tropical Atlantic and Indian Oceans by gathering information from navigators. Figure 4-1 reproduces the Atlantic portion of Halley's wind map that depicts the steady trade winds in the Northern and Southern Hemispheres. Remarkably, the southeasterly and northeasterly trades meet north of, instead of on, the equator as one might expect from equatorial symmetry.

---

1 The latitude of the sinking branch of the Hadley circulation is not directly determined by solar radiation either. Instead, it is determined by dynamic requirements like angular momentum conservation and baroclinic instability (Held and Hou 1980; Lindzen and Hou 1988).

The ITCZ—the modern term for the region where the trade winds meet—is displaced to the Northern Hemisphere in the annual mean. Halley wrote about the ITCZ: "it were improper to say there is any Trade Winds, or yet a Variable; for it seems condemned to perpetual Calms, attended with terrible Thunder and Lightning, and Rains so frequent, that our Navigators from thence call this part of the Sea the *Rains*".

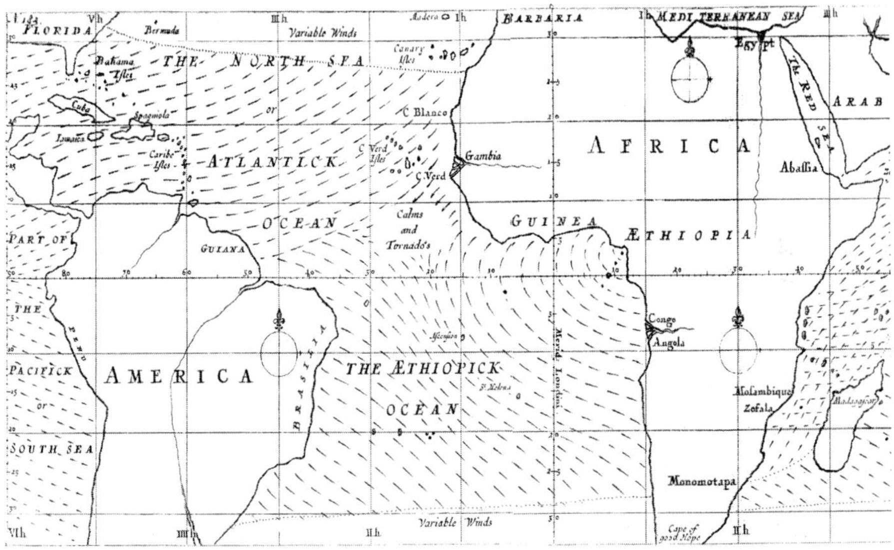

Figure 4-1. Halley's (1686) map of surface wind streamlines. The southeasterly trade winds are shown to converge onto the Northern Hemisphere.

In the ITCZ, surface air rises and in the process, the water vapor it carries condenses, resulting in the frequent rains and thunderstorms Halley noted and releasing a huge amount of latent heat that drives the Hadley and global circulation of the troposphere. Hereafter we will use the terms ITCZ,[2] convective zone, and precipitation band interchangeably. The ITCZ resides in a zone of "perpetual calms" in Halley's words, and is now called

[2] More precisely, our definition of ITCZ refers to those surface convergence zones over warm oceans with sea surface temperatures (SSTs) greater than the convective threshold (26–27°C). There are surface convergence zones over cool sea surface that are not associated with deep convection and significant precipitation. Instead, they are associated with shallow boundary-layer circulation.

the Equatorial Doldrums in textbooks. As will become clear in Section 4.1, the collocation of the Doldrums with the ITCZ is the key to the mystery of their northward displacement from the equator.

Before the late seventeenth century, the vast Pacific Ocean was much less navigated than the Atlantic and Halley had little information on its wind distribution other than accounts that "there is great conformity between the winds of this Sea and those of the Atlantic." This lack of observations forced Halley to draw an "analogy between" the Pacific winds "and those of the Atlantic." Interestingly, Halley did not draw a perfect analogy with the Atlantic winds; in his map, the Pacific trades converge on the geographical equator, rather than on northern latitudes as in the Atlantic. Perhaps Halley or his contemporaries had no reason to believe that the Pacific wind system should depart from equatorial symmetry. By the late nineteenth century, Köppen's (1899) atlas showed that the similarity between the Pacific and Atlantic is greater than Halley thought; as in the Atlantic, the Pacific trades also converge onto the Northern Hemisphere even in boreal winter.

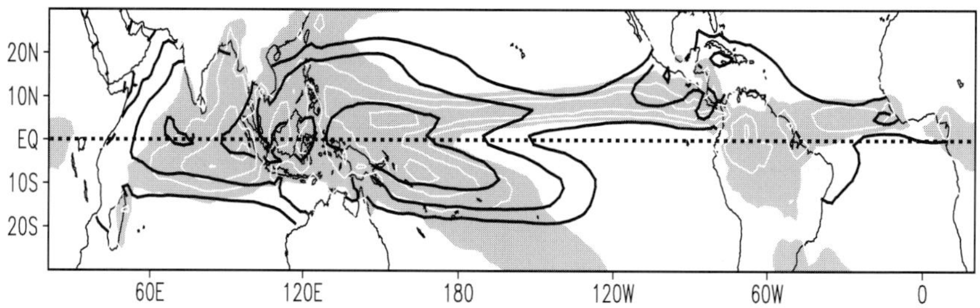

Figure 4-2. Annual-mean climatological precipitation (white contours at 2 mm/day intervals; shade > 4 mm/day), and SST (black contours at 1°C intervals; only contours for 27°C and above are plotted), based on the Climate Prediction Center Merged Analysis of Precipitation (CMAP; Xie and Arkin 1996) and the Reynolds and Smith (1994) data set, respectively.

A reliable precipitation climatology proves more difficult to obtain because of the sporadic nature of rains. In Bartholomew and Herbertson's (1899) map of annual rainfall, the Pacific Ocean was left blank. (One can nevertheless infer a strong equatorial asymmetry from the depicted rainfall on the Pacific coast that is over 160 inches/year in Colombia north of the equator but less than 10 inches/year on the Peruvian coast.) For the mid-twentieth century, Möller's (1951) map of annual mean rainfall is very similar to modern climatology (Fig. 4-2), showing that the ITCZ rain band is clearly displaced to the Northern Hemisphere over both the eastern Pacific and the Atlantic.

3. OBSERVATIONAL BACKGROUND

The advent of satellite remote sensing opened the door for global observations of clouds in the 1960s and, somewhat later, for observations of precipitation. In an early climatology of reflectivity (U.S. Air Force and U.S. Department of Commerce 1971), the Pacific and Atlantic ITCZ appears on the dark ocean background as a silver belt that is north of the equator in both boreal summer and winter and one of the most visible and striking features in such satellite images. Since 1979, outgoing long-wave radiation (OLR) measurements by satellite infrared sensors are often used as a proxy of precipitating deep convection that reaches great heights. A paradox arises: Over the eastern Pacific and Atlantic, the OLR-based estimate of rainfall is too low compared to ship reports, which indicate substantial precipitation accompanied by strong surface wind convergence there (Fig. 4-3). It turns out that this underestimation by the OLR-based method in the eastern Pacific and Atlantic ITCZ is due to the fact that the sea surface temperature (SST ~27°C) there is significantly lower than it is in the Indo-western Pacific warm pool (SST > 28°C). As a result, convection in the eastern Pacific and Atlantic does not reach as high as in the western Pacific, yielding higher OLR values (Thompson et al. 1979; J.M. Wallace 1994, personal communication.). More recent satellite microwave sensors, measuring quantities more directly related to precipitation than the infrared ones, observe similar rain rates in the eastern and in the western Pacific.

Figure 4-2 shows the annual mean precipitation climatology based on combined infrared and microwave satellite observations. Over the continents and the Indo-western Pacific sector, the annual mean precipitation distribution in the tropics is more or less symmetric about the equator, consistent with solar radiation distribution. On the seasonal time scale, the maximum rainfall in these regions moves back and forth across the equator following the sun (Mitchell and Wallace 1992). This solar control of tropical convection breaks down over the eastern half of the Pacific and entire Atlantic, where deep convection is confined to the ITCZ north of the equator. This climatic asymmetry persists for most of the year, even during boreal winter when the solar radiation south of the equator exceeds that to the north (Fig. 4-4). Only for a brief period during March and April, a double ITCZ appears with a rain band on each side of the equator.

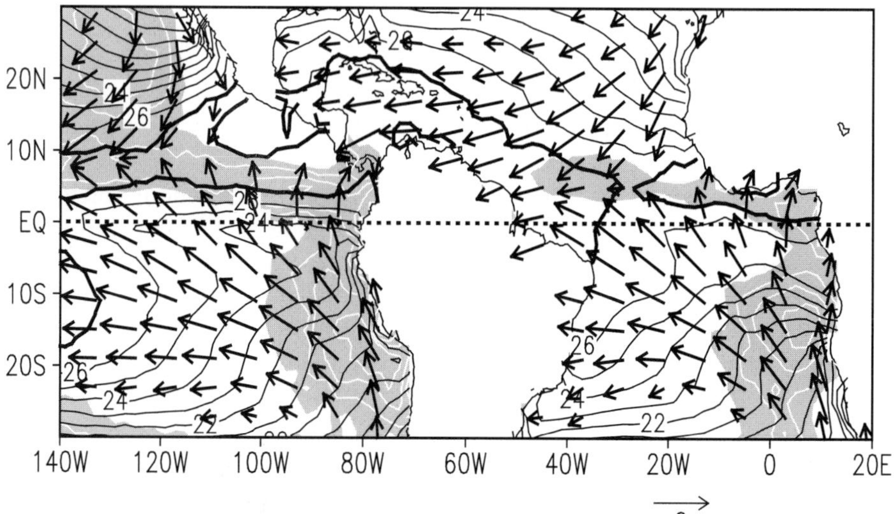

Figure 4-3. Climatological SST (contours in °C), surface wind vectors (m/s), and cloud cover (white contours at 5% intervals; shade > 60%), based on the Comprehensive Ocean-Atmospheric Data Set (COADS; Woodruff et al. 1987).

Located in the region where a great amount of latent heat is released to the atmosphere, the ITCZ is sometimes called the thermal equator. The peculiar location of the thermal equator in the eastern Pacific and Atlantic begs answers to the following questions. (1) Why is the ITCZ not on the equator where the annual mean solar radiation is the maximum? (2) Given that annual mean solar radiation is roughly symmetric about the equator, why is the ITCZ displaced north of the equator? and (3) What effect does this northward displacement of the thermal equator have on climate variability? Two schools of thought exist regarding the first two questions. One points to the strong hemispheric asymmetry in the landmass and its distribution and suggests that this continental asymmetry causes climatic asymmetry. The other camp proposes an SST control, countering the former with the fact that climatic asymmetry is weak in the Indian Ocean, yet the region has the greatest interhemispheric distribution in landmass. We will show that these schools of thought are not mutually exclusive and both are necessary for the complete answers. Let us first look at the arguments of SST control.

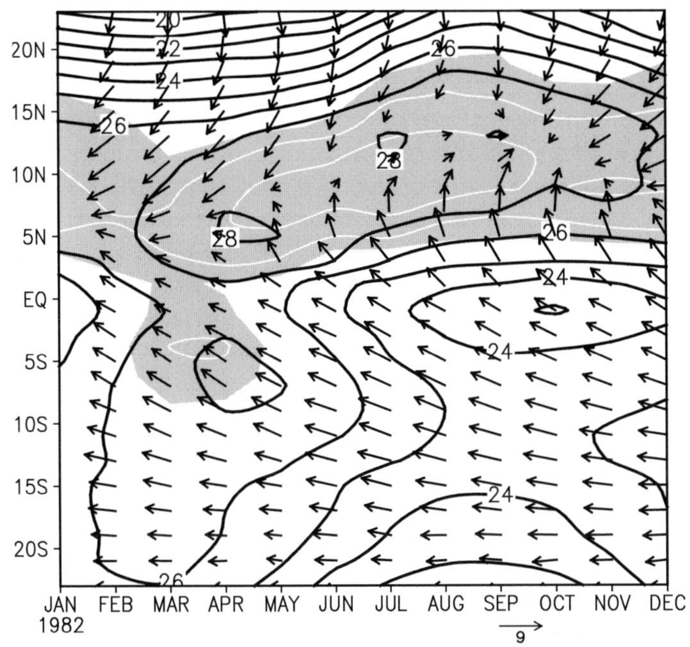

Figure 4-4. Time-latitude section of climatological SST (black contours in °C), surface wind vectors (m/s), based on COADS; and CMAP precipitation (white contours at 5 mm/d intervals; shade > 2.5 mm/d). All zonally averaged in 120°W–115°W.

Sea surface temperature affects tropical convection through its effect on moist static stability and its gradient that drives moisture-laden winds in the marine boundary layer (MBL). An empirical SST threshold for deep convection exists at 26°C–27°C (e.g., Waliser and Graham 1993). Indeed, major tropical precipitation is confined within the 27°C SST contours (Fig. 4-2). The correspondence between SST and tropical convection is not perfect, however. This is especially the case in the Indo-western Pacific region where, for example, SST has a broad equatorial maximum, yet the precipitation maximum occurs off the equator on either side. Atmospheric general circulation model (AGCM) experiments under aqua-planet conditions suggest that depending on the convection scheme used, precipitation shows a single maximum on the equator or a pair of maxima on either side of the equator in response to such a broad equatorial maximum of SST (Numaguti and Hayashi 1991). The physics of such subtlety is not well understood, and it remains unclear how SST affects convection over the Indo-western Pacific warm pool, where the SST gradient is weak.

The SST control over precipitation is much stronger in the eastern Pacific and in the Atlantic, where the SST gradient is strong. This SST control is perhaps best illustrated by its coevolution with convection and surface winds (Fig. 4-4). Most of the time, warm water with SSTs > 26°C is confined to north of the equator, and so is convection. From April to September, the precipitation maximum moves northward from 5°N to 10°N, apparently dragged by the northward shift of the SST maximum. Briefly in March and April, as the equatorial cold tongue relaxes, the meridional SST gradient weakens substantially between 10°S and 15°N, and SST south of the equator rises above the 26°C threshold, reaching as high as 27°C. Over this Southern Hemisphere warm water, considerable precipitation takes place and a double ITCZ symmetric about the equator is often observed during these months[3]. Surface wind convergence follows the same seasonal cycle of and is tightly coupled with SST and precipitation. While some details remain to be worked out, such as the equatorward displacement of the precipitation maximum from the SST maximum in the Northern Hemisphere (Hastenrath 1991), their joint seasonal cycle in Figure 4-4 illustrates the strong SST control of convection in this part of the world.

4. AIR-SEA FEEDBACK

The SST control mechanism offers a partial solution to the problem of climatic asymmetry. From such a meteorological point of view, the ITCZ remains north of the equator over the eastern Pacific and Atlantic because SST is higher north of the equator than south. From an oceanographic point of view, on the other hand, SST is higher north of the equator because the ITCZ stays in the Northern Hemisphere. This circular argument suggests that the northward-displaced ITCZ and high SST band are just two sides of the same coin and understanding both phenomena requires an air-sea interaction approach[4] .

---

3 The discovery of such a double ITCZ from satellite cloud imagery in March 1967 (Kornfield et al. 1967) led to a brief excitement that it vindicates Charney's (1971) then unpublished theory of Ekman CISK (conditional instability of the second kind) that predicts such a double ITCZ. Based on atmospheric GCM results, Manabe et al. (1974) show that SST effect is more important in controlling the eastern Pacific ITCZ.

4 Pike (1971) used a coupled ocean-atmosphere model to study the meridional configuration of the ITCZ. His results answer the first question in the previous section. Namely, under the prevailing easterlies, wind-induced upwelling reduces SST on the equator to a level that deep convection is no longer possible. At the end of his 88-day integration, a single ITCZ forms away from the cold equator. Curiously, however, SST under the ITCZ is 0.5°C lower than on the other side of the equator, in contrary to the observed SST-precipitation relation (Fig. 4-3).

In recognition of their importance for the El Niño/Southern Oscillation (ENSO) and its global impact, the eastern Pacific ITCZ and equatorial annual cycle were extensively discussed at several of the National Oceanic and Atmospheric Administration's (NOAA) Equatorial Pacific Ocean Climate Studies (EPOCS) Program principal investigators meetings in the early 1990s. Stimulated by these discussions, investigators proposed several air-sea feedback mechanisms for maintaining the observed climatic asymmetry characterized by the northward-displaced ITCZ.

4.1. Wind-Evaporation-SST Feedback

Surface evaporation, a function of both SST and wind speed, is the major means for tropical oceans to balance incoming solar radiation. Surface wind speed reaches a minimum at the ITCZ in both the eastern Pacific and the Atlantic (Figs. 4-3 and 4-4), a fact Halley (1686) documented[5]. Based on this observation, Xie and Philander (1994) propose the following mechanism for breaking the equatorial symmetry set by solar radiation. Suppose that somehow SST north of the equator becomes slightly warmer than to the south (Fig. 4-5). The sea level pressure (SLP) gradient will drive southerly winds across the equator. The Coriolis force acts to turn these southerlies westward south and eastward north of the equator. Superimposed on the background easterly trades south of the equator, these anomalous southeasterlies increase surface wind speed and hence evaporative cooling. Conversely, north of the equator wind speed and surface evaporation decrease, amplifying the initial northward SST gradient. This wind-evaporation-SST (WES) feedback offers an explanation for the observed cross-equatorial differences in both wind speed and SST in Figure 4-3. If one assumes that everything else is the same at 10°N and 10°S, a 25% wind speed difference leads to an SST difference of 3°C according to the Clausius-Clapeyron equation for saturated water vapor content (for a typical wind speed of 7–8

Why Pike's model ITCZ chooses to form in the colder hemisphere is unclear, possibly because the integration is too short to filter out chaotic variability in the tropical convection. Manabe (1969) clearly recognizes the effect of equatorial upwelling on tropical convection, but somehow the oceanic ITCZ stays on the equator in his one-year integration of a coupled GCM.

[5] In the ITCZ, annual mean scalar wind speed is considerably greater than vector speed because of the rectification by westward-traveling easterly waves (e.g., Gu and Zhang 2002) and the seasonal march of the Doldrums (Section 7.1). In the 4-year (August 1999–July 2003) observations by the QuikSCAT satellite scatterometer, the mean scalar wind speed is 6–7 m/s in the ITCZ as opposed to ~8 m/s on the other side of the equator along 10°S.

m/s). To balance the net radiative flux, SST must rise (fall) under weak (strong) winds at 10°N (S).

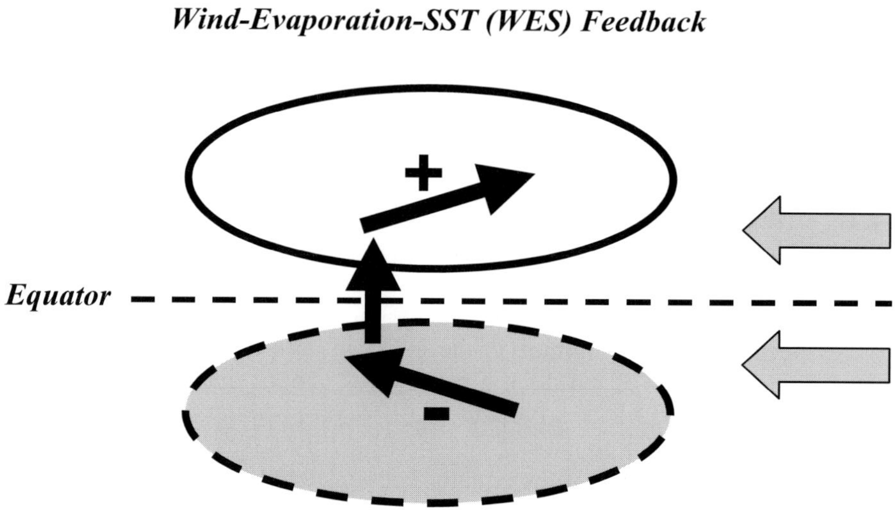

Figure 4-5. Schematic of the WES feedback: anomalies of SST in contours (negative dashed) and surface wind velocity in black vectors. The gray vectors on the right signify the background easterly trades.

Xie and Philander (1994) demonstrate the symmetry-breaking effect of this WES feedback with a zonally symmetric coupled model. The model convection is linearly proportional to SST above a threshold and vanishes when SST falls below it. The surface wind is computed based on a linear model. The model SST is computed based on a slab mixed-layer model cooled by upwelling centered on the equator. Because of the WES feedback, the symmetric solution with a double ITCZ becomes unstable and the model settles into an asymmetric steady state. Under forcing and boundary conditions that are perfectly symmetric, the coupled model develops a single ITCZ on one side of the equator that is collocated with the surface wind speed minimum and SST maximum (Fig. 4-6).

When equatorial upwelling is removed, SST reaches the maximum at the equator and so does the model convection. Only when the equator is kept colder than the convective threshold by ocean upwelling, the coupled system has a choice among the double-ITCZ symmetric solution and two asymmetric ones with a single off-equatorial ITCZ. Because of the WES feedback, the symmetric solution with a double ITCZ is unstable and the model settles into an asymmetric steady state. This necessary condition of equatorial upwelling for the development of climatic asymmetry is consistent with the observation that the ITCZ is nearly symmetric over the Indo-western Pacific warm pool but kept to the north of the equator over the eastern Pacific and the Atlantic where ocean upwelling maintains a cold equator.

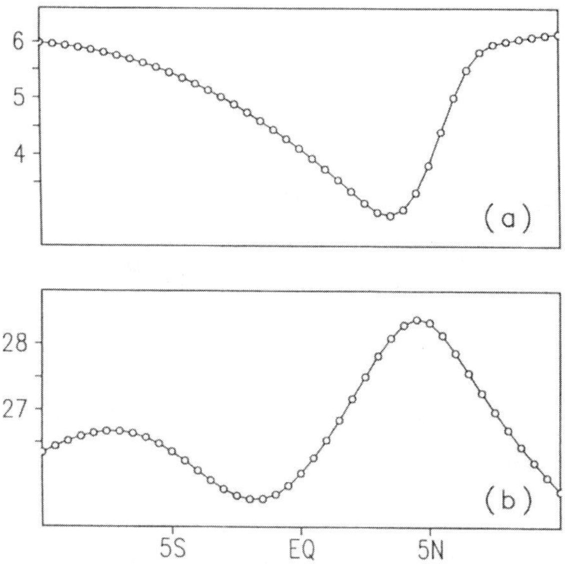

Figure 4-6. Asymmetric solution to the Xie and Philander (1994) model under equatorially symmetric conditions: (a) wind speed and (b) SST.

4.2. Stratus-SST Feedback

While solar radiation at the top of the atmosphere is nearly symmetric about the equator, its distribution at the sea surface is not, because of the reflection by clouds. Figure 4-3 shows the observed cloudiness climatology (shade). The narrow band of convective clouds in the ITCZ is a negative feedback on SST, reducing the climatic asymmetry. Toward the east end of both the Pacific and Atlantic basins, an extensive low cloud deck with annual mean cloud cover exceeding 60% helps cool the southern tropical oceans by reducing insolation at the sea surface. Klein and Hartmann (1993) report that the low-level stratus cloud cover is highly correlated and increases with atmospheric stability above the sea surface. The Peruvian cloud cover peaks in boreal fall when the local air temperature in the lower troposphere is at its minimum. Philander et al. (1996) suggest that this cloud cover peak off Peru results from seasonal cooling of local SST. They further propose a positive feedback between stratus clouds and SST: An initial SST cooling increases the atmospheric stability and hence the stratus cloud cover, which acts to amplify the initial sea surface cooling by reflecting solar radiation into space. This positive stratus-SST correlation is observed in interannual variations over the South Pacific (Klein and Hartmann 1993) and South Atlantic (Tanimoto and Xie 2002) stratus cloud regions. For the South Atlantic cloud deck, Tanimoto and Xie estimate that cloud cover increases by 10% in response to a 1°C drop in local SST, twice as much as Klein and Hartmann's estimate for the cloud cover response near the equatorial southeast Pacific. These different findings presumably arise because the authors focus on different SST and atmospheric circulation anomaly patterns—ENSO is an important player in South Pacific cloud cover variations while the anomalies of South Atlantic stratus clouds are associated with the meridional shift of the Atlantic ITCZ as well as local SST changes on the basin scale.

Using a coupled GCM, Philander et al. (1996) demonstrated the effect of this stratus-SST feedback on climatic asymmetry. They reported that climatic asymmetry is markedly strengthened if a stratus cloud parameterization based on Klein and Hartmann's (1993) observations is implemented, in which the cloud cover increases with static stability of the lower atmosphere. This stratus effect on the eastern Pacific ITCZ is found in other coupled GCMs (Ma et al. 1996; Kimoto and Shen 1997; Yu and Mechoso 1999; Gordon et al. 2000; Fu and Wang 2001).

Philander et al.'s (1996) original feedback concept considers only the indirect effect of stratus clouds on the ITCZ through SST. The intense upward long-wave radiation at the top of these clouds also cools the marine boundary layer, strengthening the South Pacific subtropical high and the cross-equatorial southerly winds that converge on the northern ITCZ (Nigam 1997). This direct effect on the atmosphere is confirmed in a full physics model but the effect is modest; a complete removal of the cloud radiative effect south of the equator leads to a 10%–20% decrease in the intensity of the eastern Pacific ITCZ (Wang et al., submitted).

4.3. Upwelling-SST Feedback

Southerly to southeasterly winds dominate the eastern Pacific and Atlantic equator. In response to these cross-equatorial winds, surface ocean currents flow downwind near the equator but become perpendicular to the wind direction 2°–3° away from the equator as the Coriolis effect becomes important. In response to a southerly wind forcing, this change in flow regime generates ocean upwelling south and downwelling north of the equator. The resultant northward SST gradient strengthens the cross-equatorial winds, completing a positive feedback loop (Chang and Philander 1994). This feedback explains the observation that the center of the equatorial cold tongue is consistently shifted south of the equator in the eastern Pacific and the Atlantic. Since the upwelling effect on SST becomes less important in the off-equatorial open ocean (say poleward of 3°), this mechanism is probably secondary in generating the broader latitudinal asymmetry between 10°S and 10°N.

The narrower meridional asymmetry, as characterized by a strong SST front at 2°N and weak SST gradients at 2°S, is receiving much attention lately. Satellite scatterometer measurements indicate a strong wind deceleration along the axis of the cold tongue and a strong acceleration as the air flows across the equator (Chelton et al. 2001). Attributed to SST-induced adjustment in vertical wind shear (Wallace et al. 1989), this deceleration of wind on the equator leads to strong wind curl that favors upwelling south of the equator (Chelton et al. 2001). It also maintains a surface wind convergence south of the equator (Liu and Xie 2002), which is not associated with deep convection because a strong temperature inversion caps the marine boundary layer except for a brief period during March–April when local SST exceeds 26°C (Fig. 4-4). Quasi-periodic (monthly) tropical instability waves produce spectacular meanders at the equatorial front centered at 1°N–2°N, inducing covariations in the atmospheric boundary layer. In particular, an increase in SST along the front is associated with an increase in boundary-layer cloud cover (Deser et al. 1993; Hashizume et al. 2001), an association opposite to the one observed over the stratus cloud decks west of South America and South Africa. The cross-equatorial flow in the MBL and its interaction with the ITCZ are foci of a recent field campaign over the eastern Pacific (Cronin et al. 2002; Raymond et al., in press; Small et al., submitted).

5. CONTINENTAL FORCING AND ITS WESTWARD CONTROL

Air-sea feedbacks are important in keeping the ITCZ north of the equator. They do not fully explain, however, why the Northern, not the Southern Hemisphere is favored in the Pacific and Atlantic. A long-held belief is that hemispheric asymmetry in area, shape, and orography of continents ultimately gives rise to the climatic asymmetry. This must be generally true, for the northward-displaced ITCZ has existed for a long time—at least since Europeans began sailing in the Atlantic many hundred years ago. Unanswered until very recently have been the following specific questions. Which continental features and how do they move the ITCZ away from the equator? For the Pacific, is the shape of the Asian-Australian continents to the west or that of the Americas more important? Given that the direct atmospheric response to continental asymmetry is likely to be confined near the coast, what sustains the climatic asymmetry in the middle of the vast Pacific, far away from any continents?

5.1. Westward Control

Xie (1996a) develops a simple theory to address these questions. The nondimensional equation for SST difference between meridional SST maxima north and south of the equator may be cast as

$$\frac{\partial(T_N - T_S)}{\partial t} = \sigma V - (T_N - T_S), \tag{1}$$

where V is the meridional wind velocity on the equator and σ is a positive quantity called the WES coefficient. The right-hand side (rhs) is obtained by linearizing the surface latent heat flux. The first term reflects its wind-speed dependence and represents the WES effect: Southerly cross-equatorial winds will reduce (increase) wind speed and evaporation north (south) of the equator (Fig. 4-5). The second term on the rhs reflects the SST dependence of surface evaporation and acts as Newtonian cooling. The local cloud-SST feedback may be absorbed in the second term. Time has been normalized with the resultant effective Newtonian cooling coefficient.

Cross-equatorial wind velocity is modeled with a quasi-steady Rossby wave equation that is forced by the meridional SST gradient

$$(1 - \frac{\partial}{\partial x})V = (T_N - T_S), \tag{2}$$

where the east-west distance x has been normalized by the e-folding scale of the damped long Rossby wave. Combining (1) and (2) yields an equation for cross-equatorial wind velocity that serves as a measure of climatic asymmetry

$$(\frac{\partial}{\partial t}+1)(1-\frac{\partial}{\partial x})V = \sigma V . \tag{3}$$

For the axisymmetric case, (3) reduces to

$$\frac{\partial}{\partial t}V = (\sigma - 1)V . \tag{4}$$

If the WES effect is strong enough to overcome the Newtonian cooling, the solution to (4) becomes unstable and small latitudinal asymmetry grows into amplitudes large enough to push the ITCZ to one side of the equator and eliminate convection on the other, as reported in Xie and Philander (1994).

In general, Equation (3) may be solved by imposing an eastern boundary condition

$$V\,|_{x=0} = V_E . \tag{5}$$

V_E is positive in both the Pacific and Atlantic (Fig. 4-3) and as part of the westward-traveling Rossby wave, results solely from continental forcing to the east. The steady-state solution to (3) and (5) is

$$V = V_E e^{(1-\sigma)x} . \tag{6}$$

Namely, the oceanic climate asymmetry is controlled by continental asymmetry on the ocean's eastern boundary. Positive air-sea feedback increases the e-folding zonal scale $(1 - \sigma)^{-1}$, allowing the influence of continental asymmetry to penetrate far into the west, over nearly 10,000 km in the Pacific. Observed meridional wind on the equator peaks on the South American coast and decays westward, indicating that the real Pacific Ocean-atmosphere is subcritical ($\sigma < 1$).

The continent's westward control over oceanic climate stems from the fact that under the long-wave approximation, all asymmetric signals in the ocean and atmosphere have to propagate westward as Rossby waves. (The Kelvin wave can propagate eastward, but is symmetric about the equator.) Figure 4-7 is an explicit demonstration of this westward control in a coupled model in which a northern land bulge creates the latitudinal asymmetry. Basin-wide northward displacement of the ITCZ occurs only when the continental forcing is placed on the eastern continent. This westward co-propagation of ocean-atmospheric anomalies is consistent with the coupled GCM results that a localized radiative cooling off the Peruvian coast strengthens climatic asymmetries across the Pacific (Ma et al. 1996; Kimoto and Shen 1997; Yu and Mechoso 1999). An immediate implication of this westward control mechanism is that the search for the symmetry-breaking forces can be narrowed down to the eastern continent—the Americas for the Pacific and Africa for the Atlantic.

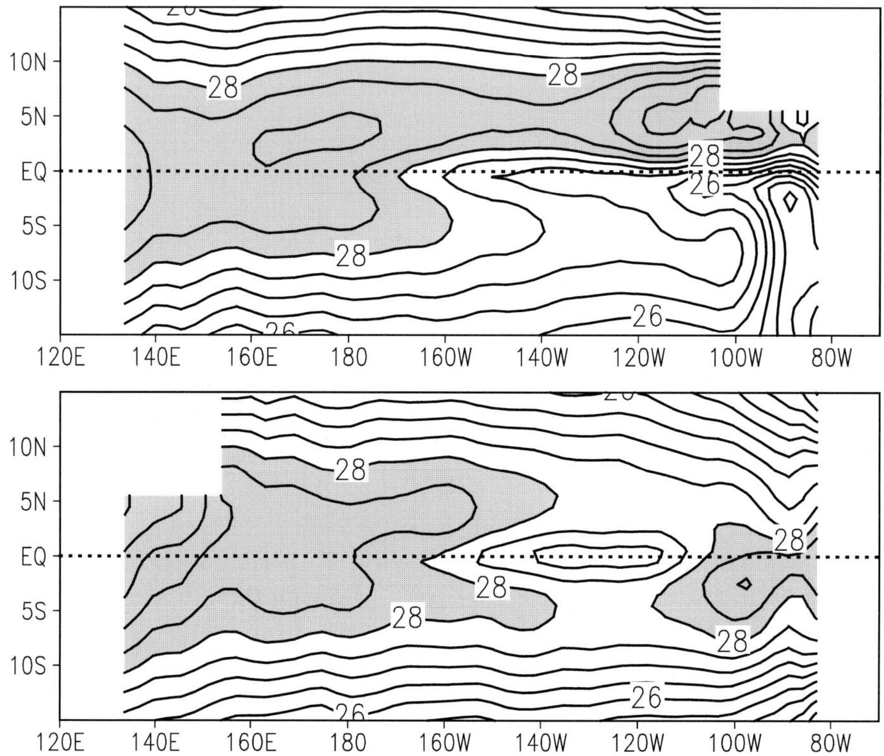

Figure 4-7. Time-mean SST in a coupled model where a northern land bulge is added to the eastern (upper panel) and western (lower panel) continent. From Xie and Saito (2001).

5.2. Continental Forcing and Basin-Wide Adjustment

Using a coupled GCM, Philander et al. (1996) were the first to demonstrate that the shape of the west coast of the Americas and Africa leads to basin-wide displacement of the ITCZ into the Northern Hemisphere in the Pacific and Atlantic, respectively. First, they run their atmospheric GCM under zonally uniform and latitudinally symmetric SST to isolate the effect of continental geometry on ocean winds. Regarding the Pacific, they point to the northwest tilt of the coast of the American continents. South of the equator, winds are roughly parallel to the coast, which would induce coastal upwelling and cool SSTs there (Fig. 4-8 [top]). North of the equator, by contrast, winds are nearly perpendicular to the coast with little effect on upwelling and as a result, SST remains high there. Then based on their coupled model runs, Philander et al. show that though confined near the coast, this initial hemispheric difference in coastal upwelling and SST leads to a basin-wide shift in Pacific climate when air-sea interaction, the stratus-SST feedback in particular, kicks in (Figs. 4-8 [middle], 4-8 [bottom]).

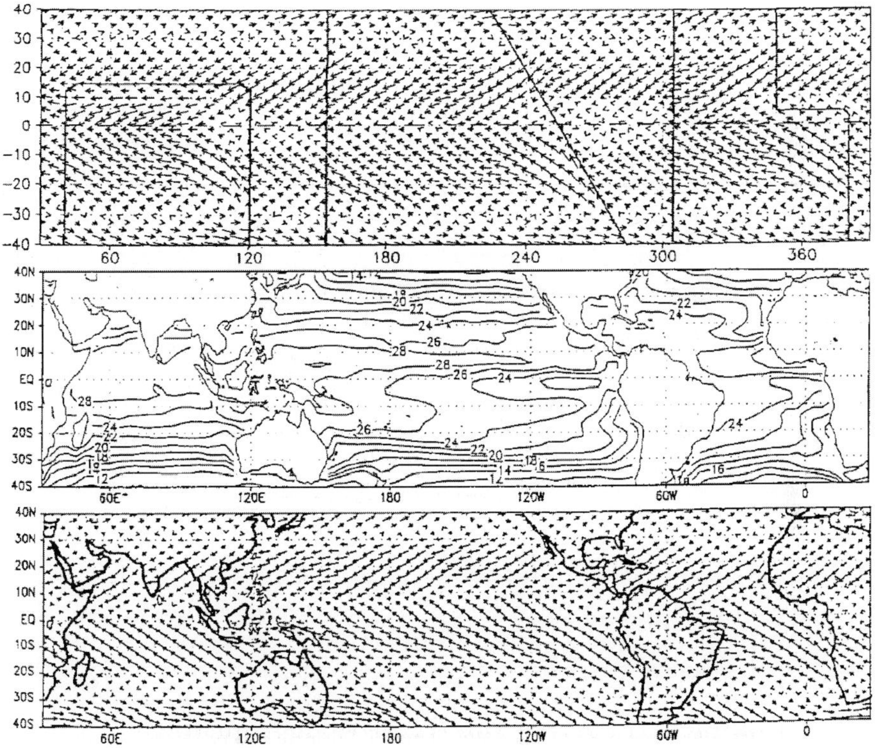

Figure 4-8. (top) Surface wind velocity in an atmospheric GCM forced with a specified SST distribution that is zonally uniform and equatorially symmetric. (middle) SST and (bottom) surface wind velocity in a coupled ocean-atmosphere GCM. The equinox insolation distribution is used and all the orography on land is removed. From Philander et al. (1996).

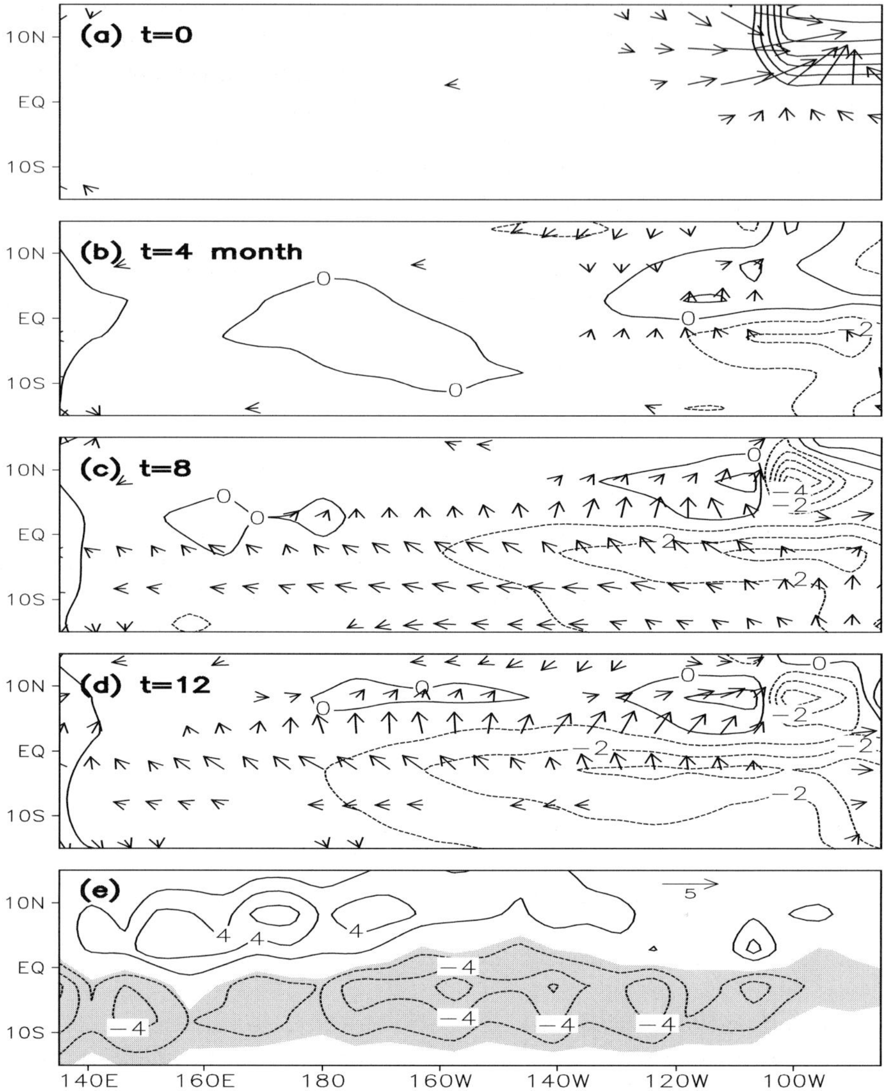

Figure 4-9. (a) Direct response of an atmospheric GCM to the addition of a northern land bulge: Anomalies of surface temperature (contours) and wind velocity. The evolution of surface temperature (contours in °C) and SST-induced surface wind velocity (m/s) changes 4, 8, and 12 months after the ocean model is turned on. (e) Changes in precipitation (mm/s) after the coupled model reaches a statistical steady state. From Xie and Saito (2001).

Regarding the Atlantic, the authors point to the bulge of West Africa, which is hotter than the ocean to the south and induces the southerly winds in the Gulf of Guinea. The winds in turn cause upwelling both off the South African coast and in the open ocean south of the equator. Xie and Saito (2001) suggest that the weakening of the northeasterly trades east of West Africa, in contrast to strong southeasterlies south of the equator, is an additional symmetry-breaking effect of the West African bulge by reducing local evaporation and hence warming the sea surface north of the equator.

Initial value problems shed further light on the role of air-sea interaction in establishing basin-wide climatic asymmetry. Figure 4-9 shows the coupled ocean-atmosphere adjustment to a sudden addition of a land bulge on a symmetric land strip north of the equator. The direct effect of this continental forcing is rather limited in space (Fig. 4-9a). When the ocean is allowed to interact with the atmosphere, however, it triggers a coupled wave front that moves quickly westward, pushing the ITCZ north of the equator on its way. In a matter of less than a year, strong climatic asymmetry resembling that observed in the Pacific and Atlantic is established basin-wide. Similar initial value problems are studied by Xie (1996a) with simpler models and by Ma et al. (1996) and Kimoto and Shen (1997) with coupled GCMs.

6. CONSEQUENCES OF ITCZ ASYMMETRY

6.1. Atmosphere and Ocean Circulation

The ITCZ supplies much of the heat that drives the global atmospheric circulation, and the latitude of its position is an important parameter for the global atmosphere and climate. A northward-displaced ITCZ induces hemispheric asymmetry in the Hadley circulation, with the southern cell being stronger than the northern one (Lindzen and Hou 1988). This difference leads to a stronger subtropical westerly jet in the Southern Hemisphere upper troposphere (Hou 1993; Chang 1995). This asymmetry in the subtropical jets furthermore results in temperature and storm activity differences.

In the Pacific, upper ocean circulation develops strong hemispheric asymmetries in response to wind curls associated with the northward-displaced ITCZ. The North Equatorial Countercurrent (NECC) is one such asymmetry, flowing eastward against the easterly trades and in geostrophic balance with a thermocline ridge beneath the ITCZ[6] (Wyrtki and Koblinsky 1984; Kessler and Taft 1987). Such an eastward countercurrent is not observed south of the equator.

Temperature stratification in the equatorial oceans is maintained by cold water subducted during the winter in the subtropics and transported into the equator along the thermocline by the so-called subtropical cells (STCs; see Liu and Philander [2001] for a review). The subtropical water reaches the equator under the surface via an interior pathway in the South Pacific, while it goes mostly through the western boundary current in the North Pacific (Lu and McCreary 1995). This asymmetry in STCs, along with that in precipitation, moreover gives rise to pronounced hemispheric differences in thermocline salinity distribution (Nonaka and Takeuchi 2001). Given strong mixing/dissipation in the western boundary current, this hemispheric difference in the thermocline water pathway may have important consequences for decadal variability (Gu and Philander 1997; Nonaka et al. 2002; McPhaden and Zhang 2002).

6.2. Equatorial Annual Cycle

The disparity between solar forcing and climatic response is also seen on the seasonal time scale, along the equator in the eastern Pacific and the Atlantic. Despite little seasonal change in insolation on the equator, SST displays a pronounced annual cycle at the Galapagos Islands (90°W, equator), rising to a maximum of 27°C in March and dropping to a minimum of below 22°C in September. For comparison, 20° to the north, near Hawaii, with much larger annual variations in insolation, the annual range in SST is only about half as large, between 24°C and 27°C.

The annual cycle of SSTs on the equator displays a pronounced westward propagation in the eastern Pacific (Horel 1982) and the Atlantic, which Mitchell and Wallace (1992) show is a result of its interaction with zonal wind. Xie (1994b) suggests that the northward displacement of the ITCZ is the ultimate cause of this peculiar annual cycle on the equator (see also Giese and Carton [1994]). In the eastern Pacific and Atlantic, the ITCZ stays north of the equator throughout the year, intensifying in boreal summer and weakening in winter in response to annual solar forcing off the equator (Fig. 4-4). As a result, the cross-equatorial southerlies that converge

[6] The collocation of the NECC with the atmospheric ITCZ led many investigators to speculate that its advection of warm water from the western Pacific is a major reason for the high-SST band in 5°N–10°N. NECC's advective effect turns out to be of secondary importance for SST asymmetry since the zonal SST gradient is very weak along it (Fig. 4-4; Xie 1994a).

onto the ITCZ strengthen in boreal summer and fall, causing equatorial SST to fall by enhancing the upwelling/entrainment of cold water and surface evaporation. Conversely, the relaxed southerlies in boreal spring warm up the equatorial ocean mixed layer by reducing cold upwelling and evaporation. If the ITCZ were displaced to the south of the equator, the equatorial seasonal variations would still be dominated by the annual cycle, but its phase would be opposite—warm in September and cold in March at the Galapagos.

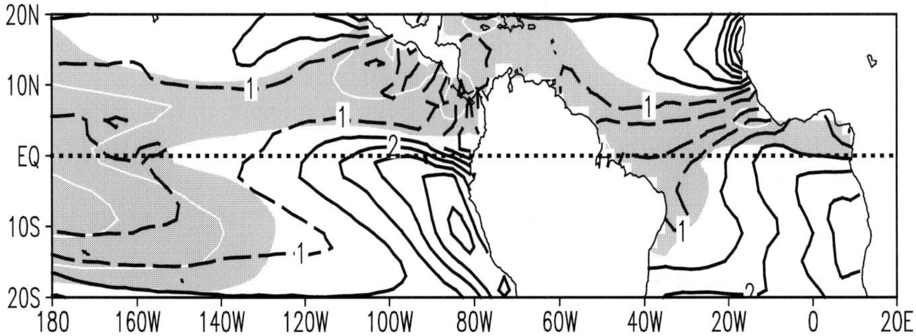

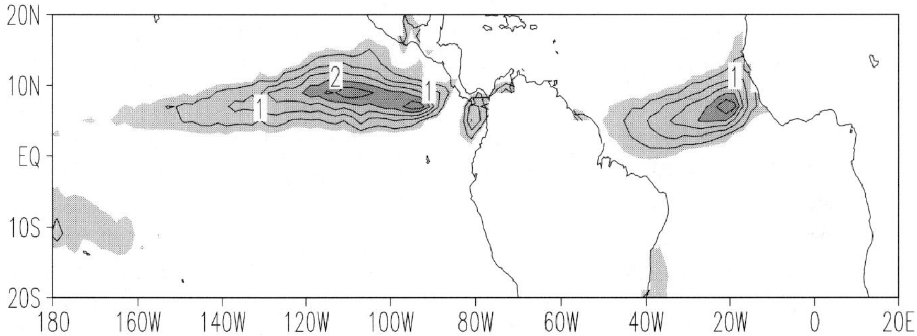

Figure 4-10. (top) Annual harmonic of SST (contours at 0.5°C intervals; values at or less than 1°C dashed), superimposed on annual mean SST (white contours at 1°C intervals; shade > 27°C). (bottom) Effect of seasonal cycle on scalar wind speed (contours at 0.5 m/s intervals; light shade > 0.25 m/s and dark shade >2 m/s). Based on COADS data.

Thus the ITCZ is the climatic equator from the standpoint of seasonal variations. South of the ITCZ the seasonal cycle bears the Southern Hemisphere characteristics of being warm in March and cold in September. To its north, SST peaks in September and is coldest in March. Figure 4-10a shows the annual harmonic of observed SST, with shaded areas indicating regions of high SST. Over most of the Atlantic and east of 140°W in the Pacific, where the ITCZ stays north of the equator and the annual mean southerlies are maintained on the equator, the SST annual harmonic reaches a local maximum on or slightly south of the equator because of the shallow thermocline and strong air-sea interaction there. The annual SST harmonic is at its minimum along the climatological ITCZ, consistent with the notion of its being the climatic equator. This collocation of the annual harmonic minimum and ITCZ is particularly conspicuous in the Atlantic. The negative feedback between convective clouds and local SST seems partly responsible for the weak annual variations in SST under the ITCZ over both the Pacific and the Atlantic.

6.3. Tropical Atlantic Variability

The position of the Atlantic ITCZ displays considerable variability in latitude on interannual to decadal time scales, resulting in droughts in some years and floods in other years in northeast Brazil (Hastenrath 1991). Empirically, such an anomalous meridional displacement of the ITCZ is known to be associated with anomalies in cross-equatorial gradients in SST and zonal wind (Nobre and Shukla 1996; Chiang, Chapter 16, "Present-Day Climate Variability in the Tropical Atlantic: A Model for Paleoclimatic Changes," this volume), very similar to the coupled pattern in Figure 4-5. Chang et al. (1997) show that air-sea interaction and the WES feedback in particular help to give rise to the observed association between anomalies of ITCZ rainfall, surface winds, and SST. Subsequent studies suggest that both the stratus-SST (Tanimoto and Xie 2002) and upwelling-SST (Xie and Saito 2001) feedbacks discussed in Section 4 also contribute. Results of Okumura et al.'s (2001) atmospheric GCM experiments suggest that an anomalous northward shift of the Atlantic ITCZ, forced by an SST pattern like the one in Figure 4-5, weakens the atmospheric Azores high in the subtropics by weakening the local Hadley cell over the North Atlantic.

The northward displacement of the climatological ITCZ over the Atlantic may affect the space-time structure of tropical variability. The WES feedback owes its positive sign to the sign change in the Coriolis parameter across the equator and is most effective under the north-south symmetric mean state. The departure of the mean ITCZ from the geographic equator weakens the WES feedback and hence the interhemispheric interaction (Okajima et al. 2003). This weakening of interaction is possibly responsible for the observed lack of significant correlation between tropical North and South Atlantic SSTs (Houghton and Tourre 1992).

7. CHALLENGES AHEAD FOR REALISTIC CLIMATE SIMULATION

Despite all of the progress outlined above, many state-of-the-art global coupled GCMs still have a problem keeping the Pacific and the Atlantic ITCZ north of the equator. (Mechoso et al. [1995] dub this deficiency as the double ITCZ syndrome.) In these GCMs, deep convection lingers for too long south of the equator, and in some models, the southern ITCZ persists throughout the year over the eastern Pacific, resulting in a double ITCZ in the annual mean climatology instead of the observed northward-displaced single ITCZ. This deficiency in simulating the observed climatic asymmetry has been identified as a major challenge at a recent workshop devoted to discussion of the tropical biases of GCMs (http://pod.tamu.edu/~bias/). These biases have important implications for climate simulation and prediction. For example, the failure to keep the eastern Pacific ITCZ north of the equator is certain to affect the simulation of the equatorial annual cycle, which is known to affect the properties of ENSO, most notably its seasonal phase-locking (Jin et al. 1994; Li and Hogan 1999). The properties of ENSO are highly sensitive to even slight changes in the mean state; the eastern equatorial ocean is considerably cooler with the ITCZ displaced to the north of the equator than with a double symmetric ITCZ because of the upwelling induced by the cross-equatorial southerlies (Fig. 4-7). The meridional configuration of the mean ITCZ also affects the WES feedback and the cross-equatorial mode of climate variability (Okajima et al. 2003) that is most pronounced in the Atlantic.

Ad hoc flux adjustments are often used in simulation and prediction models to prevent them from drifting quickly away from the realistic climate. Because the same air-sea feedbacks contribute both to the mean state and to causing temporal variability (e.g., the Bjerknes feedback for the Walker circulation/ENSO and the WES for climatic asymmetry/Atlantic cross-equatorial variability), the failure of a model to maintain a realistic mean climate indicates that it is representing these feedbacks poorly and it may be severely distorting variability. Dijkstra and Neelin (1999) show that by requiring a coupled ocean-atmosphere model to simulate a realistic cold tongue, they can narrow the parameter space for ENSO and reduce the ambiguity in parameter choice compared with a flux-adjusted version of the same model.

In light of the importance of the climatic asymmetry as discussed above and the difficulty maintaining it in GCMs, we discuss some remaining issues in this section.

7.1. Effect of Seasonal Forcing

We have implicitly assumed so far that latitudinal asymmetry of long-term annual mean climate is independent of seasonal variations, an assumption that is generally untrue. Here we consider the effect of seasonal cycle on scalar wind speed (W), which exerts a strong influence on SST via evaporation and is a nonlinear function of zonal and meridional wind velocities. The crudest estimate of W is obtained by using the annual mean climatological wind components (u_c, v_c), namely,

$$W_C = \sqrt{u_c^2 + v_c^2},$$

In the ITCZ in the far eastern Pacific and the Atlantic, both (u_c, v_c) and hence W_C approach zero, which would require a very high SST to balance the insolation. A better estimate of scalar wind speed results from using monthly climatological wind velocity (u_m, v_m) instead, with m denoting calendar month,

$$W_M = \sum_{m=1}^{12} \sqrt{u_m^2 + v_m^2}\,/12.$$

Figure 4-10b shows the difference between these two estimates, W_M–W_C, which measures the effect of seasonal-varying winds on scalar wind speed and hence SST. The effect of seasonal variations on scalar wind is large along the climatological ITCZ and mostly due to the seasonal migration of the ITCZ and its weak-wind zone, the latter being around the equator in March but moving to 10°N in August (Fig. 4-4).

Amounting to a 2 m/s increase in wind speed along the ITCZ north of the equator, this effect of seasonal forcing reduces cross-equatorial differences in wind speed and hence the latitudinal asymmetry of SST and ITCZ generated by the WES and stratus-SST feedbacks. This weakening of annual mean climatic asymmetry by seasonal forcing has been demonstrated in a simple model (Xie 1996b). While different views exist regarding the role of the annual cycle (Wang and Wang 1999), a realistic simulation of the seasonal migration of the ITCZ is very important for maintaining a realistic degree of climatic asymmetry and vice versa.

7.2. Stratus Cloud Deck

Several coupled GCM studies consistently point to the poor representation of marine boundary layer clouds in the South Pacific as the major cause of the double ITCZ syndrome (Philander et al. 1996; Ma et al. 1996; Kimoto and Shen 1997; Gordon et al. 2000). The loosely used term "stratus" in part of the air-sea interaction community (Klein and Hartmann 1993; Philander et al. 1996) includes at least two distinct cloud types: stratocumulus and trade cumulus. Stratocumulus clouds form in the surface mixed layer, which is tightly coupled with the sea surface, are more area extensive in space and more persistent in time, and hence record larger values of cloud cover than trade cumulus clouds, which rise higher but cover smaller areas and are of shorter duration. The trade cumulus clouds are cumulus clouds formed in the so-called decoupled boundary layer where a stable layer exists between the surface mixed layer and MBL top. Norris (1998) analyzes vertical soundings made at ocean weather stations and shows that the decrease in "stratus" cloud cover with increasing SST (or decreasing static stability of the lower atmosphere) in Klein and Hartmann (1993) is associated with the change in cloud type from stratocumulus over low SSTs to cumulus over higher SSTs (his Fig. 4). Because this cloud type transition is associated with a large change in cloud-layer coupling with the sea surface, Norris suggests that adequate modeling of the decoupled MBL is the key to simulating the sensitivity of cloud type and cloud cover to changes in SST. This sensitivity gives rise to the stratus-SST feedback discussed in Section 4.2.

Typical GCMs have only a few levels in the first 2 km above the sea surface, a resolution that is insufficient to represent the aforementioned transition in MBL and cloud types. Regional atmospheric models (RAMs), affording higher resolutions and more sophisticated physics for turbulence and clouds, can be a useful tool bridging global GCMs on one hand and field observations and large-eddy cloud simulations on the other. Figure 4-11 shows an example of an RAM simulation, with a vertical transect in the South Pacific. In general, MBL clouds are capped by a temperature inversion whose strength decreases westward. Over the cold water off the coast of Peru, the MBL is shallow and stratocumulus clouds form in the surface mixed layer. Both cloud base and top rise toward the west with SST increasing. West of 110°W, the MBL becomes decoupled from the sea surface, with cumulus clouds transporting moisture through the stable layer into a cloud layer below the inversion[7]. Sensitivity experiments show that de-

[7] Lasting for short time, cumuli do not leave a strong signal in the time-mean cloud water field.

trainment of cloud water at the cloud top is important for maintaining the inversion-cloud couplet (McCaa and Bretherton 2003; Wang et al., submitted). In the above RAM, the intense radiative cooling at the cloud top also induces a strong downdraft that is confined to the cloud layer. The simulated westward transition from stratocumulus in a coupled MBL to cumulus in an uncoupled MBL is consistent with the few observations that exist for this region (Garreaud et al. 2001).

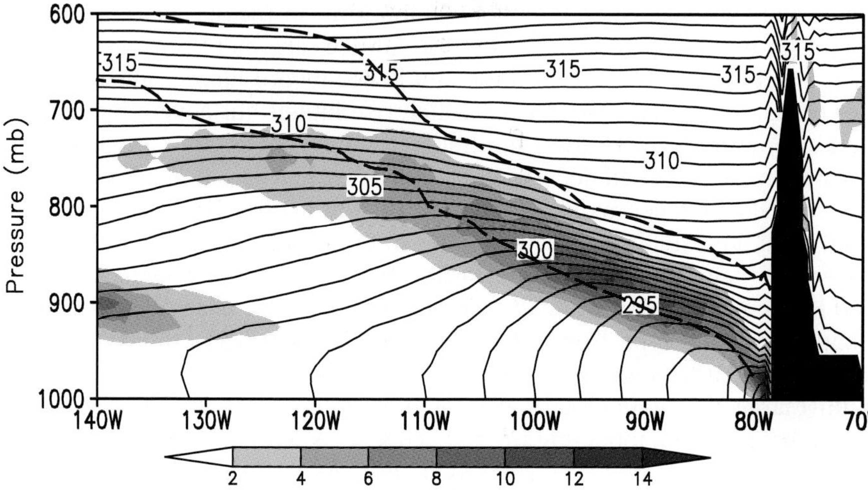

Figure 4-11. Simulated cloud liquid water content (shaded in 10^{-2} g/kg) and virtual potential temperature (solid contours in K) in Xu et al.'s (2004) regional atmospheric model, averaged for August–October 1999 along 10°S. Dashed lines denote the boundaries of the inversion layer. The Andes are shaded in black.

7.3. Andes

The steep and narrow Andes are an overlooked continental forcing. At the equator, the mountain range is only 200 km wide, rising from near sea level to 3.5 km high in less than 100 km. At typical resolutions of 300 km, global climate models severely smooth the Andean mountains to less than 1 km high between 10°S and 10°N, while in reality they are rarely lower than 3 km. Blocking the easterly winds, the Andes induce downward motion on the leeside and thereby help maintain the temperature inversion and the stratocumulus cloud deck off South America. In a RAM experiment that reduces the height of the Andes, the resultant anomalous convergence offshore weakens the inversion and reduces the stratocumulus cloud cover in the August–October season while prolonging the southern ITCZ in March and April (Xu et al. 2004). While these mountain effects on the atmosphere are confined to the coastal region, they can have a basin-wide impact on Pacific climate with the help of coupled ocean-atmosphere waves (Section 5). Thus, smoothing the Andes in global GCMs may contribute to the double-ITCZ syndrome and too weak climatic asymmetry.

8. SUMMARY

Halley recognized the effect of the West African bulge on the winds in the Gulf of Guinea. He wrote: "if a Country lying near the Sun . . . such as the Deserts of Lybia [sic] are usually reported to be, the heat occasioned by the reflection of the Sun's Beams, and the retention there of in the Sand, is incredible to those that have not felt it; whereby the Air being exceedingly rarified, it is necessary that this cooler and more dense Air should run thitherwards to restore the Equilibrium: This I take to be the cause, why near the Coast of Guinea the Wind always sets in upon the Land, blowing Westerly instead of Easterly." Amazingly, the wind map Halley drew 320 years ago contains many of the elements necessary to build a modern solution to the age-old riddle of climatic asymmetry. The bulge of West Africa causes cross-equatorial southerlies in the Gulf of Guinea, initiating an air-sea coupled wave front that pushes the southeast trades to cross the equator into the Northern Hemisphere where they meet the northeast trades. Over the warm waters where these two trade wind systems meet, convection takes place, producing the thunderstorms and rains Halley recorded. On average, winds are calm in the Doldrums of the ITCZ, allowing water to stay warm and maintaining the climatic asymmetry.

It is not the atmosphere, nor the ocean alone, not even their coupling, but the collective effort of the ocean, atmosphere, and the land that gives rise to the long silver band of clouds, stretching north of the equator by half the globe in the Pacific and Atlantic (Fig. 4-2). Like the ENSO phenomenon, this climatic asymmetry attests yet again to the importance of air-sea interaction in making the response of earth's climate deviate considerably in space and time from the pattern expected by the solar forcing alone. Given that the Pacific and Atlantic ITCZ has remained north of the equator for hundreds if not thousands of years, it is quite natural to suggest that the land-sea distribution is the ultimate cause. This answer, however, lacks crucial details such as which continental features are the cause and how they influence climate thousands of kilometers away. Recent studies of air-sea coupling have enabled us to narrow the search of continental forcing to the eastern side of an ocean basin. Furthermore, they have demonstrated that, perhaps to the surprise of many, the direct cause of the great climatic asymmetry over the Pacific and Atlantic is found in regional features of the continents such as the coastal line, not the overall distribution of landmass, such as Asia in the Northern Hemisphere versus Australia in the Southern Hemisphere. This is not to say that the global landmass distribution is of no consequence, but its role is rather indirect. The global distribution of land and seas, for example, is the reason that easterlies prevail in the equatorial Pacific and Atlantic and westerlies in the equatorial Indian Ocean (e.g., Philander et al. 1996), allowing the equatorial cold tongue and the northward-displaced ITCZ to develop in the former oceans, but not in the latter. Much needs to be studied as to what shapes and anchors the climatological rain bands over the Indian and western Pacific where SST gradients are weak and continental influences are strong.

On the geological time scale, continents constantly move and change their shape, orientation, and relative position. For example, in the middle Eocene 50 million years ago, the Andes were lower, and North America was more than 10° north of its present position and unattached to South America (www.scotese.com). Connected with the Pacific through the gap between North and South America, the Atlantic was narrower, with the south coast of West Africa closer to the equator. These changes in landmass distribution are certain to affect the meridional configuration of the ITCZ and other features of the tropical and global climate. Coupled dynamic modeling of continental effects as presented here offers a physical basis for interpreting paleoclimate records and a useful framework for inferring how earth's climate has evolved with drifting continents and changing landscapes.

9. ACKNOWLEDGMENTS

The author would like to thank J. Matsumoto for access to the library of the Department of Geography, University of Tokyo; C.-H. Chang for a literature search; and G. Speidel and H. Xu for helpful comments. This work is supported by NOAA, NSF, NASA, NSFC, and JASTEC. IPRC contribution #279 and SOEST contribution #6384.

10. REFERENCES

Bartholomew, J.G., and A.J. Herbertson. 1899. In, Buchan, A. (ed.). *Atlas of Meteorology*. Edinburgh: Royal Geographical Society.

Chang, E.K.M. 1995. The influence of Hadley circulation intensity changes on extratropical climate in an idealized model. *Journal of Atmospheric Science* 52: 2006–2024.

Chang, P., and S.G.H. Philander. 1994. A coupled ocean-atmosphere instability of relevance to the seasonal cycle. *Journal of Atmospheric Science* 51: 3627–3648.

Chang, P., L. Ji, and H. Li. 1997. A decadal climate variation in the tropical Atlantic Ocean from thermodynamic air-sea interactions. *Nature* 385: 516–518.

Charney, J.G. 1971. Tropical cyclogenesis and the formation of the intertropical convergence zone. In, Reid, W.H. (ed.). *Mathematical Problems of Geophysical Fluid Dynamics*. Providence, Rhode Island: American Mathematics Society.,.

Chelton, D.B., S.K. Esbensen, M.G. Schlax, N. Thum, M.H. Freilich, F.J. Wentz, C.L. Gentemann, M.J. McPhaden, and P.S. Schoph. 2001. Observations of coupling between surface wind stress and sea surface temperature in the eastern tropical Pacific. *Journal of Climate* 14: 1479–1498.

Cronin, M.F., N. Bond, C. Fairall, J. Hare, M.J. McPhaden, and R.A. Weller. 2002. Enhanced oceanic and atmospheric monitoring underway in Eastern Pacific. *Eos Trans.* American Geophysical Union, 83(19): 205, 210–211.

Deser, C., J.J. Bates, and S. Wahl. 1993. The influence of sea surface temperature on stratiform cloudiness along the equatorial front in the Pacific Ocean. *Journal of Climate* 6: 1172–1180.

Dijkstra, H.A., and J.D. Neelin. 1999. Coupled processes and the tropical climatology. Part III: Instabilities of fully coupled climatology. *Journal of Climate* 12: 1630–1643.

Fu, X., and B. Wang. 2001. A coupled modeling study of the seasonal cycle of the Pacific cold tongue, Part I: Simulation and sensitivity experiments. *Journal of Climate* 14: 756–779.

Garreaud, R.D., J. Rutllant, J. Quintana, J. Carrasco, and P. Minnis. 2001. CIMAR-5: A snapshot of the lower troposphere over the subtropical Southeast Pacific. *Bulletin of the American Meteorological Society* 82: 2193–2207.

Giese, B.S., and J.A. Carton. 1994. The seasonal cycle in a coupled ocean-atmosphere model. *Journal of Climate* 7: 1208–1217.

Gordon, C.T., A. Rosati, and R. Gudgel. 2000. Tropical sensitivity of a coupled model to specified ISCCP low clouds. *Journal of Climate* 13: 2239–2260.

Gu, D., and S.G.H. Philander. 1997. Interdecadal climate fluctuations that depend on exchange between the tropics and extratropics. *Science* 275: 805–807.

Gu, G., and C. Zhang. 2002. Westward-propagating synoptic-scale disturbances and the ITCZ. *Journal of Atmospheric Science* 59: 1062–1075.

Halley, E. 1686. A historical account of the trade winds, and monsoons, observable in the seas between and near the Tropicks, with an attempt to assign the phisical cause of the said winds. *Philosophical Transactions of the Royal Society of London*, 16: 153–168.

Hashizume, H., S.-P. Xie, W.T. Liu, and K. Takeuchi. 2001. Local and remote atmospheric response to tropical instability waves: A global view from the space. *Journal of Geophysical Research-Atmosphere*, 106: 10173–10185.

Hastenrath, S. 1991. *Climate Dynamics of the Tropics*. Boston: Kluwer Academic, 488 pp.

Held, I.M., and A.Y. Hou. 1980. Nonlinear axially symmetric circulations in a nearly inviscid atmosphere. *Journal of Atmospheric Science* 37: 515–533.

Horel, J.D. 1982. On the annual cycle of the tropical Pacific atmosphere and Ocean. *Monthly Weather Review*, 110: 1863–1878.

Hou, A.Y. 1993. The influence of tropical heating displacement on the extratropical climate. *Journal of Atmospheric Science* 50: 3553–3570.

Houghton, R.W., and Y.M. Tourre. 1992. Characteristics of low-frequency sea surface temperature fluctuations in the tropical Atlantic. *Journal of Climate* 5: 765–771.

Jin, F.-F., J.D. Neelin, and M. Ghil. 1994. El Niño on the devil's staircase: Annual subharmonic steps to chaos. *Science* 264: 70–72.

Kessler, W.S., and B.A. Taft. 1987. Dynamic heights and zonal geostrophic transports in the central tropical Pacific during 1979–84. *Journal of Physical Oceanography* 17: 97–122.

Kimoto, M., and X. Shen. 1997. Climate variability studies using general circulation models. In, Sumi, A. (ed.). *The Frontiers of Climate Research II.* Center for Climate System Research, University of Tokyo, pp. 91–116.

Klein, S.A., and D.L. Hartmann. 1993. The seasonal cycle of low stratiform clouds. *Journal of Climate* 6: 1587–1606.

Köppen, W. 1899. Winds over the oceans. In, Buchan, A. (ed.). *Atlas of Meteorology*. Edinburgh: Royal Geographical Society, Plate 14.

Kornfield, J., A.F. Hasler, K.J. Hanson, and V.E. Suomi. 1967. Photographic cloud climatology from ESSA III and V computer produced mosaics. *Bulletin of the American Meteorological Society* 48: 878–883.

Li, T., and T.F. Hogan. 1999. The role of the annual-mean climate on seasonal and interannual variability of the tropical Pacific in a coupled GCM. *Journal of Climate* 12: 780–792.

Lindzen, R.S., and A.Y. Hou. 1988. Hadley circulation for zonally averaged heating centered off the equator. *Journal of Atmospheric Science* 45: 2416–2427.

Liu, W.T., and X. Xie. 2002. Double intertropical convergence zones—A new look using scatterometer. *Geophysical Research Letters* 29: 2072, doi:10.1029/2002GL015431.

Liu, Z., and G. Philander. 2001. Tropical-extratropical oceanic exchange pathways. In, Siedler, G., J. Church, and W.J. Gould (eds.). *Ocean Circulation and Climate: Observing and Modeling the Global Ocean.* San Diego: Academic Press, pp. 247–254.

Lu, P., and J.P. McCreary. 1995. Influence of the ITCZ on the flow of the thermocline water from the subtropical to the equatorial Pacific Ocean. *Journal of Physical Oceanography* 25: 3076–3088.

Ma, C.-C., C.R. Mechoso, A.W. Robertson, and A. Arakawa. 1996. Peruvian stratus clouds and the tropical Pacific circulation—A coupled ocean-atmosphere GCM study. *Journal of Climate* 9: 1635–1645.

Manabe, S. 1969. Climate and the ocean circulation. Part 2: The atmospheric circulation and the effects of heat transfer by ocean currents. *Monthly Weather Review* 97: 775–805.

Manabe, S., D.G. Hahn, and J.L. Holloway. 1974. The seasonal variation of the tropical circulation as simulated by a global model of the atmosphere. *Journal of Atmospheric Science* 31: 43–83.

McCaa, J.R., and C.S. Bretherton. 2003. A new parameterization for shallow cumulus convection and its application to marine subtropical cloud-topped boundary layers: Part II: Regional simulations of marine boundary layer clouds. *Monthly Weather Review* 132: 883–896.

McPhaden, M.J., and D. Zhang. 2002. Slowdown of the meridional overturning circulation in the upper Pacific Ocean. *Nature* 415: 603–608.

Mechoso, C.R., A.W. Robertson, and Coauthors. 1995. The seasonal cycle over the tropical Pacific in general circulation models. *Monthly Weather Review* 123: 2825–2838.

Mitchell, T.P., and J.M. Wallace. 1992. The annual cycle in equatorial convection and sea surface temperature. *Journal of Climate* 5: 1140–1156.

Möller, F. 1951. Viertel Jahrs Karten des Niederschlags fur die Ganze Erde. *Petermanns Geograph. Mitt.*, 95: 1–7.

Nigam, S. 1997. The annual warm to cold phase transition in the eastern equatorial Pacific: Diagnosis of the role of stratus cloud-top cooling. *Journal of Climate* 10: 2447–2467.

Nobre, P., and J. Shukla. 1996. Variations of sea surface temperature, wind stress, and rainfall over the tropical Atlantic and South America. *Journal of Climate* 9: 2464–2479.

Nonaka, M., and K. Takeuchi. 2001. Tropical subsurface salinity and tritium distributions in the Pacific: Their differences and formation mechanisms. *Journal of Physical Oceanography* 31: 1388–1395.

Nonaka, M., S.-P. Xie, and J.P McCreary. 2002. Decadal variations in the subtropical cells and equatorial Pacific SST. *Geophysical Research Letters* 29: 1116, doi: 10.1029/2001GL013676.

Norris, J.R. 1998. Low cloud type over the ocean from surface observations. Part I: Relationship to surface meteorology and the vertical distribution of temperature and moisture. *Journal of Climate* 11: 369–382.

Numaguti, A., and Y.-Y. Hayashi. 1991. Behaviors of cumulus activity and the structures of circulations in an aqua-planet model. *Journal of the Meteorological Society of Japan* 69: 563–579.

Okajima, H., S.-P. Xie, and A. Numaguti. 2003. Interhemispheric coherence of tropical climate variability: Effect of climatological ITCZ. *Journal of the Meteorological Society of Japan* 81: 1371–1386.

Okumura, Y., S.-P. Xie, A. Numaguti, and Y. Tanimoto. 2001. Tropical Atlantic air-sea interaction and its influence on the NAO. *Geophysical Research Letters* 28: 1507–1510.

Philander, S.G.H., D. Gu, D. Halpern, G. Lambert, N.-C. Lau, T. Li, and R.C. Pacanowski. 1996. Why the ITCZ is mostly north of the equator. *Journal of Climate* 9: 2958–2972.

Pike, A.C. 1971. Intertropical convergence zone studied with an interacting atmosphere and ocean model. *Monthly Weather Review* 99: 469–477.

Raymond, D.J., S.K. Esbensen, C. Paulson, M. Gregg, C.S. Bretherton, W.A. Petersen, R. Cifelli, L.K. Shay, C. Ohlmann, and P. Zuidema. In press. EPIC2001 and the coupled ocean-atmosphere system of the tropical east Pacific. *Bulletin of the American Meteorological Society*.

Reynolds, R.W., and T.M. Smith. 1994 Improved global sea surface temperature analyses using optimal interpolation. *Journal of Climate* 7: 929–948.

Small, R.J., S.-P. Xie, Y. Wang, S.K. Esbensen, and D. Vickers. Submitted. Numerical simulation of boundary layer structure and cross-equatorial flow in the eastern Pacific. *Journal of Atmospheric Science*.

Tanimoto, Y., and S.-P. Xie. 2002. Inter-hemispheric decadal variations in SST, surface wind, heat flux and cloud cover over the Atlantic Ocean. *Journal of the Meteorological Society of Japan* 80: 1199–1219.

Thompson, R.M., S.W. Payne, E.E. Recker, and R.J. Reed. 1979. Structure and properties of synoptic-scale wave disturbances in the intertropical convergence zone of the eastern Atlantic. *Journal of Atmospheric Science* 36: 53–72.

U.S. Air Force and U.S. Department of Commerce. 1971. *Global Atlas of Reflective Cloud Cover, 1967–1970*. Washington, D.C.

Waliser, D.E., and N.E. Graham. 1993. Convective cloud systems and warm-pool surface temperatures: Coupled interactions and self-regulation. *Journal of Geophysical Research* 98: 12881–12893.

Wallace, J.M., T.P. Mitchell, and C. Deser. 1989. The influence of sea surface temperature on surface wind in the eastern equatorial Pacific: Seasonal and interannual variability. *Journal of Climate* 2: 1492–1499.

Wang, B., and Y. Wang. 1999. Dynamics of the ITCZ-equatorial cold tongue complex and causes of the latitudinal asymmetry. *Journal of Climate* 12: 1830–1847.

Wang, C., S.-P. Xie, and J. A. Carton (eds.). In press. *Earth Climate: The Ocean-Atmosphere Interaction*, Geophysical Monograph. Washington D.C.: American Geophysical Union.

Wang, Y., S.-P. Xie, B. Wang, and H. Xu. Submitted. Large-scale atmospheric forcing by southeast Pacific boundary-layer clouds: A regional model study. *Journal of Climate.*

Woodruff, S.D., R.J Slutz, R.L. Jenne, and P.M. Steurer. 1987. A comprehensive ocean-atmosphere dataset. *Bulletin of the American Meteorological Society* 68: 521–527.

Wyrtki, K., and B. Koblinsky. 1984. Mean water and current structure during the Hawaii to Tahiti shuttle experiment. *Journal of Physical Oceanography* 14: 242–254.

Xie, P., and P.A. Arkin. 1996. Analyses of global monthly precipitation using gauge observations, satellite estimates, and numerical model predictions. *Journal of Climate* 9: 840–858.

Xie, S.-P. 1994a. Oceanic response to the wind forcing associated with an ITCZ in the northern hemisphere. *Journal of Geophysical Research-Oceans* 99: 20393–20402.

Xie, S.-P. 1994b: On the genesis of the equatorial annual cycle. *Journal of Climate* 7: 2008–2013.

Xie, S.-P. 1996a. Westward propagation of latitudinal asymmetry in a coupled ocean-atmosphere model. *Journal of Atmospheric Science* 53: 3236–3250.

Xie, S.-P. 1996b. Effects of seasonal solar forcing on latitudinal asymmetry of the ITCZ. *Journal of Climate* 9: 2945–2950.

Xie, S.-P., and S.G.H. Philander. 1994. A coupled ocean-atmosphere model of relevance to the ITCZ in the eastern Pacific. *Tellus* 46A: 340–350.

Xie, S.-P., and K. Saito. 2001. Formation and variability of a northerly ITCZ in a hybrid coupled AGCM: Continental forcing and ocean-atmospheric feedback. *Journal of Climate* 14: 1262–1276.

Xu, H., Y. Wang, and S.-P. Xie. 2004. Effects of the Andes on eastern Pacific climate: A regional atmospheric model study. *Journal of Climate* 17: 589–602.

Yu, J.-Y., and C.R. Mechoso. 1999. Links between annual variations of Peruvian stratocumulus clouds and of SST in the eastern equatorial Pacific. *Journal of Climate* 12: 3305–3318.

Chapter 5

YEAR-TO-YEAR VARIABILITY IN THE HADLEY AND WALKER CIRCULATIONS FROM NCEP/NCAR REANALYSIS DATA

Shoshiro Minobe[1, 2]
[1]*Division of Earth and Planetary Sciences, Graduate School of Science, Hokkaido University, Sapporo, 060-0810, Japan*
[2]*Frontier Research System for Global Change, Yokohama, Japan*
E-mail: minobe@ep.sci.hokudai.ac.jp

Abstract The year-to-year variability in the local Hadley and Walker circulations is studied by empirical orthogonal function (EOF) analysis of zonal and meridional divergent winds at 200 and 850 hPa in the NCEP/NCAR reanalysis data. The first mode of the EOF analysis for the period since 1979 is closely related to the El Nino/Southern Oscillation (ENSO). The corresponding vertical velocity structure at the middle of the troposphere is characterized by the combination of a horseshoe pattern in the western tropical Pacific and an oval pattern in the central-eastern equatorial Pacific, consistent with satellite-derived precipitation correlations. A streamline analysis for horizontal divergent winds and vertical winds revealed that the dominant local Hadley and Walker circulation anomalies connect the oval and the horseshoe, while the other clusters of the local Hadley circulation anomalies rotating in opposing directions emanate from the Maritime Continent region.

The first EOF mode of the data for 1949 through 2002 is characterized by a trend-like increase from the 1960s to the 1980s that is consistent with a previous study by Goswami and Thomas (2000). This mode is accompanied by an increase of downward vertical wind anomalies over Sahel and over the central equatorial Pacific Ocean and by an increase in upward anomalies over the Maritime Continent and the Amazon. A consistent decrease in precipitation is observed over Sahel. Precipitation also decreased over the central Pacific, but the region of the decrease is located to the south of the center of the downward wind anomalies. Precipitation increases consistent with the local upward motions are not observed over the Maritime Continent or over the Amazon.

H.F. Diaz and R.S. Bradley (eds.), The Hadley Circulation: Present, Past and Future, 153–171.

1. INTRODUCTION

The Hadley and Walker circulations are fundamental structures in the earth's atmosphere, and hence understanding their year-to-year variability is clearly of scientific value. The interannual changes of these circulations associated with the El Niño/Southern Oscillation (ENSO) were intensively studied after pioneering works of Bjerkness (1966, 1969) by a number of researchers (e.g., Arkin 1982; Webster et al. 1998; Philander 1990; Oort and Yienger 1996; Wang et al. 2000; Wang 2002ab). In order to understand these circulation anomalies, schematic diagrams for the anomalous Walker circulation (e.g., McPhaden et al. 1998), and the combined Walker and local Hadley circulations (Wang 2002ab) were proposed. However, previous schematics may be insufficient to understand the three-dimensional nature of the anomalous Hadley and Walker circulations.

Recent advances in the atmosphere reanalysis data allow us to study more closely the Hadley and Walker circulations (e.g., Trenberth et al. 2000). Using NCEP/NCAR reanalysis data (Kalnay et al. 1996), Goswami and Thomas (2000) examined the three-dimensional decadal changes of the Hadley and Walker circulations, and found that the most dominant changes are characterized by a trend from the 1960s to the 1980s. Focusing on El Niños on interannual time scales and using NCEP/NCAR reanalysis data, Wang (2002a) proposed that anomalous local Hadley cells occur in the eastern and western Pacific rotating in opposing directions to each other. Although these previous studies have provided useful information on interannual to interdecadal Hadley and Walker circulations, a number of questions should still be explored: Is the three-dimensional structure of Hadley and Walker circulation anomalies associated with ENSO well represented by the previous schematics? Is the trend robust regardless of analysis methods, and reasonably related to other fields, i.e., precipitation? To address these questions in this chapter, a combined empirical orthogonal function (EOF) analysis of the 200 and 850 hPa divergent winds is performed.

The rest of the chapter is organized as follows: In Section 2, the data and methodology are described. The results of EOF analysis of the data for the period from 1979 through 2002, for which reanalysis data are more reliable than those for the earlier period, are explained in Section 3. Using the entire record from 1949 through 2002, EOF analysis is repeated and the results are shown in Section 4. In Section 5, conclusions and a discussion are presented.

2. DATA AND METHOD

We mainly analyzed the NCEP/NCAR reanalysis data for January 1949 to April 2002 for divergent winds, upward velocities in pressure coordinates (minus ω), and surface temperatures (sea surface temperatures [SSTs] over the ocean and land surface temperatures otherwise [Kalnay et al. 1996; Kistler et al. 2001]). In order to obtain representative circulation changes related to the Hadley and Walker circulations, we calculate combined EOFs of four scalar variables (zonal and meridional divergent winds at 200 and 850 hPa) over the globe. This method would capture anomalies in *local* Hadley circulations and the Walker circulations. Structures of other fields, such as vertical wind speeds or surface temperatures, relating with the divergent wind EOFs are extracted by calculating correlation and regression coefficients of other fields onto the principal components (PCs) or the temporal coefficients of the EOFs.

The NCEP/NCAR reanalysis data are available for the period from 1949, but the reliability of the earlier period of the record may be questionable due to the smaller number of observations[1]. In particular, inclusion of satellite data obtained since 1979 may substantially improve the reanalysis data. Thus, one of the reasonable choices for the analysis period is from 1979 to the end of the record in 2002 as will be described in Section 3. However, this choice can be too conservative so that we cannot obtain any information for the period before 1979. Therefore, we will first examine EOF analysis using data obtained after 1979 in Section 4.

In order to compare with the reanalysis vertical winds, we use a couple of precipitation data sets, since the reliability of the vertical velocities can be questionable (Kalnay et al. 1996; Newman et al. 2000). One data set is from the Climate Prediction Center satellite/gauge Merged Analysis of Precipitation (referred to as CMAP data) (Xie and Arkin 1997). The CMAP data set used in this chapter is a version in which reanalysis precipitation data were not incorporated. The other data set is gridded land precipitation from gauge measurements (referred to as gauge data) (Dai et al. 1997). The periods of the record are 1979–2002 for the CMAP data and 1850–1995 for the gauge data, respectively, though the gauge precipitation data are used for the period overlapping with the reanalysis data; i.e., from 1949 to 1995.

In addition to the above gridded data, we use monthly El Niño SST indices, i.e., NIÑO3, NIÑO3.4, and NIÑO4 indices, for the period since

[1] Number of sonde observations can be seen at http://wesley.wwb.noaa.gov/cgi-bin/pdisp_m_obscnt.sh (Sept., 2003)

1950. These NIÑO indices are obtained at the Web site of the Climate Prediction Center[2]. The NIÑO SST indices are area-averaged SST anomalies for 1950 to 2002; The longitudinal ranges are 150°W–90°W for NIÑO3, 160°E–150°W for NIÑO4, and 170°W–120°W for the NIÑO3.4 index, with a common 5°S–5°N latitudinal range.

Anomalies of parameters are calculated as deviations of monthly climatologies, which are defined for the whole available record of respective data. The anomalies are averaged seasonally (sampling rate of 3 months) for smaller numbers of computations for EOF analysis. Unless otherwise stated, raw seasonal anomalies are used.

3. EOF ANALYSIS FOR THE PERIOD SINCE 1979

Figures 5-1a and b show the spatial patterns of the first EOF mode for the 200 and 850 hPa divergent zonal and meridional wind speeds. This mode explains 21% of the total variance of the divergent winds. The spatial structure isequivalent to the regression coefficients onto principal component 1 (PC-1). Also, accompanied regressions of vertical wind speeds at the middle of the troposphere are shown in Figure 5-1c. Prominent divergences and convergences occur at the 200 hPa level over the tropical central-to-eastern Pacific and over the Maritime Continent, respectively. The divergence (convergence) in the upper troposphere is accompanied by convergence (divergence) in the lower troposphere, connected by the upward (downward) motions in the middle of the troposphere. The overall pattern of the vertical velocities can be viewed as a combination of a horseshoe pattern prevailing over the Maritime Continent and the subtropical western Pacific and an oval in the central-eastern tropical Pacific with opposing polarities.

Figure 5-1d shows correlation coefficients of surface temperatures onto PC-1. The surface temperature chart is reminiscent of the typical SST anomaly pattern during the mature phase of ENSO (e.g., Rasmusson and Carpenter 1982). Consistently, PC-1 captures the El Niño events in 1982–83, 1986–88, the long-lasting El Niño in the early 1990s (Trenberth and Hoar 1996), and the latest event in 1997–98 (Fig. 5-2). The decorrelation time scale, which is given by the smallest lag with an autocorrelation smaller than e^{-1}, for PC-1 is three seasons. Conservatively, assuming that the data at 1-year lag are independent, we find that an absolute correlation coefficient larger than 0.42 is significant at the 95% confidence limit. Correlation coefficients between the PC-1 and NIÑO SST indices are generally

[2] http://www.cpc.ncep.noaa.gov/data/indices/

high, and the highest correlation (r = 0.92) is found with the NIÑO3.4 SST index followed by the second highest correlation (r = 0.88) with the NIÑO3 SST and by the correlation with the NIÑO4 SST (r = 0.82). This result indicates that dominant variability of the divergent winds, which are the most energetic in the tropics and hence represent the local Hadley and Walker circulations, is tightly related to the ENSO for the last two decades.

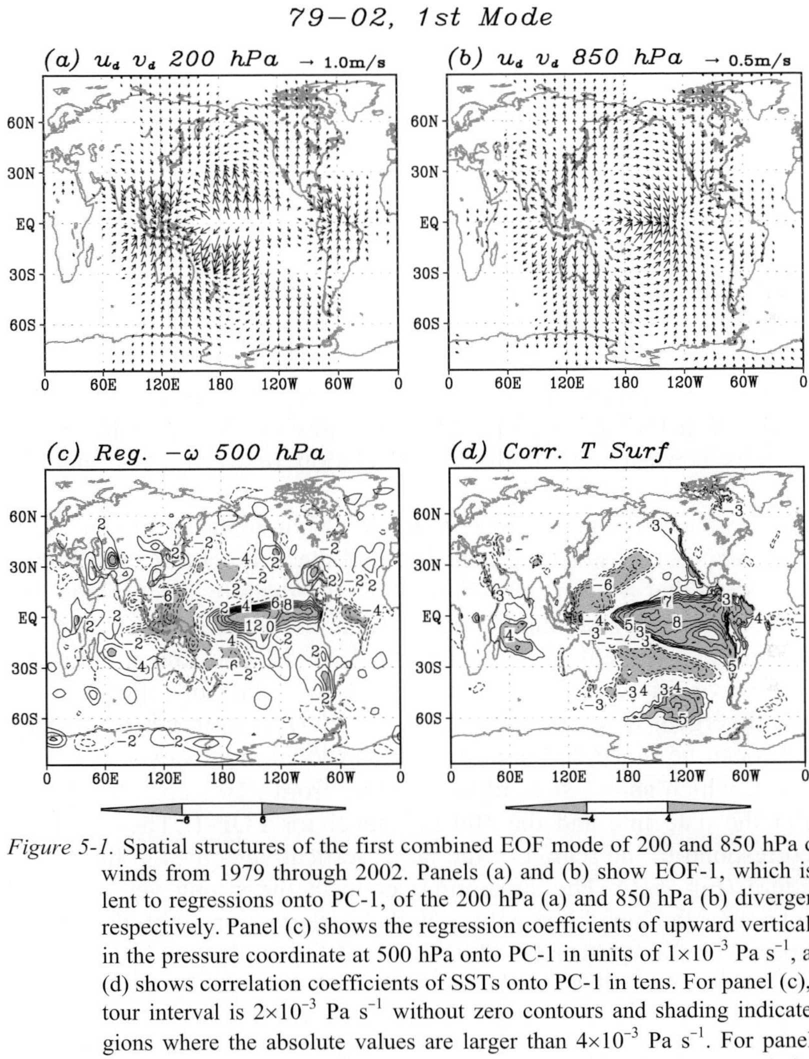

Figure 5-1. Spatial structures of the first combined EOF mode of 200 and 850 hPa divergent winds from 1979 through 2002. Panels (a) and (b) show EOF-1, which is equivalent to regressions onto PC-1, of the 200 hPa (a) and 850 hPa (b) divergent winds, respectively. Panel (c) shows the regression coefficients of upward vertical velocity in the pressure coordinate at 500 hPa onto PC-1 in units of 1×10^{-3} Pa s^{-1}, and panel (d) shows correlation coefficients of SSTs onto PC-1 in tens. For panel (c), the contour interval is 2×10^{-3} Pa s^{-1} without zero contours and shading indicates the regions where the absolute values are larger than 4×10^{-3} Pa s^{-1}. For panel (d), the contour interval is 1 in tens (0.1 for correlations) for absolute correlations larger than 0.3, and shading indicates the regions where the absolute values of correlations are larger than 0.4.

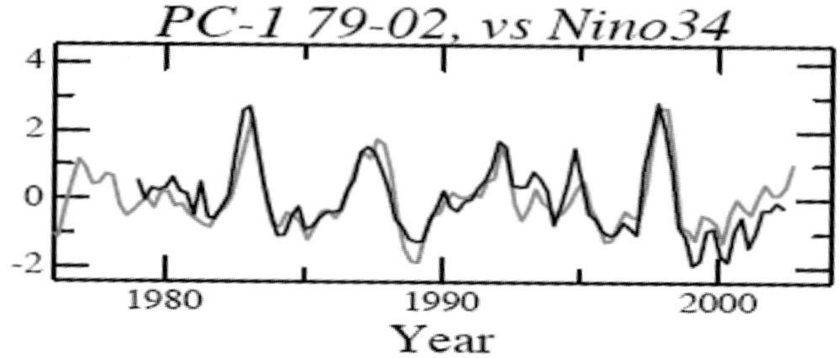

Figure 5-2. PC-1 or the time coefficient of EOF-1 shown in Fig. 5-1 (black line) and NIÑO3.4 SST index in arbitrary units (gray line).

In order to obtain some idea of the reliability of the wind structures shown in Figure 5-1, we compare the correlation coefficients of vertical winds onto PC-1 and those of the CMAP precipitations (Fig. 5-3). The combination of the horseshoe and oval is commonly found in these correlation maps, but the extra-equatorial horseshoe correlations in the Northern Hemisphere are weaker than those in the Southern Hemisphere. The correlation distribution for precipitation is similar to the CMAP precipitation anomalies associated with the NIÑO3.4 SST variability shown by Wang et al. (2000). This result is consistent with the close relation between PC-1 and the NIÑO3.4 SST index.

Given the three-dimensional structure of horizontal divergent winds and vertical winds, it is not easy to understand these structures from Figure 5-1 to the extent that we can confidently illustrate them by a schematic diagram. For a better understanding of those structures, we estimate streamlines according to the divergent winds and vertical winds. The streamline is defined as a curve for which the velocity field consisting of regressions of horizontal divergent and vertical winds onto PC-1 is everywhere tangent. Streamlines can be illustrated in a three-dimensional plot as exemplified in Figure 5-4a, which shows streamlines starting from ±20°, 25°, and 30° in latitude at the date line and the 500 hPa level for EOF-1. These starting points correspond to the region of the large vertical velocities in the horseshoe pattern (Fig. 5-1). The streamlines connect the strong vertical wind regressions over the horseshoe pattern with the regressions with the opposing polarity over the equatorial oval pattern.

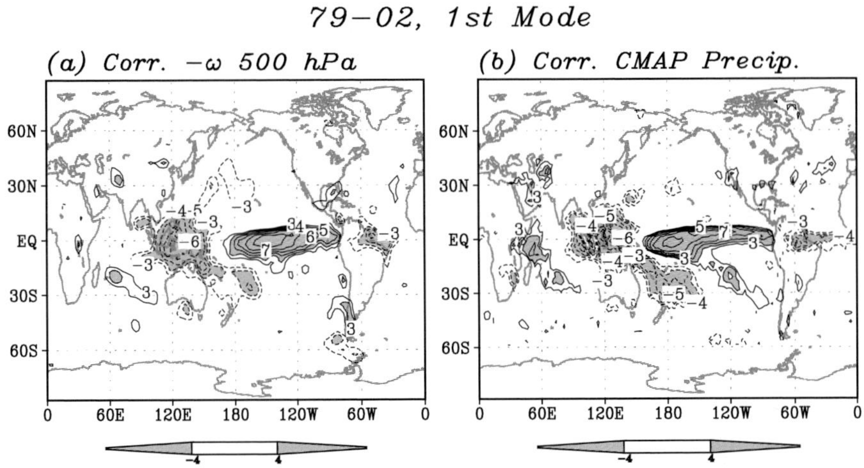

Figure. 5-3. (a) Correlations of the upward vertical velocity in the pressure coordinate at 500 hPa onto PC-1, which is shown in Fig. 5-2, and (b) correlations of the CMAP precipitation onto PC-1. Units are tens, and contour interval is 0.1 for absolute correlations larger than 0.3.

Although the three-dimensional view such as in Figure 5-4a is a powerful tool for exploring the shape of the streamlines in detail, it is not possible to draw a large number of streamlines (i.e., more than 10) in a panel. Thus, to display an overall pattern of the streamlines, we employ a plain view like Figure 5-4b, but with a much larger number of streamlines.

Figure 5-5 shows the plain view of streamlines associated with EOF-1 starting at every 10° in longitude and 5° in latitude between 40°S and 40° N and at 500 hPa. To avoid making the figure panel too busy and to focus on important streamlines, we plot only streamlines that are accompanied by absolute values of vertical velocities larger than 0.002 Pa s^{-1} at the starting points, and reach the 300 or 700 hPa level. The streamlines that meet these conditions are mostly limited to the Indian Ocean and the Pacific Ocean, and exhibit a well-organized, meridionally symmetric pattern. In the Pacific Ocean, the streamlines run over the tropical North (South) Pacific in a northwest-southeast (northeast-southwest) direction, and these streamlines connect the region of active vertical motion in the central Pacific Ocean and those in the subtropical western North and South Pacific. From another equatorial divergent center located over the Maritime Continent, streamlines emanate westward, northwestward, southwestward, and eastward. The eastward branch of streamlines from the Maritime Continent towards the date line represents the Walker circulation anomalies. Also, the Walker circulation anomalies are expressed by the bending of streamlines in the central Pacific; in this region, strong zonal wind anomalies trapped near the equator bend streamlines, which extend from off-equator toward the equator according to the local Hadley circulation, in the zonal direction near the equator. Thus, the equatorial bending indicates that substantial Walker and local Hadley circulation anomalies coexist over the central equatorial Pacific. The other divergence center over the eastern Amazon and western equatorial Atlantic is connected to the eastern Pacific by streamlines directly. That is, in this region, Walker circulation anomalies are much more pronounced than local Hadley circulation anomalies around the equator, consistent with Wang (2002b; Chapter 6, "ENSO, Atlantic Climate Variability, and the Walker and Hadley Circulations," this volume). These features are summarized in the schematic diagram in Figure 5-6.

The diagram in Figure 5-6 is similar to the diagram of Wang (2002a) with respect to the fact that the two anomalous local Hadley circulations occur in the central-eastern Pacific and in the Maritime Continent with opposing polarity. However, the present diagram emphasizes the meridionally slanted nature of the divergent winds, and the dominance of the local Hadley circulations over the central-eastern Pacific Ocean. The overwhelming local Hadley circulations over the central-eastern Pacific Ocean compared to the opposing local Hadley circulations over the Maritime Continent results in the symmetric structures of the global Hadley circulation anomalies associated with ENSO reported by previous studies (Oort and Yienger 1996; Goswami and Thomas 2000).

In order to know whether the streamlines (Fig. 5-5), which are the basis of the schematic diagram of Figure 5-6, are robust with respect to the details of the analysis method, we calculated streamlines based on regression coefficients with the NIÑO3.4 index using data for the period since 1979 (not shown) and using the data for the period since 1950 (Fig. 5-7a). As was expected from the high correlation between PC-1 and NIÑO3.4, the streamlines based on regression coefficients with the NIÑO3.4 index are quite similar to those with PC-1. (No essential difference was found according to the different periods of regression calculations.)

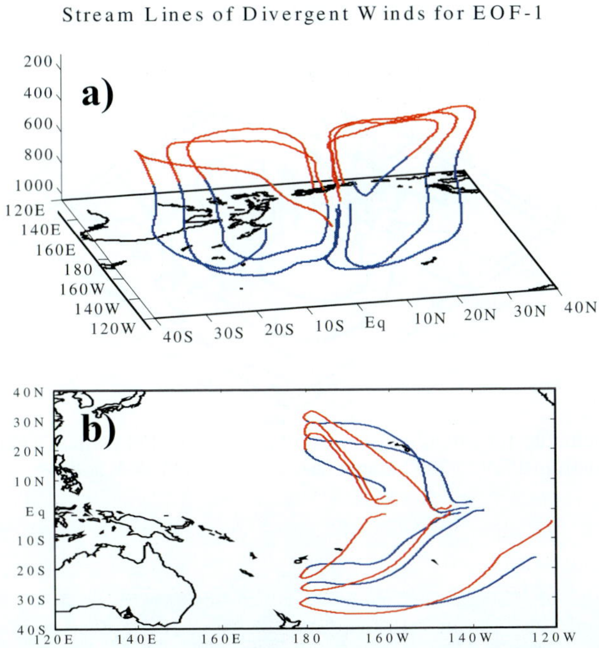

Figure. 5-4. Bird's-eye view (a) and plain view (b) of streamlines according to the regression coefficients of the divergent horizontal winds and vertical winds (see text), starting from ±20°, 25°, and 30° in latitude at the date line and at the 500 hPa level. Vertical axis in (a) is pressure height in hPa. Red and blue curves indicate streamlines lo-

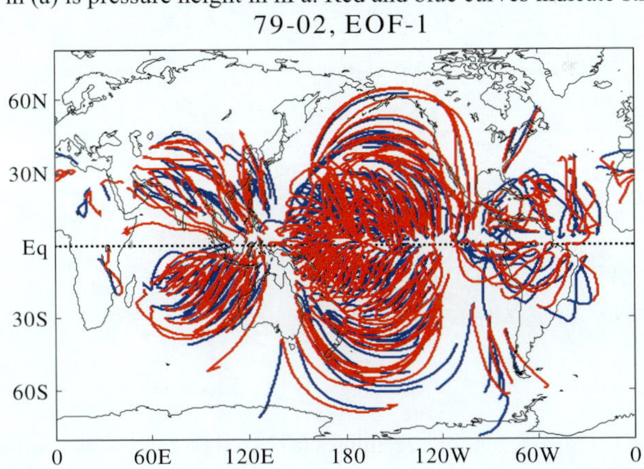

Figure 5-5. Streamlines corresponding to the horizontal divergent winds and vertical winds shown in Figs. 5-1. The streamlines start at the 500 hPa level at every 10° in longitude and 5° in latitude, traced until the streamline again crosses 500 hPa. Streamlines are not shown, if the streamline that is accompanied by the absolute vertical wind regression at the starting point is smaller than 0.002 Pa s^{-1} or the streamline does not reach the 300 or 700 hPa level. Red and blue curves indicate streamlines located above and below the 500 hPa height, respectively.

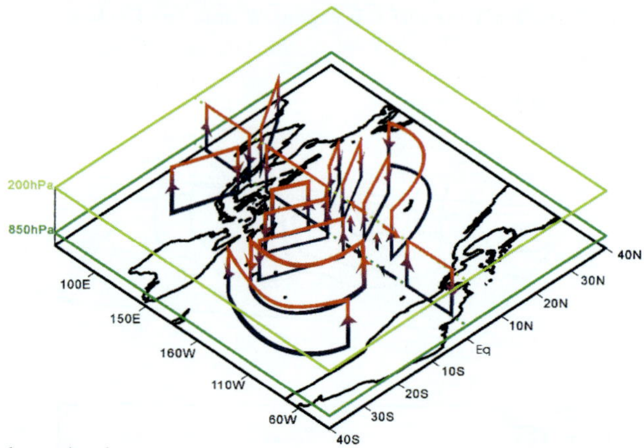

Figure 5-6. Schematics for structures of horizontal divergent and vertical wind speeds for EOF-1 from 1979 to 2002 drawn based on Figs. 5-1, 5-4, and 5-5.

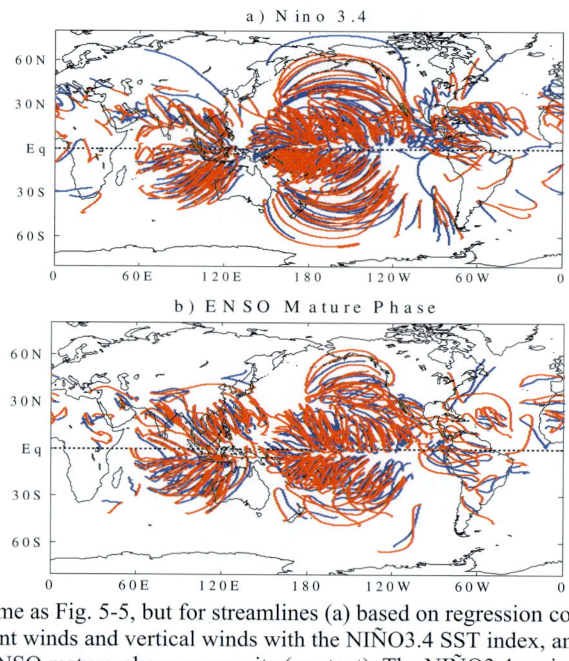

Figure. 5-7. Same as Fig. 5-5, but for streamlines (a) based on regression coefficients of divergent winds and vertical winds with the NIÑO3.4 SST index, and (b) based on the ENSO mature-phase composite (see text). The NIÑO3.4 region is bounded by 120°W–170°W and 5°S–5°N.

Furthermore, streamlines are calculated based on a composite of the ENSO mature phase (Fig. 5-7b). The composite is calculated from November through the next January for seven significant El Niño events (1957–58, 1965–66, 1972–73, 1982–83, 1986–87, 1991–92 and 1997–98) according to Wang et al. (2002a). The streamlines for the ENSO mature phase are less symmetric about the equator than those for EOF-1, with more (less) prominent zonal streamlines in the Northern (Southern) Hemisphere. Except for this difference, the overall structure is common between the streamlines for the ENSO composite and those for EOF-1. Therefore, it can be concluded that the schematic diagram shown in Figure 5-6 is not substantially dependent on the details of the analysis methods, and captures the dominant structures of the Hadley and Walker circulation anomalies associated with ENSO.

4. EOF ANALYSIS FOR THE PERIOD SINCE 1949

When we calculated the EOFs using divergent winds for the period since 1949 instead of since 1979 in the previous section, we found that PC-1 for the period since 1949 is entirely different from the aforementioned PC-1 for the period since 1979, with a correlation coefficient as low as 0.05 for the overlapping period from 1979 to 2002. The PC-1 since 1949 is characterized by a long-term increase from the 1960s to the 1980s (Fig. 5-8). The EOF-1 explains 24% of the total variance of the 200 and 850 hPa divergent winds.

Figures 5-9a–c show that the spatial pattern of the divergent winds for EOF-1 is characterized by the lower-level divergence (upper-level convergence) over the tropical Pacific, the Amazon, central Africa, and the Maritime Continent, with alternating polarities. Again, the lower-level divergence (convergence) is accompanied by the upper-level convergence (divergence) and downward (upward) vertical winds in the middle of the troposphere. The strong correlations of surface temperatures are found over the tropical Indian Ocean–Maritime Continent with smaller correlations over Africa and the tropical South Atlantic (Fig. 5-9d).

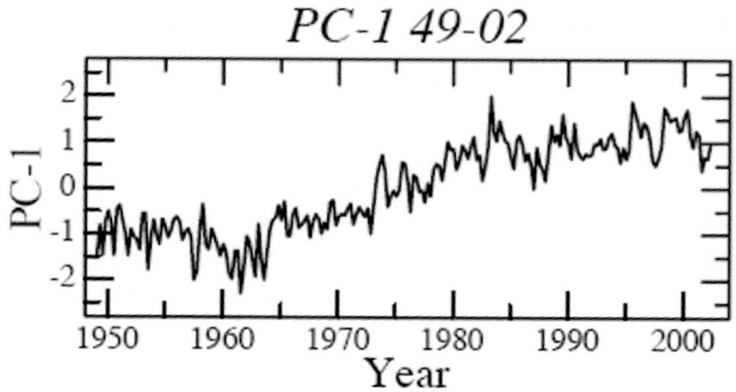

Figure 5-8. PC-1 of the EOF analysis of divergent winds from 1949 to 2002.

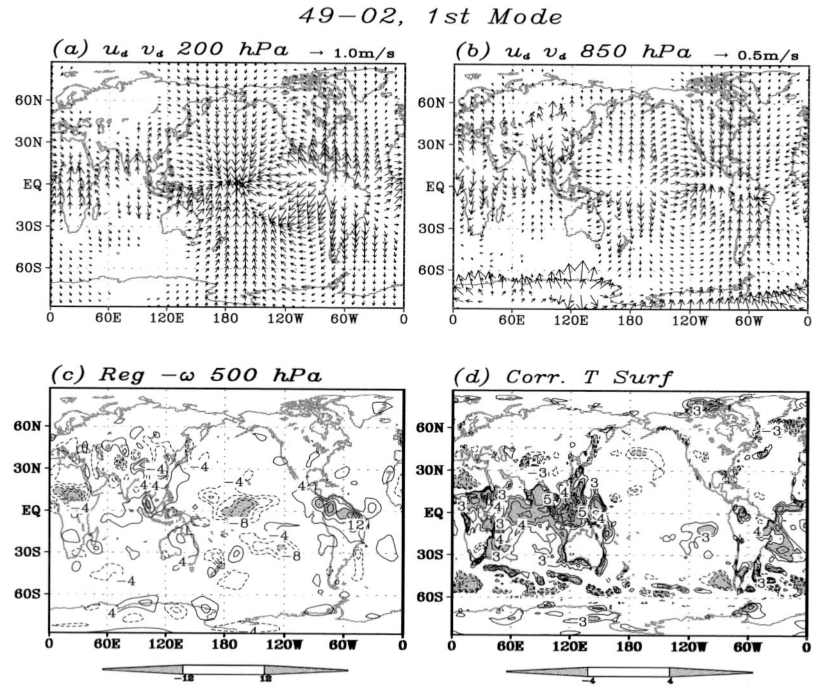

Figure 5-9. Same as Fig. 5-1, but for the EOF analysis using data obtained from 1949 to 2002.

Figure 5-10a shows a time series comparison between PC-1 and the tropical Indian Ocean–Maritime Continent SST. The Indian Ocean–Maritime Continent SST appears as an abrupt warming in the 1970s rather than a continuous increase as in PC-1. The abrupt warming is known as a part of the 1970s climatic regime shift (Nitta and Yamada 1989; Trenberth 1990). Minobe (1997) reported the occurrences of two other climatic regime shifts in the 1920s and 1940s, and that these two regime shifts are also accompanied by consistent SST changes over the tropical Indian Ocean–Maritime Continent.

The present result is consistent with the analysis of Goswami and Thomas (2000), who examined EOFs of total wind fields (not divergent winds) combined with other fields. Their temporal coefficient also exhibited a prominent increase from the 1960s to the 1980s, and the spatial structure of the total wind anomalies is similar to those of the present study (not shown). It is noteworthy, however, that PC-1 of Goswami and Thomas (2000) returned to about zero at the end of their record (1996) from the maximum in the late 1980s, but that PC-1 in the present study stays high from the 1980s to the end of the record (2002). This discrepancy may arise from an end effect of the low-pass filtering employed by Goswami and Thomas (2000), or partly from the longer record used in the present study.

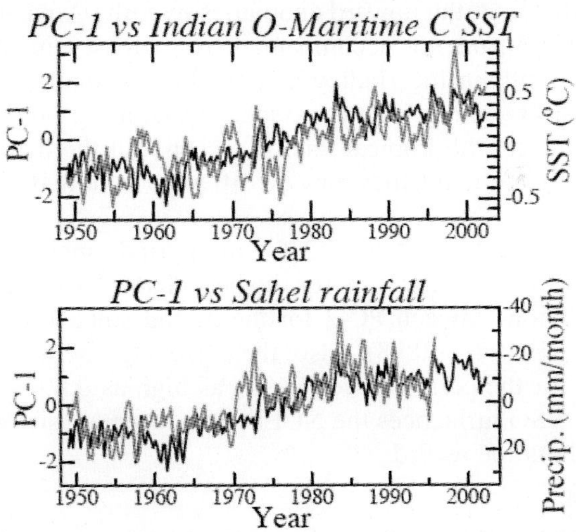

Figure 5-10. Time series comparison of tropical Indian Ocean–Maritime Continent SST anomalies (60°E–150°E, 10°S–10°N) (gray line, top panel), Sahel precipitation anomalies (20°W–30°E, 5°N–20°N) (gray line, bottom panel) and PC-1 for the period since 1949, which was also shown in Fig. 5-8 (black line, both panels). The Sahel precipitation time series is smoothed by a three-point running mean and the other time series are unfiltered (raw seasonally sampled anomalies). The axis of the precipitation is reversed for an easier comparison.

Figure 5-11 shows the correlations of vertical winds with PC-1 and those of the gauge precipitations. As was described in Section 2, the reliability of the earlier record of the NCEP/NCAR reanalysis data can be questionable, and hence we need to be careful about the trend found in PC-1 by the EOF analysis for the period since 1949. The increasing downward wind anomalies expressed by negative correlations around Sahel are consistent with the negative correlations of precipitation, which are related to previously reported decreasing Sahel rainfall (e.g., Zeng et al. 1999). Figure 5-10b shows that the Sahel rainfall decrease occurred mainly from the 1970s to the 1980s, roughly the same period of the prominent change of PC-1. The increasing downward anomalies over the equatorial central Pacific might be related to the negative correlations in precipitation over the tropical South Pacific, though the location of the latter is shifted to the south of the former (Fig. 5-11). Very recently, examining the observed data in the Comprehensive Ocean-Atmosphere Data Set (COADS), Evans and Kaplan (Chapter 8, "The Pacific Sector Hadley and Walker Circulations in Historical Marine Wind Analyses: Potential for Reconstruction from Proxy Data," this volume) showed that quantitatively similar trends are observed both in the NCEP/NCAR reanalysis data and COADS data for over the Pacific Ocean. In contrast to some support from the precipitation for those downward wind anomalies of the trend, the upward anomalies over the Amazon and over the Maritime Continent are not accompanied by increases in rainfall. Consequently, some trends in the Hadley and Walker circulations may have occurred in the late twentieth century associated with the precipitation trends over Africa and over the tropical Pacific Ocean, but the detail of the structure of the circulation anomalies may be different from that shown in Figure 5-10.

It is noteworthy that EOF-2 for the period since 1949 is virtually identical to EOF-1 of the analysis for the period since 1979 with a remarkably high correlation between PC-2 for the period since 1949 and PC-1 for the period since 1979 ($r = 0.98$). Also, the correlation between the NIÑO3.4 SST and PC-2 for the period since 1949 is as high as 0.89. This result indicates that the ENSO influences the NCEP/NCAR reanalysis data in a similar manner throughout the record.

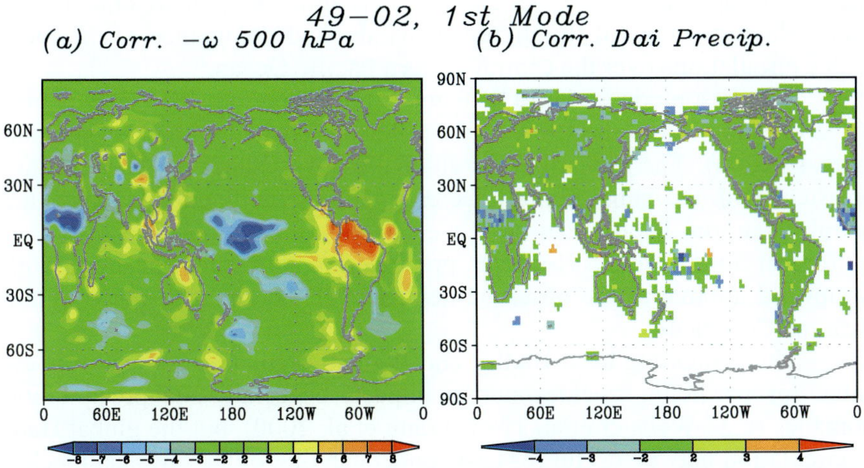

Figure. 5-11. (a) Correlations of the upward vertical velocity in the pressure coordinate at 500 hPa onto PC-1 for the period since 1949, which is shown in Fig. 5-8, and (b) correlations of gauge precipitation onto PC-1.

5. CONCLUSIONS AND DISCUSSION

The year-to-year variability of local Hadley and Walker circulations are studied. A combined EOF analysis of 200 and 850 hPa zonal and meridional divergent winds of the NCEP/NCAR reanalysis data has shown a close relation between the ENSO and local Hadley and Walker circulation anomalies, depicted as the first mode by the EOF analysis for the period since 1979 and as the virtually identical second mode by the analysis for the period since 1949. The vertical wind anomalies in the middle of the troposphere are characterized by the oval pattern in the central-eastern equatorial Pacific and the horseshoe pattern with opposing polarity over the tropical western Pacific. The regression fields of the vertical winds and horizontal divergent winds are analyzed with streamlines of these wind fields; it turned out that the streamlines connect the oval and horseshoe patterns. The major features of the streamlines are summarized in a schematic diagram (Fig. 5-6). The robustness of the streamline analysis is confirmed by a regression analysis with the NIÑO3.4 index and by a composite analysis for ENSO events. The present schematic is consistent with Wang's (2002a) schematic for the ENSO mature phase with respect to the two anomalous local Hadley circulations occurring in the central-eastern Pacific and in the western Pacific with opposing polarity, but the present diagram emphasizes the meridionally slanted nature of the divergent winds, and dominance of the local Hadley circulations over the central-eastern Pacific Ocean.

The first mode of the EOF analysis for the period since 1949 exhibits a strong trend from the 1960s to the 1980s, consistent with Goswami and Thomas (2000). The upward wind anomalies over Sahel of the trend are consistent with the Sahel rainfall decrease, and upward anomalies over the central equatorial Pacific can be related to the precipitation decrease over the tropical South Pacific. However, precipitation anomalies consistent with the upward anomalies are observed neither over the Amazon nor over the Maritime Continent.

Understanding local Hadley and Walker circulation anomalies is essential for predicting socioeconomic impacts especially via precipitation anomalies (e.g., Webster et al. 1998; Wang et al. 2000), but the global Hadley circulation may be useful for basic understanding of the physical mechanism. For example, Seager et al. (2003) recently proposed that ENSO events cause global Hadley circulation anomalies symmetric to the equator due to tropical heating and mid-latitude wave mean-flow interactions, based on a zonally averaged momentum flux analysis. The result in this chapter indicates the dominance of the local Hadley circulation anomalies over the central-to-eastern tropical Pacific, which is related to the global Hadley circulation anomalies studied by Oort and Yienger (1996).

The Hadley circulations are important for connecting the equator and the mid-latitudes. This chapter has focused on the ENSO and the trend, but the influences of the equatorial variability on the mid-latitudes were also proposed for decadal variability over the Pacific Ocean (e.g., Graham 1994; Gu and Philander 1997; Mantua et al. 1997; Luo and Yamagata 2001; Bratcher and Giese 2002; Newman et al. 2003). One of the most interesting examples of decadal variability in the Pacific Ocean may be the Pacific (inter-)Decadal Oscillation (PDO) (Mantua et al. 1997; Mantua and Hare 2002). Zhang et al. (1997) suggested that the ENSO-like decadal mode, which is closely related to the PDO, is accompanied by less pronounced Hadley circulation anomalies than the ENSO mode. Furthermore, for the PDO, two dominant time scales were proposed; one is a 50- to 70-year oscillation (Minobe 1997) and the other is about a 20-year oscillation (Mann and Park 1996; Minobe 1999, 2000; Minobe et al. 2002). Recently, Minobe and Nakanowatari (2002) showed that the 20-year variability exerts substantial influence on precipitation over the Pacific Ocean, and suggested that the Hawaii winter droughts are strongly related to the 20-year oscillation. Further studies are necessary to understand the roles of the Hadley and Walker circulations in the climate system on various time scales.

6. ACKNOWLEDGMENTS

The author thanks H. Diaz and R. Bradley for inviting me to be a participant in this book and in the closely related workshop; C. Wang, M. Evans, and I. Dima for invaluable discussion; and N. Harnik for a preprint. This study is supported by grants from the Japanese Ministry of Education, Culture, Sports, Science and Technology.

7. REFERENCES

Arkin, A. 1982. The relationship between interannual variability in the 200mb tropical wind field and the Southern Oscillation. *Monthly Weather Review* 110: 1393–1404.

Bjerkness, J. 1966. A possible response of the atmospheric Hadley circulation to equatorial anomalies of ocean temperature. *Tellus* 18: 820–829.

Bjerkness, J. 1969. Atmospheric teleconnections from the equatorial Pacific. *Monthly Weather Review* 97: 164–172.

Bratcher, A.J., and B.S. Giese. 2002. *Geophysical Research Letters* 29: 1918, doi: 10.1029/2002GL015191.

Dai, A., I.Y. Fung, and A.D.D. Genio. 1997. Surface observed global land precipitation variations during 1900–1988. *Journal of Climate* 10: 2943–2962.

Deser, C., and M.L. Blackmon. 1995. On the relationship between tropical and North Pacific sea surface temperature variations. *Journal of Climate* 8: 1677–1680.

Goswami, B.N., and M.A. Thomas. 2000. Coupled ocean-atmosphere inter-decadal modes in the tropics. *Journal of the Meteorological Society of Japan* 78: 765–775.

Graham, N.E. 1994. Decadal scale variability in the 1970"s and 1980's: Observations and model results. *Climate Dynamics* 10: 135–162.

Gu, D.F., and S.G.H. Philander. 1997. Interdecadal climate fluctuations that depend on exchanges between the tropics and extratropics. *Science* 275: 805–807.

Kalnay, E., M. Kanamitsu, R. Kistler, W. Collins, D. Deaven, L. Gandin, M. Iredell, S. Saha, G. White, J. Woollen, Y. Zhu, A. Leetmaa, B. Reynolds, M. Chelliah, W. Ebisuzaki, W. Higgins, J. Janowiak, K.C. Mo, C. Ropelewski, J. Wang, R. Jenne, and D. Joseph. 1996. The NCEP/NCAR 40-Year Reanalysis Project. *Bulletin of the American Meteorological Society* 77: 437–472.

Kistler, R., E. Kalnay, W. Collins, S. Saha, G. White, J. Woollen, M. Chelliah, W. Ebisuzaki, M. Kanamitsu, V. Kousky, H. van den Dool, R. Jenne, and M. Fiorino. 2001. The NCEP-NCAR 50-year reanalysis: Monthly means CD-ROM and documentation. *Bulletin of the American Meteorological Society* 82: 247–268.

Luo J.J., and Yamagata T. 2001. Long-term El Nino–Southern Oscillation (ENSO)-like variation with special emphasis on the South Pacific. *Journal of Geophysical Research* 106(C10): 22211–22227.

Mann, M.E., and J. Park. 1996. Joint spatiotemporal modes of surface temperature and sea level pressure variability in the Northern Hemisphere during the last century. *Journal of Climate* 9: 2137–2162.

Mantua, N.J., S.R. Hare, Y. Zhang, J.M. Wallace, and R.C. Francis. 1997. A Pacific interdecadal climate oscillation with impacts on salmon production. *Bulletin of the American Meteorological Society* 76: 1069–1079.

McPhaden M.J., A.J. Busalacchi, R. Cheney, J.-R. Donguy, K.S. Gage, D. Halpern, M. Ji, P. Julian, G. Meyers, G.T. Mitchum, P.P. Niiler, J. Picaut, R.W. Reynolds, N. Smith, and K. Takeuchi. 1998. The tropical ocean–global atmosphere observing system: A decade of progress. *Journal of Geophysical Research* 103(C7): 14169–14240.

Minobe, S. 1997. A 50–70 year climatic oscillation over the North Pacific and North America. *Geophysical Research Letters* 24: 683–686.

Minobe, S. 1999. Resonance in bidecadal and pentadecadal climate oscillations over the North Pacific: Role in climatic regime shifts. *Geophysical Research Letters* 26: 855–858.

Minobe, S. 2000. Spatio-temporal structure of the pentadecadal variability over the North Pacific. *Progress in Oceanography* 47: 99–102.

Minobe, S., and T. Nakanowatari. 2002. Global structure of bidecadal precipitation variability in boreal winter. *Geophysical Research Letters* 29(10): 10.1029/2001GL014447.

Minobe, S., T. Manabe, and A. Shouji. 2002. Maximal wavelet filter and its application to bidecadal oscillation over the Northern Hemisphere through the 20th century. *Journal of Climate* 15: 1064–1075.

Newman, M., G.P. Compo, and M.A. Alexander. 2003. ENSO-forced variability of the Pacific Decadal Oscillation. *Journal of Climate* 16: 3853–3857.

Newman, M., P.D. Sardeshmukh, and J.W. Bergman. 2000. An assessment of the NCEP, NASA and ECMWF reanalyses over the tropical West Pacific warm pool. *Bulletin of the American Meteorological Society* 81: 41–48.

Nitta, T., and S. Yamada. 1989. Recent warming of tropical sea surface temperature and its relationship to the Northern Hemisphere circulation. *Journal of the Meteorological Society of Japan* 67: 375–383.

Oort, A.H., and J.J. Yienger. 1996. Observed interannual variability in the Hadley circulation and its connection to ENSO. *Journal of Climate* 9: 2751–2767.

Philander, S.G. 1990. *El Niño and La Niña and the Southern Oscillation.* San Diego: Academic Press, 289 pp.

Rasmusson, E.M., and T.H. Carpenter. 1982. Variations in tropical sea surface temperature and surface wind fields associated with the Southern Oscillation–El Nino. *Monthly Weather Review* 110: 354–384.

Seager, R., N. Harnik, Y. Kushnir, W. Robinson, and J. Miller. 2003. Mechanisms of hemispherically symmetric climate variability. *Journal of Climate*, 16: 2960–2978.

Trenberth, K.E. 1990. Recent observed interdecadal climate changes in the Northern Hemisphere. *Bulletin of the American Meteorological Society* 71: 988–993.

Trenberth, K.E., and T.J. Hoar. 1996. The 1990–1995 El Nino–Southern Oscillation event: Longest on record. *Geophysical Research Letters* 23: 57–60.

Trenberth, K.E., D.P. Stepaniak, and J.M. Caron. 2000. The global monsoon as seen through the divergent atmospheric circulation. *Journal of Climate* 13: 3969–3993.

Wang, B., R. Wu, and X. Fu. 2000. Pacific–East Asian teleconnection: How does ENSO affect East Asian climate? *Journal of Climate* 13: 1517–1536.

Wang, C. 2002a. Atmospheric circulation cells associated with the El Nino–Southern Oscillation. *Journal of Climate* 15: 399–419.

Wang, C. 2002b. Atlantic climate variability and its associated atmospheric circulation cells. *Journal of Climate* 15: 1516–1536.

Webster, P.J., V.O. Magaña, T.N. Palmer, J. Shukla, R.A. Tomas, M. Yanai, and T. Yasunari. 1998. Monsoons: Processes, predictability, and the prospects for prediction. *Journal of Geophysical Research* 103(C7): 14451–14510.

Zeng, N., J.D. Neelin, K.-M. Lau, and C.J. Tucker.1999. Enhancement of interdecadal climate variability in the Sahel by vegetation interaction. *Science* 286: 1537–1540.

Zhang, Y., J.M. Wallace, and D.S. Battisti. 1997. ENSO-like interdecadal variability: 1900–93. *Journal of Climate* 10: 1004–1020.

Chapter 6

ENSO, ATLANTIC CLIMATE VARIABILITY, AND THE WALKER AND HADLEY CIRCULATIONS

Chunzai Wang
NOAA Atlantic Oceanographic and Meteorological Laboratory, 4301 Rickenbacker Causeway, Miami, Florida 33149, U.S.A.
E-mail: Chunzai.Wang@noaa.gov.

Abstract This chapter describes and discusses the Walker and Hadley circulations associated with the El Niño/Southern Oscillation (ENSO), the Atlantic "Niño", the tropical Atlantic meridional gradient variability, the Western Hemisphere warm pool (WHWP), and the North Atlantic Oscillation (NAO). During the warm phase of ENSO, the Pacific Walker circulation, the western Pacific Hadley circulation, and the Atlantic Hadley circulation are observed to be weakened, whereas the eastern Pacific Hadley circulation is strengthened. During the peak phase of the Atlantic Niño, the Atlantic Walker circulation weakens and extends eastward and the Atlantic Hadley circulation strengthens. The tropical Atlantic meridional gradient variability corresponds to a meridional circulation in which the air rises over the warm sea surface temperature (SST) anomaly region, flows toward the cold SST anomaly region aloft, sinks in the cold SST anomaly region, then crosses the equator toward the warm SST region in the lower troposphere. During periods when the NAO index is high, the atmospheric Ferrel and Hadley circulations are strengthened, consistent with surface westerly and easterly wind anomalies in the North Atlantic and in the middle to tropical Atlantic, respectively. The chapter also discusses a tropospheric bridge by the Walker/Hadley circulation that links the Pacific El Niño with warming of the tropical North Atlantic (TNA) and the WHWP.

H.F. Diaz and R.S. Bradley (eds.), The Hadley Circulation: Present, Past and Future, 173–202.

1. INTRODUCTION

Pacific and Atlantic climate phenomena include the El Niño/Southern Oscillation (ENSO), the Atlantic "Niño", the tropical Atlantic meridional gradient variability, the Western Hemisphere warm pool (WHWP), and the North Atlantic Oscillation (NAO). All have unique variations in time that impact Western Hemisphere climate and that are interlinked (or not) in various ways. The warm phase of ENSO, El Niño, shows positive sea surface temperature (SST) anomalies in the central/eastern equatorial Pacific and an east-west seesaw in the tropical sea level pressure (SLP) between the Western and Eastern Hemispheres (see ENSO overviews by Philander [1990] and Neelin et al. [1998]). ENSO alters the sources of atmospheric heating, which then affect atmospheric circulation and climate on a global scale (e.g., Oort and Yienger 1996; Klein et al. 1999; Wang 2002a).

An interannual phenomenon similar to but weaker than the Pacific El Niño also occurs in the Atlantic, sometimes known as the Atlantic Niño (or the Atlantic equatorial mode). The warm events reach their maximum strength in the second half of the year, with manifestations focused primarily near the equator (e.g., Zebiak 1993; Carton and Huang 1994; Latif and Grotzner 2000; Wang 2002b). During a warm phase, trade winds in the equatorial western Atlantic are weak and SST is high in the equatorial eastern Atlantic. The reverse occurs during a cold phase. The Atlantic Niño is mostly independent of the Pacific ENSO variability; it has a shorter characteristic time scale and is not to be confused with the tropical Atlantic response to the Pacific ENSO.

The Atlantic also displays tropical Atlantic meridional gradient variability that results from the contrasting behaviors of SST in the tropical North Atlantic (TNA) and tropical South Atlantic (TSA), respectively. Some researchers have claimed that this behavior is a characteristic, antisymmetric "dipole" mode (e.g., Weare 1977; Hastenrath 1978; Moura and Shukla 1981; Servain 1991; Nobre and Shukla 1996; Chang et al. 1997; Xie 1999), while studies such as Houghton and Tourre (1992), Enfield and Mayer (1997), Mehta (1998), Enfield et al. (1999), Dommenget and Latif (2000), Wang (2002b), and Melice and Servain (2003) show that the TNA and TSA SST fluctuations vary independently and have fundamentally different time scales. Whatever the case, the tropical Atlantic meridional gradient variability, defined by the SST anomaly difference between the TNA and TSA, is correlated with north-south displacements of the Atlantic Intertropical Convergence Zone (ITCZ) and with strong climate anomalies over the surrounding land regions.

The tropical Western Hemisphere warm pool (WHWP), defined by Wang and Enfield (2001) as the ocean region covered by water warmer than 28.5°C, is composed of the eastern North Pacific west of Central America; the Intra-Americas Sea (IAS), i.e., the Gulf of Mexico and the Caribbean; and the western tropical North Atlantic (see also Wang and Enfield 2003). The WHWP is the second largest tropical warm pool on earth. Unlike the Eastern Hemisphere warm pool in the western Pacific, which straddles the equator, the WHWP is entirely north of the equator. The WHWP has a large seasonal cycle and the interannual fluctuations of its areal extent are comparable to the annual variation, although it does not undergo large anomalous zonal excursions such as occur in the western Pacific. The WHWP is a critical component of the boreal summer climate of the Caribbean and surrounding land areas and also appears to influence the tropical and subtropical southeast Pacific (Wang and Enfield 2003).

Another important Atlantic climate phenomenon is the North Atlantic Oscillation (e.g., Hurrell 1995, 1996). Much of the climate variability over the North Atlantic and surrounding continents has been correlated with changes in the intensity of the NAO. The NAO is associated with the variations of surface westerly wind in the middle and high latitudes of the North Atlantic. It is characterized by a surface pressure seesaw between the Icelandic low and the subtropical anticyclone centered near the Azores. The SST signature obtained by correlating SST with the NAO (Azores minus Iceland) SLP index is a characteristic tripole pattern in the North Atlantic, while lacking significance in the South Atlantic.

This chapter reviews and examines climate phenomena of the Pacific ENSO, the Atlantic Niño, the tropical Atlantic meridional gradient variability, the WHWP, and the NAO and their associated regional atmospheric circulations. The rest of the chapter is organized as follows. Section 2 introduces the data used in the chapter. Section 3 shows the annual variability of the atmospheric circulation patterns over the Pacific and Atlantic. Sections 4, 5, 6, and 7 document atmospheric circulations associated with ENSO, the Atlantic Niño, the tropical Atlantic meridional gradient variability, and the NAO, respectively. Section 8 discusses the Walker and Hadley circulations that serve as a tropospheric bridge, transferring the Pacific ENSO effects to the Atlantic sector. Section 9 provides a summary and discussion.

2. DATA

The major data sources in this study are the NCEP/NCAR reanalysis field and the NCEP SST from January 1950 to December 1999.

The NCEP/NCAR reanalysis field uses a state-of-the-art global data assimilation system on a 2.5° latitude by 2.5° longitude grid (see Kalnay et al. [1996] for details). Variables used in this study are monthly SLP and monthly atmospheric horizontal wind velocity, vertical velocity, and velocity potential at levels of 1,000 mb, 925 mb, 850 mb, 700 mb, 600 mb, 500 mb, 400 mb, 300 mb, 250 mb, 200 mb, 150 mb, and 100 mb. The vertical component of the wind field in the reanalysis field is pressure vertical velocity. We multiply the pressure vertical velocity by –1, so positive values of the vertical velocity indicate an upward movement of air parcels.

Horizontal wind velocity can be divided into a nondivergent (or rotational) part and a divergent (or irrotational) part (e.g., Mancuso 1967; Krishnamurti 1971; Krishnamurti et al. 1973):

$$\vec{v} = \vec{v}_{\psi} + \vec{v}_{\chi} = \vec{k} \times \nabla \psi + \nabla \chi,$$

where ψ is stream function and χ is velocity potential. The first part does not contribute to atmospheric divergent fields associated with atmospheric vertical motion (it is nondivergent). It is well known that the Walker and Hadley circulations are thermally driven, associated with geographical foci of atmospheric convergence and divergence. In the tropics, atmospheric heating associated with convection induces atmospheric convergence and divergence that drive atmospheric vertical motion and circulation. This direct circulation, comprised of zonal (Walker) and meridional (Hadley) circulations, is therefore best characterized by the divergent component of flow and vertical motion. We will focus mainly on the distributions of atmospheric vertical motion and the divergent component of the wind when we discuss atmospheric circulations.

Monthly SST data are also used in this study. SST data are taken from the NCEP SST data set on a 2° latitude by 2° longitude grid from January 1950 through December 1999. These SST fields were produced by using a spatial interpolation method employing empirical orthogonal function analysis (see Smith et al. [1996] for the detailed description). With all of these data, we first calculate monthly climatologies based on the full record period (1950–99) and then anomalies are obtained by subtracting the monthly climatologies for each data set from the data.

3. ANNUAL VARIABILITY

Before we show anomaly variations of atmospheric circulations associated with ENSO and Atlantic climate variability, we first examine the seasonal cycle of tropospheric circulation patterns over the Pacific and Atlantic. Figure 6-1 shows the boreal winter (January) climatologies of tropospheric circulation. Centers of low (high) velocity potential are associated with divergent outflow (convergent inflow) winds. Notice that divergence (convergence) in the upper troposphere corresponds to convergence (divergence) in the lower troposphere (e.g., Wang 2002a, b); however, we only show velocity potential and divergent wind in the upper troposphere of 200 mb in this chapter. Figures 6-1a and b show that centers of upper tropospheric divergence over the western Pacific, the Amazon, and tropical Africa are characterized by middle tropospheric upward motions. These patterns are manifestations of, or are consistent with, three of the known major global heat sources over the "maritime continent" of the western Pacific, the Amazon, and tropical Africa. The equatorial eastern Pacific and Atlantic are associated with upper tropospheric convergence and middle tropospheric downward vertical motion. All of these are consistent with the features of warm water in the western Pacific and western Atlantic, and the equatorial cold tongue in the eastern Pacific and eastern Atlantic.

Figure 6-1b shows manifestation of the Intertropical Convergence Zone in the Pacific and Atlantic—in particular, a narrow band of middle tropospheric upward vertical motion in the eastern Pacific around 10°N. The South Pacific Convergence Zone is also indicated around 10°S near the date line. An east-west band of downward motion in the middle troposphere along 20°N manifests the subtropical high. In the mid-latitudes, upward vertical motion is in the central North Pacific and Atlantic, and downward motion is located in the east of Asia and in the United States. Figure 6-1c shows the east-west equatorial circulation: the Pacific and Atlantic Walker circulations. The air ascends in the west, flows eastward in the upper troposphere, sinks in the east, and returns toward the west in the lower troposphere. For the Atlantic Walker circulation, the sinking and westward flows do not reach the lower troposphere during the boreal winter.

As is suggested by Figures 6-1a and b, the meridional circulations behave differently in different regions. Thus, we separately plot meridional-vertical circulations in the western Pacific, eastern Pacific, and Atlantic as shown in Figures 6-1d–f. In the western Pacific, the Hadley circulation shows the air rising in the tropical region, flowing poleward in the upper troposphere in both hemispheres, and returning to the tropics in the lower troposphere (Fig. 6-1d). In the eastern Pacific, the tropical circulation shows moist air rising in the ITCZ, then diverging northward and southward in the upper troposphere, and descending over the regions of the subtropical high and the equatorial cold tongue (Fig. 6-1e). The extratropics of the Northern Hemisphere (NH) shows the classical Ferrel circulation, with upward motion in the high latitudes and downward motion in the mid-latitudes (Fig. 6-1e). In the Atlantic, the meridional circulation shows both the Hadley circulation and the Ferrel circulation with upward motion near the equator (and south of the equator) and in the high latitudes, and downward motion in the northern subtropical Atlantic (Fig. 6-1f). Figure 6-1 shows that the Pacific and Atlantic circulations are similar, except that the upward motion over land heat sources like South America is strongest near the ground, while the ascent over the ocean regions is strongest at higher levels.

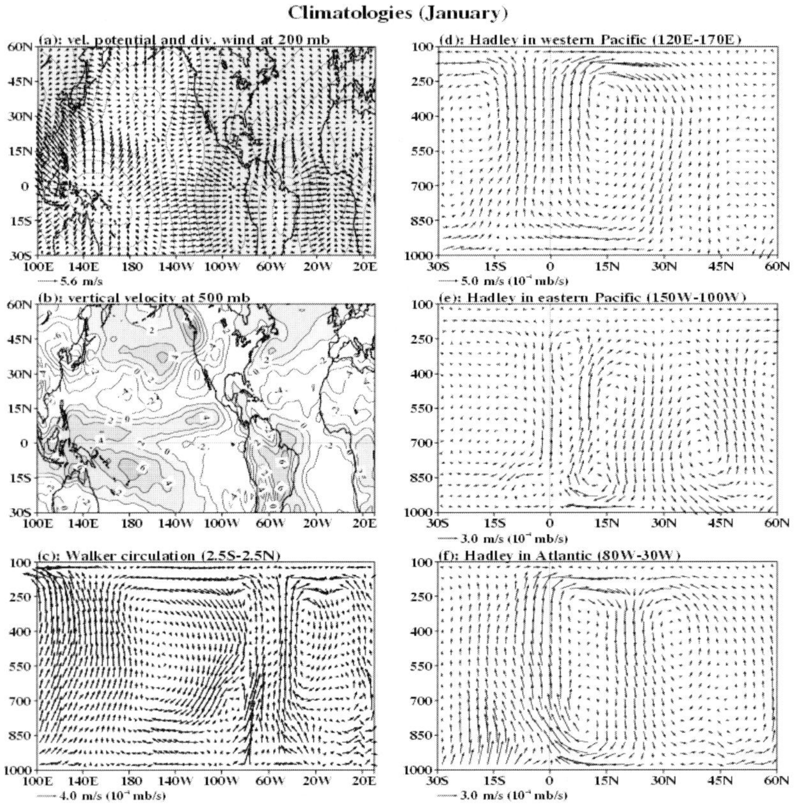

Figure 6-1. The boreal winter (January) climatologies of tropospheric circulation patterns. (a) 200 mb velocity potential (10^6 m^2/s) and divergent wind (m/s); (b) 500 mb vertical velocity (10^{-4} mb/s); (c) Walker circulation by averaging divergent wind and vertical velocity between 2.5°S and 2.5°N; (d) Hadley circulation in the western Pacific by averaging divergent wind and vertical velocity between 120°E and 170°E; (e) Hadley circulation in the eastern Pacific by averaging divergent wind and vertical velocity between 150°W and 100°W; and (f) Hadley circulation in the Atlantic by averaging divergent wind and vertical velocity between 80°W and 30°W. The vertical velocity is taken to be the negative of the pressure vertical velocity in the reanalysis; i.e., positive values indicate an upward movement of air parcels. Positive values are shaded.

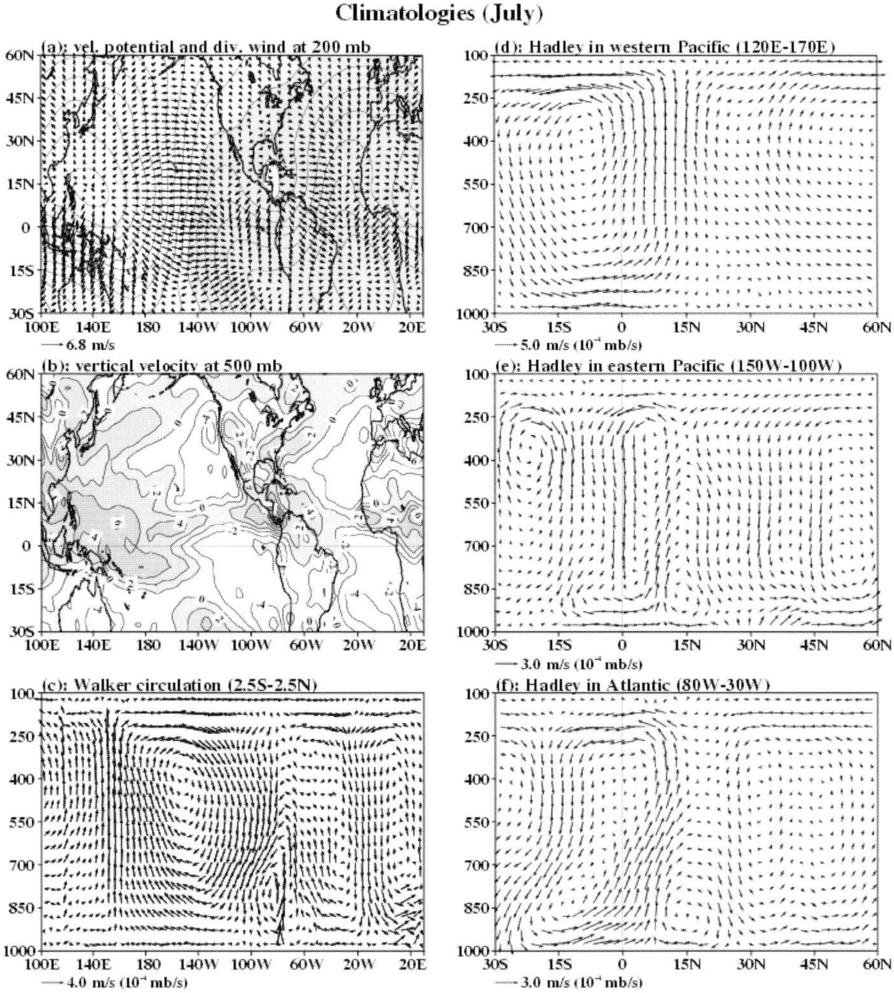

Figure 6-2. As in Fig. 6-1, but for the boreal summer (July).

The boreal summer (July) climatologies of tropospheric circulation are shown in Figure 6-2. The centers of upper tropospheric divergence associated with middle tropospheric ascent shift to the NH. For example, the ascent over the Amazon (during the boreal winter) now shifts to the Intra- Americas Sea[1]. These shifts are associated with seasonal variability of the tropical Western Hemisphere warm pool that extends from the eastern North Pacific to the Gulf of Mexico, the Caribbean, and the western tropical North Atlantic (Wang and Enfield 2001, 2003). As was shown and discussed by Wang and Enfield (2001, 2003), the WHWP is dominated by SSTs in excess of 28.5°C and it is associated with eastern North Pacific and Atlantic hurricane activities and rainfall from northern South America to the southern tier of the United States. Figure 6-2c shows the Pacific and Atlantic zonal Walker circulations, with the strong ascent near the ground over the isthmus of Central America.

The Hadley circulation in the western Pacific shows a northward shift during the boreal summer, with a strong upward vertical motion north of the equator (Fig. 6-2d). The Atlantic Hadley circulation shows a downward motion over South America and an upward motion in the region of the WHWP enclosed by the 28.5°C SST (Wang and Enfield 2003), suggesting that the WHWP is a heating source of the summer Atlantic Hadley circulation (also Figs. 6-2a, b). All of the meridional circulations in Figures 6-2d–f show cross-equatorial flows from the Southern Hemisphere (SH) to the NH in the lower troposphere. It is also noteworthy that associated with the ascent over the region of the WHWP is the decent in the southeast Pacific (Fig. 6-2b), suggesting that the WHWP may link to the southeast Pacific. Figures 6-1 and 6-2 also show that the Hadley flows are always predominantly into the winter hemisphere.

4. ENSO

During 1950–99, there were seven most significant El Niño events (1957–58, 1965–66, 1972–73, 1982–83, 1986–87, 1991–92, and 1997–98) for which the SST anomalies in the NIÑO3 region (5°S–5°N, 150°W–90°W) exceeded 1°C (e.g., Wang 2002a). The maximum NIÑO3 SST anomalies for each warm event occurred during the calendar months from November to January except for the 1986–87 event, which had double peaks with the major one in the boreal summer. This result indicates a robust tendency for the mature phase of El Niño to occur toward the end of the calendar year (e.g., Rasmusson and Carpenter 1982). In order to understand the nature of atmospheric circulation patterns during El Niño, we calculate

---

[1] The Intra-Americas Sea is broadly defined to include the Caribbean Sea, the Gulf of Mexico, the Florida-Bahamas area of the Atlantic Ocean, Bermuda and the northeast coast of South America (see Maul 1993).

El Niño composites as shown in Figure 6-3 (from November of Niño year [0] to January of Niño year [+1]).

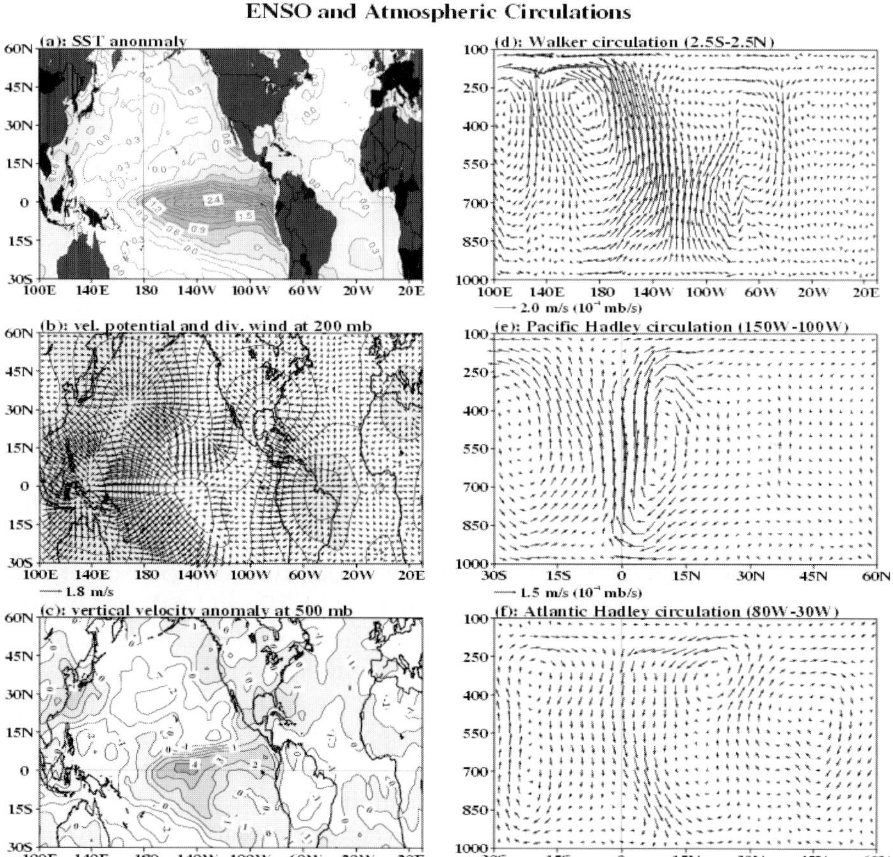

Figure 6-3. SST and atmospheric circulation anomaly composites during the mature phase of El Niño (November of Niño [0] to January of Niño [+1]). (a) SST anomalies; (b) 200 mb velocity potential anomalies (10^6 m^2/s) and divergent wind anomalies (m/s); (c) 500 mb vertical velocity anomalies (10^{-4} mb/s); (d) Walker circulation anomalies by averaging divergent wind and vertical velocity anomalies between 2.5°S and 2.5°N; (e) Hadley circulation anomalies in the eastern Pacific by averaging divergent wind and vertical velocity anomalies between 150°W and 100°W; (f) Hadley circulation anomalies in the Atlantic by averaging divergent wind and vertical velocity anomalies between 80°W and 30°W. The vertical velocity is taken to be the negative of the pressure vertical velocity in the reanalysis; i.e., positive values indicate an upward movement of air parcels. Positive values are shaded.

In the mature phase of El Niño, the equatorial eastern Pacific shows maximum anomalous warming, whereas the off-equatorial western Pacific and the central North Pacific show cold SST anomalies (Fig. 6-3a). Warm SST anomalies are found along the east coast of Asia and the west coast of North America. The TNA and WHWP also start to warm up and will reach their peaks during the spring and summer (e.g., Enfield and Mayer 1997; Wang and Enfield 2001, 2003). Upper tropospheric velocity potential anomalies show three centers near the equator: in the far equatorial western Pacific, the equatorial eastern Pacific, and the equatorial Atlantic; and three centers in the mid-latitudes: in the northwest Pacific, near Mexico, and overEurope (Fig. 6-3b). Associated with these divergent and convergent centers are middle tropospheric anomalous vertical motions (Fig. 6-3c). The equatorial eastern Pacific shows anomalous ascending motion, whereas the tropical western Pacific and the tropical Atlantic display anomalous descending motions. Anomalous descending motion is in the central North Pacific, and anomalous ascending motions are near the east coast of Asia and in the west coast of North America and over the United States.

The anomalous Walker circulation shows the air rising in the equatorial eastern Pacific, flowing westward and eastward aloft, sinking in the equatorial western Pacific and the equatorial Atlantic, and returning back to the eastern Pacific in the lower troposphere (Fig. 6-3d). Comparison of Figures 6-3c, d with Figures 6-1b, c shows that the Pacific and Atlantic Walker circulations are weakened during El Niño. The anomalous Hadley circulation in the eastern Pacific shows the air rising in the tropical region, flowing northward in the upper troposphere, descending in the mid-latitudes, and returning to the tropics in the lower troposphere (Fig. 6-3e). Notice that the anomalous Hadley circulation in the western Pacific has an opposite rotation to that of the anomalous Hadley circulation in the eastern Pacific (not shown; see Wang 2002a). The anomalous Hadley circulation in the Atlantic displays descending motion in the western equatorial Atlantic and eastern South America, and ascending motion in the subtropical North Atlantic and over subtropical South America (Figs. 6-3c and f).

Indices of the Pacific Walker circulation, the Hadley circulation in the eastern Pacific, the Hadley circulation in the western Pacific, and the Hadley circulation in the Atlantic (defined in the figure caption) are compared with the NIÑO3 SST anomalies, as shown in Figure 6-4. As is explained in the figure caption, these indices are defined based on the El Niño composite patterns in Figure 6-3. Variations of these circulations are evident in every El Niño/La Niña event.

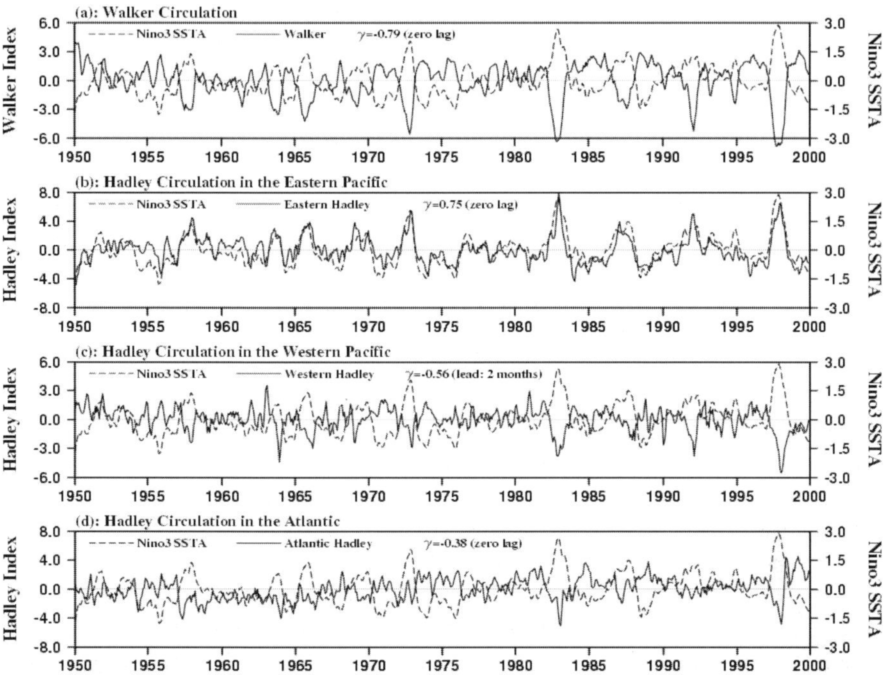

Figure 6-4. Comparison of the NIÑO3 SST anomalies with (a) the Walker circulation index; (b) the eastern Pacific Hadley circulation index; (c) the western Pacific Hadley circulation index; and (d) the Atlantic Hadley circulation index. The Walker index is defined by 500 mb vertical velocity anomaly difference between the equatorial eastern Pacific (5°S–5°N, 160°W–120°W) and the equatorial western Pacific (5°S–5°N, 120°E–160°E). The eastern Pacific Hadley index is defined by 500 mb vertical velocity anomaly difference between the central North Pacific (25°N–35°N, 170°E–150°W) and the equatorial eastern Pacific (5°S–5°N, 160°W–120°W). The western Pacific Hadley index is defined by 500 mb vertical velocity anomaly difference between the western North Pacific (25°N–35°N, 110°E–150°E) and the equatorial western Pacific (5°S–5°N, 120°E–160°E). The Atlantic Hadley index is defined by 500 mb vertical velocity anomaly difference between the tropical North Atlantic (20°N–30°N, 90°W–70°W) and the equatorial Atlantic (5°S–5°N, 50°W–30°W). All of the time series are 3-month running means. The value γ represents correlation coefficient.

The maximum correlations of the Pacific Walker circulation and the eastern Pacific Hadley circulation with the NIÑO3 SST anomalies are –0.79 and 0.75, respectively at zero lag. For the western Pacific Hadley circulation, the maximum correlation is –0.56, with the NIÑO3 SST anomalies leading the Hadley index by 2 months. The maximum correlation of the Atlantic Hadley circulation with the NIÑO3 SST anomalies is

–0.38 (above 95% significant level) at zero lag. Thus, during the warm phase of ENSO, the Pacific Walker circulation, the western Pacific Hadley circulation, and the Atlantic Hadley circulation are weakened, whereas the eastern Pacific Hadley circulation is strengthened. Weakening and strengthening of these circulations are manifestations of atmospheric responses to heating sources during ENSO (Figs. 6-3a, c).

From Figures 6-3 and 6-4, three points or comments can be mentioned. First, the higher (lower) correlation of the eastern Pacific (Atlantic) Hadley circulation with the NIÑO3 SST anomalies is due to the fact that NIÑO3 is a direct reflection of the heating anomalies in the Pacific, whereas it is only a teleconnected (indirect) proxy for the Atlantic. Second, the lower correlation of the Atlantic Hadley circulation with the NIÑO3 SST anomalies suggests that local heating anomalies also may be responsible for the Atlantic Hadley circulation. That is, the heating anomalies in the tropical Atlantic may also contribute to the Atlantic Hadley circulation (see next section for a discussion of the correlation between the Atlantic Hadley circulation and the ATL3 SST index—this index is defined as the mean sea surface temperature anomalies in the region 3°S–3°N, 20°W–0°). Third, the Amazon heating, which is a continental heat source in the boreal winter, not a maritime heating source, may also relate to the Atlantic Hadley circulation. The Atlantic Hadley circulation responds to the Amazon heating anomalies, which in turn are affected by ENSO and Atlantic climate variability.

5. THE ATLANTIC NIÑO

As is discussed by Zebiak (1993), Carton and Huang (1994), and Latif and Grotzner (2000), an interannual phenomenon similar to but weaker and more frequent than the Pacific El Niño also occurs in the Atlantic. During the Atlantic Niño, the largest near-equatorial SST anomalies are in the equatorial eastern Atlantic. Wang (2002b) showed that, during the 50-year period 1950–99 there were 11 significant warm events in which the ATL3 SST anomalies exceeded 0.7°C and lasted more than 1 month. The maximum ATL3 SST anomalies for these 11 warm events were centered in July 1963, July 1968, January 1973, November 1981, August 1984, August 1987, July 1988, June 1995, July 1996, January 1998, and July 1999. Among these 11 warm events, the peak phase of the ATL3 SST anomalies occurred in the boreal summer for 8 events and in the boreal winter for 3 events.

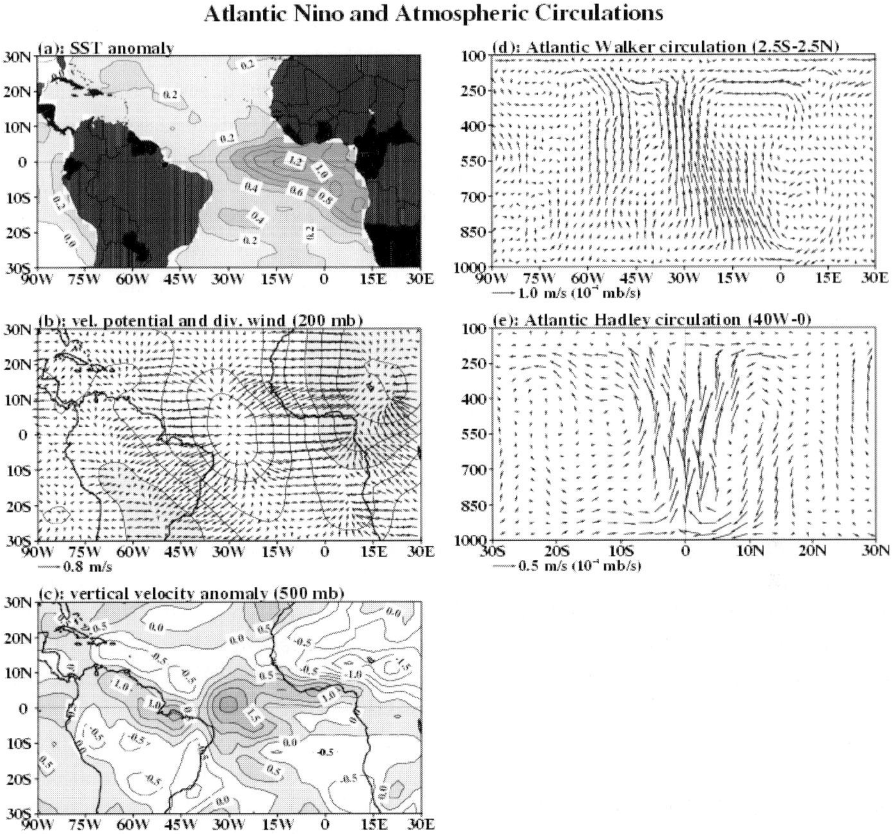

Figure 6-5. SST and atmospheric circulation anomaly composites during the peak phase of the Atlantic Niño. (a) SST anomalies; (b) 200 mb velocity potential anomalies (10^6 m^2/s) and divergent wind anomalies (m/s); (c) 500 mb vertical velocity anomalies (10^{-4} mb/s); (d) Atlantic Walker circulation anomalies by averaging divergent wind and vertical velocity anomalies between 2.5°S and 2.5°N; and (e) Atlantic Hadley circulation anomalies in the Atlantic by averaging divergent wind and vertical velocity anomalies between 40°W and 0°. The vertical velocity is taken to be the negative of the pressure vertical velocity in the reanalysis; i.e., positive values indicate an upward movement of air parcels. Positive values are shaded.

Atmospheric circulation patterns associated with the Atlantic Niño can be examined by compositing the 11 warm events mentioned above. Figure 6-5 shows the distribution of SST anomalies and atmospheric circulation anomalies during the peak phase of the Atlantic Niño. The entire tropical and subtropical Atlantic show positive SST anomalies, with maximum SST anomalies confined to the equatorial eastern

Atlantic (Fig. 6-5a). The upper tropospheric anomalous divergent outflow is in the tropical western Atlantic. The equatorial eastern Atlantic, specifically, the Gulf of Guinea, and the eastern tropical Africa land region show the upper tropospheric anomalous convergent inflow (Fig. 6-5b). Associated with these anomalous divergent/convergent flow fields are middle tropospheric anomalous ascending motion near the region of the equatorial Atlantic, and anomalous descending motion in the subtropics along 15°N (Fig. 6-5c). The equatorial zonal circulation shows anomalous ascending motion in the equatorial western Atlantic (Fig. 6-5d) near 30°W, whereas the associated ascending motion in the lower troposphere is skewed eastward to about 10°W. Although Figure 6-5d does not seem to show a clear anomalous zonal cell, we know that the Atlantic Walker circulation is weakened and extended eastward during the peak phase of the Atlantic Niño if we consider the mean state of the Atlantic Walker circulation in Figure 6-2c. The anomalous Hadley circulation shows ascent near the equator and descent in the subtropical region near 15°N. Thus, corresponding to the Atlantic Niño is a weakening of the Atlantic Walker circulation and a strengthening of the Atlantic Hadley circulation (Figs. 6-2a, b and Figs. 6-5b, c).

The indices of the Atlantic Walker circulation and the Atlantic Hadley circulation (defined in the figure caption) are shown in Figure 6-6 for comparison with the time series of the ATL3 SST anomalies that measures variability of the Atlantic Niño. Both the Atlantic Walker and Hadley circulations show high correlations with the ATL3 SST anomalies (–0.66 and 0.67, respectively). The ATL3 SST anomalies lead the Atlantic Walker index by 1 month, suggesting a response of the Walker circulation to warming in the equatorial eastern Atlantic. The Atlantic Niño is associated with a weakening of the Atlantic Walker circulation and a strengthening of the Atlantic Hadley circulation. This feature is similar to the relationship between the Pacific El Niño and the Pacific Walker and Hadley circulations, which is a key of the positive ocean-atmosphere feedback noted by Bjerknes (1969) for ENSO. During the Atlantic Niño, ascending motion associated with the IAS-Amazon heat source extends eastward. This eastward extension weakens the Atlantic Walker circulation and thus decreases surface equatorial easterly wind in the western Atlantic, which further increases SST in the equatorial eastern Atlantic. Thus, the positive ocean-atmosphere interaction associated with the Pacific Walker circulation, being responsible for the Pacific El Niño, is also operating in the Atlantic. However, in nature, both local air-sea coupling and the remote forcing may play a role in the Atlantic SST anomalies (e.g., Servain et al. 1982; Latif and Grotzner 2000).

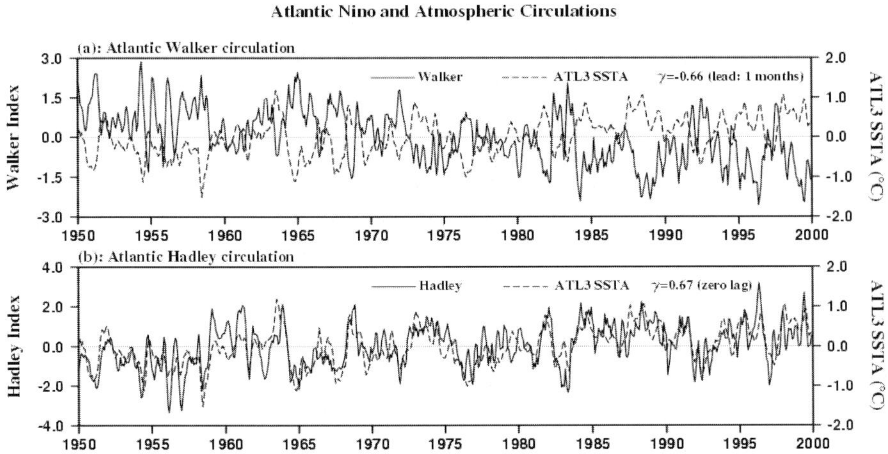

Figure 6-6. Comparisons of the ATL3 SST anomalies with (a) the Atlantic Walker circulation index and (b) the Atlantic Hadley circulation index. The Atlantic Walker index is defined by 500 mb vertical velocity anomalies in the region of 2.5°S–2.5°N, 40°W–20°W. The Atlantic Hadley index is defined by 500 mb vertical velocity anomaly difference between the regions of 12.5°N–17.5°N, 40°W–0° and 2.5°S–2.5°N, 40°W–0°. All of the time series are 3-month running means. The γ represents correlation coefficient.

6. THE TROPICAL ATLANTIC MERIDIONAL GRADIENT VARIABILITY

Many studies have investigated the tropical Atlantic meridional gradient variability (e.g., Moura and Shukla 1981; Folland et al. 1986; Servain 1991; Nobre and Shukla 1996; Rajagopalan et al. 1998; Xie and Tanimoto 1998; Enfield et al. 1999; Wang 2002b). Wang (2002b) defines a tropical North Atlantic index (5°N–25°N, 55°W–15°W) and a tropical South Atlantic index (0°–20°S, 30°W–10°E), and calculates the SST anomaly difference between the TNA and TSA regions to measure the tropical Atlantic meridional gradient variability. The main feature of the meridional (or interhemispheric) SST gradient variability is a slow variation on a decadal time scale. The decadal phases in the TNA-TSA time series are mainly due to the phased interaction between the separate and unrelated decadal variations of TNA and TSA, each with a different time scale. The meridional SST gradient is mainly positive for the periods of pre-1970, 1976–83, and after 1990 but prevailingly negative during 1971–75 and 1984–89.

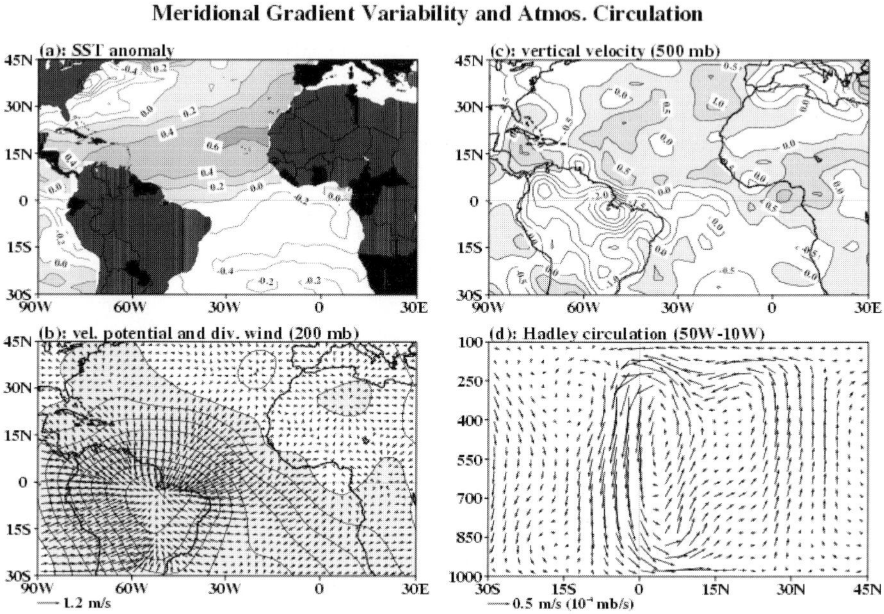

Figure 6-7. SST and atmospheric circulation anomalies for the tropical Atlantic meridional gradient variability. (a) SST anomalies; (b) 200 mb velocity potential and divergent wind anomalies; (c) 500 mb vertical velocity anomalies; and (d) Hadley circulation anomalies by averaging divergent wind and vertical velocity anomalies between 50°W and 10°W. The structures are calculated by the anomaly difference between the positive phase period of 1966–70 and the negative phase period of 1971–75. Positive values are shaded.

The structures of SST anomalies and atmospheric circulation anomalies for the meridional gradient variability can be seen by calculating the anomaly difference between a positive phase and a negative phase. Figure 6-7 shows the anomaly difference between the positive period of 1966–70 and the negative period of 1971–75. The tropical Atlantic SST anomalies show opposite signs: the TNA is warm and the TSA is cold, with the TNA being stronger than the TSA (Fig. 6-7a). The upper troposphere shows anomalous convergent inflow in the equatorial region and anomalous divergent outflow in the TNA and middle latitudes (Fig. 6-7b). Correspondingly, the middle troposphere is associated with anomalous descending motion in the TSA and over the Amazon, and anomalous ascending motion in the TNA and in the middle latitudes (Fig. 6-7c). The meridional anomalous circulation shows that the air rises over the TNA warm waters, diverges southward aloft, converges in the upper troposphere to feed the strong subsidence in the equatorial TSA, then crosses the equator toward the TNA in the lower troposphere (Fig. 6-7d).

The time series of the Atlantic Hadley circulation index (defined in the figure caption) associated with the tropical Atlantic meridional gradient variability are shown in Figure 6-8. The maximum correlation of the Hadley circulation index with the meridional SST gradient variability is 0.59 (above the 99% significance level) when the meridional SST gradient variability lags the Hadley circulation by 1 month. Thus, the tropical Atlantic meridional gradient variability is associated with the variations of the Atlantic Hadley circulation.

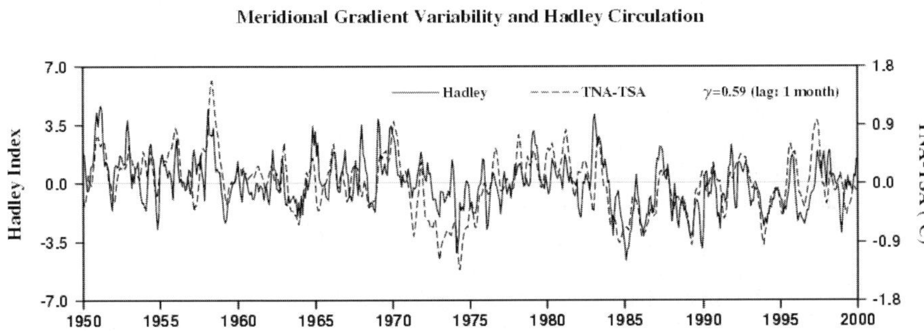

Figure 6-8. Comparisons of the tropical Atlantic meridional SST gradient with the Atlantic Hadley circulation index. The Hadley index is defined by 500 mb vertical velocity anomaly difference between the regions of 2.5°S–7.5°S, 40°W–20°W, and 25°N–30°N, 40°W–20°W. The meridional gradient mode is defined by the SST anomaly difference between the tropical North Atlantic region (5°N–25°N, 55°W–15°W) and the tropical South Atlantic region (0°–20°S, 30°W–10°E). All of the time series are 3-month running means. The γ represents correlation coefficient.

As was discussed in the Introduction, some studies have claimed that the tropical Atlantic meridional gradient variability is an antisymmetric "dipole" mode, while others show that the TNA and TSA vary independently and have different time scales. Therefore, we separately calculate the positive phase of 1966–70 and the negative phase of 1971–75. Although atmospheric circulation shows ascent over warm SST anomalies and descent over cold SST anomalies for either phase, there are differences between the positive and negative phases of the meridional gradient variability. For example, the upper tropospheric anomalous divergent outflow center for the negative phase of 1971–75 is further eastward in comparison with the anomalous convergent inflow center over the Amazon for the positive phase of 1966–70. For both the positive and negative phases, southern Africa shows middle tropospheric anomalous ascending motion. If the tropical Atlantic meridional gradient variability were an antisymmetric "dipole" mode, the tropospheric circulation patterns of the positive phase would be opposite to those of the negative phase. Our calculations thus do not seem to support an antisymmetric "dipole" mode of the tropical Atlantic meridional gradient variability.

7. THE NORTH ATLANTIC OSCILLATION

Another important climate phenomenon in the Atlantic is the NAO, whose positive phase is characterized by strong westerly air flow between the Icelandic low and the Azores high, particularly in winter. Hurrell (1995, 1996) defined an NAO index as the difference of winter (December–March) SLP anomalies between Lisbon, Portugal (38.43°N, 9.08°W) and Stykkishólmur, Iceland (65.06°N, 22.48°W). A striking feature of the NAO index has been the reversal from the negative index (weak meridional SLP gradient) to predominately positive index (high gradient) values starting near 1970. The winters of 1972–73, 1982–83, 1988–89, 1989–90, 1992–93, and 1994–95 are marked by high positive values of the NAO index.

The structures of SST and atmospheric variables for the high NAO index can be calculated by compositing anomalies of the 1972–73, 1982–83, 1988–89, 1989–90, 1992–93, and 1994–95 winters. Figure 6-9 shows the composites of SST and SLP anomalies, and anomalous atmospheric circulation patterns. The Atlantic Ocean shows an alternating pattern of zonally oriented positive-negative SST anomalies (Fig. 6-9a). The coolings appear in the North Atlantic and in the TNA. The warmings occur in the middle Atlantic with centers located at the east coast of the United States and the west coast of Europe, and in the TSA. However, the TSA SST anomalies do not significantly correlate with the NAO index (Visbeck et al. 1998; Wang 2002b). The reason for this difference (about the TSA) may be that the composite considers only the positive phase, whereas correlation deals with both the positive and negative phases.

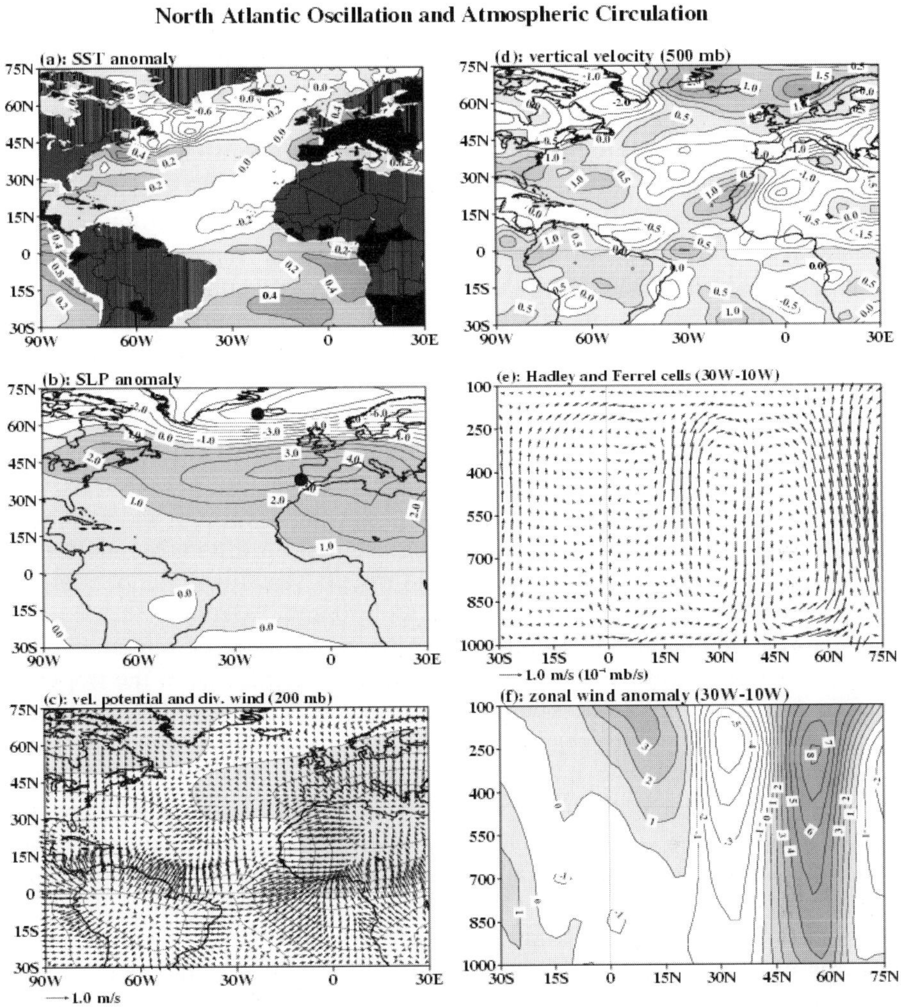

Figure 6-9. The structures of SST and atmospheric circulation anomalies for the high NAO index. (a) SST anomalies; (b) SLP anomalies; (c) 200 mb velocity potential and divergent wind anomalies; (d) 500 mb vertical velocity anomalies; (e) Atlantic Hadley and Ferrel circulation anomalies by averaging divergent wind and vertical velocity anomalies between 30°W and 10°W; and (f) zonal wind anomalies between 30°W and 10°W. The two dots in (b) are Lisbon, Portugal (38.43°N, 9.08°W) and Stykkishólmur, Iceland (65.06°N, 22.48°W), which are used to calculate the NAO index. Positive values are shaded.

Thus, in spite of seemingly being a quadrupole pattern, the spatial Atlantic SST pattern of the NAO is a tripole (e.g., Rodwell et al.

1999; Seager et al. 2000). The SLP anomalies simply show a meridional seesaw pattern, with low SLP anomalies in the high latitudes and high SLP anomalies south of about 55°N, including the Azores, which form the southern pole of the NAO index (Fig. 6-9b).

During periods when the NAO index is high, upper tropospheric anomalous divergent outflow centers are over the Amazon and the North Atlantic, whereas upper tropospheric anomalous convergent inflow centers are located in northern Africa and Europe and in the region of Greenland (Fig. 6-9c). Corresponding to these upper tropospheric anomalous divergent and convergent centers are the middle tropospheric anomalous ascending motion in the North Atlantic, in the southeast coast of the United States to the subtropical Atlantic, and over the Amazon, and anomalous descending motion in the region of Greenland and in Europe and Africa. The distribution of the vertical motion is consistent with the precipitation pattern of Hurrell et al. (2003) for high-low NAO index years (their Fig. 16). Drier conditions occur over much of Greenland, and much of central and southern Europe, the Mediterranean and parts of the Middle East, whereas more precipitation than normal falls from Iceland through Scandinavia and from the southeast coast of the United States to the subtropical Atlantic.

The anomalous meridional circulation shows a counterclockwise circulation in the North Atlantic and a clockwise circulation in the tropical-middle Atlantic (Fig. 6-9e). These two circulations correspond to the Ferrel cell and the Hadley cell, respectively. These anomalous circulations have the same rotations as the mean Ferrel and Hadley circulations, indicating that during the high NAO index periods, both the Ferrel and the Hadley circulations are strengthened. Zonal wind anomalies display westerly wind anomalies in the North Atlantic and easterly wind anomalies in the subtropical Atlantic (Fig. 6-9f), consistent with the strength of the Icelandic low and the Azores high during the high NAO index periods. These wind anomaly patterns extend through the whole troposphere, with the maximum zonal wind anomalies occurring in the upper troposphere. The distribution of the wind patterns seems to suggest that wind speed associated with latent heat flux is responsible for the SST anomaly patterns shown in Figure 6-9a (Visbeck et al. 2003; we will discuss it in Section 9).

The change of the wind and atmospheric circulations associated with the NAO can be further seen from the time series. Figure 6-10 shows a comparison of the NAO index with zonal wind anomalies in the North Atlantic, zonal wind anomalies in the middle to tropical Atlantic, the Ferrel circulation index, and the Hadley circulation index (defined in the figure caption). All of these indices are correlated with the NAO index. Every NAO event is associated with westerly wind anomalies between the Icelandic low and the Azores high, and easterly wind anomalies south of the Azores high. Both the Ferrel circulation and the Hadley circulation are also closely related to the NAO index, suggesting that the meridional circulations may be important for the NAO.

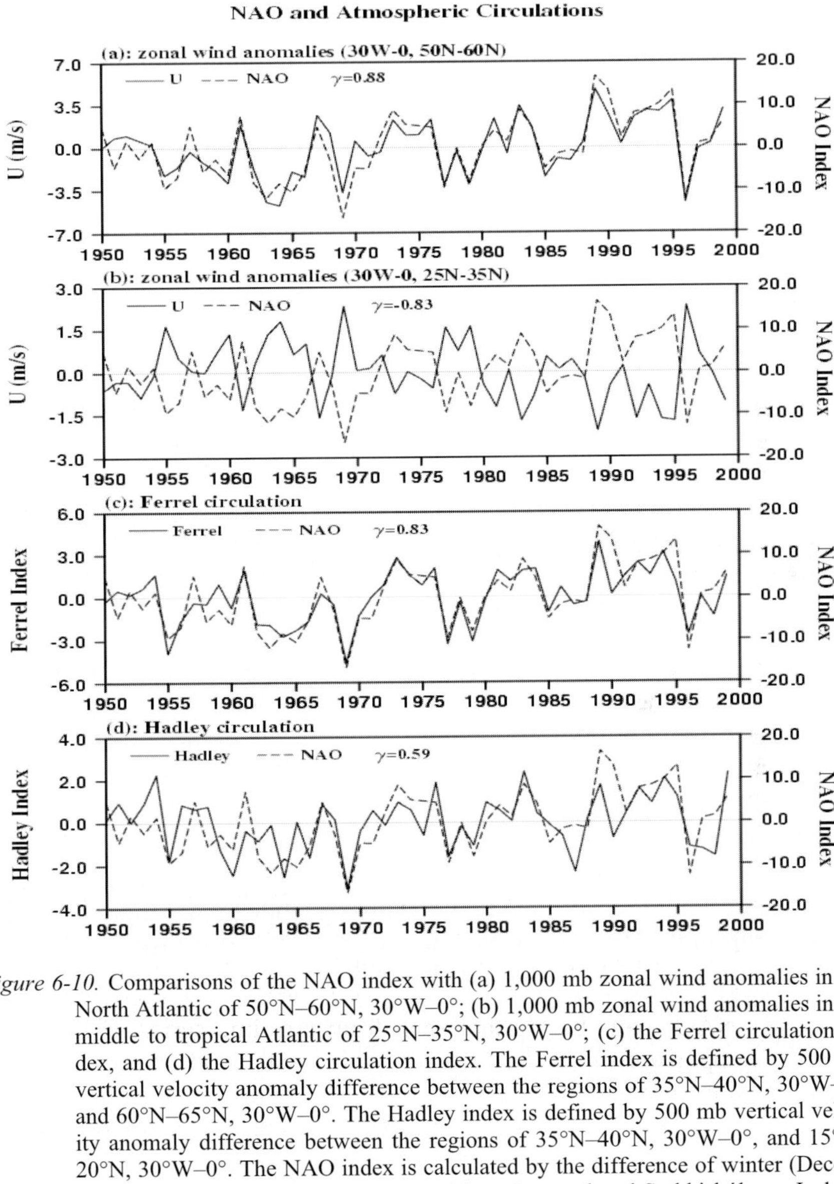

Figure 6-10. Comparisons of the NAO index with (a) 1,000 mb zonal wind anomalies in the North Atlantic of 50°N–60°N, 30°W–0°; (b) 1,000 mb zonal wind anomalies in the middle to tropical Atlantic of 25°N–35°N, 30°W–0°; (c) the Ferrel circulation index, and (d) the Hadley circulation index. The Ferrel index is defined by 500 mb vertical velocity anomaly difference between the regions of 35°N–40°N, 30°W–0°, and 60°N–65°N, 30°W–0°. The Hadley index is defined by 500 mb vertical velocity anomaly difference between the regions of 35°N–40°N, 30°W–0°, and 15°N–20°N, 30°W–0°. The NAO index is calculated by the difference of winter (December–March) SLP anomalies between Lisbon, Portugal and Stykkishólmur, Iceland. The γ represents correlation coefficient at zero lag.

8. A TROPOSPHERIC BRIDGE FOR TRANSFERRING THE EFFECTS OF THE PACIFIC ENSO TO THE ATLANTIC SECTOR

Previous studies have shown that the TNA SST anomalies are related to the Pacific El Niño (e.g., Curtis and Hastenrath 1995; Enfield and Mayer 1997; Klein et al. 1999; Hastenrath 2000). The lagged cross-correlation between the TNA and NIÑO3 SST anomalies shows that the maximum positive correlation of 0.47 occurs when the TNA SST anomalies lag the NIÑO3 SST anomalies by 5 months (Wang 2002c). Recently, Wang (2002b, c) and Wang and Enfield (2003) showed that the Pacific ENSO variability is related to variations of the TNA and the WHWP. Wang and Enfield (2003) showed that during the 50-year period since 1950 there were five large warm pools, which on average reached their maximum anomaly in July. All warm pools occurred in the summer following recognized El Niño events and they also coincided with strong warming of the TNA. These warm pools were about twice as large as the climatological average for July (Fig. 6-11). Four of these large warm pools occurred in the boreal summer of Niño [+1] years (1958, 1983, 1987, 1998), consistent with Hastenrath et al. (1987) and Enfield and Mayer (1997). Note, however, that large warm pools did not follow four other recognized El Niño events (1966, 1973, 1977, 1992). The occurrence of Pacific El Niño is no guarantee of a large ensuing warm pool. Further studies are needed for understanding why extremely anomalous warm pools develop the year following certain Pacific El Niño events, but not others.

How does the Pacific El Niño affect the TNA and the WHWP in the Atlantic sector? There are two possible ways for the Pacific El Niño to affect the Atlantic sector: (1) through the Pacific–North American (PNA) pattern; and (2) through the Walker and Hadley circulations. Wallace and Gutzler (1981) and Horel and Wallace (1981) found that equatorial Pacific warming is accompanied by a teleconnection PNA pattern that shows alternating positive and negative geopotential height anomalies emanated from the Pacific, directed poleward, and curved eastward and then equatorward. It is possible that the Pacific El Niño affects the northern Atlantic subtropical high through the PNA pattern. Changes of the Atlantic subtropical high induce variations of the northeast trade winds on its southern flank and then affect the TNA SST anomalies.

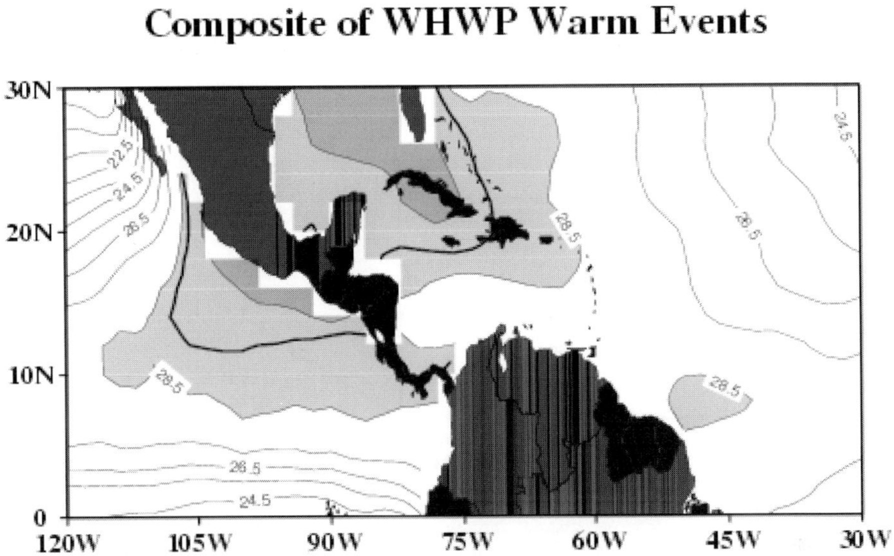

Figure 6-11. SST composites in July for the WHWP interannual warm events. The composites are calculated by averaging over the warm years of 1958, 1969, 1983, 1987, and 1998. The shading represents water warmer than 28.5°C. The dark contour is July climatological SST warmer than 28.5°C.

The Walker and Hadley circulations may also link the Pacific El Niño and warming in the tropical Atlantic. Pacific El Niño conditions were established during the winter seasons prior to warming of TNA and the WHWP. In Figure 6-3 we see composite-averaged maps of the velocity potential, divergent wind, vertical velocity, and circulation departures for the mature phase of El Niño. Most prominent are large convergent and divergent areas over northern South America and the Gulf of Mexico, respectively, at 200 mb (Fig. 6-3b). The upper convergence is fed by an anomalous northerly flow from the north, which in turn diverges from the Caribbean and subtropical North Atlantic. Figure 6-3d shows an anomalous zonal Walker circulation. The Atlantic Hadley circulation is weakened during the mature phase of the Pacific El Niño (Fig. 6-3f). Associated with these circulations are anomalous descending over northern South America and anomalous ascending motion in the region of the North Atlantic subtropical high-pressure system (Fig. 6-3c).

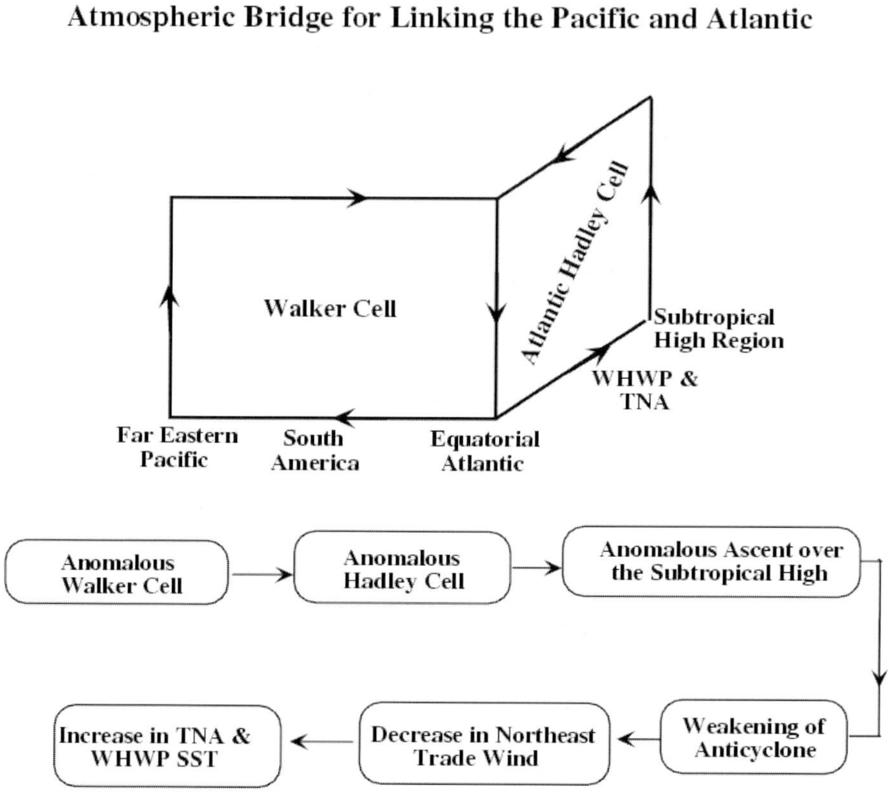

Figure 6-12. Schematic diagram showing linkage of the Pacific El Niño with the tropical North Atlantic and the Western Hemisphere warm pool by the Walker and Hadley circulations.

The connection associated with the Walker and Hadley circulations, seen from the data, is schematically summarized in Figure 6-12. The anomalous descending over northern South America is consistent with reduced rainfall observed over parts of Colombia, Venezuela, and northern Brazil (Ropelewski and Halpert 1987). The anomalous subtropical ascending motion corresponds to a late winter weakening of the North Atlantic anticyclone and the associated northeast (NE) trade winds over its southern limb in the TNA region. With the weaker NE trades come reduced evaporation and entrainment (from below the oceanic mixed layer) during late winter and early spring, leading to warmer SST anomalies over the TNA region by late spring and early summer (Enfield and Mayer 1997 and others). The TNA warming along 5°–15°N extends well into the region of the Outer Antilles that by May sees SSTs above 27°C (Wang and Enfield 2003), required for large-scale tropical convection at the start of the Caribbean rainy season. Thus, the Walker and Hadley circulations can serve as a "tropospheric bridge" for transferring the Pacific El Niño SST anomalies to the Atlantic sector and inducing the TNA SST anomalies just at the time of year when the warm pool is developing.

9. SUMMARY AND DISCUSSION

There are three major localized tropical heat sources: (1) over the maritime continent of the western Pacific; (2) migrating between the Amazon and the Intra-Americas Sea; and (3) over tropical Africa. It is widely known that zonal excursion of the western Pacific heat source is associated with the Pacific ENSO phenomenon that affects climate variations on a global scale. The seasonally varying IAS-Amazon heat source affects climate from South to North America and tropical storms and hurricanes on both sides of Central America. The heat source moves seasonally, generally, being most north and west over the WHWP in the boreal summer, and south and east over northern South America in the boreal winter. Associated with the seasonal movements of the heat sources are the seasonal variations of the equatorial zonal Walker circulation, the tropical meridional Hadley circulation, and the extratropical meridional Ferrel circulation.

ENSO shifts the western Pacific heat source and atmospheric convective activity and then affects global atmospheric circulation. During El Niño, the equatorial Pacific Walker circulation is observed to be weakened. The anomalous meridional Hadley circulation in the eastern Pacific shows the air rising in the tropics, flowing poleward in the upper troposphere, sinking in the subtropics, and returning to the tropics in the lower troposphere. The anomalous Hadley circulation in the western Pacific is opposite to that in the eastern Pacific, indicating a weakening of the western Pacific Hadley circulation during El Niño. The NCAR/NCEP reanalysis field also shows that El Niño weakens the Atlantic Hadley circulation, consistent with an earlier result of Klein et al. (1999) that is inferred from correlation maps of satellite observations, and with the direct circulation analyses of Mestas-Nuñez and Enfield (2001) and Wang (2002a). Wang (2002b, c) and Wang and Enfield (2003) suggest that following El Niño winters in which the Atlantic Hadley circulation is strongly weakened, the decreased subsidence over the subtropical North Atlantic results in the late winter weakening of the NE trades off Africa, the associated spring TNA warming (Enfield and Mayer 1997 and others), and the large summer warm pools (Wang and Enfield 2001).

The Atlantic Niño, similar to but weaker and more frequent than the Pacific El Niño and unrelated to it, shows positive SST anomalies in the equatorial eastern Atlantic. During the warm phase of the Atlantic Niño, ascending motion associated with the IAS-Amazon heat source extends eastward. This eastward extension weakens the Atlantic Walker circulation and thus decreases surface equatorial easterly wind in the western Atlantic, which in turn further increases SST in the equatorial eastern Atlantic. Thus, the positive ocean-atmosphere interaction associated with the Pacific Walker circulation, being responsible for the Pacific El Niño (Bjerknes 1969), seems to be also operating in the Atlantic. The Atlantic Hadley circulation is observed to be strengthened during the warm phase of the Atlantic Niño.

The Atlantic is unique in having the tropical meridional gradient variability, which has a strong impact on rainfall over surrounding land areas due to the associated migrations of the Atlantic ITCZ. The meridional gradient anomaly is significant either when the TNA is anomalous or when the TSA is anomalous or when both conditions exist nearly simultaneously and are opposite in sign. Corresponding to this meridional variability is an atmospheric meridional circulation in which the air rises over the warm SST anomaly region, flows toward the cold SST anomaly region aloft, converges in the upper troposphere to feed the strong subsidence and lower tropospheric divergence in the cold SST anomaly region, then crosses the equator toward the warm SST anomaly region in the lower troposphere. Since the lower tropospheric air always crosses the equator toward the warm SST anomalies, the Coriolis force will deflect air to the east (the west) over the warm (cold) SST anomaly regions. Tropical zonal wind anomalies are thus westerly (easterly) over the warm (cold) anomaly regions in the lower troposphere.

The SST pattern during the high NAO index periods shows a tripole, with cold SST anomalies in the northwest Atlantic and in the TNA and warm SST anomalies in the middle Atlantic. Accompanied by this SST anomaly pattern are a stronger Ferrel circulation and a stronger Hadley circulation. The zonal wind anomalies display westerly wind anomalies in the North Atlantic and easterly wind anomalies in the middle Atlantic, consistent with the strength of the Icelandic low and the Azores high during the high NAO index periods. The zonal wind anomaly distribution is also consistent with the strengthening of the meridional Ferrel and Hadley circulations. The strengthening of the Ferrel and Hadley circulations is associated with anomalous descending motion over the middle Atlantic. Near the sea surface, sinking air is divergent and flows both northward and southward as low branches of the Ferrel and Hadley circulations, respectively. The northward (southward) flowing air is deflected to the east (west) by the Coriolis force, resulting in westerly (easterly) wind anomalies in the North (middle) Atlantic. Since the mean zonal winds are westerly in the North and middle Atlantic, this zonal wind anomaly distribution increases (decreases) wind speed in the North (middle) Atlantic. The increasing (decreasing) of wind speed result in an increase (decrease) of latent heat flux that cools (warms) the North (middle) Atlantic (see Visbeck et al. [2003] for review). The TNA cooling when the NAO index is high (Fig. 6-9a) is also consistent with the zonal wind anomalies in the TNA (Fig. 6-9f). The negative values of the surface zonal wind anomalies in the TNA indicate the strength of the northeast trade winds, which will increase latent heat flux and then cool the TNA.

The Walker and Hadley circulations can serve as a "tropospheric bridge" for transferring the Pacific El Niño SST anomalies to the Atlantic sector and inducing the TNA SST anomalies just at the time of year when the warm pool is developing. As the Pacific El Niño warming culminates near the end of the calendar year, an alteration of the low-latitude direct circulation occurs, featuring (1) an anomalous weakening of the convection over northern South America, (2) Walker circulation anomalies along the equatorial strip to the east and west, and (3) a weakened northward Hadley flow aloft. The Hadley weakening results in less subsidence over the subtropical North Atlantic, an associated breakdown of the anticyclone, and a weakening of the NE trades in the TNA. The wind weakening leads to less evaporative surface cooling and entrainment of colder water from below the shallow mixed layer, resulting in positive SST anomalies. The TNA anomalies thus expand the WHWP area and increase the WHWP SST anomalies, just when warm pool development is taking place. The opposite is presumed to occur during boreal winters with unusually cool SSTs in the equatorial Pacific.

The Hadley circulation in some studies is referred to as a global zonal mean meridional circulation (e.g., Oort and Yienger 1996; Trenberth et al. 2000; and references there). Since ENSO is characterized as a strong east-west contrast phenomenon and this chapter focuses on atmospheric circulations associated with individual climate phenomena, we herein show the regional atmospheric circulations by using the atmospheric vertical velocity and the divergent component of wind. The divergent wind and vertical velocity also show a mid-latitude zonal circulation over the North Pacific (see Wang 2002a). However, we should keep in mind that the atmospheric circulations shown in this chapter may not be closed cells, in particular if we also consider the rotational component of wind.

10. ACKNOWLEDGMENTS.

This work was supported by a grant from the National Oceanic and Atmospheric Administration (NOAA) Office of Global Programs and by the NOAA Environmental Research Laboratories through their base funding of the Atlantic Oceanographic and Meteorological Laboratory. Discussions with and comments by Dave Enfield are appreciated. Two anonymous reviewers provided useful comments that helped improve the manuscript.

11. REFERENCES

Bjerknes, J. 1969. Atmospheric teleconnections from the equatorial Pacific. *Monthly Weather Review* 97: 163–172.

Carton, J.A., and B. Huang. 1994. Warm events in the tropical Atlantic. *Journal of Physical Oceanography* 24: 888–903.

Chang, P., L. Ji, and H. Li. 1997. A decadal climate variation in the tropical Atlantic Ocean from thermodynamic air-sea interactions. *Nature* 385: 516–518.

Curtis, S., and S. Hastenrath. 1995. Forcing of anomalous sea surface temperature evolution in the tropical Atlantic during Pacific warm events. *Journal of Geophysical Research* 100: 15835–15847.

Dommenget, D., and M. Latif. 2000. Interannual to decadal variability in the tropical Atlantic. *Journal of Climate* 13: 777–792.

Enfield, D.B., and D.A. Mayer. 1997. Tropical Atlantic sea surface temperature variability and its relation to El Niño-Southern Oscillation. *Journal of Geophysical Research* 102: 929–945.

Enfield, D.B., A.M. Mestas-Nuñez, D.A. Mayer, and L. Cid-Serrano. 1999. How ubiquitous is the dipole relationship in tropical Atlantic sea surface temperature? *Journal of Geophysical Research* 104: 7841–7848.

Folland, C.K., T.N. Palmer, and D.E. Parker. 1986. Sahel rainfall and worldwide sea temperatures, 1901–85. *Nature* 320: 602–607.

Hastenrath, S. 1978. On modes of tropical circulation and climate anomalies. *Journal of Atmospheric Science* 35: 2222–2231.

Hastenrath, S. 2000. Upper air mechanisms of the Southern Oscillation in the tropical Atlantic sector. *Journal of Geophysical Research* 105: 14997–15009.

Hastenrath, S., L.C. de Castro, and P. Aceituno. 1987. The Southern Oscillation in the Atlantic sector. *Contributions to Atmospheric Physics* 60: 447–463.

Horel, J.D., and J.M. Wallace. 1981. Planetary-scale atmospheric phenomena associated with the Southern Oscillation. *Monthly Weather Review* 109: 813–829.

Houghton, R.W., and Y. Tourre. 1992. Characteristics of low-frequency sea surface temperature fluctuations in the tropical Atlantic. *Journal of Climate* 5: 765–771.

Hurrell, J.W. 1995. Decadal trends in the North Atlantic Oscillation: Regional temperature and precipitation. *Science* 269: 676–679.

Hurrell, J.W. 1996. Influence of variations in extratropical wintertime teleconnections on Northern Hemisphere temperature. *Geophysical Research Letters* 23: 665–668.

Hurrell, J.W., Y. Kushnir, G. Ottersen, and M. Visbeck. 2003. An overview of the North Atlantic Oscillation. In, Hurrell, J.W., Y. Kushnir, G. Ottersen, and M. Visbeck (eds.). *The North Atlantic Oscillation: Climatic Significance and Environmental Impact*. AGU Geophysical Monograph Series, pp. 1–35.

Kalnay, E., and Co-authors. 1996. The NCEP/NCAR 40-year reanalysis project. *Bulletin of the American Meteorological Society* 77: 437–471.

Klein, S.A., B.J. Soden, and N.C. Lau. 1999. Remote sea surface temperature variations during ENSO: Evidence for a tropical atmospheric bridge. *Journal of Climate* 12: 917–932.

Krishnamurti, T.N. 1971. Tropical east-west circulations during the northern summer. *Journal of Atmospheric Science* 28: 1342–1347.

Krishnamurti, T.N., M. Kanamitsu, W.J. Koss, and J.D. Lee. 1973. Tropical east-west circulations during the northern winter. *Journal of Atmospheric Science* 30: 780–787.

Latif, M., and A. Grotzner. 2000. The equatorial Atlantic oscillation and its response to ENSO. *Climate Dynamics* 16: 213–218.

Mancuso, R.L. 1967. A numerical procedure for computing fields of streamfunction and velocity potential. *Journal of Applied Meteorology* 6: 994–1001.

Maul, G.A. (Ed.) 1993. *Climatic Change in the Intra-Americas Sea*. UNEP. New York: Routledge, Chapman and Hall, 389 pp.

Mehta, V.M. 1998. Variability of the tropical ocean surface temperatures at decadal-multidecadal timescales. Part I: The Atlantic Ocean. *Journal of Climate* 11: 2351–2375.

Melice, J.-L., and J. Servain. 2003. The tropical Atlantic meridional SST gradient index and its relationships with the SOI, NAO and Southern Ocean. *Climate Dynamics* 20: 447–464.

Mestas-Nuñez, A.M., and D.B. Enfield. 2001. Eastern equatorial Pacific SST variability: ENSO and non-ENSO components and their climatic associations. *Journal of Climate* 14: 391–402.

Moura, A., and J. Shukla. 1981. On the dynamics of droughts in northeast Brazil: Observations, theory, and numerical experiments with a general circulation model. *Journal of Atmospheric Science* 38: 2653–2675.

Neelin, J.D., D.S. Battisti, A.C. Hirst, F.-F. Jin, Y. Wakata, T. Yamagata, and S.E. Zebiak. 1998. ENSO theory. *Journal of Geophysical Research* 103: 14262–14290.

Nobre, P., and J. Shukla. 1996. Variations of sea surface temperature, wind stress, and rainfall over the tropical Atlantic and South America. *Journal of Climate* 9: 2464–2479.

Oort, A.H., and J.J. Yienger. 1996. Observed interannual variability in the Hadley Circulation and its connection to ENSO. *Journal of Climate* 9: 2751–2767.

Philander, S.G. 1990. *El Niño, La Niña, and the Southern Oscillation*. London: Academic Press, 289 pp.

Rajagopalan, B., Y. Kushnir, and Y.M. Tourre. 1998. Observed decadal midlatitude and tropical Atlantic climate variability. *Geophysical Research Letters* 25: 3967–3970.

Rasmusson, E.M., and T.H. Carpenter. 1982. Variations in tropical sea surface temperature and surface wind fields associated with the Southern Oscillation/El Niño. *Monthly Weather Review* 110: 354–384.

Rodwell, M.J., D.P. Powell, and C.K. Folland. 1999. Oceanic forcing of the wintertime North Atlantic Oscillation and European climate. *Nature* 398: 320–323.

Ropelewski, C.F., and M.S. Halpert. 1987. Global and regional scale precipitation patterns associated with the El Niño/Southern Oscillation. *Monthly Weather Review* 115: 1606–1626.

Seager, R., Y. Kushnir, M. Visbeck, N. Naik, J. Miller, G. Karhmann, and H. Cullen. 2000. Causes of Atlantic Ocean climate variability between 1958 and 1998. *Journal of Climate* 13: 2845–2862.

Servain, J. 1991. Simple climatic indices for the tropical Atlantic Ocean and some applications. *Journal of Geophysical Research* 96: 15137–15146.

Servain, J., J. Picaut, and J. Merle. 1982. Evidence of remote forcing in the equatorial Atlantic Ocean. *Journal of Physical Oceanography* 12: 457–463.

Smith, T.M., R.W. Reynolds, R.E. Livezey, and D.C. Stokes. 1996. Reconstruction of historical sea surface temperature using empirical orthogonal functions. *Journal of Climate* 9: 1403–1420.

Trenberth, K.E., D.P. Stepaniak, and J.M. Caron. 2000. The global monsoon as seen through the divergent atmospheric circulation. *Journal of Climate* 13: 3969–3993.

Visbeck M., and Co-authors. 1998. Atlantic climate variability experiment prospectus. A consensus document of US and European scientists reflecting the findings of workshops in Lamont (September 1997), Dallas (February 1998), and Florence (May 1998), 49 pp.

Visbeck, M., E.P. Chassignet, R.G. Curry, T.L. Delworth, R.R. Dickson, and G. Krahmann. 2003. The ocean's response to North Atlantic Oscillation variability. In, Hurrell, J.W., Y. Kushnir, G. Ottersen, and M. Visbeck (eds.). *The North Atlantic Oscillation: Climatic Significance and Environmental Impact*. AGU Geophysical Monograph Series, pp. 113–145.

Wallace, J.M., and D.S. Gutzler. 1981. Teleconnections in the geopotential height field during the Northern Hemisphere winter. *Monthly Weather Review* 109: 784–812.

Wang, C. 2002a. Atmospheric circulation cells associated with the El Niño-Southern Oscillation. *Journal of Climate* 15: 399–419.

Wang, C. 2002b. Atlantic climate variability and its associated atmospheric circulation cells. *Journal of Climate* 15: 1516–1536.

Wang, C. 2002c. ENSO and atmospheric circulation cells. *CLIVAR Exchanges* 7: 9–11.

Wang, C., and D.B. Enfield. 2001. The tropical Western Hemisphere warm pool. *Geophysical Research Letters* 28: 1635–1638.

Wang, C., and D.B. Enfield. 2003. A further study of the tropical Western Hemisphere warm pool. *Journal of Climate* 16: 1476–1493.

Weare, B.C. 1977. Empirical orthogonal analysis of Atlantic Ocean surface temperatures. *Quarterly Journal of the Royal Meteorological Society* 103: 467–478.

Xie, S.P. 1999. A dynamic ocean-atmosphere model of the tropical Atlantic decadal variability. *Journal of Climate* 12: 64–70.

Xie, S.P., and Y. Tanimoto. 1998. A pan-Atlantic decadal climate oscillation. *Geophysical Research Letters* 25: 2185–2188.

Zebiak, S.E. 1993. Air-sea interaction in the equatorial Atlantic region. *Journal of Climate* 6: 1567–1586.

Chapter 7

THE HADLEY AND WALKER REGIONAL CIRCULATIONS AND ASSOCIATED ENSO IMPACTS ON SOUTH AMERICAN SEASONAL RAINFALL

Tércio Ambrizzi,[1] Everaldo B. de Souza,[1] and Roger S. Pulwarty[2]
[1]Department of Atmospheric Sciences, University of São Paulo, Rua do Matão, 1226, São Paulo - SP, Brazil, 05508-090
[2]NOAA-CIRES Climate Diagnostics Center, 325 Broadway, Boulder, Colorado 80305, U.S.A.

Abstract

While numerous detailed studies have been conducted of the annual cycle of convection over other regions (e.g., the Asian summer monsoon and the West African summer monsoon regions), the annual cycle and its modulation in the tropical South American region has received attention only relatively recently. Most of the annual total rainfall observed over tropical South America occurs during the austral summer and autumn months. The large-scale meteorological systems that modulate rainfall during these periods are linked to the strength and movement of large-scale climatological features—in particular, the Intertropical Convergence Zone (ITCZ) and the South Atlantic Convergence Zone (SACZ). It is well known that the anomalous patterns related to the El Niño/Southern Oscillation (ENSO) influence the ITCZ and SACZ patterns, with strong interannual and seasonal variations over tropical and subtropical South America.

The goal of this chapter is to analyze the influence of ENSO events on the regional Hadley and Walker cells and their respective impacts on South American seasonal rainfall. As is well documented, ENSO events influence regional precipitation patterns over South America, with the strongest influences in the Amazon/Northeast Brazil and southern South America.

Basically, two separate responses can be composited for each phase of the ENSO cycle. El Niño (La Niña) Composite 1 is the canonical ENSO warm (cold) event with well-known impacts on large-scale atmospheric circulation and regional precipitation patterns over

H.F. Diaz and R.S. Bradley (eds.), The Hadley Circulation: Present, Past and Future, 203–235.

South America, indicating that the central-eastern Pacific sea surface temperature anomaly (SSTa) is the dominating feature in this case. On the other hand, the El Niño and La Niña Composite 2 analyses characterize the influence of the intertropical Atlantic SST gradients as being significant in modulating the influence of ENSO by intensifying the SACZ and ITCZ in some cases. For these latter composites, evidence of a completely reversed atmospheric circulation and regional precipitation patterns is found during the summer and autumn seasons. The analysis demonstrates that interaction of ENSO events with the South American monsoon produces changes in the time and space evolution of convection and circulation over northern South America, which can also be reinforced by the Atlantic. Thus, depending on conditions in the Atlantic, the South American rainy season may be strongly affected. These results suggest that some care always must be taken in producing precipitation (and impacts) forecasts based on ENSO indices and composites alone.

1. INTRODUCTION

Most of the annual total rainfall observed over the South American landmass occurs during the austral summer (December–February, DJF) and autumn (March–May, MAM) months. The large synoptic meteorological systems that modulate the summertime rainfall are linked to variations in the South Atlantic Convergence Zone (SACZ; Casarin and Kousky 1986; Figueroa et al. 1995; Nogués-Paegle and Mo 1997), Bolivian High, and upper tropospheric cyclonic vortices (Virji 1981; Kousky and Gan 1981; Kayano et al. 1997). In the subsequent period, MAM, the rainy season is located on the central-eastern Amazon and Northeast Brazil, which is modulated by the migration of the Intertropical Convergence Zone (ITCZ) south of the equator (Hastenrath and Heller 1977; Moura and Shukla 1981; Nobre and Shukla 1996; Souza et al. 1998).

The so-called El Niño/Southern Oscillation (ENSO) is one of the most prominent sources of interannual variations in weather and climate around the world (Trenberth and Caron 2000). ENSO is related to a strong and complex ocean-atmosphere coupling over the tropical Pacific basin (Cane 1992), which leads to oscillations in sea surface temperature (SST) of the equatorial Pacific, with the El Niño (warm phase) manifesting one extreme phase, and the La Niña (cold phase) the opposite extreme. The major atmospheric and oceanic features associated with El Niño episodes are: predominance of positive SST anomalies (SSTa), weakness of the trade winds in the surface, and low pressure with deep convection on the central-eastern Pacific and high pressure with subsidence movement on the western Pacific, Indonesia, and Australia. La Niña events generally feature reversed atmospheric and oceanic patterns (Kousky and Ropelewski 1989). These anomalous patterns occur over the tropical Pacific basin, including an extensive spatial area of the tropics (more than a third of the tropical belt around the globe). Hence, ENSO triggers changes in the general circulation of the atmosphere, resulting in climatic impacts in several continental areas located in the tropics and extratropics. These changes are basically related to the weakness, intensification, and/or displacements of the large-scale atmospheric circulation in the meridional and zonal planes, mainly those linked to the Hadley and Walker circulations (Kidson 1975; Kousky et al. 1984). The Walker circulation is a result of the "seesaw" in surface pressures between the Eastern and Western Hemispheres linking these action centers through an atmospheric circulation in the zonal plane, restricted in the tropical strip, with an ascending branch over the western Pacific and a descending branch over the eastern Pacific (Bjerknes 1969). On the other hand, the differential heating between the tropic-extratropics results in the formation of a meridional circulation, the Hadley circulation, with an ascending branch over equatorial areas and sinking over the subtropical latitudes (around 30° of latitude) in both the Southern (SH) and Northern (NH) Hemispheres (Hastenrath 1985).

South America is one of the continental areas that is directly influenced by the ENSO cycle (Coelho et al. 2002). Several studies have documented ENSO impacts (mainly the El Niño, or "warm," events) on South American rainfall (Aceituno 1988; Kousky et al. 1984; Rao and Hada 1990; Alves and Repelli 1992; Coelho et al. 1998; Grimm et al. 2000; Souza and Ambrizzi 2002; and others). These findings indicate, in general, that the main areas of South America influenced by ENSO are located in the west (Peru and Ecuador), north and northeast (Amazon and Brazilian Northeast), and south-southeast (southern Brazil, Uruguay, and Argentina).

Other studies have shown that the tropical Atlantic Ocean also plays an important role in the interannual variability of the rainy season of the Amazon and Northeast region of Brazil (Hastenrath and Heller 1977; Moura and Shukla 1981; Pulwarty 1994; Nobre and Shukla 1996; Souza et al. 2000; and others). During austral summer and autumn, the SST anomalies in the tropical Atlantic show a dominant large-scale mode having a dipole pattern that is at times obscured by ENSO teleconnections into the Atlantic basin. Results showed that when SST was warmer than normal over the tropical North Atlantic and colder than normal over the tropical South Atlantic, defining the positive phase of the dipole pattern (Servain 1991), there was a deficit of precipitation during the rainy season of the Brazilian Amazon and Northeast. On the other hand, when the inverse pattern occurred, a rainy season that was anomalously wetter was observed in these regions. In fact, depending on the sign of the dipole pattern, the latitudinal position of the ITCZ may be affected (Nobre and Shukla 1996; Wagner 1996; Souza and Nobre 1998). The ITCZ is dependent on SST, trade winds, and sea level pressure (SLP) variations over the tropical Atlantic (Hastenrath and Greischar 1993; Nobre and Shukla 1996; Chiang et al. 2002). Thus, during the austral summer and autumn, the quality of the rainy season in the north and northeast sectors of South America is remarkably modulated by the displacement of the cloudiness and precipitation associated with the ITCZ south of the equator.

Modeling studies by Lau and Nath (1994) and Lau (1997) showed that the changes of tropical Atlantic SST during El Niño years are forced by changes in atmospheric circulation, suggesting that both are driven by changes in Pacific SSTs. More recently, Pezzi and Cavalcanti (2001), from numerical modeling experiments, found that a positive dipole over the tropical Atlantic and El Niño conditions resulted in below normal precipitation over Northeast Brazil. A sign reversal was found during the negative phase of the Atlantic dipole. However, when La Niña conditions were tested together with a negative dipole, positive precipitation anomalies occurred in the whole Northeast region. Using the positive dipole over the tropical Atlantic, the precipitation in the same region was below average.

Observational and modeling studies have questioned the existence of the tropical Atlantic dipole (e.g., Enfield and Mayer 1997; Dommenget and Latif 2000), based on the fact that the observed correlation between SST anomalies north and south of the equator is not strongly negative as would be characteristic of a dipole. However, the atmospheric response to cross-equatorial SST gradients is quite evident from the results below (Hastenrath 2002).

Based on previous observational studies, it is quite clear that ENSO events influence precipitation patterns over South America, where Northeast Brazil seems to show one of the strongest signals. However, recent numerical studies of large-scale atmosphere-ocean climate modes have suggested that the tropical Atlantic may be important in modulating the seasonal precipitation pattern over northeastern South America. The main goal of this chapter is to analyze the changes in the regional Hadley and Walker cells and their respective impacts on South American rainfall during the ENSO episodes observed in the last 50 years. Following the same procedure as in Souza and Ambrizzi (2002), vertical cross-section analyses of atmospheric circulation in altitude, averaged in the zonal and meridional planes, will be investigated. Since our main focus will be the precipitation variability over

South America, we will show here the regional atmospheric circulation instead of a global zonal mean Hadley circulation (e.g., Oort and Yienger 1996; Trenberth et al. 2000; Trenberth and Stepaniak 2003 and references therein). Besides the well-known ENSO impacts on rainfall over the South American continent, it will be demonstrated here that there are occasions when the canonical patterns do not occur and the regional circulation can be quite different. The study is organized as follows. Section 2 introduces the data and analysis procedure used in this work. Section 3 presents the climatological aspects of South American precipitation and the large-scale circulation patterns. The ENSO composites and the analysis of the regional Hadley and Walker circulations, as well as their impact on precipitation, are discussed in Section 4. The influence on the seasonal cycle of the South American monsoon is presented in Section 5. Concluding remarks and the importance of the present study for seasonal climate prediction are provided in Section 6.

2. DATA AND ANALYSIS PROCEDURE

The data sets used for the 1950–99 period consist of a global grid of zonal, meridional, and vertical components of the wind vector in the standard pressure levels obtained from the National Centers for Environmental Prediction/National Center for Atmospheric Research (NCEP/NCAR) reanalysis project (Kalnay et al. 1996). The grid size is 2.5° latitude by 2.5° longitude. Global gridded SST produced by Smith et al. (1996) is used to analyze the spatial configuration of the SST anomalies of the tropical Pacific and Atlantic Oceans. For this data set the grid size is 2° in latitude and longitude. The 50-year gauge precipitation data set compiled by Chen et al. (2002) is also used. These data are on a 2.5° in latitude/longitude grid over the global land areas.

The seasonal climatologies are based on the full record period (1950–99) and then anomalies are obtained by subtracting them from each season. As was mentioned in the Introduction, the large-scale meteorological patterns and the associated rainfall anomalies are analyzed for the DJF and MAM seasons. To study the regional rainfall variability over South America, a seasonal standardized index is calculated at key regions over the continent. These regions are shown in Figure 7-1, which indicates the geographical locations where the spatial-average regional precipitation was calculated. The shaded regions represent: (1) the eastern Amazon, (2) Northeast Brazil, (3) the SACZ, (4) southeastern South America, (5) the Andes Altiplano, and (6) Ecuador/Peru. The index I is obtained through the equation $I = (X_i - \overline{X}_i)/\sigma_i$ where X_i is the observed seasonal variable for each year during 1950–99, $\overline{X}_i$ is the seasonal climatology, and c_i is the seasonal standard deviation where the sub-index *i* refers to each season, DJF or MAM. The calculation of *I* is applied to the precipitation data averaged over the key areas and also to the SST anomalies in the regions Niño-1-2, Niño-3, Niño-3-4 and Niño-4 in order to select the ENSO episodes in the period. The criterion of selection of the El Niño and La Niña events is based on the objective procedure suggested by Trenberth (1997). In this technique, the El Niño (La Niña) episodes are defined when monthly SSTa indices in the Niño-3 and Niño-3.4 are +0.5°C or above (–0.5°C or below) for at least six consecutive months from October through May.

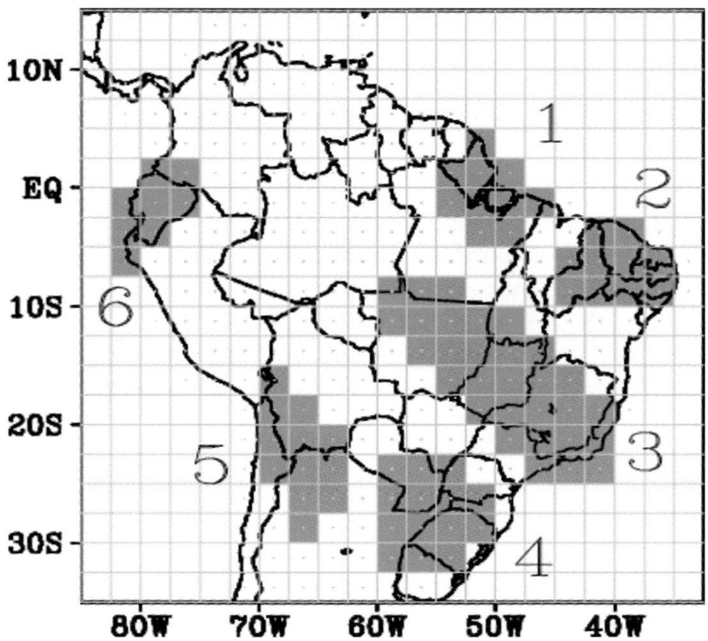

Figure 7-1. Geographical locations of the key areas over South America, where the spatial averaged regional precipitation values were computed. The shaded regions represent: 1, eastern Amazon; 2, Northeast Brazil; 3, SACZ; 4, southeastern South America; 5, Altiplano; and 6, Ecuador/Peru.

The summer and autumn austral warm events chosen are shown in Table 7-1. In order to analyze the vertical structure related to the zonally

(Walker) and meridionally (Hadley) averaged tropospheric circulation, the divergent wind components from the horizontal wind vector at all pressure levels were calculated. The divergent part of the wind is essential to study the atmospheric divergence-convergence that drives the vertical motion and circulation in the tropics (Hastenrath 2001). The large-scale atmospheric circulation patterns related to the Hadley and Walker cells were investigated with emphasis on the Pacific, South America, and Atlantic domains through analyses of the vertical cross sections of the upper divergent atmospheric circulation during the DJF and MAM seasons. For the regional Walker circulation analyses, a vertical cross section was plotted in the zonal direction (longitude x height), averaged along the equatorial area between 0° and 10°S, while for the analysis of the regional Hadley circulation, a meridional-vertical cross section (latitude x height) was calculated along the east part of South America between 70°W and 35°W as is shown in Figure 7-2.

Table 1. El Niño and La Niña episodes used for the observational composites during DJF and MAM periods. December corresponds to the Year 0 and from January to May it is the Year +1 of each ENSO event.

El Niño Episodes		**La Niña Episodes**	
Composite 1	**Composite 2**	**Composite 1**	**Composite 2**
DJF:	**DJF**:	**DJF**:	**DJF**:
1957–58	1963–64	1975–76	1950–51
1965–66	1977–78	1984–85	1970–71
1968–69	1979–80	1985–86	1995–96
1982–83			
1986–87			
1992–93			
1994–95			
1997–98			
MAM:	**MAM**:	**MAM**:	**MAM**:
1983, 1987, 1992, 1993, 1998	1977, 1995	1967, 1968, 1971, 1974, 1989, 1996	1955, 1999

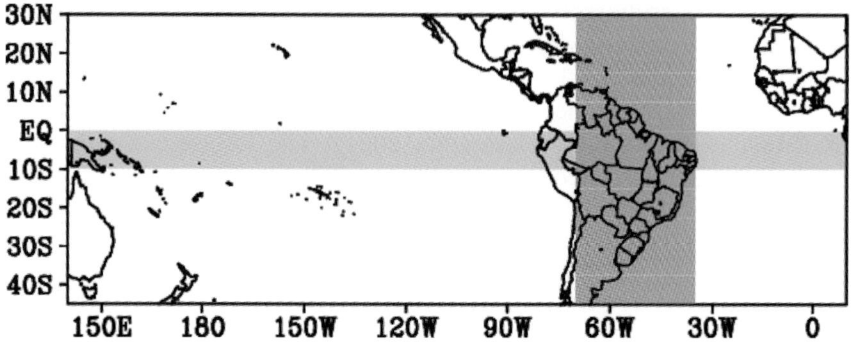

Figure 7-2. The longitudinal (Walker) and latitudinal (Hadley) bands where the vertical cross sections were calculated. The east-west direction was averaged between 0° and 10°S, and the north-south was averaged between 70°W and 35°W.

3. CLIMATOLOGICAL ASPECTS AND SEASONAL VARIABILITY

As was mentioned in the Introduction, most of the annual total rainfall observed over South America occurs during the austral summer and autumn seasons. In Figure 7-3, which illustrates the seasonal percentages of the annual total precipitation for the DJF and MAM periods, it is clearly observed that the months of DJF are the rainiest in most of the continent. During this period a maximum around 65% is observed over the Altiplano, a narrow band in the east part of the Andes, as well as over the south of Bolivia and the northwest of Argentina. A second maximum is observed around the central-west and southeast of Brazil. In fact, this rainfall maximum has a northwest-southeast orientation and it is primarily associated with the SACZ (Figueroa et al. 1995; Liebmann et al. 1999 and references therein). On the other hand, the precipitation maximum located in the central part of Northeast Brazil can be associated with upper-level cyclonic vortex events that frequently appear during the summer (Kousky and Gan 1981). During the autumn season, the precipitation maxima are located over the central-eastern Amazon and central-northern region of Northeast Brazil, where the observed accumulated percentages are 30%–45% and 35%–65%, respectively. The main meteorological system responsible for these maxima is the ITCZ (Hastenrath and Heller 1977; Moura and Shukla 1981; Souza et al. 1998), which reaches its southernmost position in the South Atlantic Ocean around March (Hastenrath and Lamb 1977; Nobre and Shukla 1996;

Chiang et al. 2002). It is interesting to note that along northern Peru and Ecuador rainfall is around 40% in both seasons (Fig. 7-3).

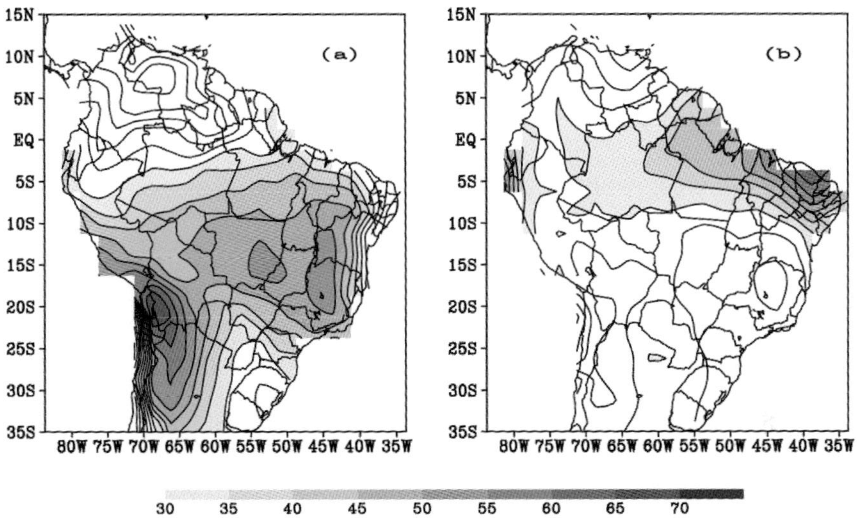

Figure 7-3. Climatological seasonal percentages of precipitation (in relation to the total annual) over South America during (a) DJF and (b) MAM. The interval is 5% and the values above 30% are shaded.

The climatological zonal-vertical (Walker) and meridional-vertical (Hadley) circulations during the austral summer and autumn seasons for the Pacific-Atlantic and South American sectors are shown in Figures 7-4 and 7-5. During the DJF and MAM periods (Figs. 7-4d, 7-5d), the ascending branch of the Walker cell is over the western Pacific Ocean around 160°E–180° and the sinking branch is located in the eastern Pacific, close to the west coast of South America (between 120°W and 90°W). Over the equatorial portion of South America (including Ecuador, northern Peru, central-southern Colombia, Venezuela, and a large part of the Brazilian Amazon), there is strong ascending motion and compensatory subsidence reaching the central-eastern tropical Atlantic Ocean. The regional Hadley circulation (Figs. 7-4e, 7-5e) shows some important differences between the seasons in comparison to the Walker cell. During DJF, ascending motion is observed from 5°N to 30°S with a maximum around 7°S. The compensatory descending branches occur over both hemispheres around 20°N and 40°S. This circulation pattern is the result of the intense convection that occurs over the continent during the summer as can be inferred from the divergent circulations at 200 hPa and 850 hPa (Figs. 7-4a, c). The vertical velocity at 500 hPa (Fig. 7-4b) also indicates an overall ascending motion during this season with a second maximum around the SACZ (20°S, 50°W). In the next season, a very different picture emerges. The maximum ascending area is concentrated between 5°N and 10°S and the descending branch is more intense over the NH between 15°N and 25°N (Figs. 7-5d, e). The presence of the ITCZ near the South American equatorial region can be observed from the vertical velocity, ω, and divergent fields in Figures 7-5a, b, and c. This pattern also agrees with the maximum precipitation seen over this region (Fig. 7-3b).

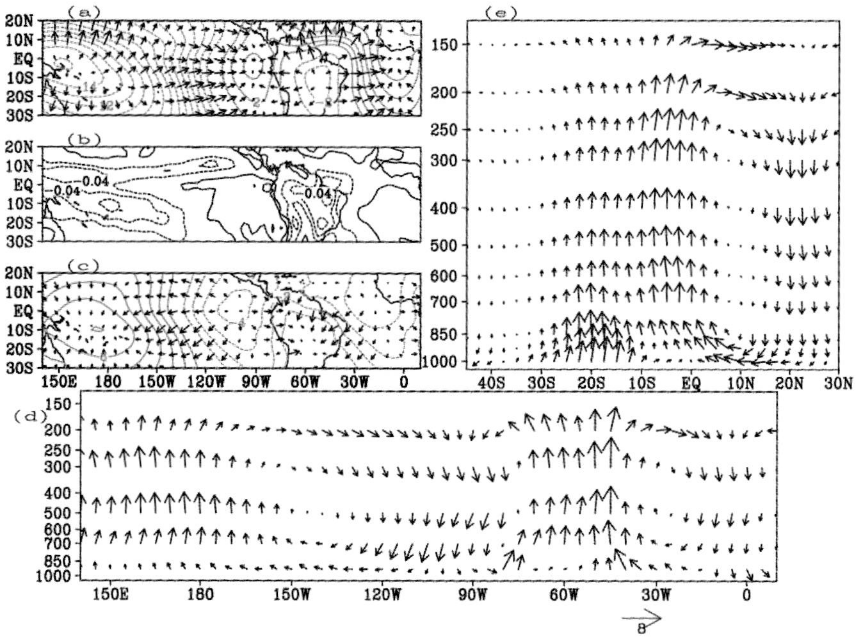

Figure 7-4. (a) Divergent circulation and velocity potential χ at 200 hPa; (b) vertical velocity at 500 hPa (in Pa/s); (c) divergent circulation and χ at 850 hPa; (d) zonal-vertical circulation (Walker) by averaging divergent wind and vertical velocity in the latitudinal band between 10°S and 0°; (e) meridional-vertical circulation (Hadley) by averaging divergent wind and vertical velocity in the longitudinal band between 70°W and 35°W. All fields represent the climatological seasonal mean for DJF.

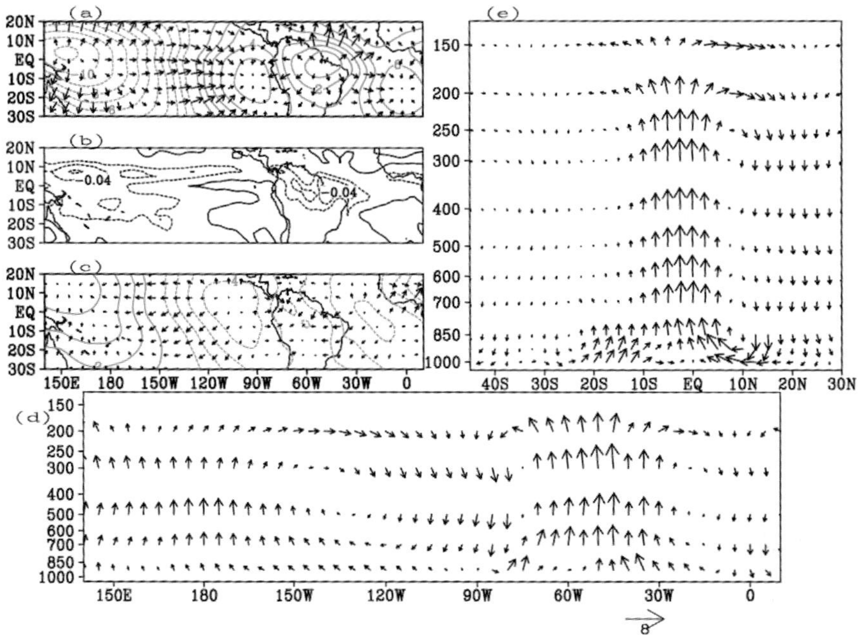

Figure 7-5. As in Fig. 7-4 but for MAM.

4. ENSO COMPOSITES

Using the seasonal standardized precipitation and SST indices (Section 2), 15 El Niño and La Niña episodes were identified between 1950 and 1999. The selected ENSO events are also in the list of Trenberth (1997) and are part of the Climate Prediction Center, NOAA classification list available at its web page (www.cpc.ncep.noaa.gov).

From the time series of the ENSO episodes (figure not shown), one can observe high precipitation variability among the events. For the regions where the El Niño and La Niña have strong signals, the pattern is often more consistent, such as in Northeast Brazil and southeastern South America. In fact, these regions are well known for experiencing rainfall deficits and excesses during warm ENSO events and La Niña episodes, respectively. However, during some years the precipitation index acquires an opposite signal and therefore an inverse pattern. In order to better analyze the implications on the atmospheric Walker and Hadley regional circulations on the precipitation patterns associated with the ENSO phases, these cases have been separated into two composites.

Based on the time series generated by the index *I* of the ENSO episodes for the six regions depicted in Figure 7-1, the first composite was constructed by using the positive precipitation index for southern South America and Ecuador/Peru and the negative index for the eastern Amazon and Northeast Brazil for El Niño (La Niña) years. For the second composite, we used the El Niño and La Niña years that essentially exhibited an inverse impact on the precipitation pattern over the same regions. Table 7-1 shows the selected years and seasons for the four composites. It should be noted that the composites are of the anomalies, which are departures from the climatological patterns presented in Figures 7-3, 7-4, and 7-5. The statistical significance of the composites is assessed through a two-tailed Student t-test (Harrison and Larkin 1998). Significant composites at the 95% confidence level are those $\geq (Z_{95}(n) \times \sigma_c)/(n)^{1/2}$, where n is the number of values used in the composite, σ_c is the corresponding standard deviation, and Z_{95} is the value of the t-distribution for *n* degrees of freedom at the 95% confidence level.

As will be demonstrated in the next section, Composite 1 shows the canonical ENSO impact on the atmospheric circulation and precipitation patterns over South America, indicating that the eastern Pacific SSTa is the dominating feature in this case. On the other hand, Composite 2 clearly indicates that the Atlantic SSTa may play a more significant role during some ENSO episodes.

4.1. El Niño Composite 1: Canonical Impacts

The Pacific SSTa patterns for the DJF and MAM periods show strong positive anomalies in both seasons, with maxima around the Niño3-4 region during its mature phase (DJF) and more significant values near the west coast of South America during its decaying period (MAM). The Atlantic SSTa patterns show a positive dipole with a northward SSTa gradient over the intertropical oceanic portion during MAM (Figs. 7-6a, b). The precipitation pattern during the austral summer shows rainfall deficits up to 30% over the northern part of South America and excess around 20% over the south of Brazil, Paraguay, and northeastern Argentina (Fig. 7-6c). In the autumn (Fig. 7-6d), the rainfall deficit over Northeast Brazil reaches up to 40%, while in southeastern South America the rainfall excess is about 30% above average.

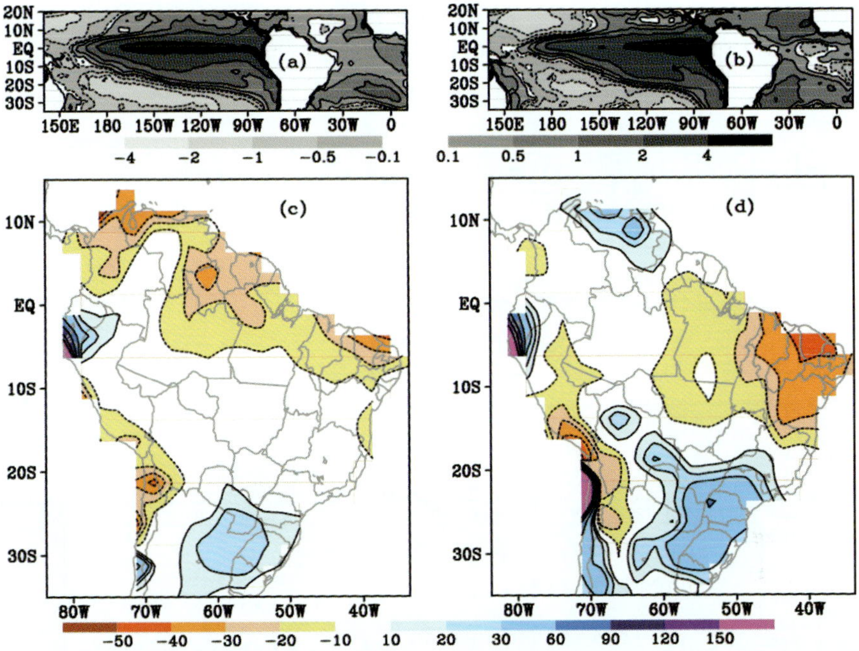

Figure 7-6. Composite 1 of El Niño episodes: The above figures show the seasonal SST anomalies (°C) over the Pacific and Atlantic Oceans during (a) DJF and (b) MAM. The seasonal precipitation percentage (%) over South America is shown in (c) DJF and (d) MAM. The interval and magnitude of the SST (precipitation) fields are shown by the gray (color) scale. The full line (dots) represents positive (negative) anomalies of the SST and precipitation percentage above (below) climatology.

The anomalous seasonal large-scale atmospheric patterns related to the Walker and Hadley cells for the El Niño composite are analyzed in Figures 7-7 and 7-8. The zonal-vertical circulation (Walker) for the austral summer and autumn seasons (Figs. 7-7d and 7-8d) shows an inverse pattern when compared to the climatology (Fig. 7-4d and 7-5d). The central and eastern Pacific is dominated by ascending motion with a maximum around 180°–170°W and descending branches over Northeast Brazil and the adjacent Atlantic Ocean, with maximum values around 50°W during DJF (Fig. 7-7a). The maximum of the upper-level convergence (Fig. 7-7a) with downward motion (positive ω anomalies in Fig. 7-7b) occur in northern South America. In the following season (MAM), the Pacific Walker maximum ascending branch has moved eastwards, around 150°W–90°W, and a significant descending cell appears over the eastern coast of Brazil and the central-eastern Atlantic (Fig. 7-8d). This pattern is clearly seen in the upper-level divergence and vertical velocity fields (Figs. 7-8a, b).

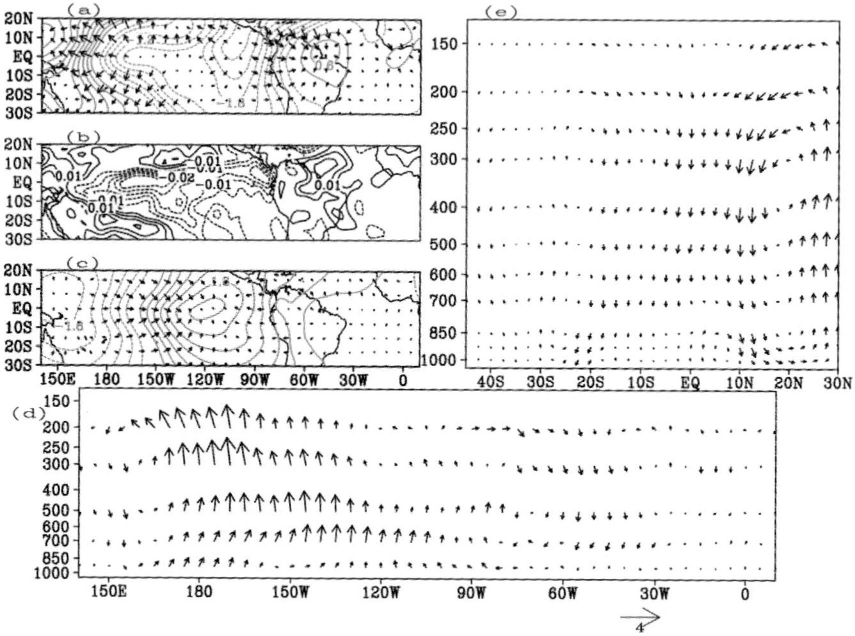

Figure 7-7. El Niño Composite 1 of the anomalous fields for DJF: (a) Divergent circulation and velocity potential χ at 200 hPa; (b) vertical velocity at 500 hPa (in Pa/s); (c) divergent circulation and χ at 850 hPa; (d) zonal-vertical circulation (Walker) by averaging divergent wind and vertical velocity in the latitudinal band between 10°S and 0°; (e) meridional-vertical circulation (Hadley) by averaging divergent wind and vertical velocity in the longitudinal band between 70°W and 35°W.

The austral summer regional Hadley cell (meridional-vertical circulation cross section) also presents a reverse pattern in comparison with the climatology (Fig. 7-4e). Anomalous subsidence is observed from 15°N to 20°S with a maximum around 10°N (Fig. 7-7e). This circulation pattern agrees with the precipitation distribution showed in Figure 7-6c. During the autumn, anomalous ascending motion dominates the subtropics of both the NH and SH, and some areas with subsidence are noticed in the lower troposphere of tropical Brazil around 10°S (Fig. 7-8e). In general, the Walker and Hadley circulation patterns observed during El Niño Composite 1 seem to be strongly coupled with the SSTa in the Pacific and the convection in the upper north of South America, respectively. The rainfall deficit over northern and northeastern South America is linked to the Walker circulation anomalies that occur during ENSO warm events. However, the above average precipitation percentages found over southern South America not only are likely to be associated with the regional Hadley circulation anomalies but are also due to modification of the quasi-stationary waves' position over the South American continent; these waves are generated by modification of the heat source over the eastern equatorial Pacific region during El Niño events (Ambrizzi and Magaña 1999, submitted; Hoerling and Kumar 2000 and references therein).

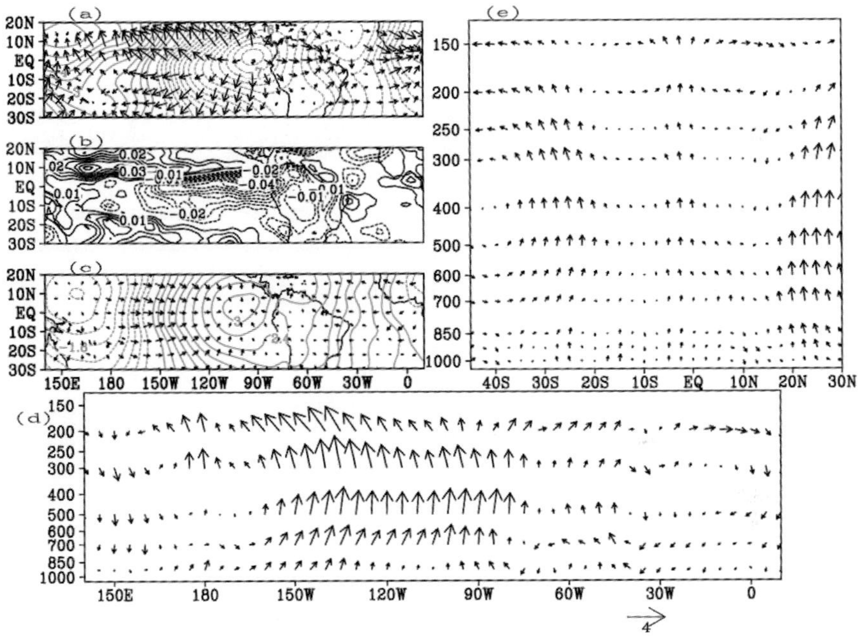

Figure 7-8. As in Fig.7-7 but for El Niño Composite 1 during MAM.

4.2. La Niña Composite 1: Canonical Impacts

The spatial configuration of the seasonal SSTa and precipitation associated with La Niña Composite 1 are shown in Figure 7-9. The typical cold ENSO event characteristics of significant negative SSTa in the central eastern Pacific can be observed during DJF and MAM periods. Values are up to –1.5°C around the Niño 3.4 region during austral summer and they are weaker (–1.0°C) during autumn. Over the tropical Atlantic there is a persistent southward SSTa gradient (negative dipole) from one season to another with absolute values around 0.5°C. The rainfall distribution follows its canonical impacts; i.e., above average precipitation in the north and northeast part of South America and below normal values in the southern part of the continent during both seasons. Positive values can reach up to 50% of the seasonal average percentage, and negative values can reach around –30% of the seasonal average. Previous observational analyses have found similar results concerning the locations of the regions of positive and negative precipitation anomalies (e.g., Ropelewski and Halpert 1987, 1989; Kiladis and Diaz 1989; Diaz et al. 1998; Uvo et al. 1998; Grimm et al. 1998, 2000).

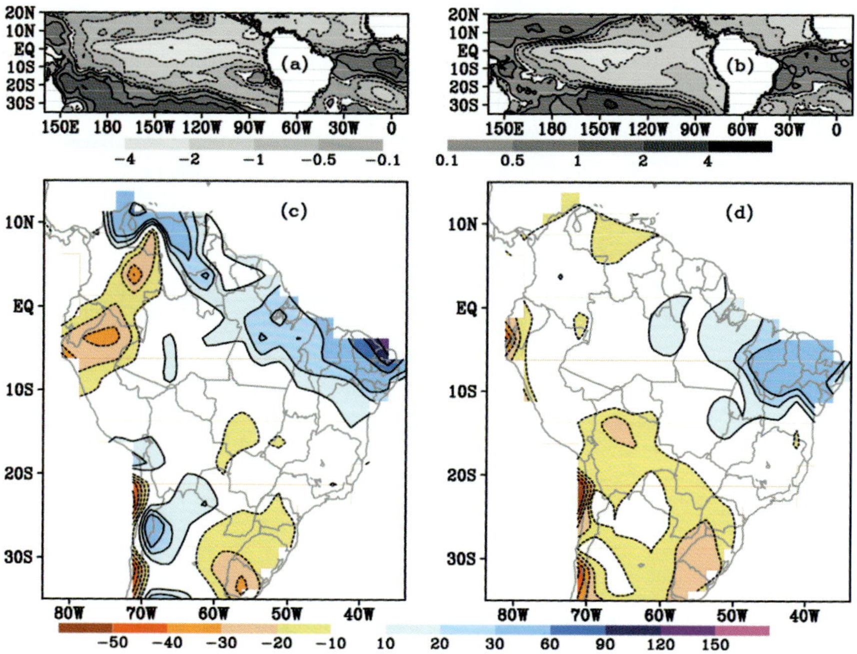

Figure 7-9. As in Fig. 7-6 but for La Niña Composite 1 during DJF (left) and MAM (right).

The vertical cross section of zonal circulation over the equator (Walker) during the austral summer shows circulation patterns similar to the climatology; however, some regional features are different. For instance, between 180° and 160°W very intense subsidence is observed. On the other hand, the downward motion in the eastern Pacific is much weaker and the ascending cell over northern South America is quite strong. It is also interesting to note that due to the positive SSTa over the tropical Atlantic, upward motion as compared to the climatology is observed along the basin (Figs. 7-4d and 10d). In the next season, the circulation shows similar patterns but with weaker amplitudes. The anomalous austral summer Hadley cell circulation depicted by Figure 7-10e has its maximum ascending branch between the equator and 5°S and the descending branch around 33°S. During the MAM period (Fig. 7-11e), the circulation is much weaker and the main upward branch of the ITCZ has almost disappeared. In general, the regional Hadley circulation patterns observed during both seasons are weaker as compared to the climatology, though the negative precipitation deviation obtained over southern South America agrees with the more intense subsidence found around this region.

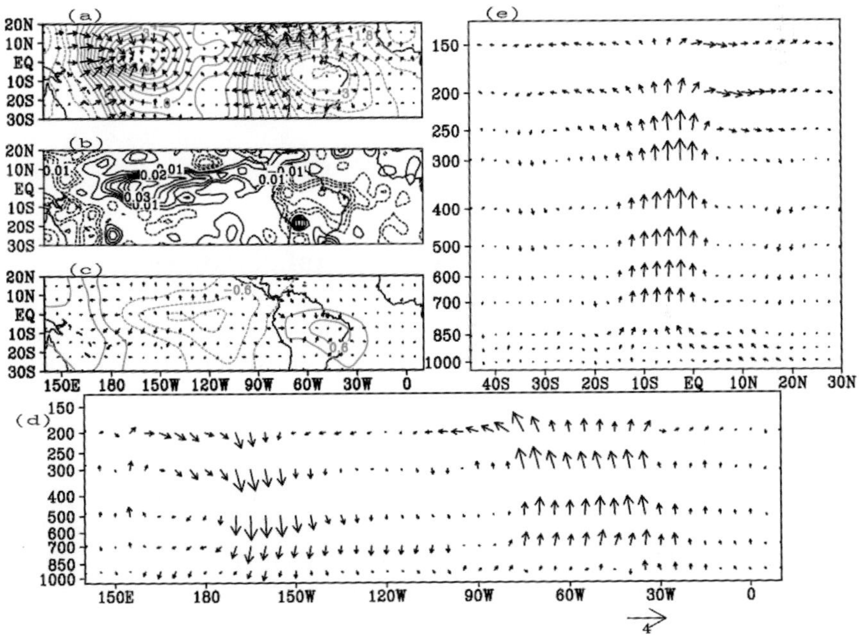

Figure 7-10. As in Fig. 7-7 but for La Niña Composite 1 during DJF.

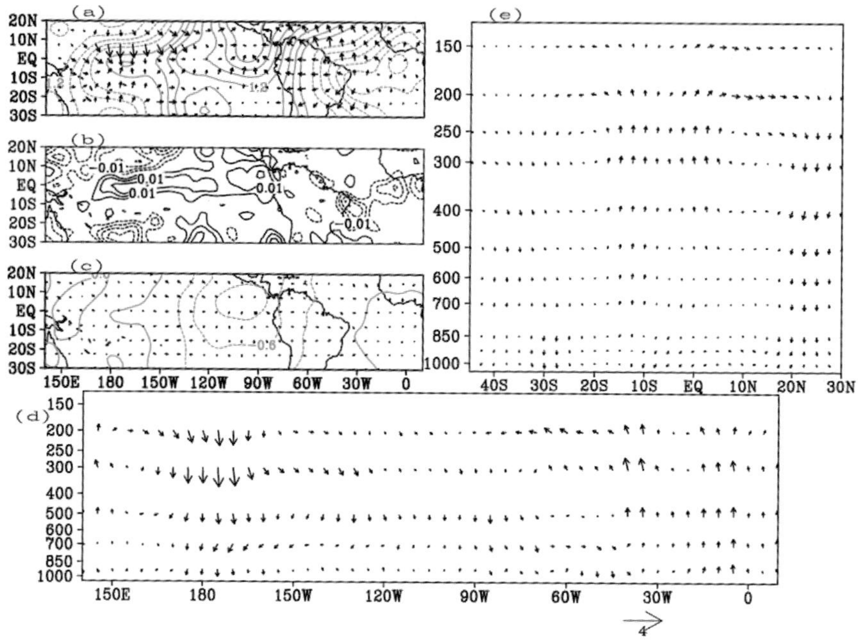

Figure 7-11. As in Fig. 7-10 but for La Niña Composite 1 during MAM.

4.3. El Niño Composite 2: Reversed Rainfall Pattern

Despite the fact that El Niño Composite 2 is based on only three events for DJF and two for MAM, the driving motivation to create such a composite was given by the two ENSO episodes (1954–55 La Niña and 1972–73 El Niño) with reversed impacts on regional precipitation over Northeast Brazil (Fig. 7-12), as analyzed by Souza et al. (in press). Based on these two case studies, the generalized question that needs to be answered is the following: What are the main atmospheric circulation and SSTa features that occur when positive rainfall anomalies are observed over Northeast Brazil and negative anomalies are found in southern South America during El Niño events? During DJF, positive SSTa in the Pacific reach up to 1.5°C in the Niño-3.4 region. Some cold waters can be seen near the west coast of South America. In the intertropical Atlantic Ocean a weak southward SSTa gradient is observed during the summer and autumn seasons. At this time, the Atlantic is dominated by positive anomalies in its tropical and southern portions. Since the composite was based on the dipole rainfall distribution between the north (above normal) and south (below normal) part of South

America, the seasonal percentage deviation of the precipitation pattern observed for both seasons is coherent with this selection. The 120% of normal precipitation observed in the northeast part of South America during DJF may have some relationship to the displacement of the SST gradient to the south of the equator and some possible interaction with the SACZ and southward displacement of the ITCZ during this period. The negative rainfall deviation in southeastern South America is mainly due to strong subsidence, as will be seen from the regional Hadley cell analysis. In the autumn, the dipole precipitation pattern is still present but is somewhat weaker. From a climatological point of view, the SACZ has retreated and therefore does not contribute to the rainfall distribution. On the other hand, the tropical Atlantic and parts of the South Atlantic still show positive SSTa and may keep the ITCZ displaced southward during the period.

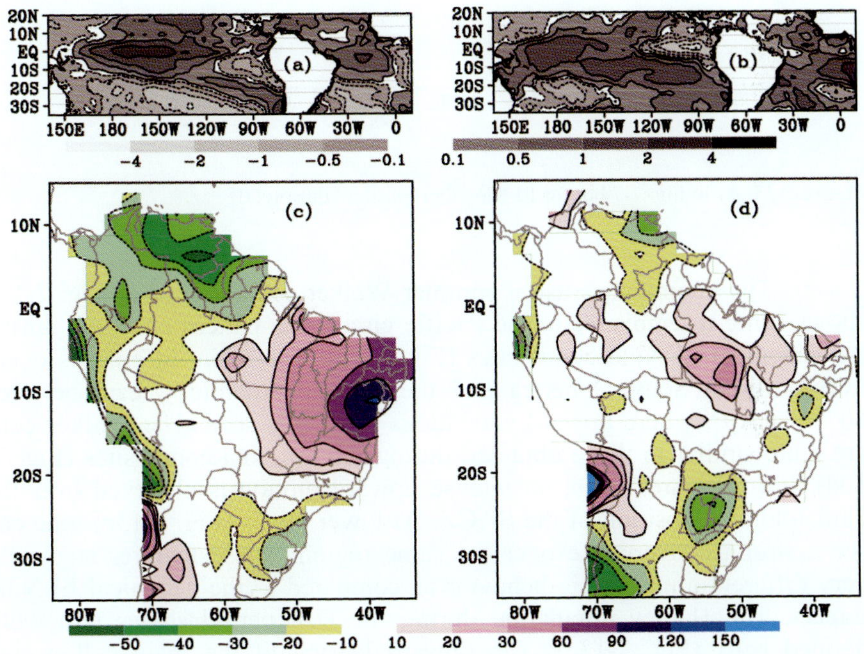

Figure 7-12. As in Fig. 7-6 but for El Niño Composite 2 during DJF (left) and MAM (right).

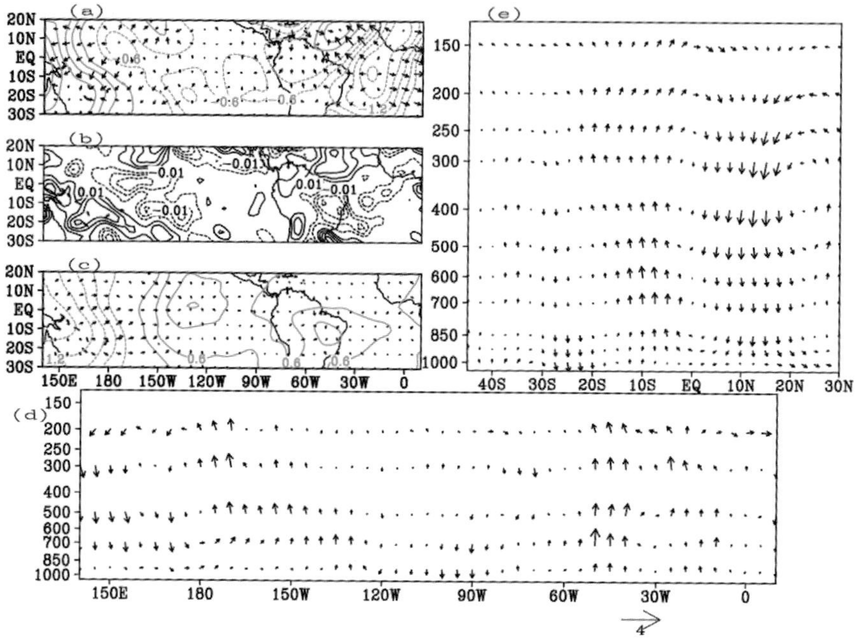

Figure 7-13. As in Fig. 7-7 but for El Niño Composite 2 during DJF.

The anomalous austral summer Walker cell depicted in Fig. 7-13d shows some resemblance to the Pacific circulation pattern observed during the canonical El Niño composites (Fig. 7-7d). The main differences occur over northern South America and the tropical Atlantic Ocean, between 60°W and 0°, where upward anomalous motion is seen. The Walker cells are quite similar to those obtained during the La Niña composites (Fig. 7-10d). From Figure 7-13b, an intense upward motion is observed over the climatological position of the SACZ. At lower levels (Fig. 7-13c), one can see a strong convergence over the same region. These features suggest a very different atmospheric behavior as compared to the canonical El Niño impact. The Hadley circulation during the DJF period shows two well-defined cells (Fig. 7-13e). The upward branch of the main cell occurs around 10°S, and the downward branches are located around 15°N and 30°S. As compared to the climatology (Fig. 7-4e) and Composite 1 (Fig. 7-7e), the circulation characteristics showed here are quite different. From this figure one can see that there is a good coincidence between the position of the Atlantic SST gradient and the convective active maximum and therefore the rainfall high-percentage values obtained. In the MAM period (Fig. 7-

14e), the maximum ascending motion occurs around 5°S, suggesting that indeed the ITCZ was quite active over this region during the period. In a recent paper, Chiang et al. (2002) showed that the Atlantic ITCZ is sensitive to anomalous SST gradients, which allow it to occupy a relatively wide range of latitudinal positions.

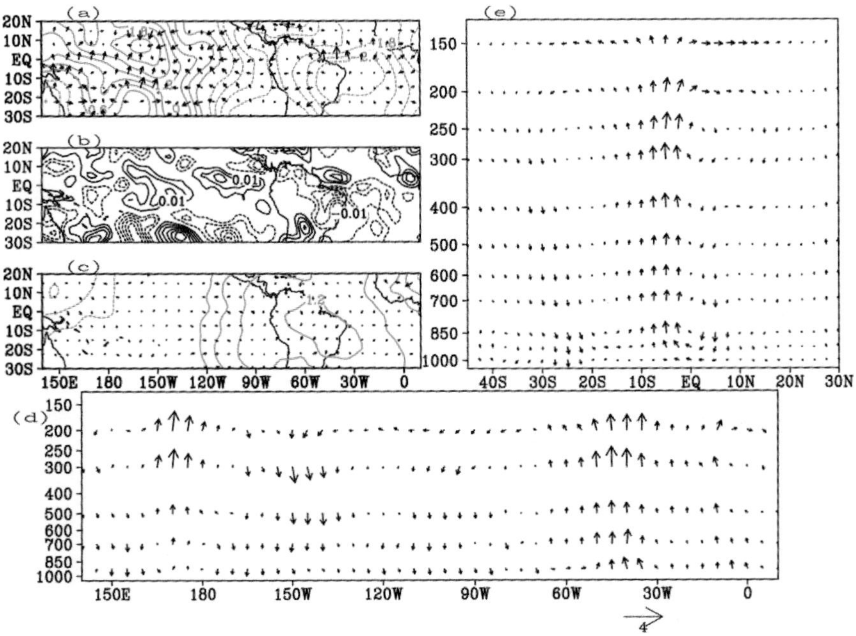

Figure 7-14. As in Fig. 7-13 but for El Niño Composite 2 during MAM.

The precipitation dipole pattern described here may be the result of the southward displacement of the Atlantic SST gradient, which in turn favors intensification of the SACZ and the ITCZ. Indeed, the Pacific SSTa was also weak and created the conditions for the Atlantic to dominate the circulation pattern over South America. The study of such interaction is out of the scope of the present diagnostic analysis, and further investigation will be needed to confirm this hypothesis.

4.4. La Niña Composite 2: Reversed Rainfall Pattern

Cooler water with up to –1.0°C SSTa dominates the tropical Pacific basin during the DJF period for Composite 2. In the Atlantic, a well-configured northward SSTa gradient with significant negative SSTa in the south and positive SSTa to the north is observed. Compared with the La Niña canonical composite, this pattern is completely reversed (Fig. 7-15a). In the following season, the negative SSTa decreases in the central-eastern Pacific as well as the SST gradient in the Atlantic.

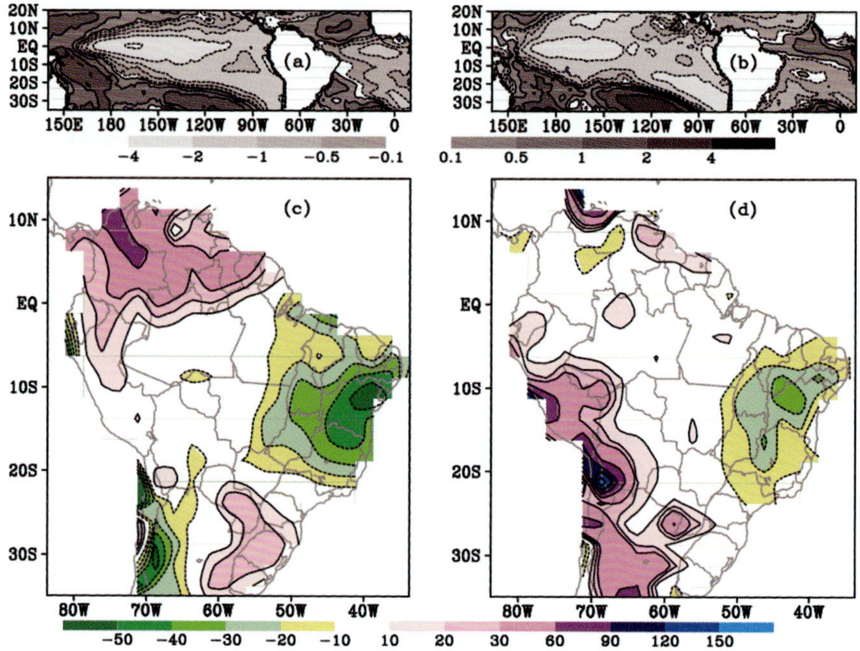

Figure 7-15. As in Fig. 7-6 but for La Niña Composite 2 during DJF (left) and MAM (right).

From the previous observational and statistical ENSO studies, one would expect that during La Niña years, the northeastern part of Brazil should have above normal rainfall during its rainy season and an inverse signal during El Niño events. Indeed, the canonical El Niño and La Niña composites presented before clearly show these patterns (Figs. 7-6 and 7-9). However, a reversed precipitation pattern is observed from Composite 2. A predominance of negative (positive) precipitation deviation varying between

–10% and –50% (10% to 20%) is seen over the northeast part of Brazil (southeastern South America). During the autumn season, the negative rainfall deviations decrease over the northeast region and a strong positive signal dominates the west part of South America, from central Argentina up to northern Peru and Bolivia. The diagnostic analyses suggest that the presence of a strong SST gradient in the Atlantic may overcome any signal from the Pacific, being the dominant feature during the rainy season.

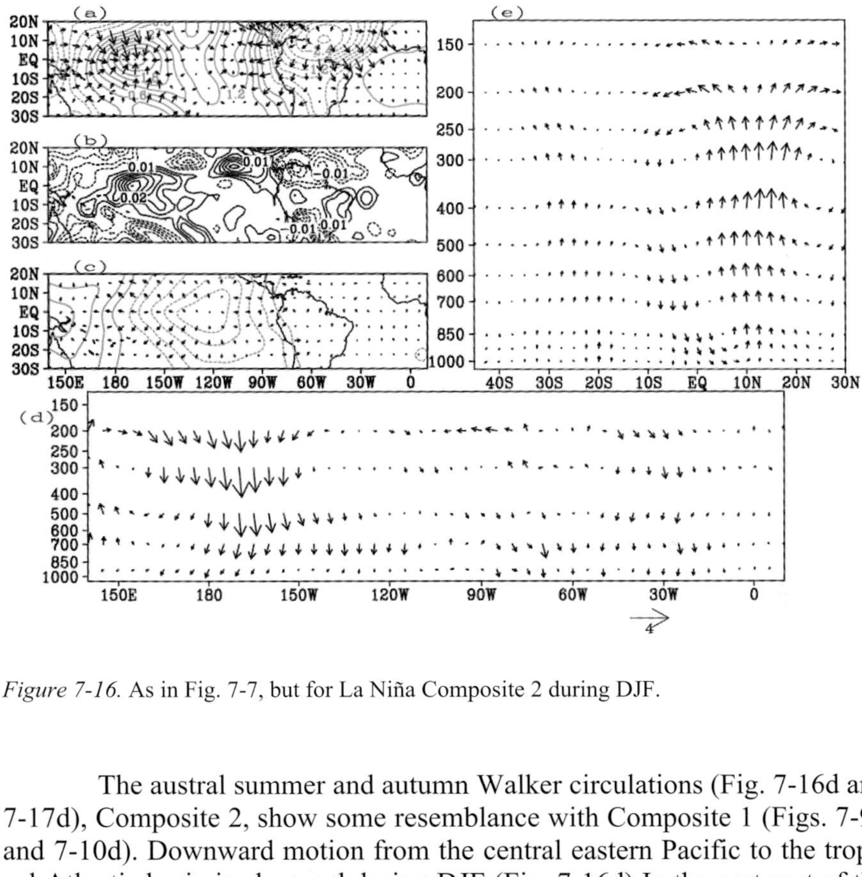

Figure 7-16. As in Fig. 7-7, but for La Niña Composite 2 during DJF.

The austral summer and autumn Walker circulations (Fig. 7-16d and 7-17d), Composite 2, show some resemblance with Composite 1 (Figs. 7-9d and 7-10d). Downward motion from the central eastern Pacific to the tropical Atlantic basin is observed during DJF (Fig. 7-16d) In the east part of the South American continent, a weak downward velocity is noticed (Fig. 7-16d) that agrees with the rainfall deficit in the region. The upper-level divergence and the lower-level convergence to the north of South America (Figs. 7-16a, c) seem to comprise the main driving circulation responsible for the anomalous precipitation patterns observed during this period. An ascending motion between 80°W and 40°W over the South American continent is observed during the MAM period (Fig. 7-17d) and completely differs from the motion observed during the previous season. This pattern is related to the lower convergence and upper-level divergence signal seen over the northwestern and northeastern parts of the South American continent (Figs. 7-17a, c). The position of the ITCZ here is off the continental coast, because the SST gradient is in the North Atlantic. Therefore, part of the precipitation deficit found over northeastern Brazil may be related to the absence of the ITCZ during this period. Indeed, the ascending branch of the regional Hadley circulation (Fig. 7-17e) clearly shows a maximum around the equator and a descending branch between 10°S and 20°S. This pattern is opposite to that observed in Composite 1, and it is similar to that in the previous season.

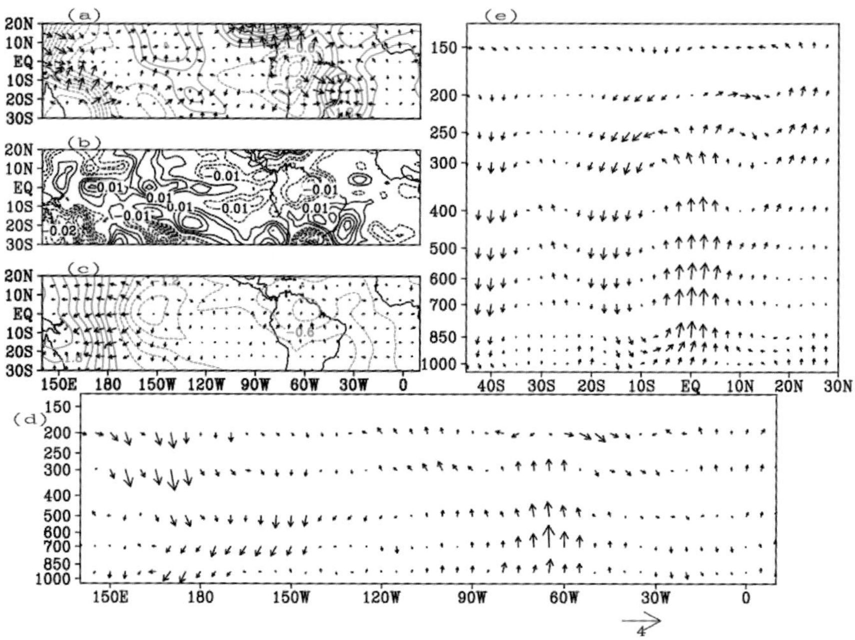

Figure 7-17. As in Fig. 7-16, but for La Niña Composite 2 during MAM

It is well known that the latitudinal position of the ITCZ is affected by the position of the Atlantic SST gradient (Nobre and Shukla 1996; Souza and Nobre 1998; Chiang et al. 2002) and therefore so are the trade winds and sea level pressure variations over the tropical Atlantic (Hastenrath and

Greischar 1993). Thus, during the austral summer and autumn the quality of the rainy season in the north and northeast sectors of South America are remarkably modulated by the displacement of the cloudiness and precipitation associated with the ITCZ to the south of the equator. The results shown here demonstrate that the regional Hadley circulation is sensitive to the position of the ITCZ over the continent. On the other hand, the ITCZ is sensitive to the position of SST gradients over the Atlantic. Thus, when the Atlantic signal dominates the Pacific cold (or warm) ENSO anomalies, the circulation generated causes a very distinct rainfall distribution over the South American continent and the canonical ENSO impact is not valid.

5. INFLUENCE ON THE SEASONAL CYCLE OF THE SOUTH AMERICAN MONSOON

Land-sea temperature contrasts in the tropics strongly affect the evolution of monsoons. In most other regions this gradient is driven primarily by land temperature changes and therefore primarily by insolation. During ENSO events, however, the contrast is created by a mix of insolation and the large SST anomalies occurring in the tropical eastern Pacific off the coast of South America. Pulwarty and Diaz (1993) observed that differences of up to 2°C between the northern South American landmass and the eastern tropical Pacific modulate the migration of MAM convection. Composites of the mean seasonal position of minima in outgoing long-wave radiation (OLR), taken as an indicator of high convective activity and heavy rainfall show westward excursions of deep convection during "warm (El Niño) events" contrasting with the northward and eastward location of convective centers during "cold (La Niña) events." Deep convection over the Peruvian and Ecuadorian coasts during this time may be driven not only by local SST increases but also by the seasonal migration from the southern Amazon Basin being displaced westward. Differing migration trajectories of the center of deep convection during Year +1 from January through September from Brazil to the Isthmus of Panama (and the eastern Pacific warm pool) are observed during (1) strong El Niño and (2) La Niña conditions. During the strong warm events, the mean January–February and MAM convective locations are found well to the west of the climatological means.

This pattern indicates that for some regions in the center of the South American landmass, a decrease in seasonal rainfall means a redistribution of convection and total regional rainfall suppression. The interaction of extreme ENSO events with the South American monsoon produces changes in the time and space evolution of convection and circulation anomalies over northern South America. Rainfall anomalies as discussed above are associated with the excitation of an anomalous east-west overturning cell with rising motions and low-level westerlies over the equatorial eastern Pacific coupled with descent and low-level easterlies over northern Brazil (Lau and Zhou 2003). Further studies of these east-west land-sea temperature contrasts and their modulation during ENSO are needed.

6. SUMMARY AND DISCUSSION

From previous observational studies, one can be assured that ENSO events influence the precipitation patterns over South America, where Northeast Brazil and southern South America seem to show the strongest signals. Besides the well-documented ENSO impacts on rainfall over the South American continent, such as dry (wet) conditions in northern and northeastern Brazil and wet (dry) conditions in southern Brazil, northern Argentina, and Paraguay during El Niño (La Niña) events, it was demonstrated here that there are occasions when this traditional pattern does not occur and the regional circulation can be quite different.

The seasonal standardized indices for precipitation and SST during 15 El Niño and La Niña episodes observed between 1950 and 1999 indicate large inter-ENSO variability.

Based on the time series generated by the index *I* of the ENSO episodes chosen, two kinds of composites were built (see Table 7-1). The first one selected the positive seasonal precipitation index for southern South America and Ecuador/Peru and the negative index for the eastern Amazon and Northeast Brazil for El Niño (La Niña) years (Composite 1). For the second composite, the El Niño and La Niña years were selected that essentially represented an inverse impact in the precipitation pattern over the same regions (Composite 2). Diagnostic analyses of seasonal precipitation percentage, SSTa, regional Walker and Hadley circulations, upper and lower divergent fields, and vertical velocity for the austral summer and autumn were done and compared to climatology. Through these analyses it is shown that the Walker and Hadley circulation patterns during the warm ENSO events for Composite 1 have in general an opposite pattern as compared to the climatology. On the other hand, during cold SSTa in the Pacific, the circulations are similar to climatology.

Some interesting features are noted from the zonal and meridional-vertical cross-section analyses. For instance, the Hadley and Walker circulation patterns observed during El Niño Composite 1 seem to be strongly coupled with the SSTa in the Pacific. The rainfall deficit over northern and northeastern South America is linked to the Walker circulation anomalies during ENSO warm events.

However, the above average precipitation percentages found over southern South America are likely to be associated not only with the regional Hadley circulation anomalies but also with modification of the quasi-stationary waves' position over the South American continent; the waves are generated by the modification of the heat source over the eastern equatorial Pacific region during El Niño events (Ambrizzi and Magaña 1999, submitted; Hoerling and Kumar 2000 and references therein).

Figure 7-18 shows a schematic diagram of Composite 1 for the regional Walker and Hadley circulations during cold and warm ENSO episodes (Figs. 7-18a, b). The red and blue arrows represent the regional Walker and Hadley circulations, respectively. Comparing the results from both circulations with the El Niño and La Niña diagrams, one can see that in general the Walker circulation is opposite to the climatology in the first case and in the same direction in the second, being in agreement with the above description. The Hadley circulation also follows this general pattern.

The El Niño and La Niña Composite 2 shows a completely reverse atmospheric circulation and precipitation pattern as compared to Composite 1. In the warm ENSO composite events, during DJF and MAM, an SST gradient in the Atlantic around 10°S seems to play the main role in the period. The above 120% positive deviation observed in the northeast part of South America during DJF may be associated with this SST gradient, which is displaced south of the equator and has some possible interaction with the SACZ and the southwards presence of the ITCZ during this period. The negative rainfall deviation in southeastern South America is mainly due to a strong subsidence as is confirmed from the regional Hadley cell circulation. In the autumn, the dipole precipitation pattern is still present but is somewhat weaker. From a climatological point of view, the SACZ is not active anymore and therefore it does not contribute to the rainfall distribution. On the other hand, the tropical Atlantic and portions of the South Atlantic still has positive SSTa that may keep the ITCZ displaced southward during the period. A schematic view of the anomalous regional Walker and Hadley circulations during both sets of ENSO composites are shown in Figure 7-19. This figure clearly indicates that the regional Hadley circulation pattern is reversed during the seasons. In particular, during the La Niña composite, the Atlantic SST gradient is displaced northward and the main upward motion occurs north of the equator, forcing a sinking movement over northern and northeastern South America, which in turn generates the rainfall deficits observed. As for the El Niño Composite 2, the Atlantic signal also completely dominates the Pacific cold ENSO anomalies.

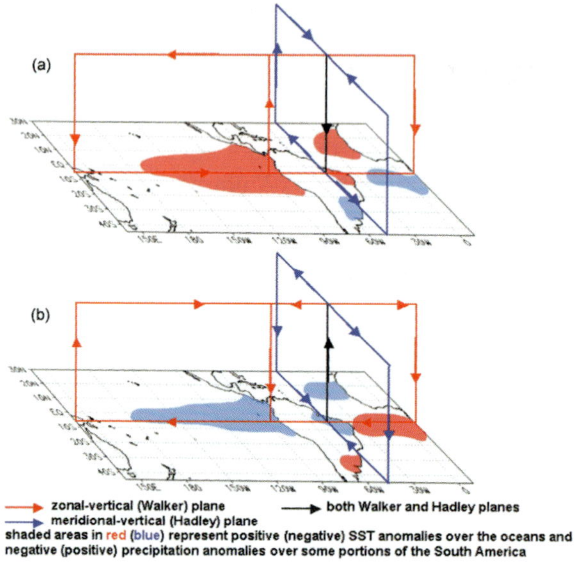

Figure 7-18. Schematic diagrams showing the anomalous regional Walker and Hadley circulations for the (a) El Niño and (b) La Niña canonical impacts (Composite 1).

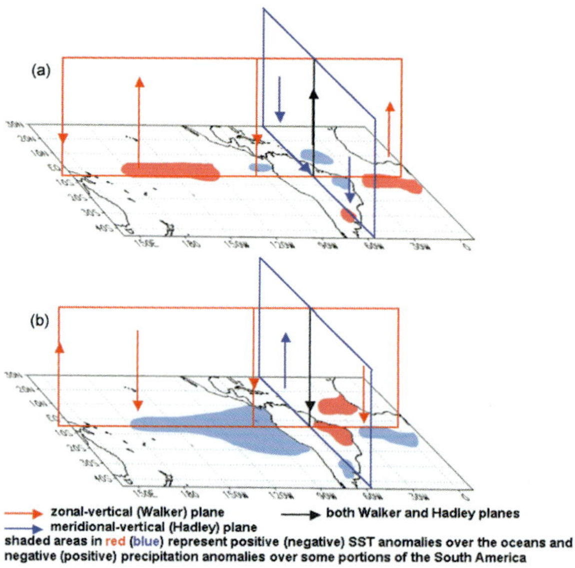

Figure 7-19. As in Fig. 7-18 but for the (a) El Niño and (b) La Niña reversed impacts (Composite 2).

In summary, the dipole precipitation pattern obtained from El Niño Composite 2 may be the result of the southward displacement of the Atlantic SST gradient, which in turn favors intensification of the SACZ and ITCZ. Indeed, the Pacific SSTa was also weak and created the conditions for the Atlantic to dominate the circulation pattern over South America. The importance of the Atlantic is further confirmed in La Niña Composite 2, in which the Atlantic SST gradient occurred north of the equator and had a direct impact on rainfall distribution. The study of such interactions is out of the scope of the present diagnostic analysis, and further investigation is needed to confirm this hypothesis.

The temporal characteristics of the precipitation relationships mentioned above are strongly modulated due the warmer SSTs and the excitation of ascending motion over the Pacific coast. These features alter the normal land-sea contrast longitudinally, which is quite different from the north-south contrasts observed in monsoon areas. These contrasts force westward displacements of convection during the northward migration of deep convection during MAM associated with strong warm events and eastward displacement during cold events. Similar but less pronounced displacements occur during the "retreat" of convection into the southern Amazon during the development of ENSO events. As has been pointed out by Pulwarty (1994), many aspects of ENSO impact on the onset and withdrawal of the South American monsoon system, and the relationships between its interannual and intraseasonal variability remain to be quantified.

Many studies have already described the climatological Walker and Hadley regional cells and their deviations during ENSO events (e.g., Wang 2002 and references therein). Here we have related the variability of these important circulations with rainfall distribution over South America. It was shown that not all ENSO events follow a canonical pattern. Depending on the conditions of the Atlantic SSTa, the position of the ITCZ can be modified and the ascending and descending branches of the related regional Hadley cell may vary and the South American rainy season may be strongly affected. These results suggest that some care always must be taken in producing precipitation (and impact) forecasts based on ENSO indices alone.

7. ACKNOWLEDGMENTS

Many thanks to the editors, Henry Diaz and Ray Bradley, and the anonymous reviewer, whose comments greatly helped in improving the original manuscript. T. Ambrizzi was partially supported by Conselho Nacional de Desenvolvimento Científico e Tecnológico (CNPq), Fundação de Amparo a Pesquisa do Estado de São Paulo (FAPESP) and the Inter-

American Institute for Global Change Research – Collaborative Research Network Program (CRN), Process No. 055 (IAI-CRN055). E.B. de Souza was supported by CNPq (process 304096/2003-2).

8. REFERENCES

Aceituno, P. 1988. On the functioning of the Southern Oscillation in the South America sector. *Monthly Weather Review* 116(2): 505–524.

Alves, J.M.B., and C.A. Repelli. 1992. A variabilidade pluviométrica no setor norte do nordeste e os eventos El Niño/Oscilação Sul. *Revista Brasileira de Meteorologia* 7(2): 583–592.

Ambrizzi, T., and V. Magaña. 1999. Dynamics of the impacts of El Niño/Southern Oscillation on the Americas climate: The December–January–February signal. In, 14th Conference on Hydrology, 79th AMS Annual Meeting, Dallas, Texas, 1: 307–308.

Ambrizzi, T., and V. Magaña. Submitted. Climate variability in the tropical and subtropical Americas associated with El Niño/Southern Oscillation (ENSO). *Climate Dynamics*.

Bjerknes, J. 1969. Atmospheric teleconnections from the equatorial Pacific. *Monthly Weather Review* 97: 163–172.

Cane, M. 1992. Tropical Pacific ENSO models: ENSO as a mode of the coupled system. In, Trenberth, K.E. (ed.). *Climate System Modeling*. Cambridge, UK: Cambridge University Press, pp. 583–614.

Casarin, D.P., and V.E. Kousky. 1986. Anomalias de precipitação no sul do Brasil e variações da circulação atmosférica. *Revista Brasileira de Meteorologia* 1: 83–90.

Chen, M., P. Xie, J.E. Janowiak, and P.A. Arkin. 2002. Global land precipitation: A 50 year monthly analysis based on gauge observations. *Journal of Hydrometeorology* 3(3): 249–266.

Chiang, J.C., Y. Kushnir, and A. Giannini. 2002. Deconstructing Atlantic Intertropical Convergence Zone variability: Influence of the local cross-equatorial sea surface temperature gradient and remote forcing from the eastern equatorial Pacific. *Journal of Geophysical Research* 107(D1): 1–19.

Coelho, C.A.S., A.R.M. Drumond, T. Ambrizzi, and G. Sampaio. 1998. Estudo climatológico sazonal da precipitação sobre o Brasil em episódios extremos da oscilação sul. *Revista Brasileira de Meteorologia* 14(1): 49–65.

Coelho, C.A.S., C.B. Uvo, and T. Ambrizzi. 2002. Exploring the impacts of the tropical Pacific SST on the precipitation patterns over South America during ENSO periods. *Theoretical and Applied Climatology* 71: 185–197.

Diaz, A.F., C.D. Studzinski, and C.R. Mechoso. 1998. Relationships between precipitation anomalies in Uruguay and Southern Brazil and sea surface temperature in the Pacific and Atlantic Oceans. *Journal of Climate* 11: 251–271.

Dommenget, D., and M. Latif. 2000. Interannual to decadal variability in the tropical Atlantic. *Journal of Climate* 13: 777–792.

Enfield, D.B., and D.A. Mayer. 1997. Tropical Atlantic sea surface temperature variability and its relation to the El Niño–Southern Oscillation. *Journal of Geophysical Research* 102: 929–945.

Figueroa, S.N., P. Satyamurty, and P.L. Silva Dias. 1995. Simulations of the summer circulation over the South American region with an ETA coordinate model. *Journal of the Atmospheric Sciences* 52: 1573–1584.

Grimm, A.M., V.R. Barros, and M. Doyle. 2000. Climate variability in Southern South America associated with El Niño and La Niña events. *Journal of Climate* 13: 35–58.

Grimm, A.M., S.E.T. Ferraz, and J. Gomes. 1998. Precipitation anomalies in Southern Brazil associated with El Niño and La Niña events. *Journal of Climate* 11: 2863–2880.

Harrison, D.E., and N.K. Larkin. 1998. El Niño–Southern Oscillation sea surface temperature and wind anomalies, 1946–1993. *Reviews of Geophysics* 36(3): 353–399.

Hastenrath, S. 1985. *Climate and Circulation of the Tropics*. Dordrecht: D.Riedel, 312 pp.

Hastenrath, S. 2001. In search of zonal circulation in the equatorial Atlantic sector from the NCEP-NCAR reanalysis. *International Journal of Climatology* 21: 37–47.

Hastenrath, S. 2002. Dipoles, temperature gradients, and tropical climate anomalies. *Bulletin of the American Meteorological Society* 83(5): 735–738.

Hastenrath, S., and L. Greischar. 1993. Circulation mechanisms related to northeast Brazil rainfall anomalies. *Journal of Geophysical Research* 98(D3): 5093–5102.

Hastenrath, S., and L. Heller. 1977. Dynamics of climatic hazards in Northeast Brazil. *Quarterly Journal of the Royal Meteorological Society* 103(435): 77–92.

Hastenrath, S., and P. Lamb. 1977. *Climatic Atlas of the Tropical Atlantic and Eastern Pacific Oceans*. Madison: University of Wisconsin Press.

Hoerling, M.P., and A. Kumar. 2000. Understanding and predicting extratropical teleconnections related to ENSO. In, Diaz, H.F., and V. Markgraf (eds.). *El Niño and the Southern Oscillation: Multiscale Variability and Global and Regional Impacts*. Cambridge, UK: Cambridge University Press, pp. 57–88.

Kalnay, E., and co-authors. 1996. The NCEP/NCAR reanalysis project. *Bulletin of the American Meteorological Society* 77: 437–471.

Kayano, M.T., N.J. Ferreira, and M.C.V. Ramírez. 1997. Summer circulation patterns related to the upper tropospheric vortices over the tropical South Atlantic. *Meteorology and Atmospheric Physics* 64: 203–213.

Kidson, J.W. 1975. Tropical eigenvector analysis and the Southern Oscillation. *Monthly Weather Review* 103: 187–196.

Kiladis, G., and H.F. Diaz. 1989. Global climatic anomalies associated with extremes in the Southern Oscillation. *Journal of Climate* 2: 1069–1090.

Kousky, V.E., and M.A. Gan. 1981. Upper tropospheric cyclonic vortices in the tropical South Atlantic. *Tellus* 33: 538–551.

Kousky, V.E., and C. Ropelewski. 1989. Extremes in the Southern Oscillation and their relationship to precipitation with emphasis on the South American region. *Revista Brasileira de Meteorologia* 4: 351–363.

Kousky, V.E., I.F.A. Cavalcanti, and M.T. Kayano. 1984. A review of the Southern Oscillation: Oceanic-atmospheric circulation changes and related rainfall anomalies. *Tellus* 36A: 490–504.

Lau, K., and J. Zhou. 2003. Anomalies of the South American summer monsoon associated with the 1997–1999 ENSO. *International Journal of Climatology* 23: 529–539

Lau, N.G. 1997. Interactions between global SST anomalies and the midlatitude atmospheric circulation. *Bulletin of the American Meteorological Society* 78: 21–34.

Lau, N.G., and M.J. Nath. 1994. A modeling study of the relative roles of tropical and extratropical SST anomalies in the variability of the global atmosphere-ocean system. *Journal of Climate* 7: 1184–1207.

Liebmann, B., G.K. Kiladis, J.A. Marengo, T. Ambrizzi, and J.D. Glick. 1999. Submonthly convective variability over South America and the South Atlantic Convergence Zone. *Journal of Climate* 12: 1877–1891.

Moura, A.D., and J. Shukla. 1981. On the dynamics of droughts in northeast Brazil: Observations, theory and numerical experiments with a general circulation model. *Journal of the Atmospheric Sciences* 38(7): 2653–2675.

Nobre, P., and J. Shukla. 1996. Variations of sea surface temperature, wind stress and rainfall over the tropical Atlantic and South America. *Journal of Climate* 10(4): 2664–2479.

Nogués-Paegle, J., and K.C. Mo. 1997. Alternating wet and dry conditions over South America during summer. *Monthly Weather Review* 107: 279–291.

Oort, A.H., and J.J. Yienger. 1996. Observed interannual variability in the Hadley circulation and its connection to ENSO. *Journal of Climate* 9: 2751–2767.

Pezzi, L.P., and I.F.A. Cavalcanti. 2001. The relative importance of ENSO and tropical Atlantic SST anomalies for seasonal precipitation over South America: A numerical study. *Climate Dynamics* 17: 205–212.

Pulwarty, R.S. 1994. Annual and interannual variability of convection over tropical South America. Ph.D. thesis, University of Colorado, Boulder, 220 pp.

Pulwarty, R., and H. Diaz. 1993. A study of the seasonal cycle and its perturbation by ENSO events over the tropical Americas. In, Fourth International Conference on Southern Hemisphere Meteorology and Oceanography, March 29–April 2, 1993, Hobart, Australia.

Rao, V.B., and K. Hada. 1990. Characteristics of rainfall over Brazil: Annual variation and connections with the Southern Oscillation. *Theoretical and Applied Climatology* 42: 81–91.

Rao, V.B., I.F.A. Cavalcanti, and K. Hada. 1996. Annual variations of rainfall over Brazil and water vapor characteristics of South America. *Journal of Geophysical Research* 101: 36350–36551.

Ropelewski, C., and M.S. Halpert. 1987. Global and regional scale precipitation patterns associated with the El Niño–Southern Oscillation. *Monthly Weather Review* 115(2): 1606–1626.

Ropelewski, C., and M.S. Halpert. 1989. Precipitation patterns associated with the high index phase of the Southern Oscillation. *Journal of Climate* 2: 268–284.

Servain, J. 1991. Simple climate indices for the tropical Atlantic Ocean and some applications. *Journal of Geophysical Research* 96: 15137–15146.

Smith, T.M., R.W. Reynolds, R.E. Livezey, and D.C. Stokes. 1996. Reconstruction of historical sea surface temperature using empirical orthogonal functions. *Journal of Climate* 9: 1403–1420.

Souza, E.B., and T. Ambrizzi. 2002. ENSO impacts on the South American rainfall during 1980s: Hadley and Walker circulation. *Atmósfera* 15: 105–120.

Souza, E.B., and P. Nobre. 1998. Uma revisão sobre o Padrão de Dipolo no Oceano Atlântico tropical. *Revista Brasileira de Meteorologia* 13(1): 31–44.

Souza, E.B., T. Ambrizzi, and C.A.S. Coelho. In press. Two ENSO episodes with reversed impacts on the regional precipitation of the northeastern South America. *Meteorológica.*

Souza, E.B., J.M.B. Alves, and P. Nobre. 1998. Anomalias de precipitação no norte e leste do Nordeste Brasileiro em associação aos eventos do Padrão de Dipolo observados no Atlântico tropical. *Revista Brasileira de Meteorologia* 13(1): 31–44.

Souza, E.B., M.T. Kayano, J. Tota, L. Pezzi, G. Fisch, and C. Nobre. 2000. On the influences of the El Niño, La Niña and Atlantic dipole pattern on the Amazonian rainfall during 1960–1998. *Acta Amazonica* 30(2): 305–318.

Trenberth, K.E. 1997. The definition of El Niño. *Bulletin of the American Meteorological Society* 78: 2771–2777.

Trenberth, K.E., and J.M. Caron. 2000. The Southern Oscillation revisited: Sea level pressures, surface temperatures and precipitation. *Journal of Climate* 13: 4358–4365.

Trenberth, K.E., and D.P. Stepaniak. 2003. Seamless poleward atmospheric energy transports and implications for the Hadley circulation. *Journal of Climate* 16: 3706–3722.

Trenberth, K.E., D.P. Stepaniak, and J.M. Caron. 2000. The global monsoon as seen through the divergent atmospheric circulation. *Journal of Climate* 13: 3969–3993.

Uvo, C.R.B., C.A. Repelli, S. Zebiak, and Y. Kushnir. 1998. The relationships between tropical Pacific and Atlantic SST and northeast Brazil monthly precipitation. *Journal of Climate* 11(4): 551–562.

Virji, H. 1981. A preliminary study of summertime tropospheric circulation patterns over South America estimated from cloud winds. *Monthly Weather Review* 109: 599–610.

Wagner, R.G. 1996. Mechanisms controlling variability of the interhemispheric sea surface temperature gradient in the tropical Atlantic. *Journal of Climate* 9(7): 2010–2019.

Wang, C. 2002. Atmospheric circulation cells associated with the El Niño–Southern Oscillation. *Journal of Climate* 15: 399–419.

SECTION B

Variability of the Hadley Circulation in the Past

Chapter 8

THE PACIFIC SECTOR HADLEY AND WALKER CIRCULATION IN HISTORICAL MARINE WIND ANALYSES

Potential for Reconstruction from Proxy Data

Michael N. Evans[1] and Alexey Kaplan[2]

[1]*University of Arizona, Laboratory of Tree-Ring Research, Tucson, Arizona 85721 U.S.A.*
[2]*Lamont-Doherty Earth Observatory of Columbia University, Palisades, New York 10964 U.S.A.*

Abstract We investigate the historical variation of the wintertime Pacific marine sector meridional atmospheric circulation, using simple diagnostics calculated from a statistical analysis of 140 years of surface wind data. Intensity of the wintertime expression of the Hadley circulation, as expressed by a wind divergence index, varies interannually and secularly. In agreement with previous studies, interannual variation is associated with variations in the Walker circulation; e.g., El Niño/Southern Oscillation (ENSO) activity. The secular variation, most likely affected by systematic measurement biases, is nevertheless consistent with results from simulation of the Indo-Pacific-sector Hadley circulation variability in the NCEP/NCAR reanalysis (see Chapter 3, "Change of the Tropical Hadley Cell since 1950," Quan et al., this volume; and Chapter 5, "Interannual to Interdecadal Variations of the Hadley and Walker Circulations," Minobe, this volume) and model simulations of the global atmospheric response to anthropogenic forcing (see Chapter 14, "The Response of the Hadley Circulation to Climate Changes, Past and Future," Rind and Perlwitz, this volume; and Chapter 17, "Mechanisms of an Intensified Hadley Circulation in Response to Solar Forcing in the Twentieth Century," Meehl et al., this volume). A proxy network tracking Hadley intensity as mirrored in sea surface temperature (SST), precipitation, surface winds, and/or ocean upwelling might be used to further study processes underlying long-term variability in the Hadley circulation over the past several hundred years.

H.F. Diaz and R.S. Bradley (eds.), The Hadley Circulation: Present, Past and Future, 239–258.

1. INTRODUCTION

How steady is the zonal mean tropical atmospheric circulation in the presence of internal and external forcing? The mean circulation of the atmosphere is now well described by the large number of observational sites established in the past 50 years and the development of remote-sensing instruments in the satellite era. One of the most prominent features of the general circulation is the thermally direct, seasonally varying tropospheric circulation system now known as the Hadley circulation. It is well established that interannual variability in the Hadley circulation is strongly tied to El Niño/Southern Oscillation (ENSO) activity (Bjerknes 1966; Bjerknes 1969; Oort and Yienger 1996). However, it is also increasingly clear that the past century may also be one of change and reorganization in the tropical Pacific component of the climate system on decadal to centennial time scales (e.g., Cane et al. 1997). Whether the observed variability on these longer time scales is due to processes internal to the natural climate system, or is related to external factors such as anthropogenically driven change in atmospheric trace gas composition, or reflects some combination of these influences, remains a matter of debate.

There are at least two means by which the question may be addressed. We can build models that simulate the relevant aspects of the climate system, and perform experiments with and without hypothesized or known forcing. Or we can develop estimates of past variability of the meridional tropical circulation from historical data and/or from localized responses to change in the general circulation, which are preserved in proxy data from geological or biological archives. Although there are well-known strengths and weaknesses to each of these approaches (Bradley 1999; Meehl et al. 2000), intercomparison of observational and model results provides mostly independent support for the conclusion that the results are not tool-specific. Due to its intermediate spatial and temporal coverage, historical observational data are the link often used to tune or calibrate both models of climate and proxy data, and form the basis for the present investigation.

As a precursor to paleoclimatic reconstruction of the Hadley circulation, we seek an index of the surface expression of the Hadley circulation in the Pacific sector from multidecadal, marine historical observations. The zonally averaged solstitial surface divergences from the International Comprehensive Ocean-Atmosphere Data Set (I-COADS; Woodruff et al. 1998) climatology for the Pacific marine sector are illustrated in Figure 8-1. These divergences were computed from surface wind components u and v as

$$D(y) = \left[\frac{\partial u}{\partial x} + \frac{\partial v}{\partial y} \right]_{150^o W - 90^o W}$$

where the square brackets indicate zonal averaging over the indicated interval, and the meridional averaging is as described in Section 2.2. These sectorial divergences are consistent with but not analogous to the globally averaged winter meridional overturning stream functions for the Northern and Southern Hemispheres, respectively (e.g., Peixoto and Oort 1992, Fig. 7.19). For instance, Figure 8-1a shows maximum tropical convergence, implying rising motion above, from approximately 2°S to 10°N and divergence, implying subsidence, from about 18°N to 30°N, during Northern Hemisphere winter. Similarly, Figure 8-1b shows convergence, implying rising motion, from approximately 8°N to 16°N, and divergence, implying descent, from about 12°S to 8°N, during Southern Hemisphere winter. Consequently our approach to paleoclimatic reconstruction is to first develop a diagnostic index from surface wind divergences, which in turn might ultimately be mirrored in paleoproxy observations. Although interpretation of historical data sets is limited due to spatiotemporal heterogeneities in observational coverage and poorly known biases, recent analyses of such data, using modern statistical techniques, may permit reconstruction of large-scale climatic phenomena from, for example, historical marine data sets (Kaplan et al. 1997, 1998, 2000). We might also be able to reconstruct such features from the growing observational network of seasonal to annual resolution proxy climate observations using similar techniques (Evans et al. 2002).

Reconstruction of interannual and longer-term variability from spatiotemporally heterogeneous historical observations and proxy climate data is limited by observational density and poorly understood biases. In addition, our understanding of the controls on the long-term variation of proxy climate observations is limited and may not be independent of frequency (for example, see Evans et al. 2002). Furthermore, resolution of the large-scale meridional circulation may only be weakly approximated by limited availability of proxy climate observations. Hence, potential application of proxy data to this problem must make use of the ability of such data to integrate conditions over large ranges of space and/or time, yet also resolve seasonal differences. Toward this goal, multiple data sources should be used to identify and minimize errors in the proxy data, and intercomparison with modeling efforts should be used to interpolate between the sparse paleoproxy data network.

Here we investigate the potential for reconstruction of the Pacific marine sector meridional overturning circulation using surface historical and proxy climate data. The basis for this study is the development of analyzed historical surface climate data products (Kaplan et al. 2004) derived from the I-COADS project (Woodruff et al. 1998). We describe the analyzed wind product and the construction of a surface winds–based index in Section 2. Discussion of the behavior of the index over the past 140 years and some implications for reconstruction of Hadley circulation variability from paleo-proxy data are discussed in Section 3. A summary is given in Section 4.

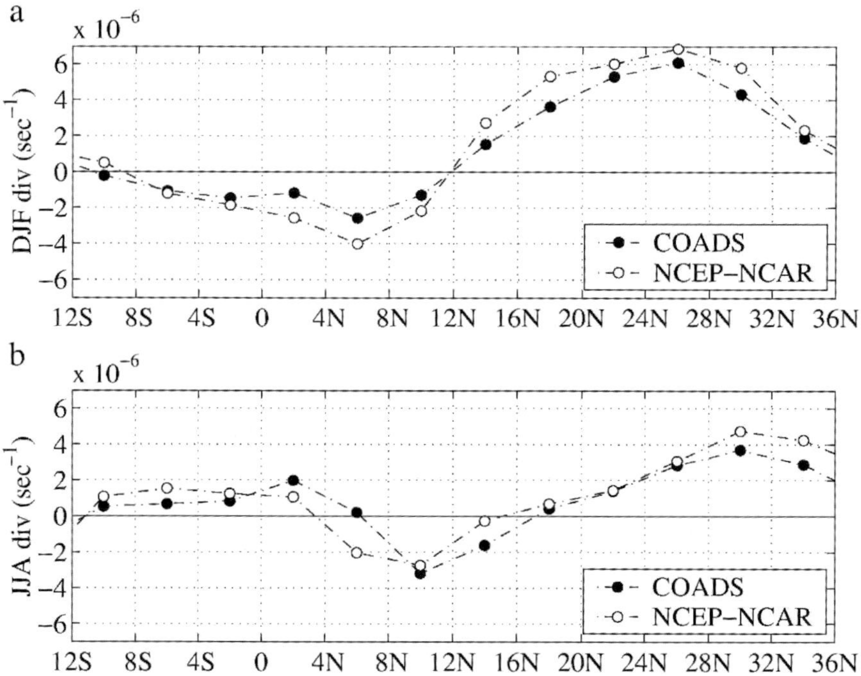

Figure8-1. **a:** Average (1951–80) December–February (DJF) divergences, zonally averaged over the Pacific basin (150°E–120°W), from I-COADS (solid circles) and NCEP/NCAR (open circles). Data are gridded as 4° x 4° averages centered between the latitude ticks indicated on the x-axis of the plot. Units are x 10^{-6} sec^{-1} as shown. **b:** As above, except for June–August (JJA) averages.

2. A SURFACE WINDS–BASED MERIDIONAL CIRCULATION INDEX

2.1. Analyzed Historical Wind Fields

Our source of historical surface wind data is the recently produced analysis (Kaplan et al. 2001, 2004) of the I-COADS gridded marine climate data set (Woodruff et al. 1987, 1998; Diaz et al. 2002). This analysis derives from the family of reduced space objective analysis procedures, which have now been used to describe and reconstruct large-scale features in sea level height, sea surface temperature (SST), and sea level pressure (SLP) from sparse gridded historical observations (Cane et al. 1996; Kaplan et al. 1997, 1998, 2000, 2003). The mechanics of the optimal interpolation analysis described here are identical to those described in detail in these preceding references.

The key to such analyses is the statistical definition of the large-scale space-time patterns of field variability from a well-observed modern period using empirical orthogonal function (EOF) analysis and filtering. The leading EOFs represent a model for the large-scale spatial and/or temporal covariance of the observations. Together with estimates of the observational and model errors, the simultaneous least-squares fit to historical observations and the model produces data field and error estimates for all locations for which the model can be defined and for all times for which observational data exist. Key prior assumptions include proper estimation of the statistical model, including its dimensionality, and accurate definition of errors in the statistical model and in the historical observations. Analysis results must be tested *a posteriori* to ensure that key prior assumptions are satisfied. These may include checking the consistency of prior and posterior error estimates (e.g., Kaplan et al. 1997) and comparison with withheld or independent observations (e.g., Evans et al. 2002).

In the case of the I-COADS winds analysis, only a spatial covariance model can be reliably defined (Kaplan et al. 2004), so the wind product is developed by using reduced space optimal interpolation (Kaplan et al. 1997). No bias corrections were applied to the I-COADS data set prior to analysis. No balanced friction force corrections (Ward and Hoskins 1996) or linear detrending to remove observational bias (Cardone et al. 1990) was used in this product. Hence in the absence of further information any interpretation of secular variability in the winds analysis or derived indices must be interpreted with caution.

The Kaplan et al. (2001, 2004) analysis of I-COADS winds potentially spans the period January 1800 to September 2001. However, severe data limitations are evident throughout the observational period in the tropical Pacific (Fig. 8-2). Data are generally fewer for the deep tropics and south of the equator; $O(10^2–10^3)$ observations per area denoted by Figures 8-2a and 8-2b are reached only in the 1850s and later, with scarcely any observations prior to the 1820s. In our opinion, severe levels of missing data preclude much use of the data or analysis for forming large-scale winds indices before about 1860 (Fig. 8-2). Intercomparison of the independent analyses of I-COADS winds, SST, and SLP for the tropical Pacific support this limitation; the three fields are tightly coupled as expected from tropical Pacific ocean-atmosphere dynamics, back to about the 1870s (Fig. 8-3).

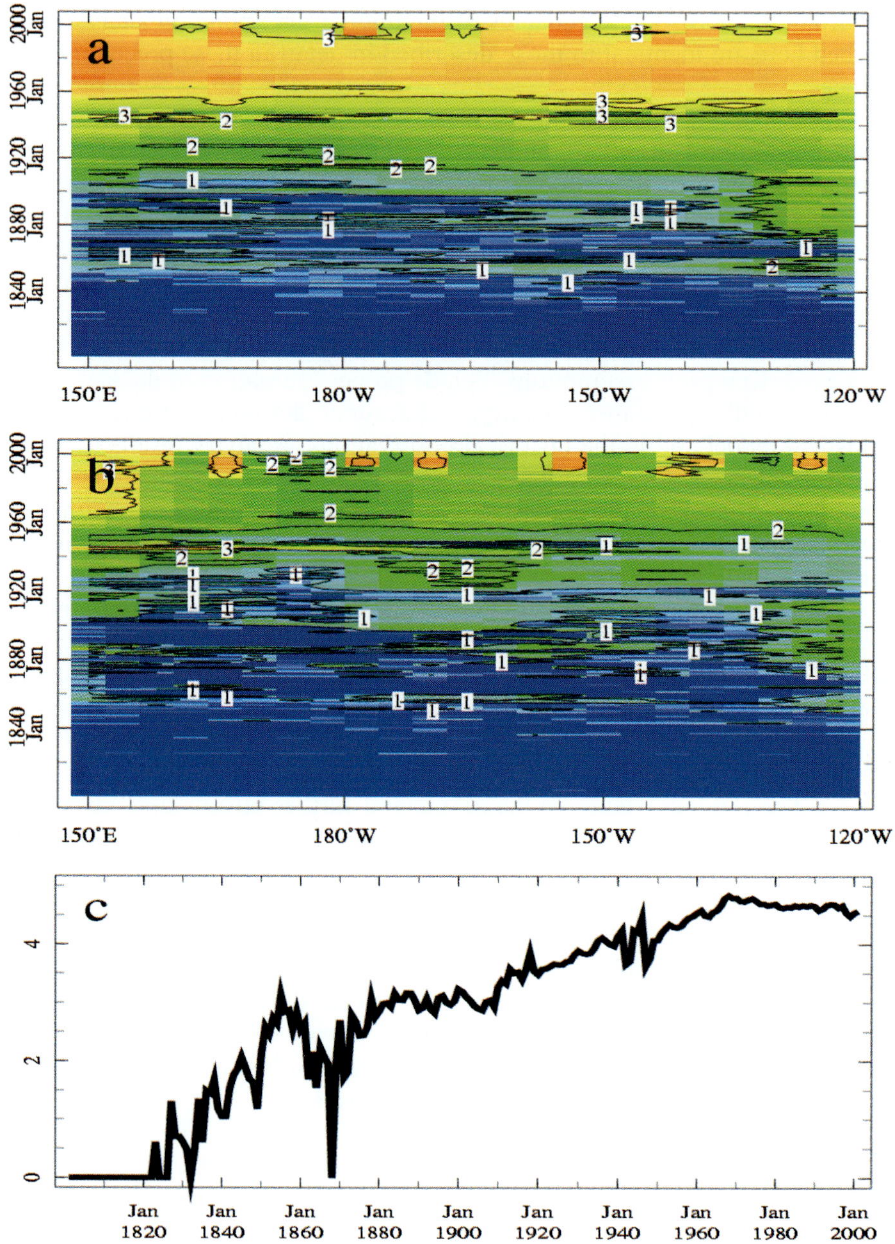

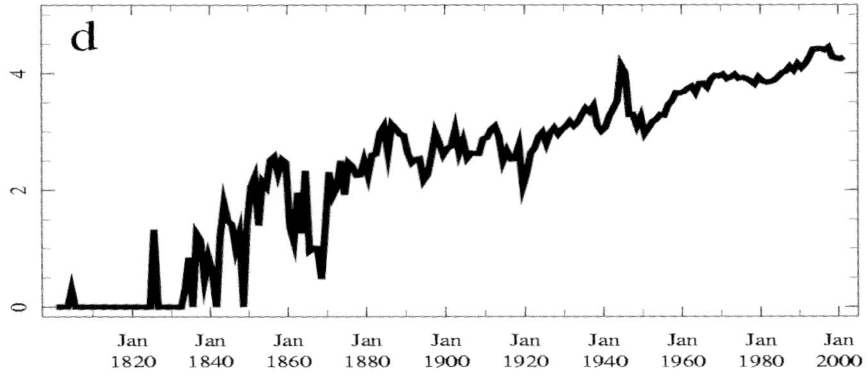

Figure 8-2. I-COADS historical meridional winds observations as a function of time (time increasing upward) and space (see Section 2). **a:** DJF averages for 150°E–120°W, 2°S–30°N, corresponding to the region over which the surface expression of the Pacific marine sector wintertime Northern Hemisphere Hadley circulation is developed (Section 2.2). **b:** JJA observations for 150°E–120°W, 12°S–16°N, corresponding to the region over which the surface expression of the Pacific marine sector wintertime Southern Hemisphere Hadley circulation is developed. **c:** DJF observations vs. time for region in panel a. **d:** JJA observations vs. time for region in panel b. In all panels, observational frequency is given on a logarithmic scale.

2.2. Pacific Marine Sector Hadley Circulation Indices

We seek a description of the surface expression of the thermally direct, zonally averaged atmospheric circulation that can be resolved in coarsely gridded historical surface marine observations and perhaps even in proxy climate observations. As a target we develop an index of the Hadley circulation over the tropical Pacific (150°E–120°W) based on the zonally averaged divergence for this region (Fig. 8-1). We define a Pacific basin, zonal mean Hadley circulation index (hereafter abbreviated HCI) for boreal and austral winters as:

$$HCI(DJF) \equiv [D]_{18^\circ N-30^\circ N} - [D]_{2^\circ S-10^\circ N}$$
$$HCI(JJA) \equiv [D]_{12^\circ S-8^\circ N} - [D]_{8^\circ N-16^\circ N}$$

where D is as defined earlier (Section 1) and averaged over December–February and June–August, respectively. It is important to note that this choice of area-averaged index is subjectively chosen to reflect the meridional circulation over the Pacific marine sector (Fig. 8-1); it may not represent a closed atmospheric circulation cell (as is guaranteed by a complete zonal average), and is therefore not analogous to the canonical zonal mean Hadley circulation (e.g., such as is described in Peixoto and Oort [1992]).

The index computed by using the analyzed I-COADS winds is denoted as "HA," for historical analysis. To assess the uncertainty in the index calculations due to temporal changes in observational coverage (Figs. 8-2, 8-3), we also compute indices using the unanalyzed I-COADS observations. We also compare the Hadley circulation index computed by using analyzed I-COADS data to that computed by using NCEP/NCAR 50-year Reanalysis Version 1 winds, available for 1949 to the present, denoted as "RA"-derived indices (Kalnay et al. 1996; Kistler et al. 2001).

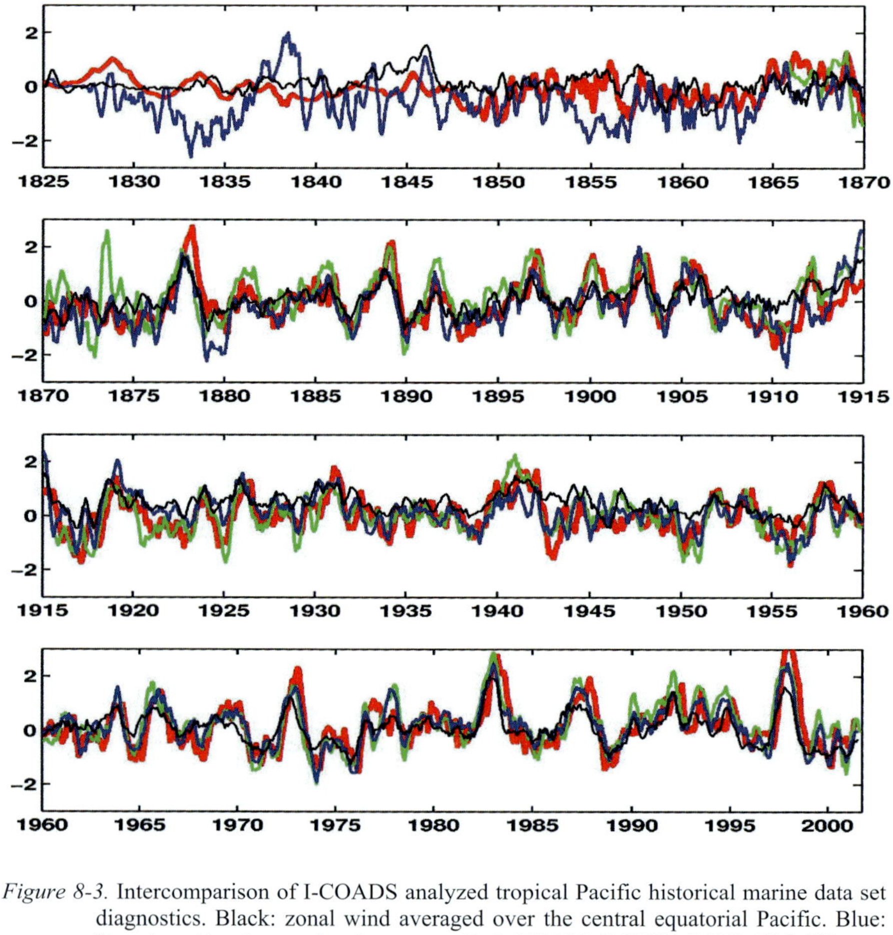

Figure 8-3. Intercomparison of I-COADS analyzed tropical Pacific historical marine data set diagnostics. Black: zonal wind averaged over the central equatorial Pacific. Blue: Darwin grid point sea level pressure. Red: NINO3 (150°W–90°W, 5°N–5°S area average) sea surface temperature anomaly. For comparison with independent observations, the green line shows Darwin station sea level pressure from Allan et al. (1991).

Scatter plots of the HC indices derived from HA and RA products are shown in Figure 8-4. In accordance with climatological observations (Fig. 8-1), the boreal circulation index is stronger than the austral circulation index. Correlation between linearly detrended HA and RA series is 0.94 and 0.82 for the 1950–2001 comparison period for DJF and JJA indices, respectively (Fig. 8-4). This is expected because COADS data are an input to the NCEP/NCAR reanalysis. However, mean differences between indices relatively to HA are 34% and 50% for DJF and JJA indices, respectively (Fig. 8-4). In addition, relative to HA, the DJF RA index is about 15%–25% larger in amplitude (Fig. 8-4a), although the JJA RA index is of comparable amplitude to the JJA HA index. Similar biases in Hadley circulation diagnostics from NCEP/NCAR reanalysis data were found by Waliser et al. (1999), and Wu and Xie (2003) argued that the NCEP/NCAR reanalysis contained seasonally dependent biases in winds for the tropical Pacific relative to COADS (see also differences between COADS and NCEP/NCEP divergence climatologies evident in Fig. 8-1).

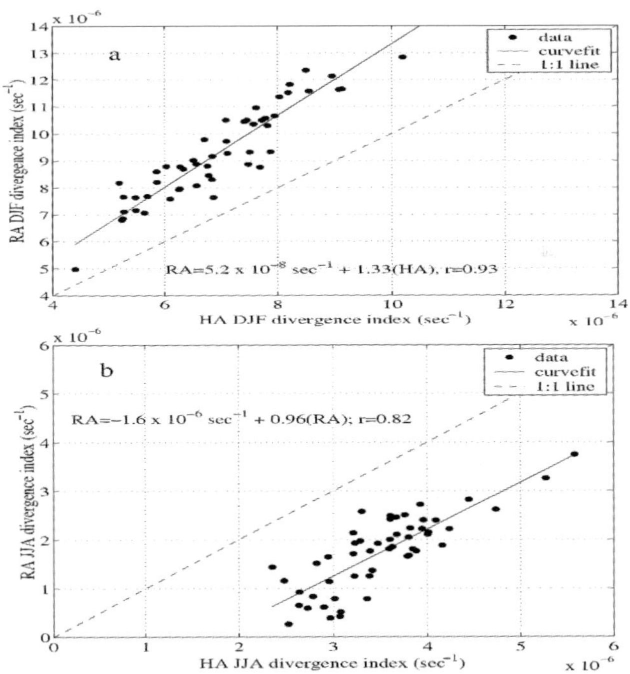

Figure 8-4. **a:** Regression of RA divergence index on HA divergence index, DJF averages, 1950–2001. Mean difference between indices is –2.4 x 10^{-6} sec^{-1}. **b:** As in panel a, except for JJA averages. Mean difference between indices is 1.8 x 10^{-6} sec^{-1}.

Following the results of Dima and Wallace (2003), we find the interannual variability evident in the boreal and austral winter indices is reasonably correlated across seasons: r = 0.49 for correlation of the DJF HA index with the JJA HA index over 1860–2001 (Fig. 8-5). We use this result to construct a combined seasonal Hadley circulation index over the marine Pacific sector by calculating the variance-weighted sum of the boreal and austral indices, and standardizing the result.

Time series of the DJF, JJA, and combined HC indices are shown in Figure 8-6. As is suggested by Figure 8-2, the drop in observational coverage results in greater noise in the I-COADS wind data, especially prior to 1870 and 1920. By general property of least-squares analysis (Kaplan et al. 2003), the wind analysis gives lower weight to scarce, noisy observations, relying more heavily on large-scale structures identified in the statistical model and producing estimates with lower variance and greater estimated uncertainty.

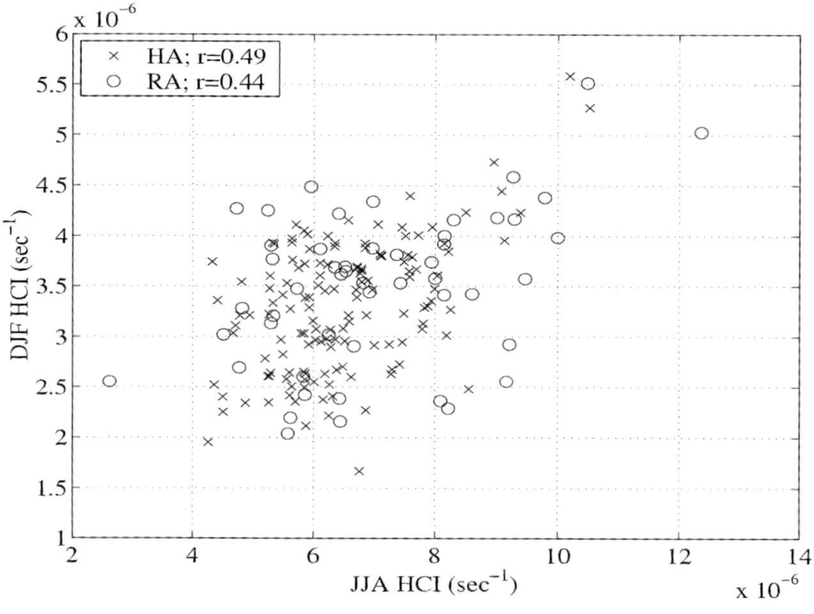

Figure 8-5. Scatter plot of DJF vs. JJA HC indices. Crosses: HA. Circles: RA. Correlations between DJF and JJA indices are 0.49 and 0.44, and are significant at the $\alpha \leq 0.01$ and 0.05 levels, respectively.

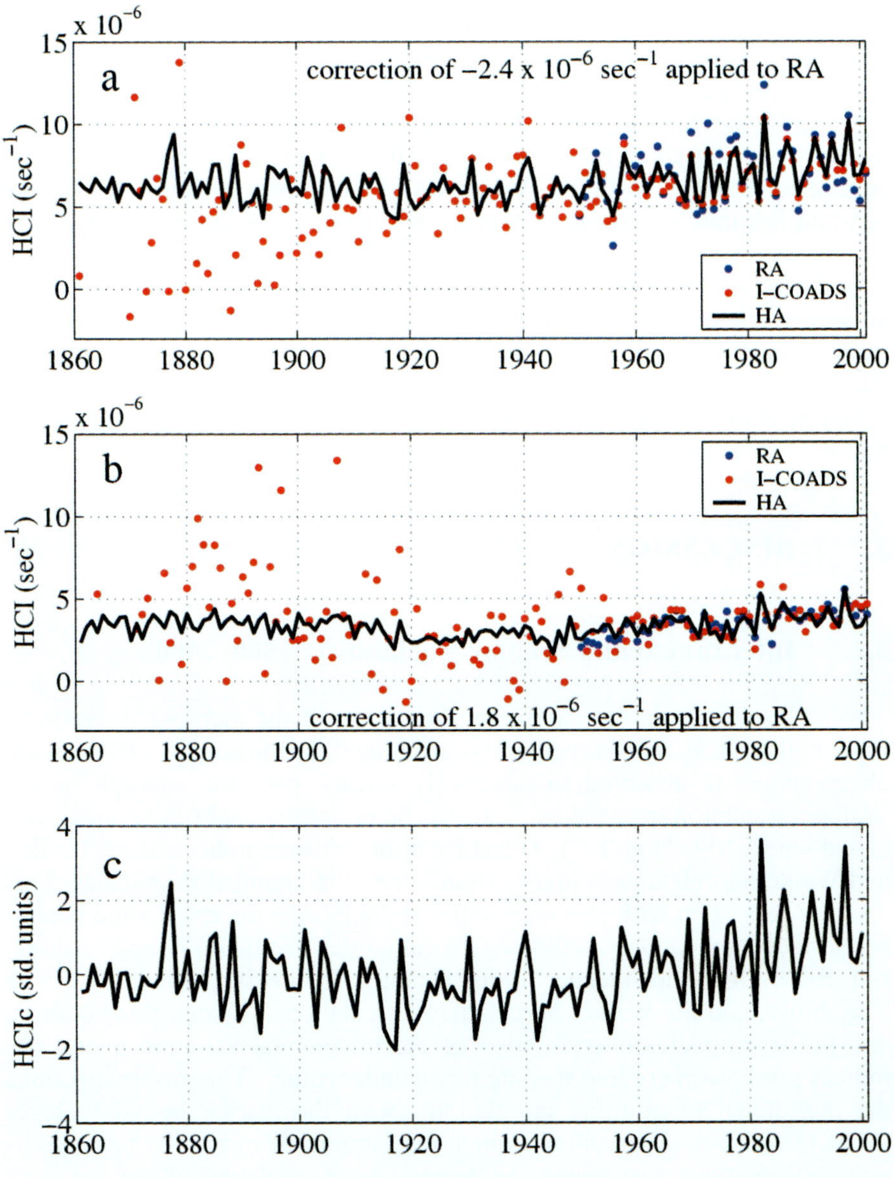

Figure8-6. **a:** Time series of HA- (solid line) and RA-derived HC indices (open circles) for the DJF season. HC indices constructed from I-COADS observations are shown as unconnected filled circles. **b:** As in panel a, except for JJA averages. **c:** The combined Hadley circulation index (HCIc), composed of the standardized, variance-weighted sum of the boreal and austral indices shown in panels a and b.

Longer-term variability is also evident in both boreal and austral indices (Fig. 8-6). This variability appears not to be an artifact of the change in observational coverage over time (Figs. 8-2, 8-6). However, since we have made no bias corrections to the I-COADS data prior to objective analysis, and no corrections were made to the COADS winds prior to assimilation into the NCEP/NCAR reanalysis, we cannot determine that the HA and RA trends since the 1940s are not due to changes in the height and manner in which wind measurements were made (see discussion by Ward and Hoskins [1996] for a review of wind bias corrections.) In addition, if some portion of the lower-frequency variation over the full period is climatic, we also cannot determine from our analysis whether this reflects a change in intensity of the Pacific marine sector Hadley circulation, a change in the position of the net convergence and divergence regions, or a combination of these two effects.

3. DISCUSSION

3.1. Interannual and Secular Variability: 1860–2000

Variability in the Pacific marine sector of the meridional overturning circulation, as reconstructed here from analyzed historical surface wind observations, is observed to negatively covary with the strength of the Walker circulation over the tropical Pacific on interannual time scales associated with ENSO (Fig. 8-7). The correlation between indices of the Hadley and Walker circulations is highly significant at interannual time scales. This result is consistent with previous analyses comparing upper air wind data to sea surface temperatures in the eastern equatorial Pacific (Oort and Rasmusson 1970; Oort and Yienger 1996; Chapter 6, "ENSO, Atlantic Climate Variability, and the Walker and Hadley Circulations," Wang, this volume), and further validates interpretation of the interannual variation in the HC indices presented here into the late nineteenth century. This result also links the meridional overturning circulation above the marine boundary layer (MBL) to surface observations, which may in turn be mirrored in proxy climate observations (see below for discussion). Correlation of the low-pass series is still significant, but there are very few effective degrees of freedom in the series (Trenberth 1984), so this correlation must be interpreted with caution. However, longer-term coherence in these indices of the Pacific sector Hadley and Walker circulations may be found in similar amplitude modulation on decadal time scales (Figure 8-8). More variance is found in the high-pass filtered series for both indices in the most recent few decades and in the late nineteenth century. Although these variance estimates will be sensitive to uncertainty and to the averaging interval chosen, and are not likely to be significantly different from the mean variance over this period, this result is consistent with previous studies of ENSO for this interval using historical data (Trenberth and Shea 1987; Trenberth and Caron 2000) and coral-based proxy observations (Cole et al. 1993).

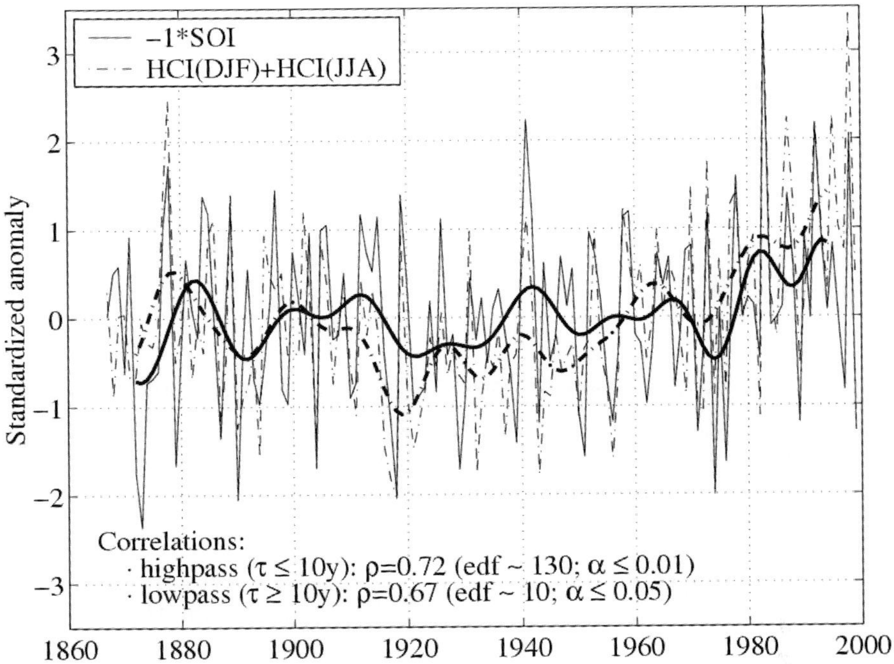

Figure 8-7. Time series plot of combined DJF + JJA HC indices (solid line) vs.–1*Southern Oscillation Index (SOI; dashed line). Thickened solid and dashed lines give respective low-pass (period (τ) ≥ 10 year) filtered data series. Correlations (ρ) and significance estimates (α) for low-pass and high-pass time series (data shown in Fig. 8-8) are indicated.

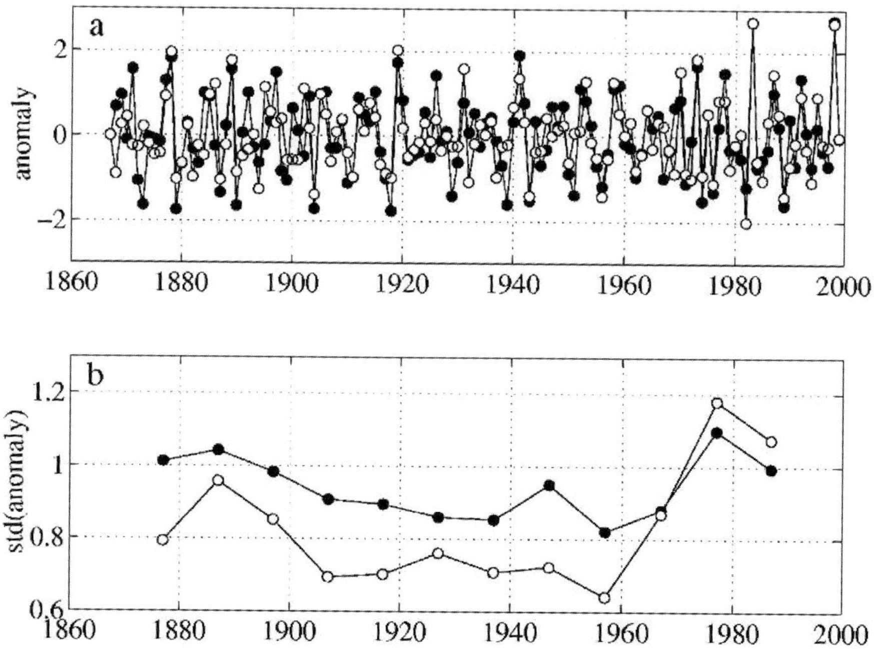

Figure 8-8. a: Time series plot of high-pass filtered combined DJF + JJA HC indices (open circles) vs. high-pass filtered–1*Southern Oscillation Index (closed circles). Correlation between series is 0.72 (Fig. 8-7). b: Estimated standard deviation of anomalies in panel a for independent 21-year windows (circles show center years).

Interpretation of the secular variability evident in the Hadley circulation indices (Fig. 8-7) must be treated with caution because assessment and correction for systematic measurement biases is difficult if not impossible (Section 2). The most likely explanation of the secular variation is in systematic measurement bias (Cardone et al. 1990; Ward and Hoskins 1996; Wu and Xie 2003). On the other hand, the tropical Pacific has shown ENSO-like variability on decadal time scales (Garreaud and Battisti 1999), so it would not be unreasonable to presume that similar interdecadal shifts in strength and/or position of the Pacific-region Hadley circulation might have also occurred (Oort and Yienger 1996). There is significant correlation with SST anomalies in the central and eastern equatorial Pacific on interannual time scales, but only a weak, nonsignificant, ENSO-like pattern on decadal time scales (Fig. 8-9).

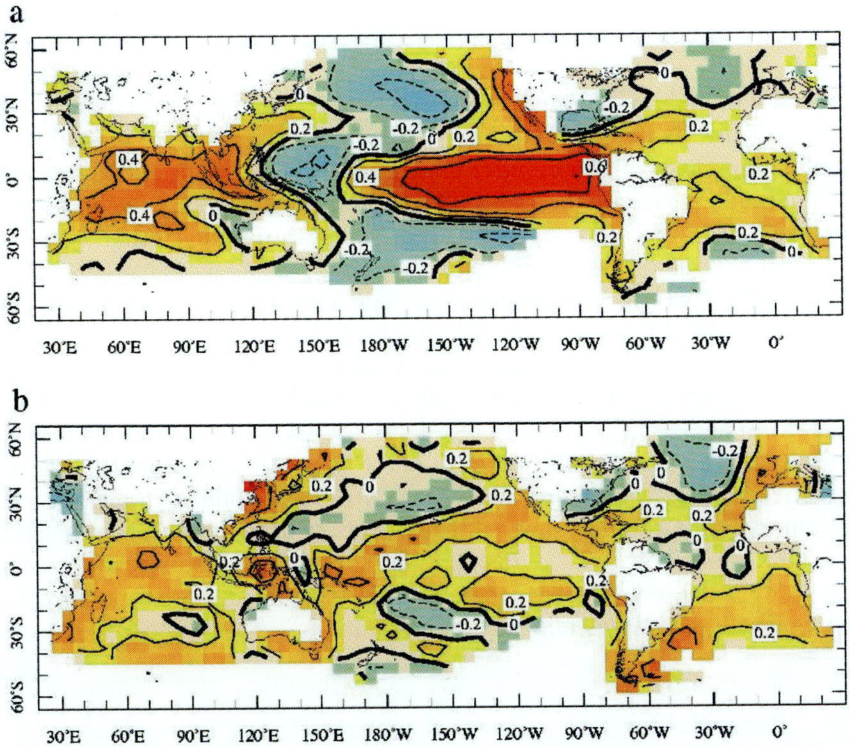

Figure 8-9. a: Pattern correlation of the high-pass ($\tau \leq 10$y) component of the combined DJF and JJA HC indices with the analyzed historical gridded SST field from Kaplan et al. (1998). Correlations of ≥ 0.2 are significant at the 95% confidence level assuming 130 degrees of freedom. b: As in panel a except for the low-pass component ($\tau \geq 10$y) of the combined DJF and JJA HC indices. Correlations of ≥ 0.6 are significant at the 95% confidence level assuming 8 effective degrees of freedom.

If some component of the trends evident in the HC indices is climatically driven, Figure 8-9 indicates that strengthening of the wintertime meridional overturning circulations has occurred over the past 50 to 80 years, which is associated with variation in the Walker circulation over the Pacific. These results appear to be consistent with model simulations reported by others in this volume. Quan et al. (Chapter 3, "Change in the Tropical Hadley Cell since 1950) has shown that the trend in a zonally averaged 850–200 hPa meridional wind index calculated from NCEP/NCAR reanalysis data is reproduced by forcing an atmospheric general circulation model (AGCM) with observed sea surface temperatures over the past 50 years. Half of the trend was attributed to an increased amplitude of ENSO activity in the most recent decades; the other half was attributed to a trend in SSTs in the tropical Indian and western Pacific Ocean. Rind and Perlwitz (Chapter 14, "The Response of the Hadley Circulation to Climate Changes, Past and Future," this volume) observed that simulated increases in the strength of the Hadley circulation in 2xCO_2 experiments were most tightly linked to increases in the tropical-subtropical SST gradient and associated precipitation increases in the tropics. Indeed, the correlation of the combined boreal and austral HC indices with the SST field shows the variability in the HC indices is associated with more vigorous ENSO activity (Figs. 8-7 through 8-9). This was also observed in a separate study of vertical structure of the NCEP/NCAR wind fields by Minobe (Chapter 5, "Interannual to Interdecadal Variations of the Hadley and Walker Circulations," this volume), who tied Hadley circulation variability on interdecadal time scales to the signature of the Pacific Decadal Oscillation (PDO) in the central tropical Pacific (also see Fig. 8-9b). Meehl et al. (Chapter 17, "Mechanisms of an Intensified Hadley Circulation in Response to Solar Forcing in the Twentieth Century," this volume) showed that a coupled ocean-atmosphere general circulation model run with realistic solar, greenhouse, aerosol, and ozone forcings over the past century produces an enhancement of the modeled intensity of both the Hadley and Walker circulations.

3.2. Potential for Paleo-Reconstructions Using Proxy Data

Further studies of long term changes in the strength and/or position of the Hadley circulation might be made using paleoclimatic proxy data. Ideally we would seek to reconstruct a measure of the zonal mean ascending and subsiding branches of the tropical atmospheric circulation from a zonally extensive network of seasonally resolved surface proxy observations. Such results might be used to further test the hypothesis that some of the change in intensity of the meridional overturning circulation is due to anthropogenic forcing, or to assess the thermodynamical and dynamical effects of changes in the seasonality of radiative forcing at various times during the Holocene.

A direct proxy-based Hadley circulation reconstruction may never be possible, because the likelihood of obtaining a dense, globally extensive network of observations is low, and the signal is relatively subtle. However, such proxies might be derived from geobiological archives influenced by related SST, precipitation, surface winds, and upwelling phenomena. The results presented here (e.g., Figs. 8-9, 8-10) suggest that proxies for the Pacific marine sector Hadley circulation, as delineated in this chapter, may be derived from the oceanographic signature of Hadley circulation variability in central and eastern tropical Pacific SST. For example, the NINO3 index calculated from reconstruction of the Pacific SST field based on statistical analysis of 65 coral proxy climate data series (Evans et al. 2001b, 2002) should reflect Hadley circulation variability, via modulation of ENSO frequency or amplitude (Evans et al. 2001a). The reconstructed NINO3 index based on the coral data is significantly correlated on interannual time scales, and shows a similar trend over the past 50–80 years (Fig. 8-10). But since the reconstruction is based on limited proxy observations and contains a time-dependent variance bias, further analysis will require additional data and intercomparison with complementary proxies (Evans et al. 2002).

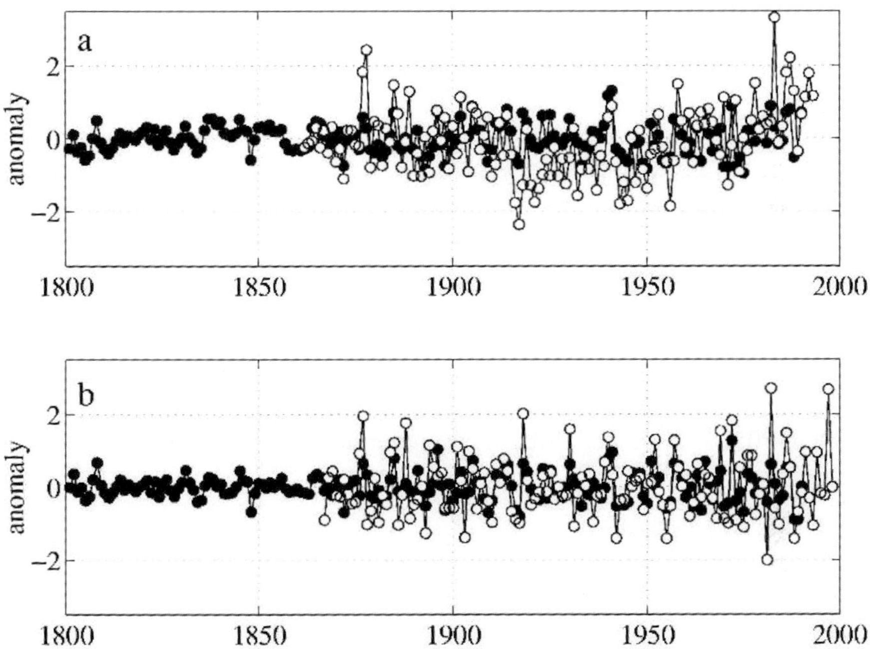

Figure 8-10. Time series intercomparison of the combined DJF + JJA HC indices (open circles) with April–March average NINO3 SST reconstructed from coral-derived proxy observations (closed circles) (Evans et al. 2001b, 2002). a: Raw series; correlation is 0.39, significant at the 95% level with 130 degrees of freedom. b: High-pass filtered series (as in Fig. 8-8); correlation is 0.59, significant at the 99% level with 130 degrees of freedom.

4. SUMMARY

We have employed a new analyzed historical surface wind product to develop proxy estimates of the boreal and austral wintertime meridional overturning circulations over the central and eastern Pacific marine sector for the past 140 years. The combined indices are negatively and significantly correlated with indices of variation in the Walker circulation—in other words, with the ENSO phenomenon—and possibly with its amplitude modulation. A trend in the HC indices is most probably due to the presence of systematic wind measurement bias. However, the secular variation over the past 50–80 years is not inconsistent with results from a number of modeling simulations described in this volume, which link an intensification of the Hadley circulation over the past 50–100 years to greenhouse and solar forcing via an enhanced tropical-subtropical SST gradient. Interdecadal shifts in Hadley circulation strength might be resolved by a network of seasonally resolved proxy observations of SST, ocean upwelling, zonal wind strength, and precipitation that describe tropical-subtropical divergence gradients and variations.

5. ACKNOWLEDGMENTS

We are grateful to H.F. Diaz for discussions and to the two reviewers whose remarks greatly improved this manuscript. This work was supported by NOAA Office of Global Programs grants NA16GP1615/6 and NAOGGP0567. LDEO contribution number 6608.

6. REFERENCES

Allan, R.J., N. Nicholls, P.D. Jones, and I. Butterworth. 1991. A further extension of the Tahiti-Darwin SOI, early SOI results and Darwin pressure. *Journal of Climate* 4: 743–749, data accessed via internet: http://www.cru.uea.ac.uk/cru/data/soi.htm.

Bjerknes, J. 1966. A possible response of the atmospheric Hadley circulation to equatorial anomalies of ocean temperature. *Tellus* 18: 820–829.

Bjerknes, J. 1969. Atmospheric teleconnections from the equatorial Pacific. *Monthly Weather Review* 97: 163–172.

Bradley, R.S. 1999. *Paleoclimatology: Reconstructing Climates of the Quaternary*, 2nd edition. New York: Academic Press.

Cane, M.A., A.C. Clement, A. Kaplan, Y. Kushnir, R. Murtugudde, D. Pozdnyakov, R. Seager, and S.E. Zebiak. 1997. 20th century sea surface temperature trends. *Science* 275: 957–960.

Cane, M.A., A. Kaplan, R.N. Miller, B. Tang, E.C. Hackert, and A.J. Busalacchi. 1996. Mapping tropical Pacific sea level: Data assimilation via a reduced state space Kalman filter. *Journal of Geophysical Research* 101: 22599–22617.

Cardone, V.J., J.G. Greenwood, and M.A. Cane. 1990. On trends in historical marine wind data. *Journal of Climate* 3: 113–127.

Cole, J.E., R.G. Fairbanks, and G.T. Shen. 1993. Recent variability in the Southern Oscillation: Isotopic results from a Tarawa Atoll coral. *Science* 260: 1790–1793.

Diaz, H.F., C. Folland, T. Manabe, D. Parker, R. Reynolds, and S. Woodruff. 2002. Workshop on Advances in the Use of Historical Marine Climate Data. *WMO Bulletin* 51: 377–380.

Dima, I., and J.M. Wallace. 2003. On the seasonality of the Hadley Cell. *Journal of Climate* 60: 1522–1527.

Evans, M.N., M.A. Cane, D.P. Schrag, A. Kaplan, B.K. Linsley, R. Villalba, and G.M. Wellington. 2001a. Support for tropically-driven Pacific decadal variability based on paleoproxy evidence. *Geophysical Research Letters* 28: 3689–3692.

Evans, M.N., A. Kaplan, and M.A. Cane. 2002. Pacific sea surface temperature field reconstruction from coral $\delta^{18}O$ data using reduced space objective analysis. *Paleoceanography* 17: 10.1029/2000PA000590.

Evans, M.N., A. Kaplan, B.K. Reichert, and M.A. Cane. 2001b. Reconstruction/deconstruction: Toward better paleoclimate estimates. In, EOS, Transactions, AGU. Vol. 82(47) Suppl. Abstract GC21A-08, accessed via online abstract database (http://www.agu.org/meetings/waisfm01.html), July 2003.

Garreaud, R.D., and D.S. Battisti. 1999. Interannual (ENSO) and interdecadal (ENSO-like) variability in the Southern Hemisphere tropospheric circulation. *Journal of Climate* 12: 2113–2122.

Kalnay, E., M. Kanamitsu, R. Kistler, W. Collins, D. Deaven, L. Gandin, M. Iredell, S. Saha, G. White, J. Woollen, Y. Zhu, M. Chelliah, W. Ebusuzaki, W. Higgins, J. Janowiak, K.C. Mo, C. Ropelewski, J. Wang, A. Leetmaa, R. Reynolds, R. Jenne, and D. Joseph. 1996. The NCEP/NCAR 40-Year Reanalysis Project. *Bulletin of the American Meteorological Society* 77: 437–471.

Kaplan, A., M.A. Cane, and Y. Kushnir. 2003. Reduced space approach to the optimal analysis interpolation of historical marine observations: Accomplishments, difficulties, and prospects. In, *Advances in the Applications of Marine Climatology: The Dynamic Part of the WMO Guide to the Applications of Marine Climatology*. Geneva, Switzerland: World Meteorological Organization WMO/TD-1081, pp. 199–216, available at: http://www.wmo.ch/web/aom/marprog/Wordpdfs/Jcomm-TR/JCOMM\%20TR13\%20Marine\%20Climatology/JCOMM_TR13.pdf.

Kaplan, A., M.A. Cane, Y. Kushnir, A.C. Clement, M.B. Blumenthal, and B. Rajagopalan. 1998. Analyses of global sea surface temperature 1856–1991. *Journal of Geophysical Research* 103(C9): 18567–18589.

Kaplan, A., M.A. Cane, Y. Kushnir, and D.L. Witter. 2004. Interannual variability of the tropical Pacific surface winds. *International Journal of Climate* (in preparation).

Kaplan, A., M.N. Evans, B.K. Reichert, and M.A. Cane. 2001. Constraints from the instrumental and paleo data on 19th century climate. In, EOS, Transactions, AGU. Vol. 82(47) Suppl. Abstract GC21A-07, accessed via online abstract database (http://www.agu.org/meetings/waisfm01.html), July 2003.

Kaplan, A., Y. Kushnir, and M.A. Cane. 2000. Analysis of historical sea level pressure 1854–1992. *Journal of Climate* 13: 2987–3002.

Kaplan, A., Y. Kushnir, M.A. Cane, and M.B. Blumenthal. 1997. Reduced space optimal analysis for historical datasets: 136 years of Atlantic sea surface temperatures. *Journal of Geophysical Research* 102: 27835–27860.

Kistler, R., E. Kalnay, W. Collins, S. Saha, G. White, J. Woollen, M. Chelliah, W. Ebusuzaki, M. Kanamitsu, V. Kousky, H. Vanden Dool, R. Jenne, and M. Fiorino. 2001.

The NCEP/NCAR 50-year Reanalysis: Monthly means CD-ROM and documentation. *Bulletin of the American Meteorological Society* 82: 247–267.

Meehl, G.A., G.J. Boer, C. Covey, M. Latif, and R.J. Stouffer. 2000. The Coupled Model Intercomparison Project (CMIP). *Bulletin of the American Meteorological Society* 81: 313–318.

Oort, A.H., and E.M. Rasmusson. 1970. On the annual variation of the monthly mean meridional circulation. *Monthly Weather Review* 98: 423–442.

Oort, A.H., and J.J. Yienger. 1996. Observed interannual variability in the Hadley Circulation and its connection to ENSO. *Journal of Climate* 9: 2751–2767.

Peixoto, J.P., and A.H. Oort. 1992. *Physics of Climate*. American Institute of Physics, New York.

Trenberth, K.E. 1984. Some effects of finite sample size and persistence on meteorological statistics. Part I: Autocorrelations. *Monthly Weather Review* 112: 2359–2379.

Trenberth, K.E., and J.M. Caron. 2000. The Southern Oscillation revisited: Sea level pressures, surface temperatures, and precipitation. *Journal of Climate* 13: 4358–4365.

Trenberth, K.E., and D.J. Shea. 1987. On the evolution of the Southern Oscillation. *Monthly Weather Review* 115: 3078–3096.

Waliser, D.E., Z. Shi, J.R. Lanzante, and A.H. Oort. 1999. The Hadley Circulation: Assessing NCEP/NCAR Reanalysis and sparse in-situ estimates. *Climate Dynamics* 15: 719–735.

Ward, M.N., and B.J. Hoskins. 1996. Near-surface wind over the Global Ocean 1949–1988. *Journal of Climate* 9: 1877–1895.

Woodruff, S.D., H.F. Diaz, J.D. Elms, and S.J. Worley. 1998. COADS Release 2 data and metadata enhancements for improvements of marine surface flux fields. *Physical. Chemistry of the Earth* 23: 517–526, accessed via Internet, May 1, 2003: http://www.cdc.noaa.gov/coads/egs_paper.html.

Woodruff, S.D., R.J. Slutz, R.L. Jenne, and P.M. Steurer. 1987. A comprehensive ocean-atmosphere data set. *Bulletin of the American Meteorological Society* 68: 521–527.

Wu, R.G., and S.-P. Xie. 2003. On equatorial Pacific surface wind changes around 1977: NCEP-NCAR reanalysis versus COADS observations. *Journal of Climate* 16: 167–173.

Chapter 9

HOLOCENE RECORDS OF RAINFALL VARIATION AND ASSOCIATED ITCZ MIGRATION FROM STALAGMITES FROM NORTHERN AND SOUTHERN OMAN

Dominik Fleitmann,[1] Stephen J. Burns,[2] Ulrich Neff,[3] Manfred Mudelsee,[4] Augusto Mangini,[3] Jan Kramers,[1] and Albert Matter[1]*
[1]Institute of Geological Sciences, University of Bern, CH -3012 Bern, Switzerland
[2]Department of Geosciences, Morrill Science Center, University of Massachusetts, Amherst, Massachusetts 01003-9297, U.S.A.
[3]Heidelberg Academy of Sciences, 69117 Heidelberg, Germany
[4]Department of Earth Sciences, Boston University, Boston, Massachusetts 02215, U.S.A.
**Present address: Department of Geological and Environmental Sciences, Stanford University, Stanford, California 94305-2115, U.S.A.*

Abstract Oxygen isotope ($\delta^{18}O$) profiles of uranium-series-dated stalagmites from Oman provide a record of Holocene Indian Ocean monsoon intensity at sub-decadal resolution over the past 10,000 years. The $\delta^{18}O$ values are a proxy for the amount of monsoon precipitation, which is controlled by the mean summer latitudinal position and convection intensity of the Intertropical Convergence Zone (ITCZ). The longest stalagmite record, derived from a stalagmite (Q5) from southern Oman, continuously covers the period from 10,300 to 2,700 BP (before present) with an average temporal resolution of 4 to 5 years. Additional contemporaneously deposited stalagmites from northern and southern Oman are used to confirm the results obtained from Q5 time series. All stalagmite records indicate that changes in monsoon precipitation between 10,300 and 8,000 BP are in phase with high-latitude temperature fluctuations recorded in Greenland ice cores during the glacial-to-interglacial transition; this result indicates that late glacial and early Holocene monsoon precipitation and tropical convection are largely controlled by glacial boundary conditions. After the final melting of Northern Hemisphere ice sheets, monsoon precipitation decreases gradually in near linear response to changing Northern Hemisphere summer solar insolation. Comparison with the Cariaco Basin precipitation record, off the Venezuelan coast, indicates that post-glacial to modern precipitation patterns in the northern tropics are controlled on a global scale by the gradual southward migration of the ITCZ due to

H.F. Diaz and R.S. Bradley (eds.), The Hadley Circulation: Present, Past and Future, 259–287.

orbitally induced changes in insolation. At the decadal and centennial scales, fluctuations in monsoon rainfall appear to be driven primarily by variations in solar irradiance, with higher solar irradiance being correlated with higher monsoon rainfall.

1. INTRODUCTION

The atmospheric circulation of the low latitudes is dominated by the meridional circulation of the Hadley cells on both sides of the equator. The Intertropical Convergence Zone (ITCZ), a belt of relatively low surface pressure, rising air, maximum cloudiness, and convergence of the trade winds, is located between both Hadley cells. The seasonal migration of the ITCZ in response to the annual solar cycle and maximum surface heating controls the start, duration, and end of the rainy season in the tropics, particularly in the African and Asian monsoon domains (Figs. 9-1a, b). Terrestrial paleoclimatic reconstructions—based almost entirely on lacustrine sequences—from both monsoon domains show that the mean latitudinal summer position of the ITCZ and the associated monsoon rainfall belt has varied considerably over the course of the Holocene, and was located much farther north of its present position between approximately 10,000 BP and 5,000–6,000 BP (e.g., Gasse and Van Campo 1994; Gasse 2000; Burns et al. 1998, 2001; Neff et al. 2001; McClure 1976; Enzel et al. 1999; deMenocal et al. 2000).

During this period, also known as the Holocene Optimum, African Humid Period, or Neolithic pluvial phase, the ITCZ and the associated monsoonal rainfall belt were located between 700 and 2,500 km north of their present position and, as a consequence, extensive lakes were formed in the now arid to hyperarid deserts (e.g., Gasse 2000; An et al. 2000). In addition, data from across northern Africa to India reveal that the early to mid-Holocene humid period was not continuously wet, but was interrupted by several century-scale drought events centered at 8,200, 6,600, and 4,000 BP (Gasse 2000). Thus, the relative position of the ITCZ may also have varied on sub-millennial time scales. However, although the number of monsoon records has greatly increased during the last two decades, significant discrepancies still exist about the onset, duration, stability, and termination of this humid period in the African and Indian monsoon domains (Fig. 9-2). In these areas most of the paleoclimate reconstructions rely on lacustrine sediments, and observed inconsistencies between individual records result mainly from chronological uncertainties, incompleteness, and/or low temporal resolution (e.g., Gasse 2000; McClure 1976; Lézine et al. 1998). For instance, because chronologies of lacustrine sediments are mainly based on radiocarbon dating of lacustrine carbonates, local hard water reservoir effects and diagenesis (Gasse and Fontes 1992) can significantly impact age determinations. In addition to chronological uncertainties, the temporal resolution of lacustrine sequences in these areas is, with only a few exceptions, generally low (>200 years). The records are in some cases also incomplete due to desiccation and subsequent erosion, particularly after the termination of the early to mid-Holocene humid period at 5,000 to 6,000 BP. Lake level depends on both the basin morphology and the balance of precipitation and evaporation, and the response of lake levels to hydrological fluctuations is often not straightforward but highly nonlinear (Fleitmann et al. 2003). Finally, in many areas continuous high-resolution Holocene climate records are largely absent, particularly on the Arabian Peninsula.

To summarize, reconstruction of ITCZ migration and insights into the mechanisms behind it in the Indian monsoon domain suffer from a paucity of precisely dated and temporally well-resolved Holocene climate records.

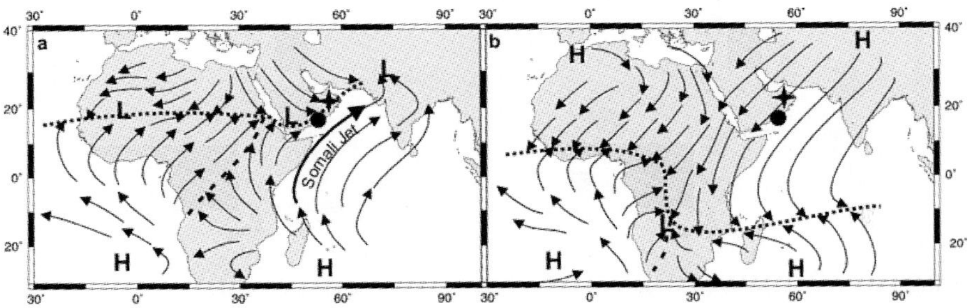

Figure 9-1. Schematic drawing of atmospheric circulation pattern during Northern Hemisphere summer (**a**, left figure) and winter (**b**, right figure). Dashed line marks the position of the Intertropical Convergence Zone. Solid black circle marks the location of Qunf Cave and Kahf Defore and the black star symbol the location of Hoti Cave.

A possible additional source of climate information is speleothems (e.g., stalagmites and flowstones) found in caves in Oman. Previously published studies on U-Th dated speleothems from Oman revealed that oxygen isotope profiles and thickness of annual growth bands can provide detailed information on monsoonal rainfall variability on annual to millennial time scales (Burns et al. 1998, 2001, 2002; Fleitmann et al. 2003, 2004). In this chapter we seek to accomplish the following three goals:

(1) To demonstrate that oxygen isotope ratios ($\delta^{18}O$) of stalagmites from northern and southern Oman are suitable climate proxies of monsoon precipitation.

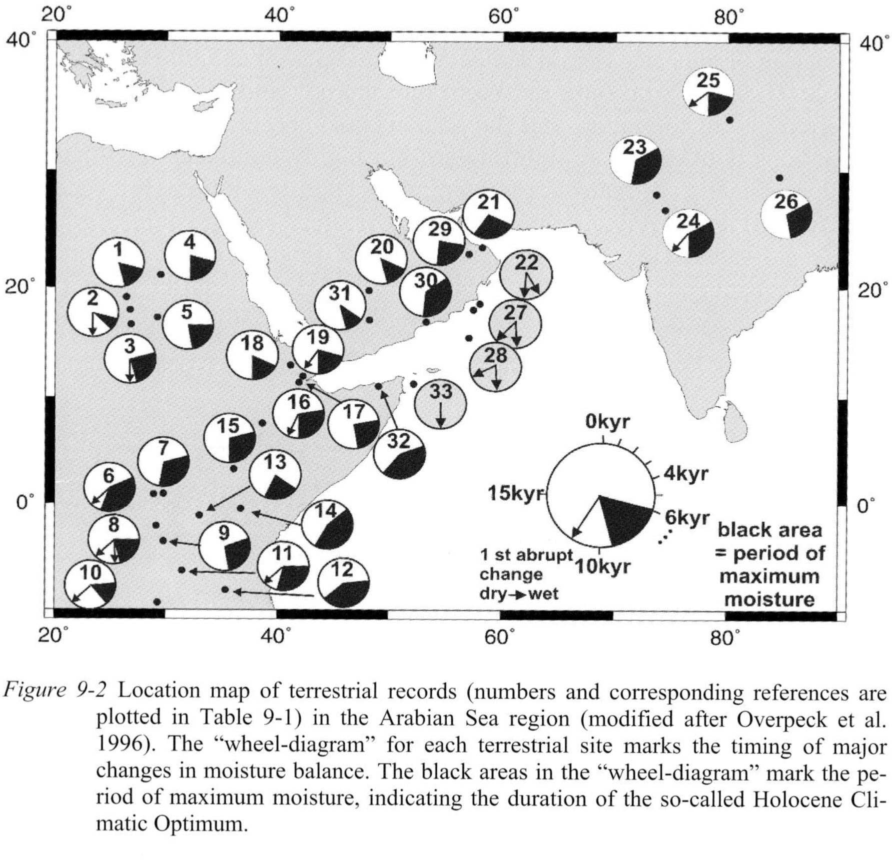

Figure 9-2 Location map of terrestrial records (numbers and corresponding references are plotted in Table 9-1) in the Arabian Sea region (modified after Overpeck et al. 1996). The "wheel-diagram" for each terrestrial site marks the timing of major changes in moisture balance. The black areas in the "wheel-diagram" mark the period of maximum moisture, indicating the duration of the so-called Holocene Climatic Optimum.

(2) To evaluate whether the records from southern Oman rather reflect regional- or broader-scale climate variability, and whether and to what extent they may reflect changes in the mean summer position of the ITCZ over the Arabian Peninsula.

Table 9-1. References for each site shown in Figure 9-2.

Site	Site or core name	Reference(s)
1	Oyo	Ritchie et al. 1985; Ritchie 1994
2	Bir Atrun (El Atrun Oasis)	Ritchie 1987; Ritchie and Haynes 1987
3	Lake Sidigh complex	Pachur and Hoelzmann 1991
4	Selima Oasis	Ritchie and Haynes 1987
5	Wadi Hoar	Pachur and Kröpelin 1987; Kröpelin and Soulie-Marshe 1991
6	Muchoya Swamp	Taylor 1990
7	Ahakagyezi Swamp	Taylor 1990
8	Lake Kivu	Kaberyan and Hecky 1987
9	Kuruyange	Bonnefille et al. 1991
10	Lake Cheshi	Stager 1988
11	Lake Tanganyika	Gasse et al. 1989; Haberyan and Hecky 1987; Vincens 1989, 1991
12	Lake Rukwa	Haberyan 1978
13	Lake Viktoria	Adamson et al. 1980
14	Mt. Satima mire	Street-Perrott and Perrott 1990
15	Lake Turkana	Owen et al. 1982
16	Lake Ziway-Shala complex	Gasse and Street 1978; Street-Perrott and Perrott 1990
17	Lake Abhé	Gasse and Street 1978; Gillespie et al. 1983
18	Lake Afrera	Gasse and Street 1978; Gillespie et al. 1983
19	Lake Asal	Gasse and Street 1978; Gillespie et al. 1983
20	Rub`al Khali	McClure 1976
21	Nizwa complex	Clark and Fontes 1990

22	Marine core RC27-23, 28, 14	Overpeck et al. 1996
23	Lunkaransar	Bryson and Swain 1981; Swain et al. 1983; Enzel et al. 1999
24	Didwana Lake	Bryson and Swain 1981; Swain et al. 1983; Singh et al. 1990
25	Sumix Co	Gasse et al. 1991; Van Campo and Gasse 1993
26	Southern Tibet lakes	Fang 1991
27	Marine core 723A	Gupta et al. 2003
28	Marine core 74KL	Sirocko et al. 1993
29	Stalagmites, northern Oman	Neff et al. 2001; Burns et al. 1998, 2001
30	Stalagmites, southern Oman	Fleitmann et al. 2003; unpublished data
31	Ramlat as-Sab'atayn	Lézine et al. 1998
32	Galweda and Hayla Cave	Brook et al. 1990
33	Marine core 905	Jung et al. 2002

(3) To compare Holocene stalagmite records from northern and southern Oman with marine and terrestrial paleoclimate records from the Indian Ocean monsoon (IOM) domain and other tropical and extratropical regions, trying to draw a more comprehensive view of the causes of ITCZ and associated monsoon rainfall belt migration over the course of the Holocene.

2. SITE LOCATIONS AND CLIMATIC SETTING

The stalagmites used for paleoclimate reconstructions were collected from three caves: Kahf Defore (17°07'N, 54°05'E; ~150 masl) and Qunf Cave (17°10'N, 54°18'E; 650 meters above sea level [masl]), in the Dhofar region in southern Oman, and Hoti Cave (23°05'N, 57°21'E; ~800 masl) from northern Oman (Figs. 9-1a, b). The present climate in the region surrounding the Arabian Sea is influenced by the seasonal latitudinal migration of the ITCZ and the associated monsoon rainfall belt, respectively. During Northern Hemisphere summer, differential heating of the Asian landmass and the southern Indian Ocean results in the establishment of a low atmospheric pressure cell (as a part of the global ITCZ) over the Tibetan Plateau and a high atmospheric pressure cell over the Indian Ocean located at about 30°S (Fig. 9-1a) (Hastenrath and Lamb 1979). The resultant pressure gradient generates a strong low-level cross equatorial airflow, which becomes a strong southwesterly flow, also known as the Somali, East African, low-level or Findlater Jet (Fig. 9-1a). The southwesterly winds transport large quantities of moisture from the southern Indian Ocean, which is then released as monsoon precipitation over some parts of the Arabian Peninsula and the Indian subcontinent. The release of latent heat from condensing moisture over the Indian subcontinent and the slopes of the Himalayas maintains and further strengthens monsoon circulation during the summer months. Therefore, both sensible and latent heating contribute to the land-sea temperature and pressure difference that drives the Indian summer monsoon. During Northern Hemisphere winter, the land-sea pressure gradient reverses and the dry northeast or winter monsoon prevails (Fig. 9-1b). At this time the ITCZ is located over the southern Indian Ocean.

In southern Oman, the most important source of precipitation is the Indian summer monsoon, locally known as the "Kahreef" season, which lasts from June to September. At this time, moist air masses, transported by the Somali Jet, condense over the cold wind-driven upwelling zones to form banks of fog and clouds, which then hit the south side of the Dhofar Mountains in southern Oman. Monsoon precipitation commonly occurs as fine drizzle, seldom exceeding more than 5 mm d^{-1}. Mean annual precipitation increases with altitude, varying between 60 and 150 mm y^{-1} in Salalah (20 masl; 1942–98) and 400–500 mm y^{-1} in Tawi Atayr (650 masl; 1977–89). In contrast to southern Oman, northern Oman is at present not directly affected by the summer monsoon and most of the mean annual precipitation is sourced from Mediterranean frontal systems that occur during the winter and spring months.

3. METHODS

3.1. Uranium Series Ages

Th-U measurements were in part performed on a multicollector thermal ionization mass spectrometer (Finnigan MAT 262 RPQ) and in part on a multicollector ICP mass spectrometer (nu Instruments). Details of the measurements are described by Neff et al. (2001) and Fleitmann et al.

(2003). It is important to note that uranium-series ages are absolute ages and in most cases no correction (except for detrital contamination) is needed.

3.2. Oxygen Isotopes

Samples (0.2–0.5 mg) for stable isotope analysis were drilled with a 0.5 mm diameter conical dentist drill. $\delta^{18}O$ and $\delta^{13}C$ isotopic composition were measured by using a VG Isocarb system attached to a VG Prism II isotope mass spectrometer. The phosphoric acid extraction was made at 90 C. All calcite values are reported relative to the Vienna Pee Dee Belemnite Standard (VPDB); the reproducibility of standard materials is ± 0.08‰ (VPDB).

3.3. Annual Growth Layers

The thicknesses of annual bands were measured from high-resolution (1200 x 1600 pixels) digital images, which were taken with a high-resolution CCD camera from polished sections and thick sections (0.5 mm thick). Multiple counts indicate an error of approximately 1% to 1.5% of the absolute age.

4. STALAGMITE OXYGEN ISOTOPE RATIOS AND THEIR PALEOCLIMATIC SIGNIFICANCE

Speleothem-based paleoclimate reconstructions commonly rely on oxygen isotope ratios, which reflect variations in cave air temperatures (e.g., Lauritzen and Lundgren 1999) and variations in $\delta^{18}O$ of seepage water forming the speleothem (e.g., Bar-Matthews et al. 1999). Previous studies of the active-water carbonate system in all three caves revealed that temperature-controlled variations in speleothem $\delta^{18}O$ are quite small in comparison to variations in $\delta^{18}O$ of the seepage water and surface precipitation, respectively (Fleitmann et al. 1999, 2004). In tropical and subtropical regions, such as in Oman, $\delta^{18}O$ values and the amount of monsoon precipitation are often inversely correlated, with higher monsoon precipitation being associated with more negative $\delta^{18}O$ values. This inverse correlation has been termed an "amount effect" and is typical for precipitation in many parts of the tropics (Dansgaard 1964; Rozanski et al. 1992), especially in those areas affected by the monsoon.

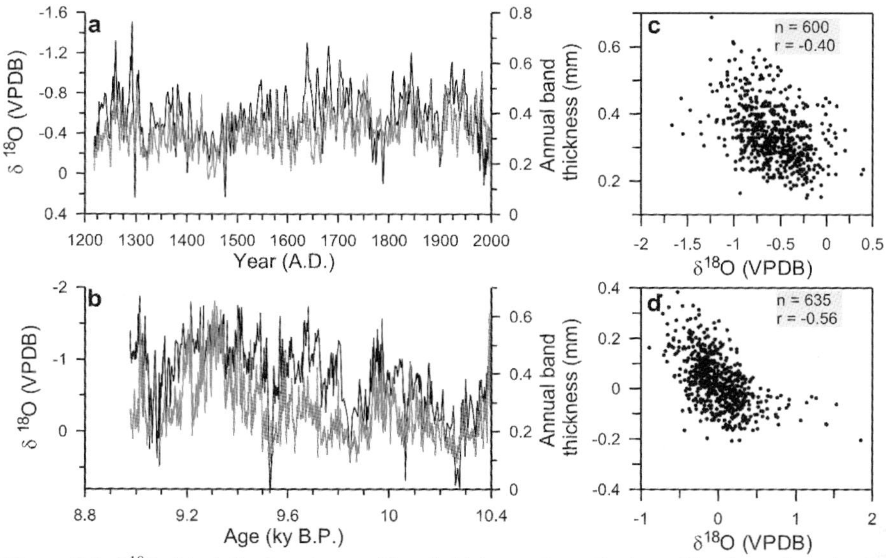

Figure 9-3. $\delta^{18}O$ (black line) and annual band thickness (gray line) profiles for stalagmite S3 (**a**) and S4 (**b**). Correlation analysis for $\delta^{18}O$ and annual band thickness for stalagmite S3 (**c**) and S4 (**d**); *n* is the number of data points, *r* is Pearson's correlation coefficient. Bootstrap simulations that take serial dependence into account (Mudelsee 2003) confirm that *r* is significantly different from zero at the 99% confidence level.

On this basis, $\delta^{18}O$ values of stalagmites from Oman are interpreted to predominantly reflect changes in the amount of monsoon precipitation (Burns et al. 1998, 2001, 2002; Fleitmann et al. 1999, 2003, 2004; Neff et al. 2001). Supporting evidence for this assumption comes from the comparison between $\delta^{18}O$ and annual band thickness profiles measured on stalagmites from Kahf Defore (Burns et al. 2002; Fleitmann et al. 2004). In shallow caves, such as in Kahf Defore, the thickness of annual growth bands is commonly controlled by the drip rate, which closely corresponds to surface precipitation. A number of studies have shown that thicker annual bands indicate higher drip rates due to higher precipitation above the cave (Genty and Quinif 1996; Polyak and Asmeron 2001). The statistically significant correlation between $\delta^{18}O$ and thickness of annual bands from both actively growing stalagmites (Figs. 9-3a, b; Fleitmann et al. 2004) and an early Holocene fossil stalagmite (S4; Figs. 3c, d) supports our interpretation that fluctuations in $\delta^{18}O$ values are inversely correlated to variations in the amount of monsoon precipitation. Such a relationship holds for time scales ranging from 100,000 years (Burns et al. 1998, 2001) to, as shown here, sub-decadal and persisted throughout the Holocene.

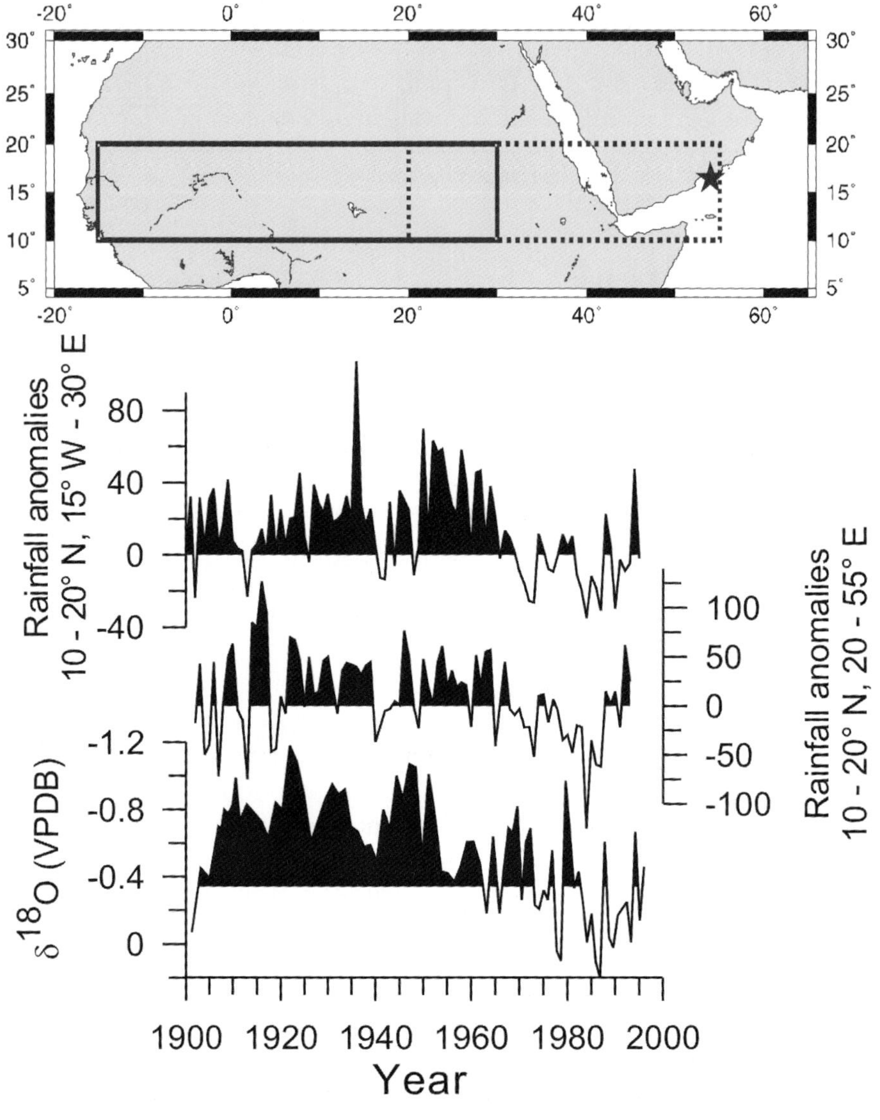

Figure 9-4 Comparison of S3 δ^{18}O (black star) with instrumental rainfall records from northern Africa (solid black line; Hulme 1996) and from East Africa and southern Arabia (dashed black line; Vose et al. 1992). Blacked-in areas indicate rainfall above the 1961–90 average.

Another important question is whether stalagmites from southern Oman reflect only local rainfall variability or are representative of much broader regional rainfall variability. To answer this question, we compared the S3 speleothem isotopic record with two gridded annual precipitation anomaly records from West Africa (Hulme 1996) and from East Africa and southern Arabia (Vose et al. 1992) (Fig. 9-4). The rainfall and speleothem records are in general agreement and show that monsoon precipitation was generally high between 1900 and ~1950 and then continuously decreased until 1985, when monsoon precipitation was at a minimum. Because of the corresponding trends in all records, we suggest that stalagmite $\delta^{18}O$ records from southern Oman likely reflect broader-scale monsoon rainfall variability.

5. HOLOCENE TIME SERIES

Stable oxygen isotope profiles from southern Oman, two from Qunf Cave (Q5 and Q11) and two from Kahf Defore (S3 and S4), are plotted versus time in Figure 9-5. Taken together, the $\delta^{18}O$ profiles cover almost the entire Holocene with a temporal resolution of between 1 and 5 years, except for a gap between 2,500 and 1,400 BP. The overall difference of ~0.6‰ in $\delta^{18}O$ between contemporaneously deposited stalagmites from Kahf Defore (~150 masl) and Qunf Cave (650 masl) is caused by an "altitude effect," whereby $\delta^{18}O$ in precipitation decreases with increasing altitude. In general, $\delta^{18}O$ decreases by approximately 0.1‰ per 100 m increase in elevation (Clark et al. 1987; Fleitmann et al. 2004). In this case the theoretical decrease of 0.4 per mil is very close to the observed overall difference in $\delta^{18}O$.

The longest $\delta^{18}O$ profile of stalagmite Q5 (Figs. 9-6 a, b) has an average temporal resolution of 5 years and covers the time intervals from 10,300 to 2,700 and from 1,400 to 400 BP (the data are presented on the ^{14}C absolute age scale where "present" is defined as AD 1950). The Q5 profile has three main features (Fig. 9-5). First, a rapid increase in precipitation between 10,300 and 9,800 BP is indicated by a sharp decrease in $\delta^{18}O$ from –0.8‰ (a value almost identical to values of actively growing stalagmites in Qunf Cave) to ~–2‰, a shift that is also recorded in the overlapping stalagmite S4 $\delta^{18}O$ record. Second, an interval of generally high monsoon precipitation lasted from 9,800 to 5,500 BP with $\delta^{18}O$ values averaging –2‰.

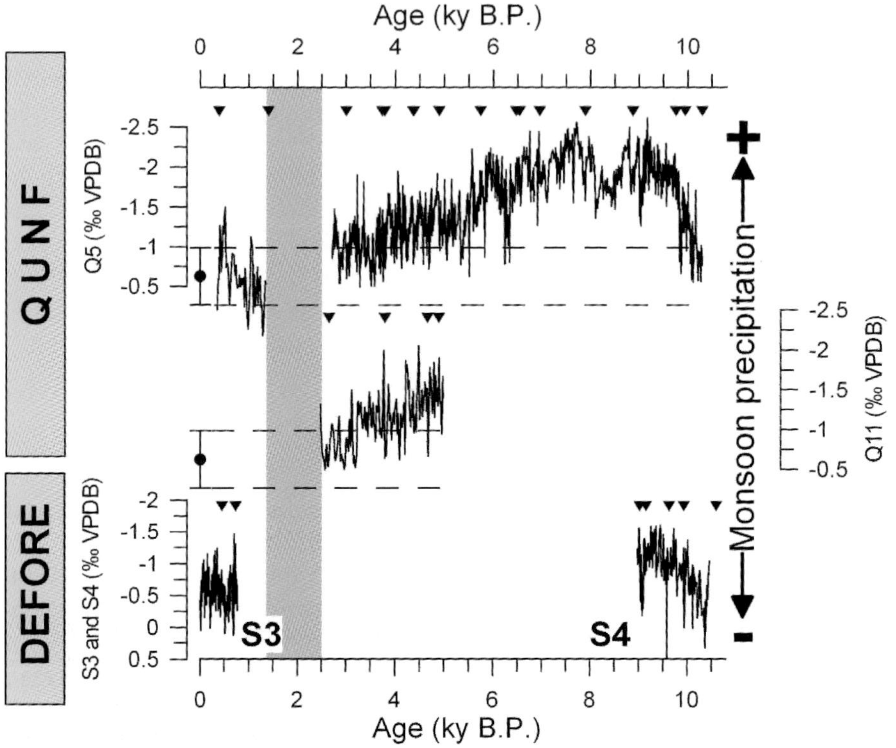

Figure 9-5. $\delta^{18}O$ profiles of stalagmites from Qunf Cave and Kahf Defore. Black triangles mark U/Th-ages and black dots with vertical error bars and gray shaded area shows the $\delta^{18}O$ range of modern stalagmites in Qunf Cave.

Third, a long-term gradual decrease in monsoon precipitation starting at about 7,000 BP is indicated by a slow shift in $\delta^{18}O$ from –2.2‰ at 7,000 BP to ~–0.9‰ (slightly more negative than $\delta^{18}O$ values of modern stalagmites) at 2,700 BP. Stalagmite Q11, deposited between 5,000 and 2,500 BP, also shows a gradual decrease in $\delta^{18}O$ over this time period and confirms the results from the Q5 record. The $\delta^{18}O$ values of the second growth phase of Q5 (after the short hiatus) are within the range of actively growing stalagmites in Qunf Cave. Superimposed on the main features of the Q5 record are distinct decadal and multi-decadal second-order variations in $\delta^{18}O$, including major positive excursions in $\delta^{18}O$ that mark sudden decreases in monsoon precipitation occurring at about 9,200–9,100, 8,500–8,100, and 6,300–6,200 BP.

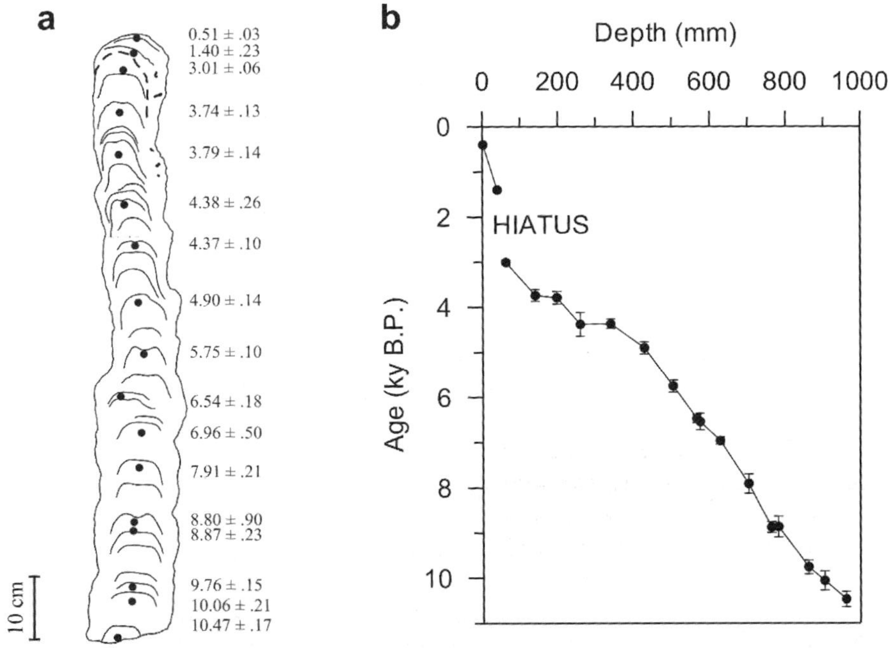

Figure 9-6. (**a**) Schematic drawing of stalagmite Q5. Black circles display the sampling location of U/Th ages. Dashed line marks hiatus; (**b**) age versus depth profile for stalagmite Q5.

6. ITCZ-CONTROLLED MONSOON PRECIPITATION IN SOUTHERN OMAN: A WORKING HYPOTHESIS

In southern Oman, strong convective cloud development is prevented by a temperature inversion, which is created by the convergence between hot, dry northwesterly winds blowing from the Arabian Desert and the relatively cool and moist low-level southwest monsoon winds (Fig. 9-7a). The height of this temperature inversion is dynamically linked to the mean latitudinal summer position of the ITCZ and to the wind pattern over southern Arabia (Sirocko et al. 1991). During the early and middle Holocene, a northward displacement of the ITCZ into the Arabian Peninsula, as indicated by lake sediments and stalagmites (McClure 1976; Lézine et al. 1998; Burns et al. 1998, 2001; Fleitmann et al. 2003) would have led to a lifting of the height of this temperature inversion, leading to stronger convective cloud development and higher monsoonal rainfall over southern Oman (Fig. 9-7b).

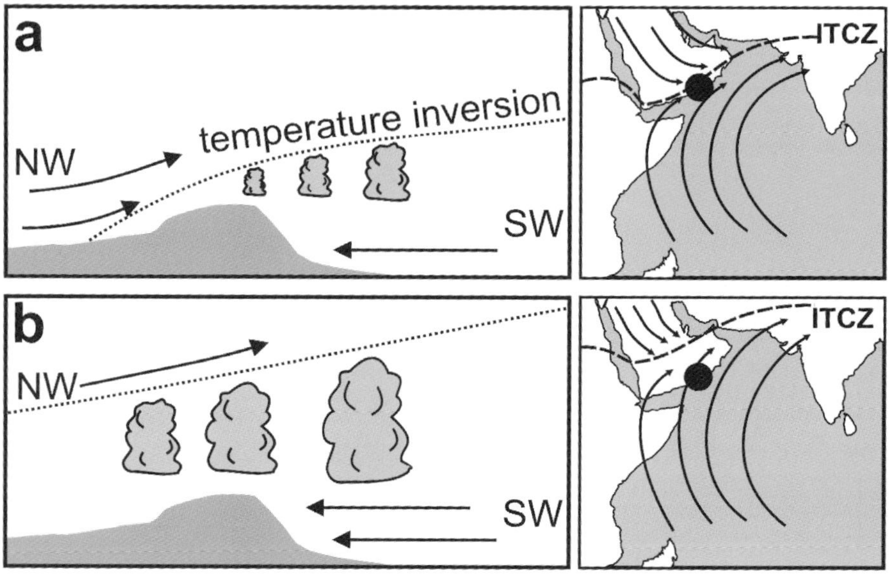

Figure 9-7. (**a**) Schematic figure of modern summer circulation pattern over southern Oman. The red star shows the location of Qunf Cave. The black dashed line shows the position of the temperature inversion and the red dashed line the location of the ITCZ. (**b**) Schematic figure of summer circulation pattern at about 7,000 BP.

However, the amount of monsoon precipitation is likely dependent not just on the mean latitudinal summer position of the ITCZ alone, but also on the overall intensity of convection of the Indian summer monsoon. As the low-level monsoon winds transport large amounts of moisture, convective cloud development may also depend on the overall moisture flux to southern Oman, and thus on the overall strength of the Indian monsoon. A weaker (stronger) Indian monsoon reduces (increases) moisture flux and intensity of convective activity and hence monsoon rainfall over southern Oman. Because of the "amount effect" (Daansgard 1964; Rozanski et al. 1992; see also Xie, Chapter 4, "The Shape of Continents, Air-Sea Interaction, and the Rising Branch of the Hadley Circulation," this volume), $\delta^{18}O$ values of precipitation become more negative with increasing precipitation. Stalagmite $\delta^{18}O$ values from southern Oman, therefore, reflect changes in the amount of monsoon precipitation, although changes in the amount of monsoonal rainfall may be controlled by the mean latitudinal summer position of the ITCZ and by the strength of the Indian monsoon and intensity of convective activity (Figs. 9-7a, b). It is, however, likely that displacement of the ITCZ and variation in monsoon intensity act in tandem, such that periods of a northward (southward) displaced ITCZ coincide with intensified (reduced) monsoon circulation and stronger (weaker) convection above southern Oman.

7. DISCUSSION

Paleoclimatic interpretations in the following discussion are based mainly on the Q5 time series (Fig. 9-5), which is currently the longest available record from Oman. Other partly overlapping stalagmite records from Oman are used to verify and refine the results obtained from the Q5 $\delta^{18}O$ record.

7.1. Early Holocene (12,000 to 8,000 BP)

In Oman, speleothem deposition starts between 10,500 and 10,100 BP. The strong decrease in $\delta^{18}O$ between 10,500 and 9,200 BP (Fig. 9-8) indicates a rapid increase in monsoon precipitation, probably due to a rapid northward displacement of the ITCZ. In northern Oman, monsoon precipitation starts and peaks approximately 400 years later than in southern Oman (Fig. 9-8). Such an abrupt increase in monsoon precipitation between 11,500 and 9,500 BP is documented in most nearby terrestrial and marine records, particularly in those located north of ~10°N (Fig. 9-2).

Although some of these records suggest a two-step increase in monsoon precipitation and wind strength occurring at ~13,000–12,500 BP and at 11,000–10,000 BP (e.g., Gasse 2000; Overpeck et al. 1996; Sirocko et al. 1993; see site numbers 19, 22, 24, 27, 28, 33 in Fig. 9-2), there is currently no speleothem evidence in Oman, neither in the southern nor in the northern part of the country, for an onset in Indian monsoon precipitation prior to ~10,500 BP In southern Oman monsoon precipitation was fully established, as estimated by RAMPFIT[1] (Mudelsee 2000), at ~9,620 ± 50 years BP (YBP; stalagmite Q5) and ~9,670 ± 50 YBP (stalagmite S4), respectively, nearly synchronous with, or slightly later than, when North Atlantic air tem-

[1] RAMPFIT (Mudelsee 2000) is an implementation of a piecewise linear regression aimed to model a climate transition. The $\delta^{18}O$ levels (before/after transition) are determined by a weighted least-squares criterion, the transition dates by a computing-intensive brute-force search. Bootstrap simulations provide 1–σ errors for estimated parameters.

perature reached its highest Holocene level at 9,740 ± 60 YBP (Fig. 9-8). As was mentioned before, in northern Oman monsoonal rainfall started at ~10,100 BP, and reached its early Holocene level at approximately 9,250 ± 50 YBP (Fig 9-8), approximately 400 years later than in southern Oman.

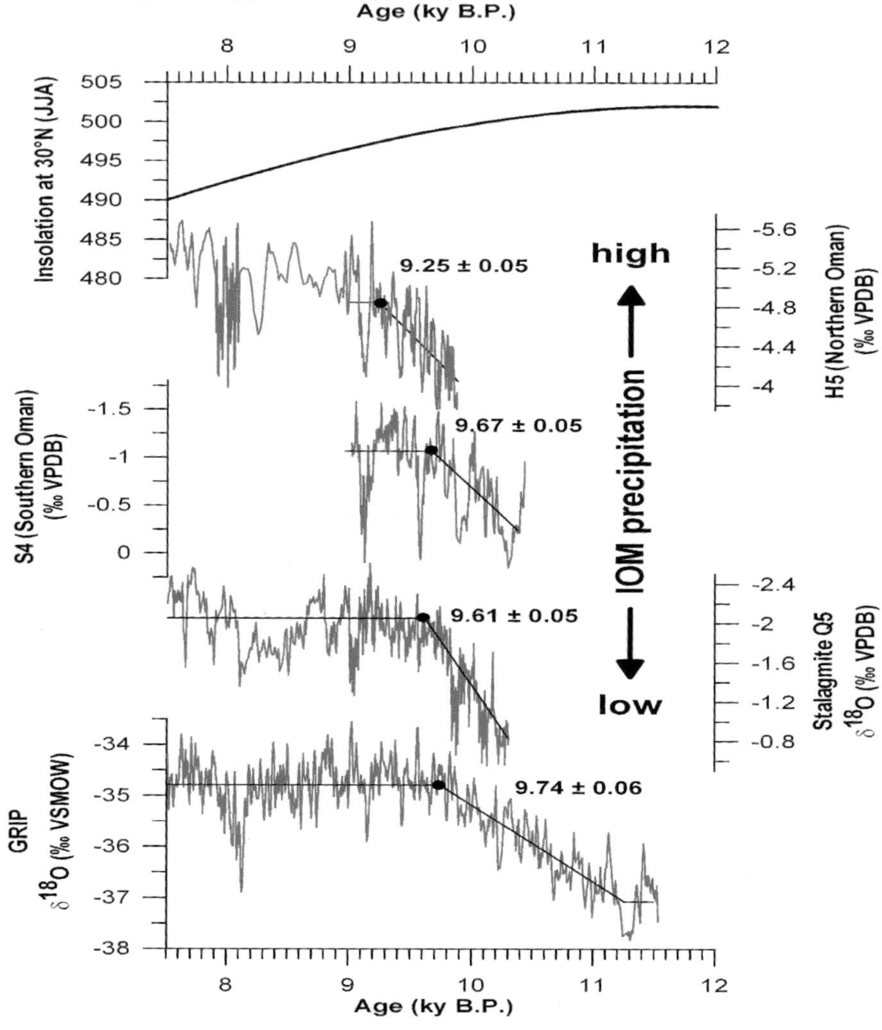

Figure 9-8. Comparison of stalagmite $\delta^{18}O$ records from northern (H5) and southern Oman (S4 and Q5) with the smoothed (5-point running average) GRIP $\delta^{18}O$ ice core record (Dansgaard et al. 1993). Lower monsoon precipitation correlates with colder North Atlantic air temperatures. The heavy black line shows "ramp" function trends (Mudelsee 2000). Change-point times are given with their statistical errors ($\pm1\sigma$), which are determined from bootstrap simulations.

As Hoti Cave (57°21'E, 23°05'N) is located approximately 600 km to the north of Qunf Cave (17°10'N, 54°18'E) and Kahf Defore (17°07'N, 54°05'E), the time lag between the onset and increase in monsoon precipitation in northern and southern Oman allows us to roughly estimate how fast the ITCZ and the associated monsoon rainfall belt were displaced northwards during the early Holocene. Figures 9-9a and b show the timing of the onset and peak in monsoon precipitation along a transect from southern to northern Oman. By assuming a northwest migration of the ITCZ over the Arabian Peninsula (Fig. 9-9a), our speleothem data suggest that the ITCZ and associated zone of maximum rainfall were displaced at an average rate of approximately 80–100 km/100 y during the early Holocene.

The onset of monsoon precipitation in Oman generally lags behind maximum boreal summer insolation (JJA), calculated for 30°N, by approximately 2,000 years and is extremely abrupt compared to the smooth increase in orbital-induced summer solar insolation (Fig. 9-8; Berger and Loutre 1991). Previously published data (e.g., Sirocko et al. 1993; Overpeck et al. 1996; Gupta et al. 2003) and the speleothem data suggest that during the early Holocene fluctuations in Indian monsoon wind strength and precipitation were closely linked to high-latitude temperature variations and boundary conditions. As the high temporal resolution and precise chronology of the stalagmite records are currently unparalleled by any other monsoon records in this area, we compared three early Holocene stalagmite records with the GRIP ice core $\delta^{18}O$ records (Fig. 9-8), which reflect fluctuations in air temperature (Daansgard et al. 1993). Taking age uncertainties of approximately 1% to 2% of the absolute age for all records into account, in southern Oman the rapid increase in monsoon precipitation from ~10,500 to 9,800 BP is in phase with continuously increasing air temperature in the northern North Atlantic (Fig. 9-8; Fleitmann et al. 2003). The comparison between the stalagmite and the GRIP $\delta^{18}O$ records further reveals that decadal-scale $\delta^{18}O$ events occur almost simultaneously in all three stalagmite records. Intervals of reduced monsoon precipitation (more positive $\delta^{18}O$ values) correlate with cooling events (more negative GRIP $\delta^{18}O$ values) in Greenland and vice versa.

The linkage between changes in North Atlantic air temperature and IOM precipitation, most apparent at ~9,100 and 8,200 BP (Fig. 9-8), suggests that both climate systems were dynamically linked during the early Holocene, probably in the following ways. First, decadal- to multi-decadal-long intervals of lower Northern Hemisphere air temperatures, perhaps triggered by a reduction in North Atlantic Deep Water (NADW) formation, caused longer and colder Eurasian winters with enhanced snow cover (e.g., Sirocko et al. 1993; Overpeck et al. 1996; Gupta et al. 2003). A greater extent of snow cover promoted a reduction in surface heating of the Tibetan Plateau during the following summer, leading to a decreased land-sea pressure gradient and weaker summer monsoon circulation (e.g., Barnett et al. 1988). A reduction in Indian monsoon wind strength and a weakening of the Somali low-level jet, respectively, would also reduce the influx of moisture from the southern Indian Ocean and the Arabian Sea and thereby weaken the intensity of convective activity along the ITCZ. Furthermore, lower influx of moisture and weaker convective activity also reduced the release of latent heat, and thereby additionally weakened IOM monsoon intensity. Second, a reduction and/or delay in surface heating of the landmasses during the summer may also have affected the latitudinal position of the ITCZ, leading to a more southerly position of the summer ITCZ.

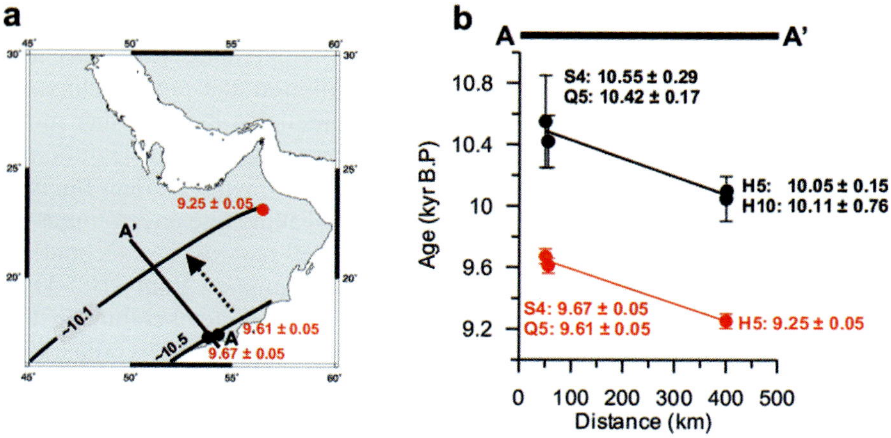

Figure 9-9. (**a**) Map showing the estimated location of the ITCZ and the associated monsoon rainfall belt. Black dot marks the location of Qunf Cave and Kahf Defore and the red dot marks the location of Hoti Cave; (**b**) timing of the onset (black dots with error bars) and peak in early Holocene monsoon precipitation (red dots with error bars) along a transect (A-A', Fig. 9-9a). To improve clarity, the age error bar (±760 y) of stalagmite H10 is not shown.

To summarize, during the early Holocene all mechanisms mentioned above would have reduced precipitation in Oman. Although summer insolation in the Northern Hemisphere was near its peak at this time, decadal- and centennial-long periods of colder winter temperatures during the early Holocene and enhanced snow cover extent over the Eurasian continent, triggered by large meltwater pulses in the North Atlantic (e.g., 8,200 BP cold event; Alley et al. 1997) and perturbations of the North Atlantic thermohaline circulation, may have directly affected the IOM as well as the convection intensity and the mean summer location of the ITCZ over the Arabian Peninsula.

7.2. Middle to Late Holocene (8,000 BP to present)

After approximately 8,000 BP, when the Northern Hemisphere ice sheets were almost entirely decayed and the North Atlantic thermohaline circulation was stabilized, monsoon precipitation in southern Oman reached its Holocene maximum at ~7,800 BP and then decreased gradually until today, as is indicated by the apparent long-term trend in $\delta^{18}O$ towards modern values (Fig. 9-5). This gradual long-term decrease in Indian monsoon precipitation parallels decreasing summer insolation (Berger and Loutre 1991) (Fig. 9-10b), which, with the exception of glacial boundary forcing, is the most important forcing of the Indian monsoon on millennial time scales (Prell and Kutzbach 1987). A long-term reduction in summer insolation reduced surface heating of the Tibetan Plateau and thereby weakened the pressure gradient between the Tibetan Plateau and the southern Indian Ocean. As a result IOM wind strength decreased and the mean summer ITCZ position over Arabia migrated continuously to the south. Furthermore, a long-term decrease in Indian monsoon wind strength led to a lower influx of moisture from the southern Indian Ocean (via the Somali Jet; Fig. 9-1a) and, thus, weaker convective activity within the ITCZ, which in turn led to a reduction in monsoon precipitation.

The long-term decrease in monsoon wind strength is further indicated by a long-term decrease in the abundance of *Globigerina bulloides* (a foraminifer associated with coastal upwelling of cold water during the Indian Ocean summer monsoon) in a marine core offshore of Oman (Figs. 9-10a, d; Site 723A; Gupta et al. 2003). Additional evidence for an insolation-controlled gradual southward migration of the ITCZ and the associated monsoon rainfall belt during the middle to late Holocene has also been found in other parts of the tropics, such as in China (An et al. 2000) and tropical South America (Haug et al. 2001; Figs. 9-10a, e). In the latter paleoclimate record, decreasing titanium content indicates a long-term decline in precipitation due to a steady southward displacement of the summer ITCZ (Fig. 9-10a). These lines of evidence may demonstrate the importance of precessional-driven changes in insolation (in the absence of glacial boundary forcing) on the mean latitude summer position of the ITCZ and convective activity within the ascending branch of the northern summer Hadley cell.

In Oman, however, the long-term gradual decrease in monsoon precipitation is in contradiction to almost all lake level and marine dust records from across northern Africa to India (e.g., McClure 1976; Enzel et al. 1999; Gasse 2000; deMenocal et al. 2000), which indicates that an abrupt change in precipitation occurred between 6,000 and 5,000 BP. How can we explain this apparent discrepancy between individual paleoclimate records? We think that the apparent differences in the timing and nature of the monsoon decrease during the Holocene between our speleothem records and most lake records is largely an artifact of archive type. First, lake levels in tropical and subtropical areas depend critically on the precipitation–evaporation (P–E) balance and the geomorphology of the lake basin. The response of lake level to climate fluctuations may be highly nonlinear. For instance, relatively small changes in monsoon season length and/or high-amplitude fluctuations in monsoon precipitation can induce a shift from positive to negative P–E balance, leading to a fast drop in lake level or even to desiccation of the lake and termination of the record of monsoon precipitation. As a result, lakes in climatically sensitive areas of the tropics and subtropics will dry out soon after the ITCZ and the summer monsoonal rainfall belt retreat south of their drainage basin. This is especially true of paleolake records located in northern Africa and on the Arabian Peninsula (e.g., site numbers 1 to 5 and 20 in Fig. 9-2). Lake levels may also depend on the relative proportion of summer and winter precipitation, such as in the Lunkaran record from northern India (site number 24 in Fig. 9-2; Enzel et al. 1999).

Another factor that may contribute to differences between our speleothem records and some African lake records is differences in the atmosphere dynamics of the African and Indian monsoon caused by the large extent of the African landmass. For the African monsoon, moisture originates solely from the eastern Atlantic and Gulf of Guinea and is transported over long distances. Therefore, land surface conditions (e.g., soil moisture, vegetation, surface hydrology) are important controls for the penetration of moisture into the North African interior, as has been revealed by fully coupled climate model simulations (ocean-atmosphere-vegetation; e.g., Kutzbach and Liu 1997; Claussen et al. 1999) and geological data (deMenocal et al. 2000; Figs. 9-10a, f).

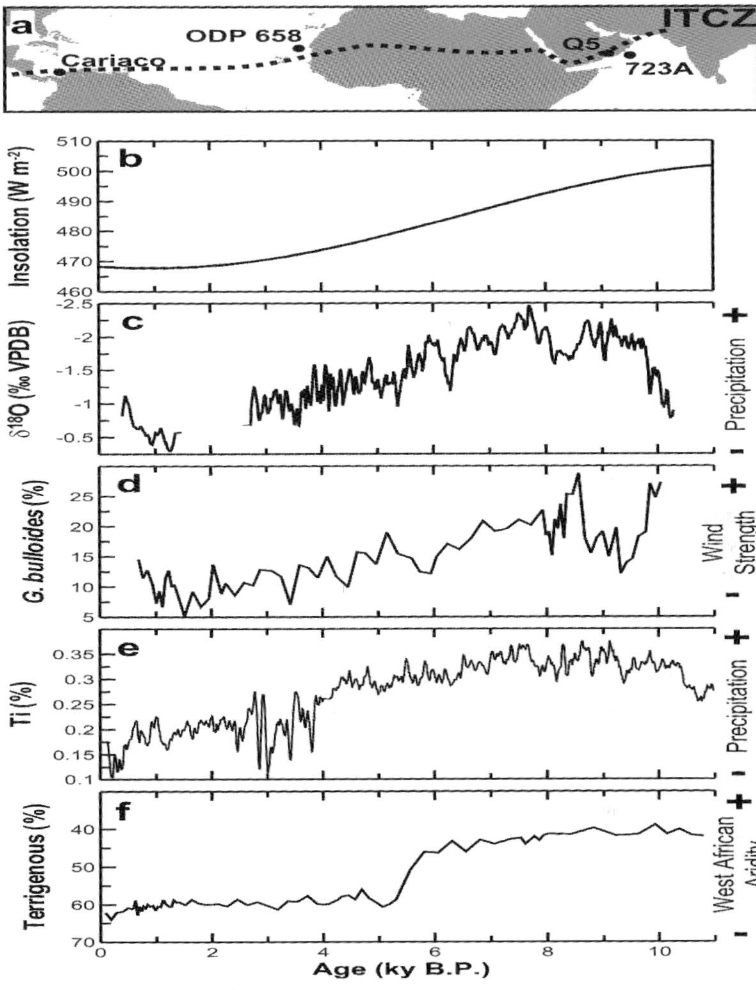

Figure 9-10. (**a**) Location map of paleoclimate records shown in Figure 9-10c to f; also shown is the mean latitudinal position of the ITCZ during modern summer. (**b**) Insolation curve at 30°N, averaged for June through August (Berger and Loutre 1991). (**c**) Smoothed (9-point running average) Q5 $\delta^{18}O$ record. (**d**) Indian monsoon upwelling record based on abundances of *Globigerina bulloides* (Gupta et al. 2003). Higher abundances of *G. bulloides* reflect more intense upwelling due to increased IOM wind strength. (**e**) Smoothed Cariaco Basin metal record (9-point running average) (Haug et al. 2001). High titanium concentrations reflect higher river discharge due to increased summer precipitation. (**f**) ODP 658 terrigenous dust record from West Africa (deMenocal et al. 1999). High terrigenous dust concentrations reflect greater aridity in West Africa.

Thus, the abrupt mid-Holocene reduction in monsoon precipitation in the northern African monsoon domain was in part the result of negative vegetation-atmosphere feedbacks due to a collapse in surface vegetation and an associated change in surface albedo. As this limited the recycling of moisture over land, the monsoon rainfall belt was displaced more rapidly to the south. In the Indian monsoon domain, moisture is transported from the Indian Ocean over much shorter distances and vegetation-atmosphere feedbacks are probably less important.

To summarize, the observed abrupt changes in monsoon precipitation that are recorded in the paleolake records in the IOM domain partly result from archive-specific and/or site-specific nonlinear responses to gradually decreasing monsoon precipitation. Additionally, the observed discrepancy between paleomonsoon records from the African and Indian monsoon domains is caused by the different atmospheric dynamics in each monsoon domain. Based on our observations from southern Oman, we suggest that the ITCZ migrated gradually and fairly continuously to the south from the beginning of the middle Holocene at approximately 8,000 BP more or less to the present.

7.3. High-Frequency Monsoon Variations

During the middle to late Holocene, high-amplitude decadal to multi-decadal variations in $\delta^{18}O$ (Fig. 9-5) are prominent in the Q5 and other stalagmite records, indicating that monsoon precipitation also varied considerably on shorter time scales. The comparison between the partly overlapping Q5 and H5 $\delta^{18}O$ profiles shows first that that decadal- to centennial-scale fluctuations in monsoon precipitation occurred almost simultaneously in northern and southern Oman and second that sample- and site-specific noise did not blur the broader-scale climatic signal (Figure 9-11). The comparison between the Q5 and H5 monsoon records and the solar irradiance $\Delta^{14}C$ record (Stuiver et al. 1998) reveals a strong sun-monsoon linkage (Fig. 9-12a), with periods of stronger (weaker) solar irradiance coinciding with periods of higher (lower) monsoon precipitation (Neff et al. 2001; Fleitmann et al. 2003). This visible correlation is supported by results of spectral analysis and cross-spectral analysis (Fig. 9-12b), which confirm the presence and correspondence of statistically significant solar cycles, centered at

205 (de Vries cycle), 132, 105, 90 (Gleissberg cycle), 60, and 55 years (Neff et al. 2001; Fleitmann et al. 2003) in both monsoon records. Additional evidence for a sun-monsoon linkage has been found in marine and lake sediments from the Arabian Sea (Agnihotri et al. 2002; Staubwasser et al. 2002) and equatorial East Africa (Verschuren et al. 2000).

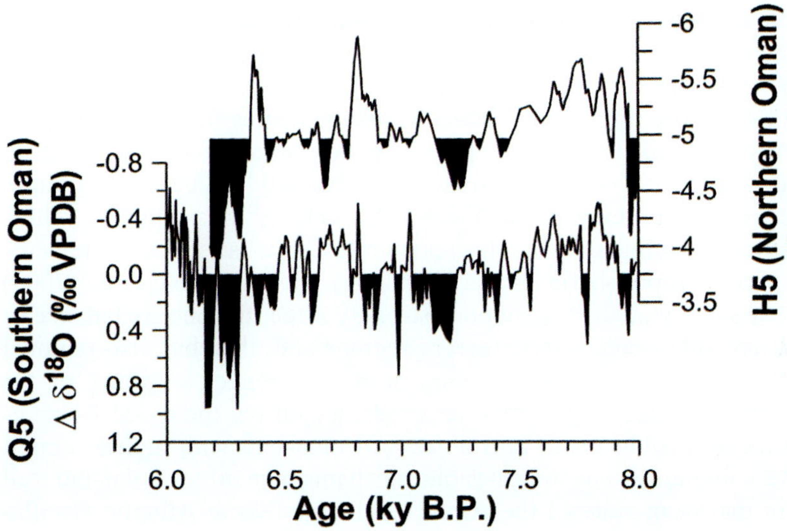

Figure 9-11. Comparison between the H5 and the detrended Q5 (see Fleitmann et al. [2003] for details) oxygen isotope profiles.

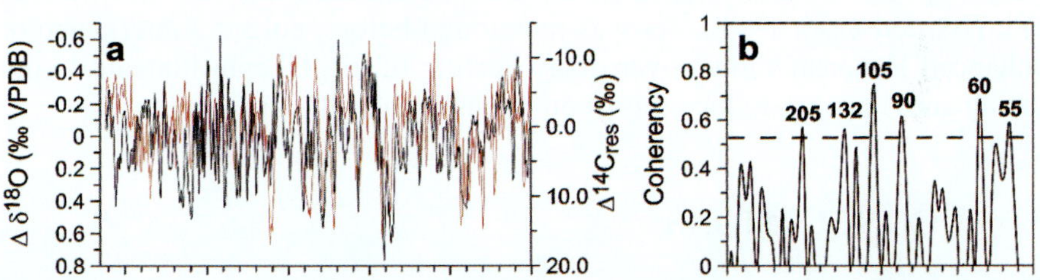

Figure 9-12. (**a**) Comparison between the smoothed (3-point running average) and detrended Q5 $\delta^{18}O$ (black line) and detrended atmospheric $\Delta^{14}C_{res}$ (red line) profiles (see Fleitmann et al. [2003] for details). (**b**) Coherency spectrum, calculated with an alignment (for details see Schulz and Stattegger [1997]) of –400 years of the $\Delta^{14}C$ time series, is compared against the 90% false-alarm-level (dashed line), resulting in significant coherent cycles at 205-, 132-, 105-, 90-, 60-, and 55-year periods, as well as 32- and 24-year periods (not shown).

Despite the increasing geological evidence for a sun-climate connection, the physical mechanism(s) behind the sun-monsoon linkage are not yet resolved and several mechanisms are conceivable. One possibility is that changes in solar irradiance directly influence monsoon intensity. An increase in solar irradiance could reinforce the meridional temperature and interhemispheric pressure gradient and, thereby, strengthen monsoon circulation (Cubasch et al. 1997). Indeed, equilibrium climate simulations reveal that centennial variations of 0.2°C to 0.5°C in surface temperatures can be explained by variations in solar irradiance (Lean et al. 1995). Supporting evidence for a more direct sun-monsoon linkage comes from recently published results of global coupled climate model simulations (Meehl et al. 2003; and Chapter 17, "Mechanisms of an Intensified Hadley Circulation in Response to Solar Forcing in the Twentieth Century," this volume), which included time-evolving solar, anthropogenic (greenhouse gas, sulfate aerosol), and ozone (tropospheric and stratospheric) forcing. Meehl et al. (2003) suggest that solar irradiance variations strongly affect the energy balance of the cloud-free subtropical high-pressure regions and, thereby, also regional precipitation regimes such as the monsoons. A stronger warming of the Eurasian landmass due to greater solar irradiance during the cloud-free premonsoon months could result in a reinforced monsoon with higher rainfall during the summer. A second possible mechanism involves solar-induced changes in the mean state of the Arctic Oscillation/North Atlantic Oscillation (AO/NAO), which lead to regionally large shifts in winter temperature (Shindell et al. 2001). Lower winter temperatures during periods of decreased solar irradiance, such as during the Maunder Minimum, could possibly impact monsoon intensity via the well-known monsoon–snow cover link (Barnett et al. 1988). As was mentioned before, colder winters, and/or enhanced Eurasian snow cover delay heating of the Tibetan Plateau during spring and, thereby, reduce monsoon intensity the following summer.

8. CONCLUSIONS

Speleothems, specifically stalagmites, from both northern and southern Oman are sensitive recorders of changes in the amount of Indian monsoon precipitation in the region. The $\delta^{18}O$ values of stalagmite calcite are inversely correlated to precipitation amount, as was demonstrated by comparisons of $\delta^{18}O$ profiles and rainfall records and also by comparisons with the thickness of annual growth layers where they are present.

During the final transition from the last glacial period to the Holocene, growth of large stalagmites began in both northern and southern Oman as a response to rapid northward migration of the ITCZ. Stalagmite growth in Qunf cave, in southern Oman, began ~400 years earlier than in Hoti Cave, 600 km to the north, suggesting an average rate of ITCZ migration of approximately 80–100 km/100 y. Stalagmite $\delta^{18}O$ values decrease quite rapidly during the first few hundred years of growth, indicating a rapid increase in monsoon precipitation. The sudden strengthening of the Indian monsoon at the beginning of the Holocene is a response to the rapid increase in temperature and decrease in ice volume in the high latitudes of the Northern Hemisphere as is demonstrated by the close correspondence between stalagmite and ice core $\delta^{18}O$ curves.

Over the course of the Holocene (a period of relatively stable climate boundary conditions) stalagmite $\delta^{18}O$ values record a steady, gradual decrease in monsoon precipitation. This decrease is interpreted as resulting from a gradual southward migration in the mean location of the ITCZ during the Northern Hemisphere summer in response to decreasing summer solar insolation. The abrupt decreases in monsoon precipitation during the mid-Holocene often seen in other records of the African and Indian monsoon are not observed. Higher-frequency variations in monsoon precipitation are superimposed on the overall Holocene trend. Most of this decadal- to century-scale variation appears to be related to variations in solar irradiance.

9. REFERENCES

Adamson, D.A., F.Gasse, F.A. Street, and M.A.J. Williams. 1980. Late Quaternary history of the Nile. *Nature* 287: 50–55.

Agnihotri, R., K. Dutta, R. Bhushan, and B.L.K. Somayajulu. 2002. Evidence for solar forcing on the Indian monsoon during the last millennium. *Earth and Planetary Science Letters* 198: 521–527.

Alley, R.B., P.A. Mayewski, T. Sowers, M. Stuiver, K.C. Taylor, and P.U. Clark. 1997. Holocene climatic instability: A prominent, widespread event 8200 yr ago. *Geology* 25: 483–486.

An, Z.S., S.C. Porter, J.E. Kutzbach, X.H. Wu, S.M. Wang, X.D. Liu, X.Q. Li, and W.J. Zhou. 2000. Asynchronous Holocene optimum of the East Asian monsoon. *Quaternary Science Reviews* 19: 743–762.

Bar-Matthews, M., A. Ayalon, A. Kaufman, and G.J. Wasserburg. 1999. The Eastern Mediterranean paleoclimate as a reflection of regional events: Soreq cave, Israel. *Earth and Planetary Science Letters* 166: 85–95.

Barnett, T.P., L. Dümenil, U. Schlese, and E. Roeckner. 1988. The effect of Eurasian snow cover on global climate. *Science* 239: 504–507.

Berger, A., and M.F. Loutre. 1991. Insolation values for the climate of the last 10 million years. *Quaternary Science Reviews* 10: 297–317.

Bonnefille, R., G. Riollet, and G. Buchet. 1991. Nouvelle Sequence pollinique d'une tourbiere de la crete Zaire-Nil (Burundi). *Reviews of Palaeobotany and Palynology* 67: 315–330.

Brook, G.A., D.A. Burney, and J.B. Cowart. 1990. Desert paleoenvironmental data from cave speleothems with examples from the Chihuahuan, Somali-Chalbi, and Kalahari desert. *Palaeogeography, Palaeoclimatology, Palaeoecology* 76: 311–329.

Bryson, R.A., and A.M. Swain. 1981. Holocene variations of monsoon rainfall in Rajasthan. *Quaternary Research* 16: 135–145.

Burns, S.J., D. Fleitmann, A. Matter, U. Neff, and A. Mangini. 2001. Speleothem evidence from Oman for continental pluvial events during interglacial periods. *Geology* 29: 623–626.

Burns, S.J., D. Fleitmann, M. Mudelsee, U. Neff, A. Matter, and A. Mangini. 2002. A 780 yr annually-resolved record of Indian Ocean Monsoon precipitation from a speleothem from south Oman, *Journal of Geophysical Research: Atmospheres*, manuscript no. 2001JD001281

Burns, S.J., A. Matter, F. Norbert, and A. Mangini. 1998. Speleothem-based paleoclimate record from northern Oman. *Geology* 26: 499–502.

Clark, I.D., and J. Fontes. 1990. Paleoclimatic reconstruction in northern Oman based on carbonates from hyperalkaline groundwaters. *Quaternary Research* 33: 320–336.

Clark, I.D., P. Fritz, O.P. Quinn, P.W. Rippon, H, Nash, and B. bin Ghalib al Said. 1987. Modern and fossil groundwater in an arid environment: A look at the hydrogeology of southern Oman. In, Use of Stable Isotopes in Water Resources Development. IAEA Symposium, 299, March 1987, Vienna. pp. 167–187.

Claussen, M., C. Kubatzki, V. Brovkin, A. Ganopolski, P. Hoelzmann, and H.-J. Pachur. 1999. Simulation of an abrupt change in Saharan vegetation in the mid-Holocene. *Geophysical Research Letters* 26: 2037–2040.

Cubasch, U., R. Voss, G.C. Hegerl, J. Waskewitz, and T.J. Crowley. 1997. Simulation of the influence of solar radiations variations on the global climate with an ocean-atmosphere general circulation model. *Climate Dynamics* 13: 757–767.

Dansgaard, W. 1964. Stable isotopes in precipitation. *Tellus* 16: 436–468.

Dansgaard, W., S.J. Johnsen, H.B. Clausen, D. Dahl-Jensen, N.S. Gundestrup, C.U. Hammer, C.S. Hvidberg, J.P. Steffensen, A.E. Sveinbjörnsdottir, J. Jouzel, and G. Bond. 1993. Evidence for general instability of past climate from a 250-kyr ice-core record. *Nature* 364: 218–220.

deMenocal, P.B., T. Ortitz, T. Guilderson, J. Adkins, M. Sarnthein, L. Baker, and M. Yarusinsky. 2000. Abrupt onset and termination of the African Humid Period: Rapid climate responses to gradual insolation forcing. *Quaternary Science Reviews* 19: 347–363.

Enzel, Y., L.L. Ely, S. Mishra, R. Ramesh, R. Amit, B. Lazar, S.N. Rajaguru, V.R. Baker, and R. Sandler. 1999. High-resolution Holocene environmental changes in the Thar desert, Northwestern India. *Science* 284: 125–128.

Fang, J. 1991. Lake evolution during the past 30 000 years in China, and its implications for environmental change. *Quaternary Research* 36: 37–60.

Fleitmann, D., S.J. Burns, and A. Matter. 1999. Stable isotope study of the active water-carbonate system in caves in Oman: A test of applicability to paleoclimate studies. *Eos Trans. AGU* 80(46), Fall Meet. Suppl., Abstract OS22B-12.

Fleitmann, D., S.J. Burns, M. Mudelsee, U. Neff, J. Kramers, A. Mangini, , and A. Matter. 2003. Holocene forcing of the Indian monsoon recorded in a stalagmite from Southern Oman. *Science* 300: 1737–1739.

Fleitmann, D., S.J. Burns, U. Neff, M.Mudelsee, A. Mangini, and A. Matter. 2004. Paleoclimatic interpretation of high-resolution oxygen isotope profiles derived from annually laminated speleothems from Southern Oman. *Quaternary Science Reviews* 23, 935–945.

Gasse, F. 2000. Hydrological changes in the African tropics since the Last Glacial Maximum. *Quaternary Science Reviews* 19: 189–211.

Gasse, F., and J.Ch. Fontes. 1992. Climatic changes in Northwest Africa during the last deglaciation (16–7 ka BP). In, Bard, E., and W.S. Broecker (eds.). *The Last Deglaciation: Absolute and Radiocarbon Chronologies*. NATO ASI series, Berlin: Springer Verlag, pp. 295–325.

Gasse, F., and F.A. Street. 1978. Late Quaternary lake-level fluctuations and environments of the northern Rift Valley and Afar region (Ethiopia and Djibouti). *Palaeogeography, Palaeoclimatology, Palaeoecology* 24: 279–325.

Gasse, F., and E. Van Campo. 1994. Abrupt post glacial climate events in West Asia and North Africa monsoon domains. *Earth and Planetary Science Letters* 126: 435–456.

Gasse, F., M. Arnold, J.C. Fontes, M. Fort, E. Gibert, A. Huc, B.Y. Li, Y.F. Li, Q. Liu, F. Melieres, E. Van Campo, F.B. Wang, and Q.S. Zhang. 1991. A 13,000-year climate record from western Tibet. *Nature* 353: 742–745.

Gasse, F., V. Ledee, M. Massault, and J. Fontes. 1989. Water-level fluctuations of Lake Tanganyika in phase with oceanic changes during the last glaciation and deglaciation. *Nature* 342(6245): 57–59.

Genty, D., and Y. Quinif. 1996. Annually laminated sequences in the internal structure of some Belgian stalagmites: Importance for paleoclimatology. *Journal of Sedimentary Research* 66: 275–288.

Gillespie, R., F.A. Street-Perrott, and R. Switsur. 1983. Post-glacial arid episodes in Ethiopia have implications for climate prediction. *Nature* 306: 680–683.

Gupta, A.K., D.M. Anderson, and J.T. Overpeck. 2003. Abrupt changes in the Asian southwest monsoon during the Holocene and their links to the North Atlantic Ocean. *Nature* 421: 354–357.

Haberyan, K.A. 1978. Fossil diatoms and the paleolimnology of Lake Rukwa, Tanzania. *Freshwater Biology* 17: 429–436.

Haberyan, K.A., and R.E. Hecky. 1987. The late Pleistocene and Holocene stratigraphy and paleolimnology of Lakes Kivu and Tanganyika. *Palaeogeography, Palaeoclimatology, Palaeoecology* 61: 169–197.

Hastenrath, S., and P.J. Lamb. 1979. *Climatic Atlas of the Indian Ocean Vol. 1, Surface Climate and Atmospheric Circulation*. Madison: University of Wisconsin Press, 116 pp.

Haug, G.H., K.A. Hughen, D.M. Sigman, L.C. Peterson, and U. Röhl. 2001. Southward migration of the Intertropical Convergence Zone through the Holocene. *Science* 293: 1304–1308.

Hulme, M. 1996. Recent climatic change in the world's drylands. *Geophysical Research Letters* 23: 61–64.

Jung, S.J.A., G.R. Davies, G. Ganssen, and D. Kroon. 2002. Decadal-centennial scale monsoon variations in the Arabian Sea during the Early Holocene. *Geochemistry Geophysics Geosystems* 3, Art. No. 1060.

Kropelin, S., and I. Soulie-Marsche. 1991. Charophyte remains from Wadi Hower as evidence for deep Mid-Holocene freshwater lakes in the Eastern Sahara of Northwest Sudan. *Quaternary Research* 36: 210–223.

Kutzbach, J.E., and Z. Liu. 1997. Response of the African Monsoon to orbital forcing and ocean feedbacks in the Middle Holocene. *Science* 278: 440–443.

Lauritzen, S.E., and J. Lundberg. 1999. Calibration of the speleothem delta function: An absolute temperature record for the Holocene in northern Norway. *Holocene* 9: 659–669.

Lean, J., J. Beer, and R. Bradley. 1995. Reconstruction of solar irradiance since 1610: Implications for climate change. *Geophysical Research Letters* 22: 3195–3198.

Lézine, A.M., J.F. Saliege, C. Robert, F. Wertz, and M.L. Inizan. 1998. Holocene lakes from Ramlat as-Sab'atayn (Yemen) illustrate the impact of monsoon activity in Southern Arabia. *Quaternary Research* 50: 290–299.

Meehl, G.A., W.M. Washington, T.M.L. Wigley, J.M. Arblaster, and A. Dai. 2003. Solar and greenhouse gas forcing and climate response in the twentieth century. *Journal of Climate* 16: 426–444.

McClure, H.A. 1976. Radiocarbon chronology of late Quaternary lakes in the Arabian Desert. *Nature* 263: 755–756.

Mudelsee, M. 2000. Ramp function regression: A tool for quantifying climate transitions. *Computers and Geosciences* 26: 293–307.

Mudelsee, M. 2003. Estimating Pearson's correlation coefficient with bootstrap interval from serially dependent time series. *Mathematical Geology* 35: 651–665.

Neff, U., S.J. Burns, A. Mangini, M. Mudelsee, D. Fleitmann, and A. Matter. 2001. Evidence for solar forcing of the Indian Ocean monsoon in a high-resolution speleothem record from Oman. *Nature* 411: 290–293.

Overpeck, J., D. Anderson, S. Trumbore, and W Prell. 1996. The southwest Indian Monsoon over the last 18 000 years. *Climate Dynamics* 12: 213–325.

Owen, R.B., J.W. Barthelme, R.W. Renaut, and A. Vincens. 1982. Palaeolimnology and archaeology of Holocene deposits northeast of Lake Turkana, Kenya. *Nature* 298: 523–529.

Pachur, H.J., and P. Hoelzmann. 1991. Paleoclimatic implications of late Quaternary lacustrine sediments in western Nubia, Sudan. *Quaternary Research* 36: 257–276.

Pachur, H.J., and S. Kröpelin. 1987. Wadi Howar: Paleoclimatic evidence from an extinct river system in the southeastern Sahara. *Science* 237: 298–300.

Polyak, V.J., and Y. Asmeron. 2001. Late Holocene climate and cultural changes in the Southwestern United States. *Science* 294: 148-151.

Prell, W.L., and J.E. Kutzbach. 1987. Monsoon variability over the past 150,000 years. *Journal of Geophysical Research* 92: 8411–8425.

Ritchie, J.C. 1987. A Holocene pollen record from Bir Atrun, Northwest Sudan. *Pollen Spores* 29: 391–410.

Ritchie, J.C. 1994. Holocene pollen spectra from Oyo, northwestern Sudan: Problems of interpretation in a hyperarid environment. *The Holocene* 4: 9–15.

Ritchie, J.C., and C.V. Haynes. 1987. Holocene vegetation zonation in the eastern Sahara. *Nature* 330: 645–647.

Ritchie, J.C., C.H. Eyles, and C.V. Haynes. 1985. Sediment and pollen evidence for an early to mid-Holocene humid period in the eastern Sahara. *Nature* 314: 352–355.

Rozanski, K., L. Araguás-Araguás,, and R. Gonfiantini. 1992. Relation between long-term trends of oxygen-18 isotope composition of precipitation and climate. *Science* 258: 981–985.

Schulz, M., and K. Stattegger. 1997. SPECTRUM: Spectral analysis of unevenly spaced paleoclimatic time series. *Computers and Geosciences* 23: 929–945.

Shindell, D.T., G.A. Schmidt, M.E. Mann, D. Rind, and A. Waple. 2001. Solar forcing of regional climate change during the Maunder Minimum. *Science* 294: 2149–2152.

Singh G., R.J. Wasson, and D.P. Agrawal. 1990. Vegetational and seasonal climatic changes since the last full glacial in the Thar Desert, northwestern India. *Reviews of Palaeobotany and Palynology* 64: 351–358.

Sirocko, F., M. Sarnthein, H. Lange, and H. Erlenkeuser. 1991. Atmospheric summer circulation and coastal upwelling in the Arabian Sea during the Holocene and the last glaciation. *Quaternary Research* 36: 72–93.

Sirocko, R.C., H. Sarnthein, H. Erlenkeuser, H. Lange, M. Arnold, and J.-C. Duplessy. 1993. Century-scale events in monsoonal climate over the past 24.000 years. *Nature* 364: 322–364.

Stager, J.C. 1988. Environmental changes at Lake Cheshi, Zambia since 40000 years BP. *Quaternary Research* 29: 54–65

Staubwasser, M., F. Sirocko, P.M. Grootes, and H. Erlenkeuser. 2002. South Asian monsoon climate change and radiocarbon in the Arabian Sea during early and middle Holocene. *Paleoceanography* 17, doi:10.1029/2000PA000608.

Street-Perrott, F.A., and R.A. Perrott. 1990. Abrupt climate fluctuations in the tropics: The influence of Atlantic Ocean circulation. *Nature* 343: 607–612.

Stuiver, M., P.J. Reimer, E. Bard, J.W. Beck, G.S. Burr, K.A. Hughen, B. Kromer, G. McCormac, J. Van der Plicht, and M. Spurk. 1998. INTCAL98 radiocarbon age calibration, 24,000–0 cal BP. *Radiocarbon* 40: 1041–1083.

Swain, A.M., J.E. Kutzbach, and S. Hastenrath. 1983. Estimates of Holocene precipitation for Rajesthan, India, based on pollen and lake-level data. *Quaternary Research* 19: 1–17

Taylor, D.M. 1990. Late Quaternary pollen records from two Ugandan mires: Evidence for environmental change in the Rukia Highlands of southwest Uganda. *Palaeogeography, Palaeoclimatology, Palaeoecology* 80: 283–300.

Van Campo, E., and F. Gasse. 1993. Pollen- and diatom-inferred climatic and hydrological changes in Sumxi Co Basin (western Tibet) since 13 000 yr BP. *Quaternary Research* 39: 300–313.

Vincens, A. 1989. Paleoenvironnements du bassin nord-Tanganyika (Zaire, Burundi, Tanzanie) au cours des 13 derniers mille ans: apport de la palynologie. *Reviews of Palaeobotany and Palynology* 61: 69–88

Vincens, A. 1991. Late Quaternary vegetation history of the South-Tanganyika Basin. Climate implications in South Central Africa. *Palaeogeography, Palaeoclimatology, Palaeoecology* 86: 207–226

Vose, R.S., R.L. Schmoyer, P.M. Steurer, T.C. Peterson, R. Heim, T.R. Karl, and J. Enscheid. 1992. The Global Historical Climatology Network: Long-term monthly temperature precipitation, sea level pressure, and station pressure data, ORNL/CDIAC-53, NDP-041. Oak Ridge, Tennessee: .Carbon Dioxide Information and Analysis Center, Oak Ridge National Laboratory.

Chapter 10

EVOLUTION OF THE INDO-PACIFIC WARM POOL AND HADLEY-WALKER CIRCULATION SINCE THE LAST DEGLACIATION

Michael K. Gagan[1] and Lonnie G. Thompson[2]

[1]Research School of Earth Sciences, The Australian National University, Canberra, ACT 0200, Australia

[2]Department of Geological Sciences and Byrd Polar Research Center, The Ohio State University, 108 Scott Hall, 1090 Carmack Road, Columbus, Ohio 43210-1002, U.S.A.

Abstract

The Indo-Pacific warm pool (IPWP), East Pacific cold tongue, and deep overturning atmospheric Hadley (meridional) and Walker (zonal) circulations form a tightly coupled system. In this chapter, we explore the concept of the Hadley circulation as the fundamental driver of changes in this system, and examine its possible impact on global climates of the past. Recent modeling studies indicate that the Hadley circulation is sensitive to Milankovitch forcing, dominated by the precession cycle (22,000 years) in the tropics. It is well established that the increasing Northern Hemisphere summer insolation during the post-glacial transition enhanced northern summer monsoon rainfall, particularly across the Asian landmass. Based on the results of modeling studies, it is probable that the northward asymmetry in tropical heating led to asymmetrical intensification of the Hadley circulation during the early Holocene.

The response of the tropical ocean to the intensification of the Hadley circulation is given by foraminiferal Mg/Ca and coral Sr/Ca sea surface temperature (SST) reconstructions, which show that ocean-atmosphere feedbacks drove the tropical Pacific into a westward-concentrated La Niña–like state (warming in the west, cooling in the east) between ~11,000 and ~4,000 years ago. At the same time, air temperatures reconstructed from Southern Hemisphere high-altitude tropical ice cores also equal or exceed late Holocene values. The widespread warming of the tropical middle troposphere during the early Holocene suggests that the additional flux of water vapor and heat from the warmer IPWP during the La Niña state overwhelmed any atmospheric cooling brought about by the expansion of the East Pacific cold tongue. However, the expanded cold tongue area could also play a role in the early Holocene warming through enhanced

H.F. Diaz and R.S. Bradley (eds.), The Hadley Circulation: Present, Past and Future, 289–312.

evasion of CO_2 to the atmosphere.

Taken together, the paleoclimate records indicate that a post-glacial strengthening of the Hadley circulation initiated ocean-atmosphere feedbacks that altered the energy budget of the tropics to amplify early Holocene warming. Synchronous warming of the Southern Hemisphere high latitudes, as indicated by Antarctic ice core records, may be the result of southward-directed dynamical heating produced by the asymmetrical Hadley circulation. The demise of the early Holocene warming in the tropics and Southern Hemisphere ~4,000 through 7,000 years ago correlates with decreasing Northern Hemisphere summer insolation, a southward migration of the Intertropical Convergence Zone (ITCZ), and the onset of modern El Niño/Southern Oscillation (ENSO) variability.

1. INTRODUCTION

The Hadley circulation is the large-scale tropical atmospheric circulation consisting of a deep overturning meridional circulation cell where regions of ascending air, near the thermal equator, are connected to subsiding branches over the subtropics (Pierrehumbert 2000). Energy associated with the equatorial maximum in solar radiation released through vigorous atmospheric convection is the ultimate driver of the mean position of the Hadley circulation (Lindzen and Hou 1988; Hou 1998; Rind 1998). In terms of the coupled ocean-atmosphere system, the Hadley circulation can be viewed as the fundamental driver of the western warm pool – eastern cold tongue configuration in sea surface temperature (SST) that is particularly prominent in the tropical Pacific Ocean (Liu and Huang 1997; Liu 1998). Surface winds converging toward the thermal equator in the lower limbs of the Hadley circulation are deflected westward by the Coriolis force. The resulting easterly trade winds drive equatorial upwelling, a cold tongue in the east, and the westward accumulation of advected warm water in the Indo-Pacific warm pool (IPWP), where average annual SSTs exceed 28°C (Fig. 10-1; Yan et al. 1992).

The zonal Walker circulation embedded in the Hadley circulation owes its origin to the cold tongue – warm pool gradient in SSTs across the Pacific Ocean. This is because deep atmospheric convection above the IPWP concentrates thermal energy in the ascending limb of Indo-Pacific section of the Hadley circulation (Bjerknes 1966), whereas the air above the cold tongue in the east makes a relatively minor contribution. Thus, relatively cool air flows westward in the lower limb of the Walker circulation where it becomes moistened and heated, and rises through deep convection above the IPWP to reinforce the Hadley circulation. This concept of the Hadley circulation as a fundamental driver of a tightly coupled Pacific cold tongue, warm pool, and Walker circulation system may apply since the final closure of the Isthmus of Panama and the establishment of thermal gradients across the Pacific about 3 million years ago (Keigwin 1978; Romine 1982; Philander and Federov 2003).

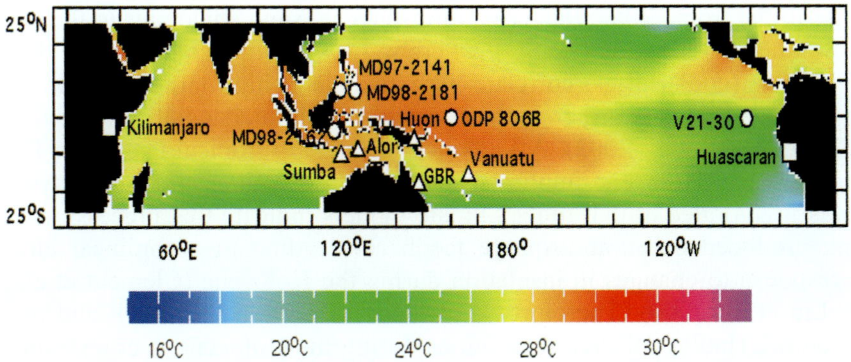

Figure 10-1. Locations of deep-sea sediment cores (circles), corals (triangles), and tropical ice core (squares) paleoclimate records described in Figures 10-2 through 10-5. Red shows the average extent of the Indo-Pacific warm pool (mean annual SST > 28°C), as defined by Yan et al. (1992).

Recent modeling studies have shown that the nonlinear response of the coupled cold tongue, warm pool, Hadley-Walker circulation system to insolation fluctuations, dominated by the precession cycle (22,000 years) in the tropics, may be sufficient to yield substantial millennial-scale climate variability (Cane and Clement 1999; Clement et al. 2001). The IPWP is a good candidate for driving this variability because changes in its temperature, size, and positioning can alter atmospheric circulation at higher latitudes (Barlow et al. 2002; Chen et al. 2002; Hoerling and Kumar 2003). Hou (1993) showed that displacements of the tropical heating maximum are reflected in changes in the winter polar temperature, and relatedly, in the heat flux between the tropics and the polar regions. The resulting changes in heat flux lead to changes in the mean temperature of the earth, even though there is no change in the globally averaged insolation. The implication is that Milankovitch variations in the seasonal distribution of insolation could have a profound effect on global climate (Lindzen and Pan 1994). Therefore, a crucial issue in global climate change research is to reconstruct tropical SSTs and the locus of atmospheric convection in the tropics.

Recently, tandem measurements of Mg/Ca and $\delta^{18}O$ in planktonic foraminifera have been used to reconstruct changes in the temperature and $\delta^{18}O$ of surface seawater over glacial-interglacial time scales, and make inferences about the surface-ocean water balance in the tropics (Lea et al. 2000; Kienast et al. 2001; Koutavas et al. 2002; Stott et al. 2002; Rosenthal et al. 2003; Visser et al. 2003). These records reveal millennial-scale changes in the tropical Pacific ocean/atmosphere system analogous to the El Niño/Southern Oscillation (ENSO), whereby El Niño–like conditions in the IPWP region (cooler/drier) correlate with stadials at high latitudes and La Niña–like conditions (warmer/wetter) correlate with interstadials (Koutavas et al. 2002; Stott et al. 2002). A La Niña–like state in tropical Pacific SSTs during the early Holocene implies that the Hadley-Walker circulation system was invigorated at that time. Climate models indicate that precessional forcing produced ocean-atmosphere feedbacks leading to a nonlinear climate response to changes in insolation during the Holocene (Clement et al. 2000; Liu et al. 2000). Therefore, knowledge of SST in the IPWP and its effect on the Hadley-Walker circulations during the Holocene is crucial for understanding the specific mechanisms by which subtle changes in insolation seasonality were converted to significant changes in tropical climate.

In this chapter, we review the most recent estimates of SST variability in the coupled Pacific cold tongue–IPWP system since the last glacial maximum (LGM). We then examine the oxygen isotopic signals recorded by high-altitude tropical ice cores over the same period to investigate interactions between tropical SSTs and the tropical troposphere. Comparisons with air temperature records reconstructed from polar ice cores provide evidence for the potential role of the coupled Pacific cold tongue–IPWP–Hadley/Walker circulation system in global climate change. Data-model comparisons provide insight into the ocean-atmosphere interactions driving post-glacial climate change. Finally, we discuss possible analogies with future climate change.

2. POST-GLACIAL EVOLUTION OF THE PACIFIC COLD TONGUE AND IPWP

Two promising new paleothermometers, Mg/Ca in surface-dwelling foraminifera and Sr/Ca in corals, have led to important findings about the temperature history of the IPWP since the LGM. Studies of foraminiferal Mg/Ca from deep-sea sediment cores provide long, continuous histories of changes in mean SST (Lea et al. 2000; Kienast et al. 2001; Koutavas et al. 2002; Stott et al. 2002; Rosenthal et al. 2003; Visser et al. 2003). Corals from the tropical Pacific provide complementary high-resolution records of Holocene climate change at monthly or better temporal resolution (Beck et al. 1992, 1997; McCulloch et al. 1996; Gagan et al. 1998; Corrège et al. 2000; Tudhope et al. 2001). Recently, these paleothermometers have been used to extract the temperature component of the oxygen isotope signal in biogenic carbonates and thereby reveal changes in seawater ^{18}O concentrations as a proxy for surface-ocean salinity (e.g., Gagan et al. 1998, 2000; Lea et al. 2000; Hendy et al. 2002; Stott et al. 2002; Rosenthal et al. 2003).

We first review the primary findings from planktonic foraminiferal Mg/Ca records from the IPWP region (Lea et al. 2000; Kienast et al. 2001; Stott et al. 2002; Rosenthal et al. 2003; Visser et al. 2003) and the East Pacific cold tongue (Koutavas et al. 2002). We then present revised temperature estimates for the IPWP region, based on coral Sr/Ca ratios, including data sets from Vanuatu (Beck et al., 1992, 1997; Corrège et al. 2000), Papua New Guinea (McCulloch et al. 1996), the Great Barrier Reef (Gagan et al. 1998), and eastern Indonesia (Gagan et al. 2004).

Paleotemperature estimates for the IPWP region derived from Mg/Ca in *Globigerinoides ruber* in the IPWP region show cooling of 2°C–4°C during the LGM (Lea et al. 2000; Stott et al. 2002; Rosenthal et al. 2003; Visser et al. 2003). In contrast, the East Pacific cold tongue exhibits a much smaller cooling of ~1°C (Koutavas et al. 2002). Comparisons of Mg/Ca and $\delta^{18}O$ measured in the same foraminifers indicates that the rise in IPWP SSTs led deglaciation by ~3,000 years (Lea et al. 2000; Stott et al. 2002; Visser et al. 2003). Interestingly, all the records show a rapid rise to modern SST values by the early Holocene (~11,000 BP; Fig. 10-2).

Estimating mean SST in the tropical western Pacific using coral Sr/Ca paleothermometry has been controversial because early estimates indicated cooling of 3°C–6°C during the early Holocene (Beck et al. 1992, 1997; McCulloch et al. 1996). A clearer picture emerges for the Southern Hemisphere portion of the IPWP when a single calibration equation derived specifically for application to continental and island arc fringing reefs (Gagan et al. 1998) is applied to fossil coral Sr/Ca data for Vanuatu (Beck et al.

1997; Corrège et al. 2000), Papua New Guinea (McCulloch et al. 1996), the inshore Great Barrier Reef (Gagan et al. 1998), and Sumba/Alor in southeastern Indonesia (Gagan et al. 2004). The revised Vanuatu coral SST estimates for the early Holocene now indicate cooling of 1°C–3°C, rather than 4°C–6°C. New coral Sr/Ca records from Alor, southeast Indonesia, show that SSTs reached modern values by ~8,500 BP, in good agreement with the foraminiferal Mg/Ca estimates of SST. This generally warm period is interrupted by a brief cold-spike centered on 8,100 BP. Mid-Holocene SSTs in Indonesia (Sumba) fall within 0.5°C of modern values, whereas corals from the inshore Great Barrier Reef, Australia, indicate SSTs ~1°C warmer than the present.

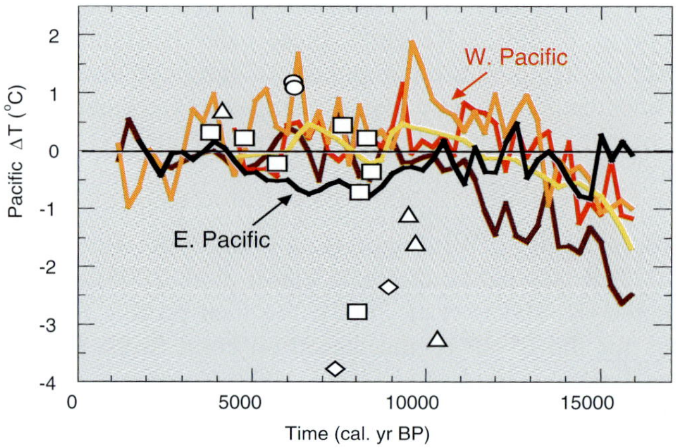

Figure 10-2. Comparison of SSTs in the equatorial eastern Pacific and IPWP regions during the last 16,000 years, as reconstructed using foraminiferal Mg/Ca and coral Sr/Ca thermometry. For the IPWP region, SST anomalies (relative to 0–5,000 BP) were reconstructed from Mg/Ca in the planktonic foraminifer *Globigerinoides ruber* from core MD98-2162 in Makassar Strait, Indonesia (4°41.33'S, 117°54.17'E; burgundy curve; Visser et al. 2003); core MD97-2141 in the Sulu Sea (8.8°N, 121.3°E; red curve; Rosenthal et al. 2003); core MD98-2181 south of the Philippines (6.3°N, 125.83°E; orange curve; Stott et al. 2002); and core ODP 806B from Ontong Java Plateau (0°19.1'N, 159°21.7'E; yellow curve; Lea et al. 2000). For the equatorial eastern Pacific, SST anomalies (relative to 1,700–4,000 BP) were reconstructed from Mg/Ca in the planktonic foraminifer *Globigerinoides sacculifer* from core V21-30 near the Galapagos Islands (1°13'S, 89°41'W; black curve; Koutavas et al. 2002). Sediment core age models are based on accelerator mass spectrometer (AMS) ^{14}C dates calibrated to calendar years before the present (cal. yr BP), and foraminiferal δ^{18}O stratigraphy. Coral Sr/Ca SST anomalies are based on fossil specimens of *Porites* from Espiritu Santo, Vanuatu (15°40'S, 167°E; triangles; Beck et al. 1997; Corrège et al. 2000); Huon Peninsula, Papua New Guinea (6°S, 147°E; diamonds; McCulloch et al. 1996); Orpheus Island, Great Barrier Reef (18°34'S, 146°29'E; circles; Gagan et al. 1998); and southern Indonesia (Alor, 8°14'S, 124°24'E; Sumba, 09°28'S, 120°06'E; squares). The relationship for converting coral Sr/Ca to SST is: T(°C) = 168.2 – [15,674*(Sr/Ca)atomic], which has been derived for modern *Porites* specimens from open-water continental and island arc fringing reefs throughout the Indo-Pacific region (Gagan et al. 1998). ^{230}Th/^{234}U calendar ages for corals were determined by thermal ionization mass spectrometry (TIMS).

Although the new coral SST estimates and foraminiferal Mg/Ca SST estimates are now in much better agreement, some of the coral records are still significantly cooler or warmer than the deep-sea sediment core estimates. Coral records showing cool SSTs could be recording real changes, particularly coastal upwelling, on time scales of decades. Cool artifacts may also be caused by early marine aragonite cements (Müller et al. 2001) and off-axis sampling of coral skeletons (de Villiers et al. 1994). Warm artifacts, on the other hand, may result from calcite diagenesis in corals sampled from uplifted coral terraces (McGregor and Gagan 2003).

The rapid late-glacial to early Holocene warming of the IPWP is consistent with a semipermanent La Niña–like state in mean SSTs (Koutavas et al. 2002; Stott et al. 2002). The La Niña state would tend to elevate the thermocline in the eastern equatorial Pacific, and deepen it in the west. Westward advection of warm water would increase surface-ocean dynamic heights in the western Pacific and increase poleward flows in western boundary currents and the export of warm water to the southeastern Indian Ocean, via the Indonesian Throughflow (Meyers 1996). Such processes would increase SSTs at Southern Hemisphere study sites such as the Great Barrier Reef, Sumba, and Alor during the early to middle Holocene, in agreement with the coral Sr/Ca SST reconstructions.

The existence of prolonged La Niña conditions with a strong focus on a warm IPWP during the early Holocene is supported by paleo-ENSO records (Fig. 10-3). The most continuous, high-resolution record of ENSO for the post-glacial and early Holocene comes from laminated clastic deposits in a high-altitude lake, Laguna Pallcacocha, in Ecuador (Rodbell et al. 1999; Moy et al. 2002). Today, these clastic laminae record anomalously high rainfall during El Niño events. However, the sedimentary record shows a clear suppression of ENSO variability, with periodicities of ~15 years, from 12,000 (the beginning of the record) to 7,000 BP.

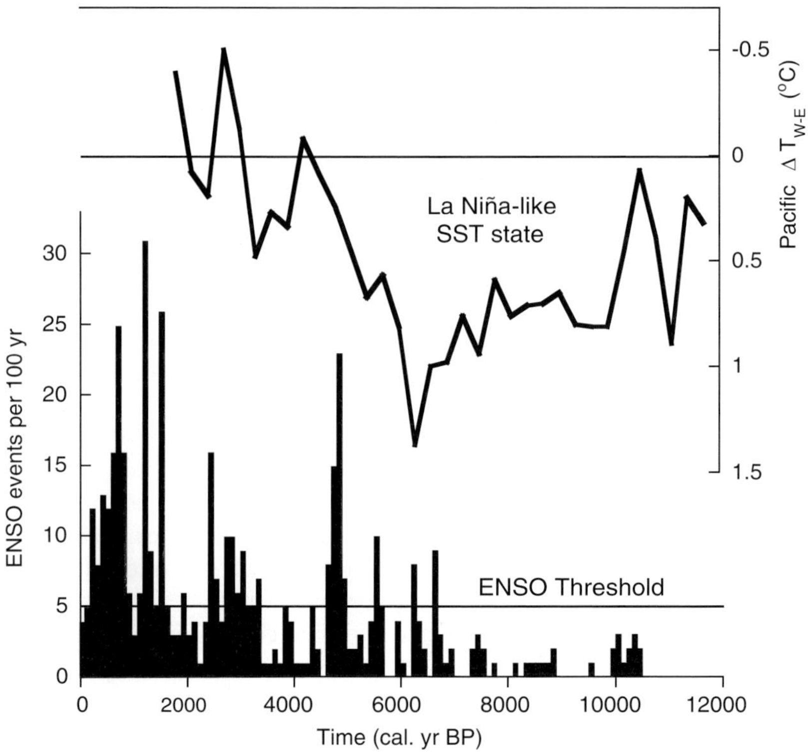

Figure 10-3. Comparison of changes in the east-west SST gradient (ΔT_{W-E}) across the tropical Pacific and the frequency of moderate to strong El Niño events over the last 12,000 years. ΔT_{W-E} is the difference (relative to the late Holocene) between the mean SST given by the four foraminiferal Mg/Ca records from the IPWP and the Galapagos foraminiferal Mg/Ca SST record (shown in Fig. 10-2). Relatively warm SSTs in the IPWP (positive values downward on Y axis) indicate a La Niña–like state in Pacific SSTs. The histogram shows the number of El Niño events in 100-year windows since 12,000 BP., based on the analysis of clastic laminae in lake Laguna Pallcacocha, southern Ecuador (after Moy et al. 2002). The solid line indicates the minimum number of events (~5) required to produce ENSO-band variance.

3. POST-GLACIAL INTENSIFICATION OF THE HADLEY-WALKER CIRCULATION

The rise of tropical SSTs in advance of the change in global ice volume (Lea et al. 2000; Stott et al. 2002; Visser et al. 2003) and early Holocene SSTs similar to, or above, modern values indicate that the tropics may have played a key role in driving post-glacial global warming. Here we explore the possibility that the early Holocene intensification of the Pacific cold tongue–IPWP SST gradient and, presumably, the Hadley-Walker circulation, initiated feedbacks that altered the energy budget of the planet to amplify post-glacial warming. It is well established that, in addition to initiating the melting of the Northern Hemisphere ice caps, the increasing Northern Hemisphere summer insolation also enhanced summer monsoon rainfall across the Asian landmass, particularly during the early Holocene (see review by Morrill et al. 2003). Our hypothesis is that this asymmetry in tropical heating also enhanced the Hadley circulation and synchronously warmed the Southern Hemisphere high latitudes via dynamical heating mechanisms (Hou 1993, 1998; Lindzen and Pan 1994). The resulting La Niña–type configuration in zonal SSTs across the Pacific could then have altered the atmospheric greenhouse effect through increased emission of CO_2 from the expanded cold tongue region, and enhanced water vapor evaporated from the warmer IPWP.

High-altitude tropical ice core records should be particularly sensitive to changes in tropical Pacific SSTs and evaporation in the IPWP region considering that the heat of condensation released during convection over the IPWP sets the temperature above ~4 km in the tropics (Broecker 1997). The relative abundance of the oxygen isotopes ^{18}O and ^{16}O (expressed as $\delta^{18}O$) is the most common parameter measured in tropical ice cores. If the initial $\delta^{18}O$ of water vapor condensing to yield snow is constant, then the resulting $\delta^{18}O$ of the snow will be a function of both condensation temperature and precipitation amount (Pierrehumbert 1999; Lawrence and Gedzelman 2003). In tropical ice cores, seasonal oscillations in $\delta^{18}O$ show an apparent inverse relationship with temperature, in contrast to $\delta^{18}O$ values in polar ice (Dansgaard 1964). This is because seasonal variations in the $\delta^{18}O$ of high-altitude tropical precipitation reflect the temperature of the mean condensation level, which is significantly higher (and colder) during the summer wet season in the tropics (Thompson et al. 2000). On the other hand, seasonal changes in the precipitation "amount effect" on $\delta^{18}O$ may overwhelm the effect of seasonal changes in temperature on the $\delta^{18}O$ of tropical ice (Bradley et al. 2003; Hardy et al. 2003). However, because

70%–80% of tropical precipitation falls during the summer wet season, tropical ice core $\delta^{18}O$ is dominated by the wet season temperature. Despite these high-frequency seasonal variations, $\delta^{18}O$ in tropical ice is positively correlated with global temperature over centennial-millennial time scales, and accurately records twentieth-century warming (Thompson et al. 2003). Therefore, although it is somewhat counterintuitive, evidently large seasonal variations in the $\delta^{18}O$ of tropical ice produced by the seasonal contrast in precipitation amount are superimposed on long-term ice core $\delta^{18}O$ trends where changes in air temperature dominate the mean climate signal.

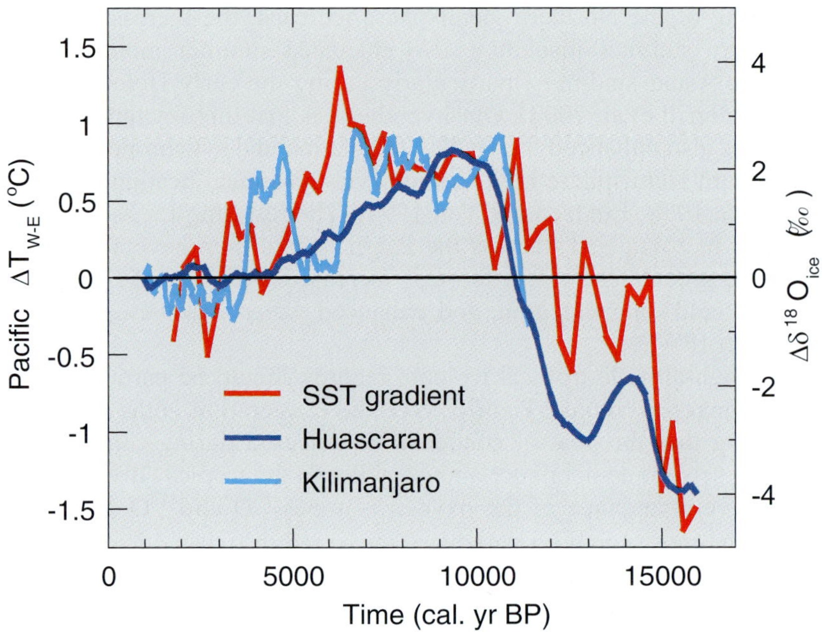

Figure 10-4. Comparison of changes in the east-west SST gradient (ΔT_{W-E}) across the tropical Pacific and $\delta^{18}O$ in high-elevation tropical ice cores from Huascarán, Peru (09°07' S, 77°37' W; 6,048 m above sea level (masl); Thompson et al. 1995) and Mt. Kilimanjaro, Tanzania (03°04.6' S, 37°21.2' E; 5,893 masl; Thompson et al. 2002). Ice core data have been smoothed with a 500-year running mean and normalized to the mean late Holocene $\delta^{18}O$ values.

Figure 10-4 shows late-glacial and Holocene $\delta^{18}O$ values in the high-altitude ice core from Huascarán, located at 9°S in the Cordillera

Blanca of Peru (Thompson et al. 1995). The striking feature in the Huascarán ice core record is the high $\delta^{18}O$ values between 11,000 and 4,000 BP. If temperature was the dominant forcing, then the $\delta^{18}O$ in the Huascarán ice core suggests that the early Holocene was ~1.5°C–2°C warmer, relative to the late Holocene (Thompson et al. 1995; Bradley et al. 2003). The recent discovery of tropical plants exposed from beneath the retreating Peruvian Quelccaya Ice Cap, with radiocarbon dates of 5.18 ± 0.45 thousand years BP, indicates that conditions were indeed warmer in the high near-equatorial Andes.

A high-altitude equatorial ice core $\delta^{18}O$ record for Mt Kilimanjaro, Tanzania (Thompson et al. 2002), provides a means for determining if, as suggested, transport of latent heat from the IPWP to the equatorial middle troposphere was a general feature of the La Niña–like state in Pacific SSTs during the early Holocene. The climate of equatorial east Africa was relatively wet during the early Holocene, so the precipitation amount effect on the $\delta^{18}O$ of Kilimanjaro ice should be opposed to any early Holocene warming signal. The results show that the early Holocene warming in the Kilimanjaro $\delta^{18}O$ record is in good agreement with that observed in the Huascarán ice core record (Fig. 10-4). Taken together, the records suggest that long-term changes in precipitation amount had a minor influence on the $\delta^{18}O$ of the ice core records and that, instead, widespread warming of the middle troposphere occurred in response to warming of the IPWP during the early Holocene.

We note that ice core $\delta^{18}O$ data for Sajama, Bolivia, at 18°S indicate that the climate was cooler/wetter when Huascarán (9°S) was warmer. The pattern is suggestive of a northward shift in the mean annual position of the Intertropical Convergence Zone (ITCZ). Geochemical records from deep-sea sediment cores from the Cariaco Basin, in the tropical North Atlantic, indicate intensified ITCZ rainfall during the early Holocene, in response to higher summer insolation in the Northern Hemisphere (Haug et al. 2001). In contrast, the level of Lake Titicaca in the Andean altiplano was at its lowest in the last 25,000 years during the early to middle Holocene (Baker et al. 2001). Therefore, the opposing signals in the Huascarán and Sajama $\delta^{18}O$ records may reflect changes in the position of the ITCZ and the location of ascending and descending air masses within the Hadley circulation (Thompson et al. 2000). If the mean position of the ITCZ was shifted towards the north, Huascarán would tend to be warm and dry beneath the descending limb of the Hadley cell. Sajama, on the other hand, could be cooler and wetter during the early Holocene under the influence of enhanced easterly airflow and precipitation, as observed during modern La Niña events (Vuille 1999; Bradley et al. 2003; Hardy et al. 2003). The mid- to late Holocene shift to modern $\delta^{18}O$ values may reflect a southward migration of the Hadley circulation that is coeval with the intensification of Southern Hemisphere summer insolation (Haug et al. 2001; Seltzer et al. 2002).

4. INFLUENCE OF THE TROPICS ON POST-GLACIAL WARMING AT HIGH LATITUDES

Widespread warming of the tropics should generate a globally synchronous climate response to Milankovitch forcing, without a net increase in global solar radiation. A tropical influence on the climate of the higher latitudes would explain why ice sheets and glaciers in the Southern Hemisphere decayed almost synchronously with those in the Northern Hemisphere, despite the decrease in Southern Hemisphere summer insolation during deglaciation (Karner and Muller 2000). Our knowledge of post-glacial air temperature changes over Antarctica has grown recently through the development of deuterium isotopic profiles (δD) in five ice cores from the East Antarctic plateau (see summary by Jouzel et al. 2001). The δD record in ice from Vostok, Antarctica, is typical of the five cores in showing a strong late glacial (~11,000 years) warming step culminating with an early to middle Holocene optimum (Petit et al. 1999). While the details are different, the timing of the early Holocene warming and the cooling trend to ~4,000 years ago is approximately synchronous with the warming observed in tropical ice cores, and the enhanced La Niña–like state in the Pacific (Fig. 10-5).

The correlation between the development of a La Niña–like state in Pacific SSTs and atmospheric warming in the Southern Hemisphere is interesting because it is opposite to the observation that La Niña years produce cooler surface temperatures in North America (Cane 1998). Moreover, model studies indicate that a La Niña–like state in Pacific SSTs should promote the growth of Northern Hemisphere ice sheets and planetary cooling (Cane and Clement 1999). The early Holocene warming in the tropics and Southern Hemisphere leads and exceeds the warming observed in ice $\delta^{18}O$ records from the GRIP and GISP2 sites in central Greenland (Fig. 10-5; Grootes et al. 1993). On the other hand, temperature profiles measured down through the GRIP borehole indicate that mean annual temperatures in the early Holocene were 2.5°C warmer, relative to the last ~500 years (Dahl-Jensen et al. 1998). While the nature of the mean air temperature trend over Greenland is still controversial, several lines of evidence show that summer air temperatures in the Arctic were indeed warmer during the early Holocene than they are today. For example, melt layers and $\delta^{18}O$ records from low-elevation Canadian Arctic ice caps (Agassiz [Fisher et al. 1995]; Devon [Koerner 1977]) show the warmest summer temperatures during the early Holocene, followed by long-term summer cooling (see reviews by Bradley et al. 2003; Fisher and Koerner 2003).

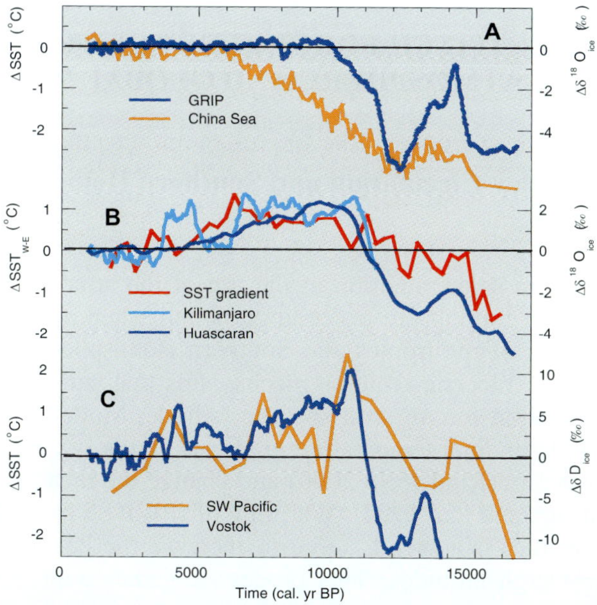

Figure 10-5. Interhemispheric comparison of the post-glacial timing of changes in SST and air temperature in the tropics, Greenland, and Antarctica. (A) Lag of change in SSTs reconstructed from alkenone unsaturation ratios in deep-sea core 17940 from the South China Sea (20°07'N, 117°23'E; Pelejero et al. 1999) relative to the change of $\delta^{18}O$ in the GRIP Greenland ice core (Grootes et al. 1993). (B) Synchronous changes in the east-west SST gradient across the tropical Pacific and $\delta^{18}O$ in ice cores from Huascarán, Peru (Thompson et al. 1995) and Mt. Kilimanjaro, Tanzania (Thompson et al. 2002). (C) Comparison of Southern Hemisphere mid-latitude SSTs reconstructed from Mg/Ca in the planktonic foraminifer *Globigerina bulloides* in deep-sea core MD97-2120 from the Chatham Rise, east of New Zealand (45°32.06'S, 174°55.85'E; Pahnke et al. 2003) and the change of δD in ice from Vostok, Antarctica (Petit et al. 1999). All data are normalized relative to late Holocene values. Ice core data have been smoothed with a 500-year running mean for comparison with relatively coarse-resolution marine records.

A picture is emerging for the tropics and Southern Hemisphere of early post-glacial warming culminating in an early Holocene climatic optimum. However, even when conventional feedbacks are considered, such as changes in ocean thermohaline circulation, increases in atmospheric CO_2, and decreasing atmospheric dust, they appear to be incapable of inducing a 3°C–5°C post-glacial rise in tropical temperatures by the early Holocene. In the next section, we examine potential mechanisms by which increasing Northern Hemisphere insolation could serve to invigorate the Hadley circulation, increase the SST contrast between the Pacific cold tongue and IPWP, and thus bring about the unanticipated warming of the tropics and Southern Hemisphere during the early Holocene.

5. MILANKOVITCH FORCING OF THE TROPICAL OCEAN-ATMOSPHERE AND GLOBAL WARMING

5.1. The Hadley Circulation and Southern Hemisphere Warming

Changes in solar radiation at the top of the atmosphere associated with Milankovitch forcing are exceedingly small when averaged annually. Thus the warming of the tropics and Southern Hemisphere during the late glacial and early Holocene requires mechanisms that can amplify seasonal changes in insolation forcing at a specific latitude. Modeling studies show that changes in latitudinal temperature gradients play a key role in the equator-to-pole heat flux (Rind 1998 2000) and change the mean temperature of the earth (Lindzen and Pan 1994). Warming of IPWP SSTs (and the tropical troposphere) would, in general, serve to increase the equator-to-pole heat flux and warm high latitudes. This process, together with the increase in Northern Hemisphere summer insolation and associated ice albedo feedback, would have warmed the Northern Hemisphere high latitudes during deglaciation. However, the ice core records indicate that post-glacial warming in the Southern Hemisphere was closely synchronized with the warming in the tropics. Thus a more direct link between the tropics and the Southern Hemisphere must be acting to transport additional heat to Antarctica.

Results of a simplified general circulation model (GCM) and heat transport calculations show that even a slight shift (2° latitude) of the mean tropical heating off the equator leads to a more intense cross-equatorial "winter" Hadley circulation accompanied by warming of the winter high latitudes (Lindzen and Hou 1988; Hou 1993). As the heating center (ITCZ) moves off the equator toward the summer hemisphere, the Hadley circulation extending into the winter hemisphere becomes more intense while the summer cell becomes much weaker. Thus, under asymmetrical heating in the tropics, poleward heat transport is significantly a winter hemisphere phenomenon. Calculations also show that the annually averaged asymmetrical Hadley circulation is stronger than the equinoctial circulation (Lindzen and Hou 1988). The results indicate that the annually averaged climate of middle to high latitudes may be dependent on the summer-winter asymmetry in insolation forcing brought about by precession of the equinoxes (Lindzen and Pan 1994), even though the insolation asymmetry does not contribute to the annually averaged heating.

This mechanism of intensification of the Hadley circulation provides interesting possibilities for heating the Southern Hemisphere synchronously with the northward movement of the ITCZ into the Northern Hemisphere under the influence of enhanced summer insolation during the early Holocene. It is now well established that the northward asymmetry in summer insolation during the early Holocene acted to shift the annual mean position of the ITCZ north of the equator (e.g. Haug et al. 2001; Morrill et al 2003). Northern Hemisphere summer heating of the Asian landmass, in particular, caused significant strengthening of the Asian monsoon and a profound northward distortion of the mean position of the center of tropical heating (see review by Morrill et al. 2003).

Thus the heat budget of the high latitudes of the Southern Hemisphere in winter during the late glacial and early Holocene may have been primarily determined by Hadley circulation dynamics while, for the Northern Hemisphere, local radiative budgets were probably more important (Webster 1982). Given that ocean surface temperatures in summer are strongly influenced by winter heat fluxes because of the large heat capacity of the oceans, year-round warming of the Southern Hemisphere could have been produced by poleward heat transport during the winter (Lindzen and Pan 1994).

5.2. Coupling of the Tropical Ocean-Atmosphere and Greenhouse Gas Concentrations

It is generally accepted that the oceans played a significant role in promoting changes in atmospheric CO_2 concentrations during glacial-interglacial cycles (Broecker 1982). The observed La Niña–like state in Pacific SSTs during the late glacial and early Holocene is consistent with the rapid rise of CO_2 trapped in polar ice cores (Indermühle et al. 1999; Petit et al. 1999). During La Niña events, the area of high pCO_2 equatorial cold tongue water expands towards the west (Feely et al. 1995). Such an SST configuration could play a role in the late glacial and early Holocene warming because the cold tongue area of the equatorial Pacific is the site of the greatest evasion of CO_2 (0.8–1.0 Pg C/yr) from the modern oceans (Takahashi et al. 2002).

Recently, Palmer and Pearson (2003) produced a boron isotope record for planktonic foraminifera from the western equatorial Pacific to reconstruct the pH of surface seawater and, by inference, pCO_2 over the last 25,000 years. The results indicate that the equatorial Pacific was a significant source of CO_2 to the atmosphere between 15,600 and 13,800 years ago. The timing of the peak in the pCO_2 is coincident with the steepest rise in atmospheric CO_2 levels during the last deglaciation (Indermühle et al. 1999; Petit et al. 1999), and the anomaly is best explained if there were more frequent and/or more intense La Niña events (Palmer and Pearson 2003). Therefore, an increase in atmospheric CO_2 potentially brought about by the intensification of the Hadley-Walker circulation and La Niña–like SST configuration could certainly have contributed to the late glacial warming and early Holocene temperature maximum. However, given that the shift in atmospheric pCO_2 from the LGM to the Holocene (~90 parts per million by volume [ppmv]) is associated with far more warming than has been observed under similar greenhouse-gas forcing during the twentieth century, additional mechanisms must be contributing to the warming.

The widespread and synchronous late glacial to early Holocene warming signal suggests that water vapor, the most important greenhouse gas, must somehow be involved. Surface-ocean evaporation becomes strongly nonlinear above SSTs of 28°C (Webster 1994), and drives tropical convection. The paleoclimate records suggest that, on long time scales at least, the flux of water vapor and heat brought about by enhanced IPWP evaporation and convection during the La Niña state overwhelmed any atmospheric cooling due to expansion of the East Pacific cold tongue. Evaporation near convective cloud tops humidifies the atmosphere at high altitudes, reduces the Earth's energy emission to space, and thus warms the atmosphere (Pierrehumbert 2000). Although still a matter of considerable debate, it has been suggested that a post-glacial rise in the absolute water vapor content of the atmosphere could produce a 2°C global warming. Taken together, the results indicate that the Hadley-Walker circulation system and the IPWP–cold tongue configuration could have conspired to increase the atmospheric concentrations of CO_2 and water vapor, resulting in widespread early Holocene warming.

6. THE 11,000 TO 4,000 BP TURN-ON OF SOUTHERN HEMISPHERE WARMING

What caused tropical SSTs and Southern Hemisphere air temperatures to equal or exceed late Holocene values from ~11,000 to 4,000 years ago? We envisage three primary processes that led to the abrupt "turn-on" of warming at ~11,000 years ago and "turn-off" at ~4,000 years ago:

(1) The timing of maximum warming observed in the IPWP SSTs and Southern Hemisphere ice core records coincides with the ~11,000-year maximum in Northern Hemisphere summer (July) insolation at 65°N. The off-equatorial heating and resulting southward-directed heat flux of the Hadley circulation at this time would have promoted the temperature maximum in the Southern Hemisphere high latitudes. A synchronous increase in the east-west SST gradient across the Pacific may have superimposed a greenhouse-gas feedback contributing to more widespread warming Thus, the initiation of the late glacial to early Holocene stage of the warming would have been driven primarily by changes in ocean-atmosphere dynamics associated with the maximum in Northern Hemisphere summer insolation.

(2) It is possible that as the Northern Hemisphere summer insolation anomaly weakened, continued invigoration of the Pacific cold tongue–warm pool SST contrast may have served to extend global warming into the mid-Holocene, as indicated by the paleo records. A semipermanent La Niña–like state during the mid-Holocene has been suggested by recent model studies of the direct effect of orbitally induced changes in the seasonal distribution of insolation on the tropical ocean-atmosphere system. Clement et al. (2000) attributed a suppression of the early to mid-Holocene ENSO to the peak in insolation on the equator during the boreal summer/fall brought about by precession of the earth's equinoxes. According to the numerical model, the additional heating of equatorial Pacific surface waters in the boreal summer/fall during the early to mid-Holocene produces an easterly wind anomaly that suppresses the development of El Niño events. Ocean-atmosphere feedbacks drive the ENSO system towards a La Niña state by increasing SST and pressure gradients across the Pacific, in good agreement with the paleoclimate records.

A similar effect has been observed in a global coupled ocean-atmosphere model, whereby the intensified Asian monsoon during the early Holocene further enhances Pacific trade winds, thus cooling the eastern equatorial Pacific and reducing ENSO interannual variability (Liu et al. 2000). Northward transport of warm SST anomalies from the subtropical South Pacific into the equatorial Pacific thermocline via the meridional circulation may also serve to subdue El Niño SST anomalies (Liu et al. 2000), and create a La Niña–like state in mean Pacific SSTs.

(3) The demise of tropical warming at ~4,000 BP appears to correlate with the onset of modern ENSO variability between ~4,000 and 7,000 years ago (Rodbell et al. 1999; Moy et al. 2002) and may signal the onset of coordinated heat removal from the tropical Pacific. Today, the periodic relaxation of the tropical Pacific ocean-atmosphere system during El Niño events provides an efficient mechanism for releasing heat accumulated in the tropical western Pacific (Sun and Trenberth 1998). Evidence for the demise of the suppressed ENSO in the mid-Holocene is most clear in the tropical eastern Pacific and northern South America. Spectral analysis of the 15,000-year high-resolution record of storm-derived clastic sedimentation in Laguna Pallcacocha, Ecuador (Rodbell et al. 1999; Moy et al. 2002) shows that the transition to modern ENSO periodicities (2–8 yr) began ~7,000–5,000 years ago (Fig. 10-3). A similar conclusion was reached by Sandweiss et al. (1996), based on their analysis of fossil mollusk assemblages and geoarcheological evidence from coastal Peru.

More indirect evidence of Holocene ENSO variability is provided by titanium concentrations in sediment from Ocean Drilling Project (ODP) site 1002 in the Cariaco Basin, off northern Venezuela (Haug et al. 2001). Titanium concentrations in Cariaco Basin sediments reflect variations in runoff associated with shifts in the position of the ITCZ. Enhanced runoff variability beginning ~3,800 years ago indicates a mean southward shift in the position of the ITCZ thought to be linked to the strengthening of El Niño events. In addition, a recent synthesis of paleoclimate records for the Asian monsoon reveals an abrupt reduction in monsoon intensity ~4,500–5,000 years ago across the entire Asian monsoon domain (Morrill et al. 2003). According to the paleo-ENSO models of Liu et al. (2000), the associated reduction in trade wind velocity across the Pacific would have served to enhance ENSO variability and reduce the east-west SST gradient across the equatorial Pacific.

7.` WILL OCEAN-ATMOSPHERE DYNAMICS CONTRIBUTE TO FUTURE WARMING?

Transient greenhouse warming simulations suggest that the distribution of global warming will not be homogeneous in the twenty-first century, and that large-scale changes in surface temperature gradients and atmospheric circulation may result (Cai and Whetton 2000). Recently, Anderson et al. (2002) argued that an increase in Asian monsoon intensity during the twentieth century is related to Northern Hemisphere air temperature changes during the past century. The effect of accelerated heating of the Asian landmass, relative to the tropical ocean, would be to pull the mean annual position of the ITCZ north of the equator. Such a northward shift of the center of tropical convection should invigorate the Hadley circulation and warm the Southern Hemisphere high latitudes.

Several recent studies have noted changes in the tropical energy budget related to a strengthening of the Hadley-Walker circulations during the 1990s (e.g., Chen et al. 2002; Hoerling and Kumar 2003). The alteration of the tropical general circulation was associated with intensified ascending motion and moistening of the equatorial convective regions and stronger sinking motion and drying of the equatorial and subtropical subsidence regions. Such a scenario is similar to that associated with the La Niña–like state in Pacific SSTs during the early Holocene. Indeed, a persistent La Niña from 1998 through 2002, together with above average SSTs in the western Pacific, have been linked with warming and drying of the mid-latitudes of both hemispheres (Barlow et al. 2002; Hoerling and Kumar 2003.).

While the ocean-atmosphere feedbacks identified for the early Holocene warming provide only partial analogues for a climate that may be influenced by greenhouse-gas forcing, both the paleo data and recent observations indicate that a strengthening of the tropical general circulation may well amplify any warming produced by enhanced levels of atmospheric CO_2.

8. ACKNOWLEDGMENTS

We thank Ray Bradley and Henry Diaz for organizing the "Hadley Circulation: Present, Past and Future" meeting in November 2002 at the University of Hawaii, which provided fuel for thought when we wrote this chapter. Peter Isdale and David Hopley kindly provided access to their suites of modern and fossil coral cores for the paleoclimate reconstructions for the Great Barrier Reef. Bambang Suwargadi, Dudi Prayudi, and Anto Sanyoto of the Indonesian Institute of Sciences are warmly thanked for dedicated assistance during coral drilling expeditions in Indonesia. Particular thanks go to Joe Cali, Graham Mortimer, and Heather Scott-Gagan for skillful mass spectrometry in the Research School of Earth Sciences laboratories. The collaborative contributions of Linda Ayliffe, Wahyoe Hantoro, and Malcolm McCulloch toward the production of the coral records reviewed in this chapter are greatly appreciated. Support for MKG's research in the Great Barrier Reef and Indonesia was provided by The Australian National University.

9. REFERENCES

Anderson, D.M., J.T. Overpeck, and A.K. Gupta. 2002. Increase in the Asian southwest monsoon during the past four centuries. *Science* 297: 596–599.

Baker, P.A., G.O. Seltzer, S.C. Fritz, R.B. Dunbar, M.J. Grove, P.M. Tapia, S.L. Cross, H.D. Rowe, and J.P Broda. 2001. The history of South American tropical precipitation for the past 25,000 years. *Science* 291: 640–643.

Barlow, M., H. Cullen, and B. Lyon. 2002. Drought in central and southwest Asia: La Niña, the warm pool, and Indian Ocean precipitation. *Journal of Climate* 15: 697–700.

Beck, J.W., R.L. Edwards, E. Ito, F.W. Taylor, J. Récy, F. Rougerie, P. Joannot, and C. Henin. 1992. Sea-surface temperature from coral skeletal strontium/calcium ratios. *Science* 257: 644–647.

Beck, J.W., J. Récy, F. Taylor, R.L. Edwards, and G. Cabioch. 1997. Abrupt changes in early Holocene tropical sea surface temperature derived from coral records. *Nature* 38: 705–707.

Bjerknes, J., 1966: A possible response of the atmospheric Hadley circulation to equatorial anomalies of ocean temperature. *Tellus* 18: 820–829.

Bjerknes, J. 1969. Atmospheric teleconnections from the equatorial Pacific. *Monthly Weather Review* 97: 163–172.

Bradley, R.S., K.R. Briffa, J.E. Cole, M.K. Hughes, and T.J. Osborn. 2003. The climate of the last millennium. In, Alverson, K., R.S. Bradley, and T.F. Pedersen (eds.). *Paleoclimate, Global Change and the Future.* Berlin: Springer-Verlag, pp. 105–141.

Bradley, R.S., M. Vuille, D. Hardy, and L.G. Thompson. 2003. Low latitude ice cores record Pacific sea surface temperatures. *Geophysical Research Letters* 30(4): 1174, doi:10.1029/2002GL016546.

Broecker, W.S. 1982. Glacial to interglacial changes in ocean chemistry. *Progress in Oceanography* 11: 151–197.

Broecker, W.S. 1997. Mountain glaciers: Recorders of atmospheric water vapor content? *Global Biogeochemical Cycles* 11: 589–597.

Cai, W., and P.H. Whetton. 2000. Evidence for a time-varying pattern of greenhouse warming in the Pacific Ocean. *Geophysical Research Letters* 27: 2577–2580.

Cane, M.R. 1998. A role for the tropical Pacific. *Science* 282: 59–61.

Cane, M.R., and A.C. Clement. 1999. A role for the tropical Pacific coupled ocean-atmosphere system on Milankovitch and millennial timescales: Part II: Global impacts. *Mechanisms of Global Climate Change at Millennial Time Scales. Geophysical Monograph* 112: 373–383.

Chen, J., B.E. Carlson, and A.D. Del Genio. 2002. Evidence for strengthening of the tropical general circulation. *Science* 295: 838–841.

Clement, A., M. Cane, and R. Seager. 2001. An orbitally driven tropical source for abrupt climate change. *Journal of Climate* 14: 2369–2375.

Clement, A.C., R. Seager, and M.A. Cane. 2000. Suppression of El Niño during the mid-Holocene by changes in the Earth's orbit. *Paleoceanography* 15: 731–737.

Corrège, T., T. Delcroix, J. Recy, W. Beck, G. Cabioch, and F. Le Cornec. 2000. Evidence for stronger El Niño–Southern Oscillation (ENSO) events in a mid-Holocene massive coral. *Paleoceanography* 14: 465–470.

Dahl-Jensen, D., K. Mosegaard, N. Gundestrup, G.D. Clow, S.J. Johnsen, A.W. Hansen, and N. Balling. 1998. Past temperatures directly from the Greenland ice sheet. *Science* 282: 268–271.

Dansgaard, W. 1964. Stable isotopes in precipitation. *Tellus* 16: 436–468.

de Villiers, S., G.T. Shen, and B.K. Nelson. 1994. The Sr/Ca-temperature relationship in coralline aragonite: Influence of variability in $(Sr/Ca)_{seawater}$ and skeletal growth parameters. *Geochimica et Cosmochimica Acta* 58: 197–208.

Feely, R.A., R. Wanninkhof, C.E. Cosca, P.P. Murphy, M.F. Lamb, and M.D. Steckley. 1995. CO_2 distribution in the equatorial Pacific during the 1991–92 ENSO event. *Deep-Sea Research* II 42: 365–386.

Fisher, D.A., and R.M. Koerner. 2003. Holocene ice-core climate history—A multivariable approach. In, Mackay, A.W., R.W. Battarbee, H.J.B. Birks, and F. Oldfield. *Global Change in the Holocene: Approaches to Reconstructing Fine-Resolution Climate Change.* London: Arnold, pp. 281–293.

Fisher, D.A., R.M. Koerner, and N. Reeh. 1995. Holocene climatic records from Agassiz Ice Cap, Ellesmere Island, NWT, Canada. *The Holocene* 5: 19–24.

Gagan, M.K., L.K. Ayliffe, J.W. Beck, J.E. Cole, E.R.M. Druffel, R.B. Dunbar, and D.P. Schrag. 2000. New views of tropical paleoclimates from corals. *Quaternary Science Reviews* 19: 45–64.

Gagan, M.K., L.K. Ayliffe, D. Hopley, J.A. Cali, G.E. Mortimer, J. Chappell, M.T. McCulloch, and M.J. Head. 1998. Temperature and surface-ocean water balance of the mid-Holocene tropical western Pacific. *Science* 279: 1014–1018.

Gagan, M.K., E.J. Hendy, S.G. Haberle, and W.S. Hantoro. 2004. Post-glacial evolution of the Indo-Pacific Warm Pool and El Niño–Southern Oscillation. *Quaternary International* 118–119: 127–143.

Grootes, P.M., M. Stuiver, J.W.C. White, S. Johnsen, and J. Jouzel. 1993. Comparison of oxygen isotope records from the GISP2 and GRIP Greenland ice cores. *Nature* 366: 552–554.

Hardy, D.R., M. Vuille, and R.S. Bradley. 2003. Variability of snow accumulation and isotopic composition on Nevado Sajama, Bolivia. *Journal of Geophysical Research* 108: D22, 4693, doi:10.1029/2003JD003623, 2003.

Haug, G.H., K.A. Hughen, D.M. Sigman, L.C. Peterson, and U. Rohl. 2001. Southward migration of the intertropical convergence zone through the Holocene. *Science* 293: 1304–1308.

Hendy, E.J., M.K. Gagan, C.A. Alibert, M.T. McCulloch, J.M. Lough, and P.J. Isdale. 2002. Abrupt decrease in tropical Pacific sea surface salinity at end of Little Ice Age. *Science* 295: 1511–1514.

Hoerling, M., and A. Kumar. 2003. The perfect ocean for drought. *Science* 299: 691–694.

Hou, A.Y. 1993. The influence of tropical heating displacements on the extratropical climate. *Journal of Atmospheric Science* 50: 3553–3570.

Hou, A.Y. 1998. Hadley circulation as a modulator of the extratropical climate. *Journal of Atmospheric Science* 55: 2437–2457.

Indermühle, A., T.F. Stocker, F. Joos, H. Fischer, H.J. Smith, M. Wahlen, B. Deck, D. Mastroianni, J. Tschumi, T. Blunier, R. Meyer, and B. Stauffer. 1999. Holocene carbon-cycle dynamics based on CO_2 trapped in ice at Taylor Dome, Antarctica. *Nature* 398: 121–126.

Jouzel, J., V. Masson, O. Cattani, S. Falourd, M. Stievenard, B. Stenni, A. Longinelli, S.J. Johnsen, J.P. Steffenssen, J.R. Petit, J. Schwander, R. Souchez, and N.I. Barkov. 2001. A new 27 ky high resolution East Antarctic climate record. *Geophysical Research Letters* 28: 3199–3202.

Karner, D.B., and R.A. Muller. 2000. A causality problem for Milankovitch. *Science* 288: 2143–2144.

Keigwin, L.D. 1978. Pliocene closing of the Isthmus of Panama based on biostratigraphic evidence from nearby Pacific Ocean and Caribbean Sea cores. *Geology* 6: 630–634.

Kienast, M., S. Steinke, K. Stattegger, and S.E. Calvert. 2001. Synchronous tropical south China Sea SST change and Greenland warming during deglaciation. *Science* 291: 2132–2134.

Koerner, R.M. 1977. Devon Island ice cap: Core stratigraphy and paleoclimate. *Science* 196: 15–18.

Koutavas, A., J. Lynch-Stieglitz, T.M. Marchitto, Jr., and J.P. Sachs. 2002. El Niño–like pattern in ice age tropical Pacific sea surface temperature. *Science* 297: 226–230.

Lawrence, J.R., and S.D. Gedzelman. 2003. Tropical ice core isotopes: Do they reflect changes in storm activity? *Geophysical Research Letters* 30(2): 1072, doi:10.1029/2002GL015906.

Lea, D.W., D.K. Pak, and H.J. Spero. 2000. Climate impact of late Quaternary equatorial Pacific sea surface temperature variations. *Science* 289: 1719–1724.

Lindzen, R.S., and A.Y. Hou. 1988. Hadley circulation for zonally averaged heating centered off the equator. *Journal of Atmospheric Science* 45: 2417–2427.

Lindzen, R.S., and W. Pan. 1994. A note on orbital control of equator-to-pole heat fluxes. *Climate Dynamics* 10: 49–57.

Liu, Z. 1998. The role of ocean in the response of tropical climatology to global warming: The west-east contrast. *Journal of Climate* 11: 864–875.

Liu, Z., and B. Huang. 1997. A coupled theory of tropical climatology: Warm Pool, cold tongue, and Walker circulation. *Journal of Climate* 10: 1662–1679.

Liu, Z., J. Kutzbach, and L. Wu. 2000. Modeling climate shift of El Niño variability in the Holocene. *Geophysical Research Letters* 27: 2265–2268.

McCulloch, M.T., G. Mortimer, T. Esat, L. Xianhua, B. Pillans, and J. Chappell. 1996. High resolution windows into early Holocene climate: Sr/Ca coral records from the Huon Peninsula. *Earth and Planetary Science Letters* 138: 169–178.

McGregor, H.V., and M.K. Gagan. 2003. Diagenesis and geochemistry of *Porites* corals from Papua New Guinea: Implications for palaeoclimate reconstruction. *Geochimica et Cosmochimica Acta* 67: 2147–2156.

Meyers, G. 1996. Variation of Indonesian throughflow and the El Niño–Southern Oscillation. *Journal of Geophysical Research* 101: 12255–12263.

Morrill, C., J.T. Overpeck, and J.E. Cole. 2003. A synthesis of abrupt changes in the Asian summer monsoon since the last deglaciation. *The Holocene* 13: 465–476.

Moy, C.M., G.O. Seltzer, D.T. Rodbell, and D.M. Anderson. 2002. Variability of El Niño/Southern Oscillation activity at millennial timescales during the Holocene epoch. *Nature* 420: 162–165.

Müller, A., M.K. Gagan, and M.T. McCulloch. 2001. Early marine diagenesis in corals and geochemical consequences for paleoceanographic reconstructions. *Geophysical Research Letters* 28: 4471–4474.

Pahnke, K., R. Zahn, H. Elderfield, and M. Schulz. 2003. 340,000-year centennial-scale marine record of Southern Hemisphere climatic oscillation. *Science* 301: 948–952.

Palmer, M.R., and P.N. Pearson. 2003. A 23,000-year record of surface water pH and pCO_2 in the western equatorial Pacific Ocean. *Science* 300: 480–482.

Pelejero, C., J.O. Grimalt, S. Heilig, M. Kienast, and L. Wang. 1999. High-resolution U^{k}_{37} temperature reconstructions in the South China Sea over the past 220 kyr. *Paleoceanography* 14: 224–231.

Petit, J.R., J. Jouzel, D. Raynaud, N.I. Barkov, J.-M. Barnola, I. Basile, M. Bender, J. Chappellaz, M. Davis, G. Delaygue, M. Delmotte, V.M. Kotlyakov, M. Legrand, V.Y. Lipenkov, C. Lorius, L. Pepin, C. Ritz, E. Saltzman, and M. Stievenard. 1999. Climate and atmospheric history of the past 420,000 years from the Vostok ice core, Antarctica. *Nature* 326: 273–277.

Philander, S.G., and A.V. Federov. 2003. Role of the tropics in changing the response to Milankovitch forcing some three million years ago. *Paleoceanography* 18(2), 1045, doi:10.1029/2002PA000837.

Pierrehumbert, R.T. 1999. Huascarán $\delta^{18}O$ as an indicator of tropical climate during the Last Glacial Maximum. *Geophysical Research Letters* 26: 1345–1348.

Pierrehumbert, R.T. 2000. Climate change and the tropical Pacific: The sleeping dragon wakes. *Proceedings of the National Academy of Science* 97: 1355–1358.

Rind, D. 1998. Latitudinal temperature gradients and climate change. *Journal of Geophysical Research* 103: 5943–5971.

Rind, D. 2000. Relating paleoclimate data and past temperature gradients: Some suggestive rules. *Quaternary Science Reviews* 19: 381–390.

Rodbell, D.T., G.O. Seltzer, D.M. Anderson, M.B. Abbott, D.B. Enfield, and J.H. Newman. 1999. An ~15,000-year record of El Niño–driven alluviation in southwestern Ecuador. *Science* 283: 516–520.

Romine, K. 1982. Late Quaternary history of atmospheric and oceanic circulation in the eastern equatorial Pacific. *Marine Micropaleontology* 7: 163–187.

Rosenthal, Y., D.W. Oppo, and B.K. Linsley. 2003. The amplitude and phasing of climate change during the last deglaciation in the Sulu Sea, western equatorial Pacific. *Geophysical Research Letters* 30(8): 1428, doi:10.1029/2002GL016612.

Sandweiss, D.H., J.B. Richardson, III, E.J. Reitz, H.B. Rollins, and K.A. Maasch. 1996. Geoarchaeological evidence from Peru for a 5000 B.P. onset of El Niño. *Science* 273: 1531–1533.

Seltzer, G.O., D.T. Rodbell, P.A. Baker, S.C. Fritz, P.M. Tapia, H.D. Rowe, and R.B. Dunbar. 2002. Early warming of tropical South America at the last glacial-interglacial transition. *Science* 296: 1685–1686.

Stott, L., C. Poulsen, S. Lund, and R. Thunell. 2002. Super ENSO and global climate oscillations at millennial time scales. *Science* 297: 222–226.

Sun, D.-Z., and K.E. Trenberth. 1998. Coordinated heat removal from the equatorial Pacific during the 1986–87 El Niño. *Geophysical Research Letters* 25: 2659–2662.

Takahashi, T., S.C. Sutherland, C. Sweeney, A. Poisson, N. Metzl, B. Tilbrook, N. Bates, R. Wanninkhof, R.A. Feely, C. Sabine, J. Olafsson, and Y. Nojiri. 2002. Global sea-air CO_2 flux based on climatological surface ocean pCO_2, and seasonal biological and temperature effects. *Deep-Sea Research* II 49: 1601–1622.

Thompson, L.G., M.E. Davis, E. Mosley-Thompson, T.A. Sowers, K.A. Henderson, V.S. Zagorodnov, P.-N. Lin, V.N. Mikhalenko, R.K. Campen, J.F. Bolzan, J.A. Cole-Dai, and B. Francou. 1998. A 25,000-year tropical climate history from Bolivian ice cores. *Science* 282: 1858–1864.

Thompson, L.G., E. Mosley-Thompson, and K.A. Henderson. 2000. Ice-core palaeoclimate records in tropical South America since the Last Glacial Maximum. *Journal of Quaternary Science* 15(4): 377–394.

Thompson, L.G., E. Mosley-Thompson, M.E. Davis, K.A. Henderson, H.H. Brecher, V.S. Zagorodnov, T.A. Mashiotta, P.-N. Lin, V.N. Mikhalenko, D.R. Hardy, and J. Beer. 2002. Kilimanjaro ice core records: Evidence of Holocene climate change in tropical Africa. *Science* 298: 589–593.

Thompson, L.G., E. Mosley-Thompson, M.E. Davis, P.-N. Lin, K.A. Henderson, J. Cole-Dai, J.F. Bolzan, and K.-b. Liu. 1995. Late glacial stage and Holocene tropical ice core records from Huascarán, Peru. *Science* 269: 46–50.

Thompson, L.G., E. Mosley-Thompson, M.E. Davis, P.-N. Lin, K. Henderson, and T.A. Mashiotta. 2003. Tropical glaciers and ice core evidence of climate change on annual to millennial time scales. *Climatic Change* 59: 137–155.

Tudhope, A.W., C.P. Chilcott, M.T. McCulloch, E. Cook, J. Chappell, R.M. Ellam, D.W. Lea, J.M. Lough, and G.B. Shimmield. 2001. Variability in the El Niño Southern Oscillation through a glacial-interglacial cycle. *Science* 291: 1511–1517.

Visser, K., R. Thunell, and L. Stott. 2003. Magnitude and timing of temperature change in the Indo-Pacific warm pool during deglaciation. *Nature* 421: 152–155.

Vuille, M. 1999. Atmospheric circulation over the Bolivian Altiplano during dry and wet periods and extreme phases of the Southern Oscillation. *International Journal of Climatology* 19: 1579–1600.

Webster, P.J. 1982. Seasonality in the local and remote response to sea surface temperature anomalies. *Journal of Atmospheric Science* 39: 41–52.

Webster, P.J. 1994. The role of hydrological processes in ocean-atmosphere interactions. *Reviews of Geophysics* 32: 427–476.

Yan, X.-H., C.-R. Ho, Q., Zheng, and V. Klemas. 1992. Temperature and size variabilities of the Western Pacific Warm Pool. *Science* 258: 1643–1645.

Chapter 11

LATE QUATERNARY HYDROLOGIC CHANGES IN THE ARID AND SEMIARID BELT OF NORTHERN AFRICA

Implications for Past Atmospheric Circulation

Françoise Gasse[1] and C. Neil Roberts[2]

[1]*CEREGE, UMR 6635, BP 80, 13545, Aix-en-Provence Cedex 4, France*
E-mail: gasse@cerege.fr
[2]*School of Geography, University of Plymouth, Drake Circus, Plymouth, PL4 8AA, UK*
E-mail: c.n.roberts@plymouth.ac.uk

Abstract The zonal climate pattern associated with the Hadley cell circulation is best exemplified in northern Africa, with its Mediterranean northern tip, subtropical Sahara desert, and belts of monsoonal and equatorial climates related to the seasonal migration of the Intertropical Convergence Zone (ITCZ). In the past, astronomical forcing has been the prime factor driving the meridional shifts of these climate belts, but feedback processes from oceans and land surfaces have amplified and modified the direct effects of insolation changes.

This chapter uses selected groundwater, paleolake, and paleobotanic data to illustrate changes in precipitation, moisture sources and trajectories, and wind intensity over the Sahara and its southern semiarid margins (~30°N–10°N) over the past 25,000 years. This time interval spans a wet Late Pleistocene phase, followed by two periods of extreme dry and wet conditions, the last glacial maximum (LGM) and the early to mid-Holocene, respectively.

During the cool Late Pleistocene wet period, data indicate a strengthening and a southward displacement of extratropical cyclonic disturbances associated with an equatorial shift of the subtropical westerly jet and of the Saharan anticyclone. During the LGM, the generally dry conditions are in good agreement with model simulations. Northern Africa climates responded to reduced summer insolation over the Northern Hemisphere, associated with a stronger northern branch of the Hadley cell circulation, tropical cooling, and a global decrease in water vapor.

Data from the early to mid-Holocene wet period show a northward migration of the tropical rainfall belts as far as 20°N–24°N. They suggest a strengthening of the ITCZ, a northward migration of 5°–6° of the core of the upper-level

H.F. Diaz and R.S. Bradley (eds.), The Hadley Circulation: Present, Past and Future, 313–345.

easterly jets, and a recycling of water vapor along the West African Monsoon (WAM) flow toward the southeastern Sahara. North of 20°N–24°N, moisture was most likely of extratropical origin, as is the case today. The best agreement between observations and simulations is found with coupled ocean-atmosphere-vegetation models. Orbital forcing enhanced the land-sea pressure gradient; feedbacks from the vegetation and the oceans amplified the intensity and the northward penetration of WAM rainfalls and the length of the monsoon season.

1. INTRODUCTION

Africa (37°N–34°5'S), with the equator in its center and a rather uniform surface, conforms better than all other continents to the zonal pattern of climates associated with the Hadley cell circulation (Fig. 11-1). Climate belts are roughly symmetric about the equator. The extratropical northern and southern tips of the continent project into the belts of mid-latitude westerlies, receiving winter precipitation from westerly cyclonic disturbances. These temperate regions are flanked by subtropical deserts, the large hyperarid Sahara desert north of the equator, and the smaller coastal Namib Desert in southwestern Africa, associated with the descending branch of the Hadley cells. A wide belt of tropical climates separates the two arid zones. A zone of maximum rainfall related to the Intertropical Convergence Zone (ITCZ) follows the overhead position of the sun with a 4- to 6-week time lag. The ITCZ migration results in an equatorial zone of humid climates with two seasonal rainfall maxima, flanked on the north and south by broad belts of monsoonal climates with a single summer rainy season and winter drought (Fig. 11-2). There are two distinct monsoonal systems. The West African Monsoon (WAM) originates from the high-pressure system over the subtropical southern Atlantic Ocean (St. Helena) and has an important zonal component in northern Africa; the complex East African monsoonal system shows reversed sub-meridional flows originating from the subtropical high cells over the southern Indian Ocean (Mascarene High) and over Arabia during the Northern and Southern Hemisphere summer, respectively.

The zonal climate pattern related to the Hadley cell circulation is well displayed north of the equator. It is much less clear southward, where a marked east-west climate asymmetry is linked to the steep topography of East Africa and to sea surface currents. This chapter thus concentrates on northern Africa, more specifically on its arid and semiarid belt, the Sahara

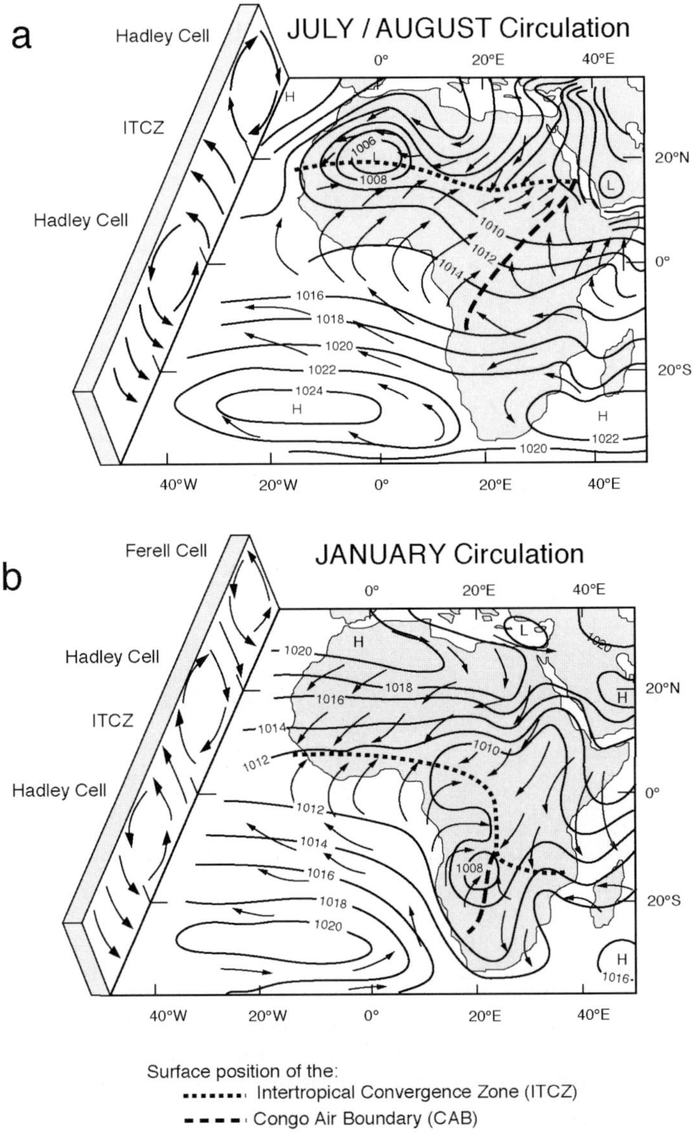

Figure 11-1. General patterns of low-level winds and pressure over Africa. (a): July/August. (b): January. From Nicholson (1996) and Griffiths (1972).

(~32°N–16°N) and its southern margins from the Atlantic Ocean to Ethiopia (~16°N–10°N). It uses selected paleohydrological and paleobotanical evidence to discuss millennial time scale changes in precipitation, moisture sources and trajectories, and wind intensity in the region over the past 25,000 years.

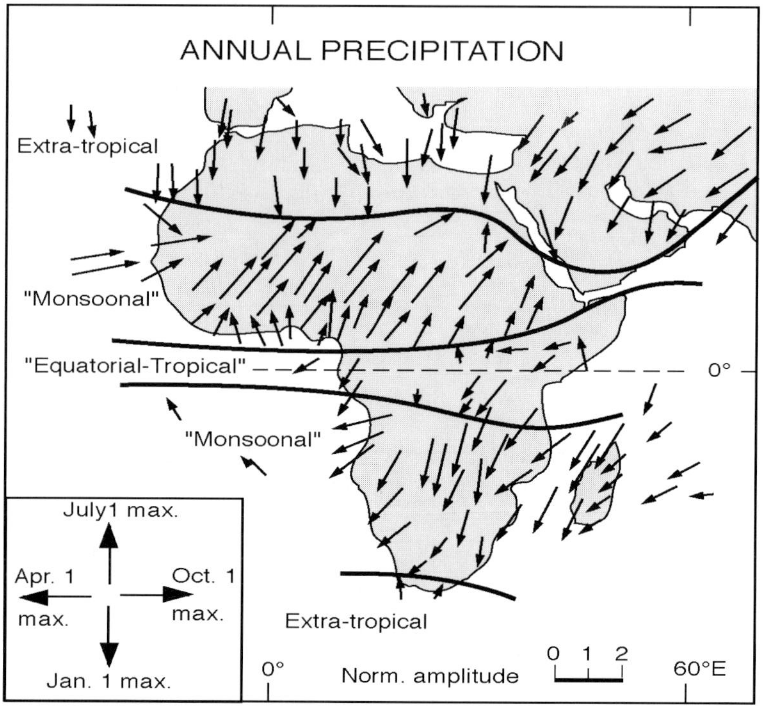

Figure 11-2. African precipitation regimes. Annual harmonics of precipitation, from Hsu and Wallace (1976), adapted by deMenocal and Rind (1996). Rainfall time series from individual climatological stations were averaged into monthly means and then subjected to harmonic analysis, which defined the phase and amplitude of the annual precipitation cycle. The vector length indicates normalized amplitude. Vector direction indicates month of maximum precipitation (e.g., southward-pointing vectors indicate January 1 rainfall maximum).

It has long been established that northern Africa has undergone large hydrologic fluctuations during the Late Quaternary (e.g., Flohn and Nicholson 1980; Kutzbach and Street-Perrott 1985; Street-Perrott and Perrott 1993; Gasse 2000; Hoelzmann et al., in press). The region experienced a late Pleistocene wet period between ~40,000 and 23,000 BP, generally dry climates during and after the last glacial maximum (LGM) at ~21,000 BP, much wetter conditions than today during the early mid-Holocene (~11,000–4,000 BP), and the establishment of modern climates in the late Holocene. The most striking event has been the wetting and greening of the Sahara between about 10,000 and 5,000 BP, the so-called "African Humid Period" (AHP). General circulation models (GCMs) have shown that these long-term changes can be primarily attributed to astronomical precession forcing, which predicts a dry LGM and strengthening of monsoon rainfalls during periods of increased summer insolation in the Northern Hemisphere (e.g., Kutzbach and Street-Perrott 1985); but feedback processes from the oceans and land surfaces are required to account for the amplitude and the abruptness of observed changes in available moisture (see, e.g., Braconnot et al., in press).

In this chapter, attention is paid to the mean hydrological conditions, compared to modern, of the Late Pleistocene wet phase, the LGM, and the AHP in the Sahara and its southern semiarid margins. Emphasis is given to the AHP, which is best documented both in data and GCM simulations.

2. MAJOR MODERN ATMOSPHERIC CIRCULATION AND CLIMATIC PATTERNS

The mean annual and seasonal distribution of rainfall in Africa results mainly from the atmospheric circulation patterns over the continent, which have marked shifts between the summer of the Northern (July–August) and Southern (January) Hemispheres (Griffiths 1972; Hastenrath 1991; McGregor and Nieuwolt 1998; Nicholson 2000).

2.1. Large-Scale Atmospheric Circulation Patterns over Africa

The temperate extremes of the continent receive winter rains from the temperate frontal disturbances embedded within the mid-latitude westerlies associated with the subtropical westerly jet (SWJ; Figs. 11-1, 11-2, 11-3a). These westerlies become more frequent in January, when those of the Northern Hemisphere are displaced far equatorward. Over the tropics and subtropics, the circulation is associated with the seasonal migrations of the Hadley cells.

At low altitude, in July–August, thermal low-pressure cells build up over the Sahara and southern Arabia. The surface position of the ITCZ—the confluence between the northeast trade winds and the monsoon flow—lies at 15°N–22°N (Figs. 11-1a, 11-3b). Along a meridional-vertical transect in

West Africa (Fig. 11-3b), the slope of the discontinuity front (sometimes called the intertropical front or the intertropical discontinuity) is very weak in summer; the zone of maximum cloudiness and rainfall is separated from the surface position of the ITCZ by 500–1,000 km.

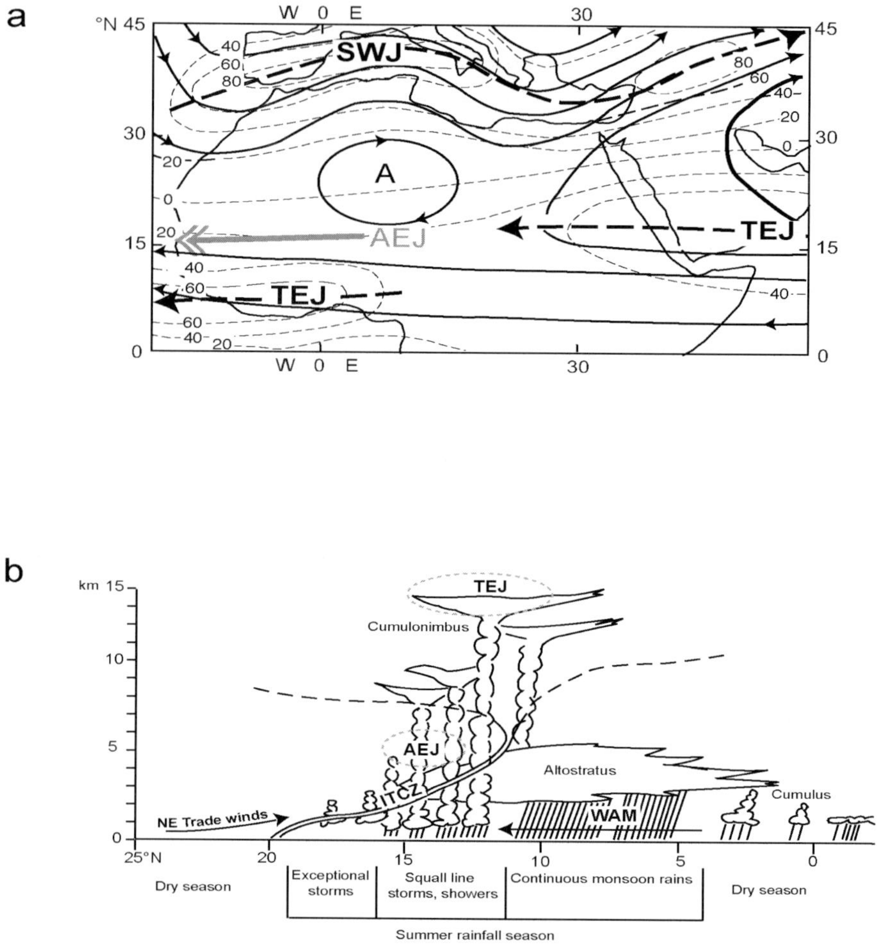

Figure 11-3. Relationships between low- and upper-level circulation patterns over Africa. (a): Upper-level winds over northern Africa in July–August (adapted from McGregor and Nieuwolt 1998). Thin dashed lines are wind speeds (isotacks) in knots (1 knot = 0.51 m s^{-1}). (b): Meridional-vertical transect through the Intertropical Convergence Zone (or Intertropical Front) over West Africa along about 0° in July–August (adapted from Leroux [983] and Janicot [1990]).
ITCZ: Intertropical Convergence Zone. WAM: West African Monsoon. TEJ: tropical easterly jet. AEJ: African easterly jet. SWJ: subtropical westerly jet. A: upper-level anticyclone.

A second convergence zone, the Congo Air Boundary, separates the flows from the southern Atlantic and Indian Oceans. The southern subtropical high-pressure systems over the Atlantic and Indian Oceans are strengthened and high pressure prevails over southern Africa. The thermally direct circulation of WAM penetrates far inland through monsoonal westerlies over the rather flat northern Africa subcontinent, but monsoonal rainfall strongly decreases along its southwest-to-northeast flow.

In East Africa, the roughly meridional Indian monsoon flow brings some moisture to the northern tropics. In January (Fig. 11-1b), the general picture is reversed, with high pressure over the Sahara and Arabia and a low pressure over southern Africa. A low-pressure area extends over the central and eastern Mediterranean, but most of northern Africa is dry, as it is dominated by the Saharan high and the strong northeasterly trade winds. An equatorial trough develops over central Africa. The convergence zones have moved southward, with the ITCZ and related rainfalls lying close to the equator over West and Central Africa and penetrating far into the Southern Hemisphere over East Africa.

At high altitude, two important jet streams are embedded in the easterlies in July–August (Fig. 11-3). The well-developed tropical easterly jet (TEJ; 200–100 mb) at 5°N–15°N originates from the upper-level anticyclone that develops over the Indian-Tibetan heat low during the Indian monsoon season. Its subsidence over subtropical Africa likely plays a role in moderating the northeastward advance of the WAM. The West African easterly jet (AEJ; 600 mb) at about. 15°N originates from the mid-tropospheric temperature gradient between the warm Sahara desert and the cold Gulf of Guinea. The TEJ and the AEJ are crucial in the development of rain-bearing disturbances and of summer rainfall. At the surface, rainfall has been found to be greatest on their equatorward side.

2.2. Major Modern Climate Features in the Sahara and Its Southern Margins

The Sahara desert stretches about 1,500 km from north to south between the ≈200 mm isohyets, with almost no measurable rainfall in its core (Fig. 11-4). Its climate is modulated by the distance from the coast and by the Saharan mountains (Ahaggar, Aïr, Tibesti, Djebel Marrah; Fig. 11-4). There is an overall decrease in mean annual rainfall from west to east. The mountains act as regional water towers, but their mean annual rainfall (P) does not rise above an average of 100–150 mm yr^{-1}. North of 20°N–24°N, most of the disturbances are basically of extratropical origin and occur during the winter and the transition seasons (Fig. 11-2). These disturbances develop from troughs in the mid-latitude westerlies (Nicholson 2000). South of 20°N–24°N, occasional summer thunderstorms are associated with the ITCZ. Some areas, e.g. the Ahaggar Mountains, in fact experience a bimodal rainfall regime, with tropical rainfall events in summer and cyclonic disturbances in winter. Winter aridity of the Sahara is accounted for by the position of the descending branch of the northern Hadley cell. In summer, subsiding air of the TEJ may contribute to aridity of the southern Sahara. Rodwell and Hoskins (1996) have suggested that the Indian monsoon heating induces a Rossby-wave pattern to its west, generating a warm thermal structure that interacts with air on the southern flank of the mid-latitude westerlies and causing it to descend; the strongest descent is centered over the eastern Mediterranean and the eastern Sahara, enhancing aridity in these regions (Rodwell and Hoskins 1996).

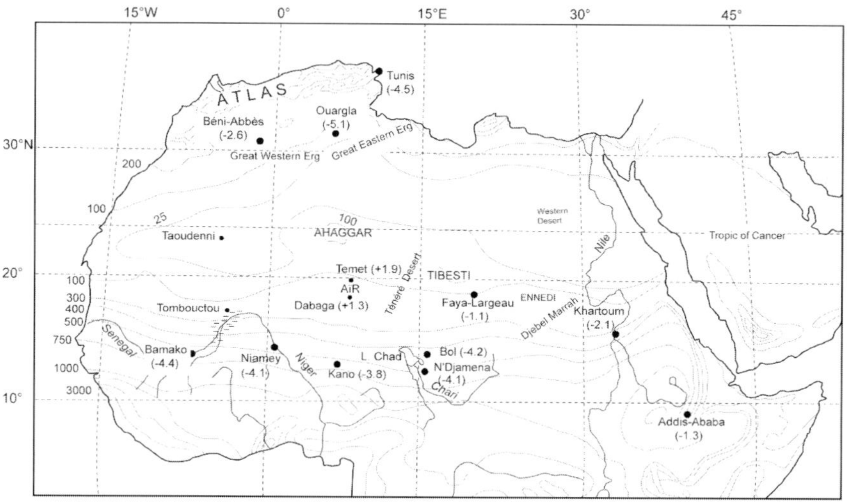

Figure 11-4. Northern Africa. mean annual rainfall (in mm yr^{-1}). Numbers between brackets represent the yearly weighted mean $\delta^{18}O$ values of modern precipitation at some African stations. After IAEA (1992), IAEA/WMO (1998); Gallaire (1995).

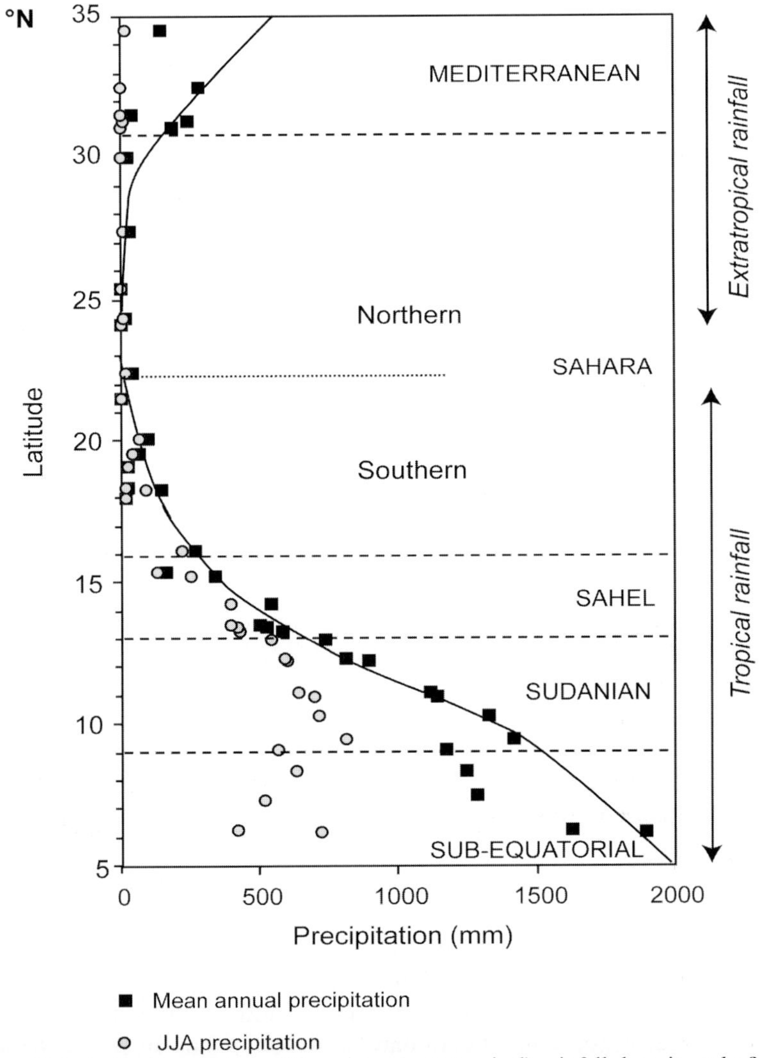

Figure 11-5. Latitudinal pattern of the winter (extratropical) rainfall domain and of the summer (tropical) rainfall domain over northern Africa. JJA: June + July + August. Diagram based on 41 meteorological stations. Data from Griffiths (1972).

Southward, the Sahara grades into a semiarid zone (P: 200–1,000 mm yr^{-1}), a region having from 3 to 6 summer months each with at least 50 mm of rain (Griffiths 1972). The semiarid zone includes the Sahel belt (~16°N–13°N) and the so-called Sudanian belt (~13°N–10°N) (Fig. 11-5). It is flanked equatorward by zones of increasing precipitation with a bimodal regime. The climatic zonation results in a broadly zonal pattern of the vegetation distribution. Schematically, the very restricted Saharan vegetation progressively turns southward into the Sahelian steppe and wooded grasslands (P: 200–500 mm yr^{-1}), which grade into the Sudanian dry forest and savanna (P: 500–1,500 mm yr^{-1}). The Sudanian vegetation is replaced by evergreen and semideciduous forests near the equator (White 1983).

The semiarid zone experiences a very steep meridional precipitation gradient (Fig. 11-5; 180 mm yr^{-1} per degree latitude). In most of the region, maximum rainfall occurs in August from heavy squall line thunderstorms (Fig. 11-3b). Squall lines are basically a part of the large-scale WAM wind system, but are initiated at much higher altitude than the continuous monsoon rains that fall south of 12°N–10°N (Fig. 11-3b). The upper-level jets, especially the AEJ, are important in initiating and driving the westward movement of these linear depression lines. Small meridional shifts in the position of the easterly jets can produce rather large rainfall anomalies. Squall lines, which may develop as high as 16 km (Fig. 11-3b), receive moisture from the low-level WAM flow in which their bases are embedded (Lamb et al. 1998).

The origin of atmospheric moisture required for West African rainfall is complex. Advection from the Atlantic Ocean is very sensitive to sea surface temperatures (SSTs) (e.g., Giannini et al. 2003). A significant part of the moisture flux that is vertically integrated in the layers 1,000–300 hPa or 1,000–925 hPa is advected into the WAM from the Mediterranean and from Central Africa (Fontaine et al. 2003). Evaporation from land (5°N–15°N; 10°W–15°E) may contribute significantly to precipitation (Gong and Eltahir 1996).

The clear climatic and vegetation zonations observed in most of the arid and semiarid belt of northern Africa are altered in its eastern part by the steep topography, which creates marked gradients in temperature, precipitation, and cloudiness over short distances. In northern East Africa south of 20°N, the sources of moisture are the Atlantic and the Indian Oceans. The WAM flow converges with the Indian monsoon winds over the strongly heated Ethiopian Plateau, generating heavy summer rains.

2.3. The Isotopic Signature of Rainfall Patterns

Data on the isotope composition of precipitation in northern Africa are scarce (IAEA 1992; Rozanski et al. 1993; IAEA/WMO 1998), but some general characteristics can, however, be outlined.

- In the monsoon domain, the relationships of rainfall isotope composition with ground temperature and altitude, the so-called "temperature effect" and "altitude effect," are attenuated compared to the Mediterranean and higher latitudes. The rainfall $\delta^{18}O$ value (δ_p) is strongly and negatively correlated to rainfall amount: The wettest months and the heaviest rainfall events within a month show the lowest δ_p values (Rozanski et al. 1993; Taupin et al. 2000). Across the semiarid zone, correlation between δ_p and the rainfall amount is in the range of –1‰ to –2‰ per 100 mm (IAEA 1992). This "amount effect" is also observed in the northern Sahara, where heavy winter rainfalls are generally depleted in heavy isotopes (e.g., Conrad and Fontes 1970).

- The stable isotope composition of the continuous monsoon rains (south of 10°N–12°N) is high compared with that of the squall line showers. The latter exhibit δ_p values as low as –11‰. This difference is attributed to the cloud height (Fontes et al. 1993): Continuous monsoon rains are generated at low altitude (1,000–4,000 m), while the formation process of squall line showers involves low-temperature condensation at high altitude (Fig. 11-3b).

- On the δ^2H versus $\delta^{18}O$ diagrams (Fig. 11-6), the yearly weighted means for stations in northern Africa fall on or close to the Global Meteoric Water Line (GMWL). Stations below the GMWL indicate an evaporative enrichment of raindrops falling in a dry atmosphere. This substantial enrichment partly explains the apparent increase of weighted mean δ_p, which reflects increasing aridity along the southwest-to-northeast flow of the WAM; e.g., from Niamey to Dabaga or Temet (Figs. 11-4 and 11-6).

- At the monthly scale, data from some stations located above the GMWL indicate an addition of reevaporated moisture from continental areas in August (Taupin et al. 2000). It was suggested that recycled moisture from the rain forest of central Africa represents an important source of water vapor in the northern tropics (Sonntag et al. 1978), in agreement with observations by Fontaine et al. (2003).

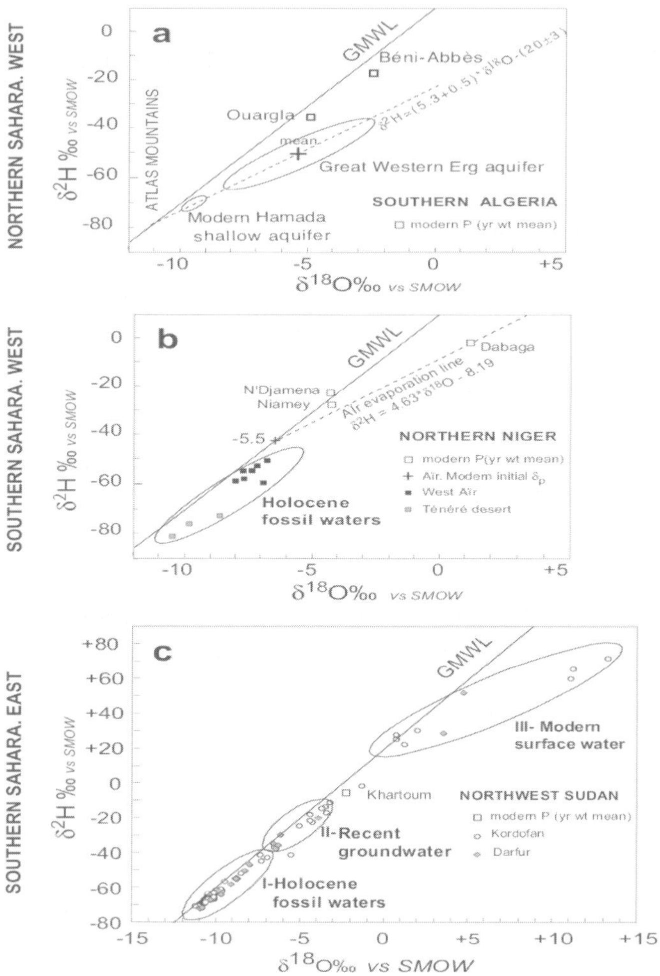

Figure 11-6. Isotopic composition of modern rainwater and of Holocene groundwaters. Data for modern rainfall are from IAEA (1992), IAEA/WMO (1998)), and from Gallaire (1995) for the Aïr region (northern Niger). GMWL: Global Meteoric Water Line. P: modern precipitation; yr wt mean: yearly weighted means. (a) Northern Sahara, southern Algeria. The Great Western Erg (GWE) aquifer and its relation with the Hamada shallow aquifer along the regional evaporation line (after Gonfiantini et al. 1974); a ^{14}C measurement of dissolved carbon in GWE water at Béni-Abbès suggests an age of ca. 6,000–7,000 YBP (unpublished data). (b) Southwestern Sahara, northern Niger. West Aïr after Fontes et al. (1993); Ténéré desert after J. Aranyossy (personal communication). Comparison with isotopic composition of modern rainfall in the Aïr massif (Dabaga) and regional rainfall evaporation line (Gallaire 1995). (c) Southeastern Sahara, northwestern Sudan (Darfur, ca. 14°N/25°E–27°E ; Kordofan: ca 15°N/31°E–32°E) compared to recent groundwater and modern surface waters (after Gröning 1994).

3. EVIDENCE FOR CHANGES IN MEAN HYDROLOGIC CONDITIONS

Late Quaternary paleoclimatic records available in the Sahara-Sahel (Hoelzmann et al., in press) and in the Horn of Africa (Umer et al., in press) have recently been summarized. Edmunds et al. (in press) provided a synthesis of available groundwater data in North Africa. We refer to these works for detailed reviews of regional hydrologic and climatic changes. This section aims at giving an overview, mainly based on published works, of mean hydroclimatic conditions that took place during the Late Pleistocene wet episode, the LGM, and the AHP as recorded from groundwater, lake, and paleovegetation data, successively.

Most of the data used here are based on ^{14}C-dated groundwater archives (major and trace element concentration, stable isotope composition, and noble gas content), lake records (shoreline studies, sediment facies, elemental and stable isotope geochemistry, aquatic microorganisms), and paleobotanical data (pollen and charcoal). Some information on continental eolian deposits and eolian dust influx to the ocean are also considered. Radiocarbon ages were translated into calendar ages using CALIB 4 (Stuiver and Reimer 1993; Stuiver et al. 1998). Record selection criteria are chronological reliability and the understanding of the relationships between proxies and climate that are archive dependent.

3.1. Groundwater Data

The Late Quaternary wet periods have been times of intensive recharge of the deep water reservoirs in the Sahara-Sahel. Fossil waters stored in the large aquifers provide information on climate and environmental conditions when and where the aquifers recharged. Time resolution is low due to the inherent characteristics of subsurface flow and the long residence time, but groundwaters register the average magnitude of long climate periods, e.g., glacial/interglacial changes, since the signal of local effects or short-term extremes is smoothed out. The groundwater $\delta^{18}O$ and $\delta^{2}H$ values

(Fig. 11-6) represent the heaviest rains contributing to groundwater recharge and can be, in favorable cases, interpreted in terms of weighted mean composition of rainfall (initial δ_p) after correction for the evaporation effect (Fontes et al. 1993). Spatial and temporal distribution of groundwater $\delta^{18}O$ values (δ_g) are illustrated in Figure 11-7. Major points arising from groundwater records (Edmunds et al., in press) are the following (Figs. 11-6 and 11-7).

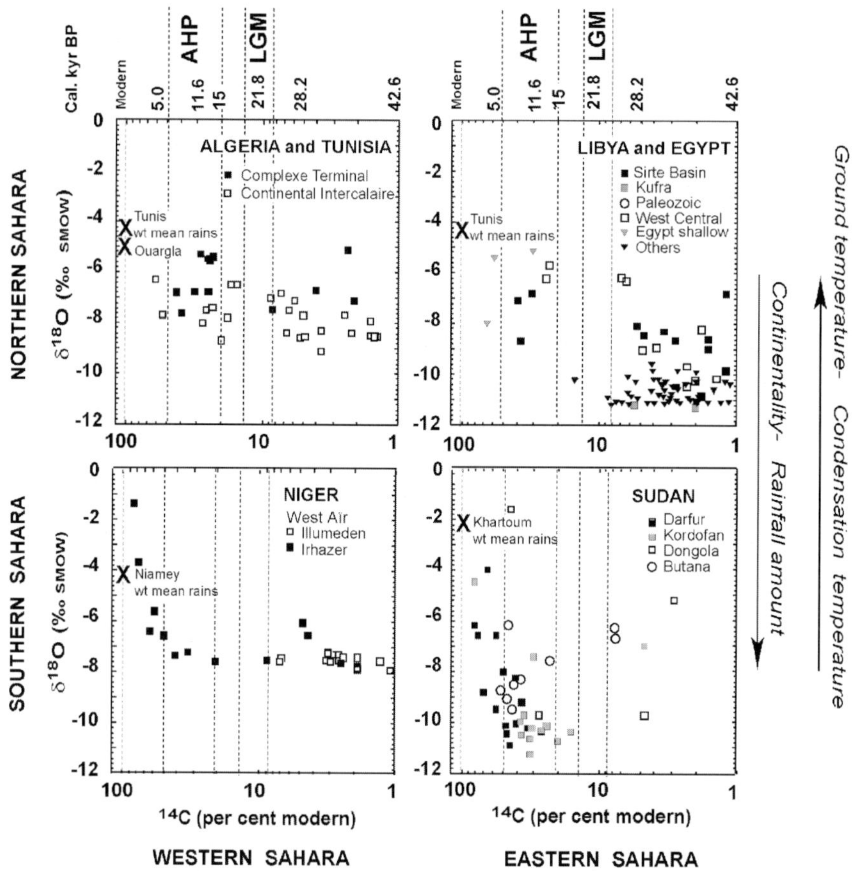

Figure11-7. Summary diagram of changes in groundwater $\delta^{18}O$ values versus ^{14}C activity (providing a time scale) for some aquifers of the Sahara and Sahel. The crosses on the left side of the graphs represent the weighted mean rainfall $\delta^{18}O$ values at neighboring stations. Adapted from Edmunds et al. (in press).

- Major groundwater recharge took place during the Late Pleistocene and early to mid-Holocene wet phases. A gap in most of the data sets during the LGM is interpreted as an arid interlude (Fig. 11-7).

- Late Pleistocene groundwaters are significantly depleted in ^{18}O compared with recent times (Fig. 11-7). Lower ground-level temperatures (ΔT, compared to modern mean annual temperature, T) may have contributed to that difference. For example, noble gas–inferred ΔT was estimated at –5°C to –6°C in northern Nigeria (Edmunds et al., in press), –3°C in northwestern Sudan (Gröning 1994), –4°C in southwestern Egypt (Sonntag et al. 1982), and –2°C to –3°C in southern Algeria (Guendouz et al. 1998).

- An overall decrease in both $\delta^{18}O$ (Fig. 11-7) and $\delta^{2}H$ values from west to east describes a "continental effect," with an Atlantic moisture source evolving according to Rayleigh fractionation from the ocean to the east.

- In the northern Sahara, this "continental effect" is extremely strong during the wet Late Pleistocene period, the major recharge period in the region. Very low δ_g values characterize the eastern Sahara (Fig. 11-7). The similarity of the present winter rain pattern of Europe with that of the northern Sahara in the past indicates that rainfall was of extratropical origin and was brought by the mid-latitude westerlies: Moist Atlantic air masses crossed the northern Sahara from Morocco to the Western Desert in southern Egypt (Sonntag et al. 1978, 1980; Sultan et al. 1997; Edmunds et al., in press). Conversely, the early to mid-Holocene has been a period of moderate recharge; in the Great Western Erg and the Great Eastern Erg aquifers (Fig. 11-4), δ_g (mean values of –5‰ to –6‰) and noble gas–inferred temperature suggest recharge conditions close to those of today (Figs. 11-6a and 11-7) (Gonfiantini et al. 1974; Guendouz et al. 1998).

- In the southern Sahara, the lowest δ_g values fall within the AHP (Fig. 11-7). In West Africa, very low early to mid-Holocene δ_g values are observed at 18°N–21°N. For example, west of the Aïr massif (Fig. 11-6b), δ_g ranges from –6‰ to –11‰. Holocene initial δ_p was estimated at –10‰ to –12‰ (Fontes et al. 1993), 5‰–6‰ lower than the modern initial δ_p in the Aïr massif (δ_p = –5.5‰, Fig. 11-6b), but close to the isotope composition of the heaviest squall line events of the modern Sahel. Similar anomalies were observed along the northwest margin of the Tibesti mountains (Dodo and Zuppi 1997, 1999) and in northern Mali (Fontes et al. 1993). These strong isotopic anomalies can hardly be explained by the "amount effect" alone. They were attributed to a northward migration of the Sahelian squall line rainfall belt of at least 5°, an intensification of squall line showers that possibly formed at higher altitudes than they do at present, thus at lower condensation temperatures (Fontes et al. 1993). A northward development of Sahel-type environments is confirmed by the high dissolved nitrogen content in Holocene groundwater of northern Mali, reflecting biologically active soils up to 20°N–21°N (Fontes et al. 1993). In the eastern Sahara, the Holocene groundwater isotope depletion and the contrast between north and south δ_g values are even more emphasized than in the western Sahara (Figs. 11-6c and 11-7). Gröning (1994) proposed that the extremely low δ_g of the Darfur and Kordofan aquifers (northwestern Sudan), east of the Djebel Marrah massif (Fig. 11-4), was due, first, to evaporation and recycling of a large amount of water from Lake Chad, which was huge at that time (the so-called "Megachad"); and second, to eastward transport of water vapor through the WAM flow, which rose at high altitude over the Djebel Marrah (>3,000 m above sea level [masl]). This moisture may then have significantly contributed (≥150 mm yr^{-1}) to the rainfall amount and isotopic signature in northwestern Sudan. An enhanced Indian monsoon flow may have also brought isotopically depleted rainfall to the region (Thorweihe 1990).

- During the AHP, the southwest- to-northeast gradient in rainfall isotopic composition over the southern Sahara and the Sahel was reversed compared with the modern (Fig. 11-4). This change was due to enhanced rainfall amounts, intensified squall line showers, and much lower evaporation rates.

3.2. Lake Records

Major changes in effective moisture over the past 25,000 years BP (YBP) are illustrated (Fig. 11-8) by the reconstruction of the precipitation minus evaporation (P–E) balance or water levels of some closed lake basins (without surface outlet), and a synthesis of "lake status" fluctuations in the region (8°N–28°N/20°W–43°E) for the past 15,000 YBP (Hoelzmann et al.

1998, in press). Changes in the eolian dust influx off Mauritania (Fig. 11-8) can also be regarded as an integrated aridity index in West Africa (deMenocal et al. 2000). Late Pleistocene lake records are extremely scarce due to intense wind deflation, especially during the LGM and the Younger Dryas event when the northeast trades were strengthened and the desert expanded to the south, as has been documented by the study of continental eolian deposits (Swezey 2001). Figure 11-9 summarizes the spatial distribution and evolution of closed lakes from 18,000 BP to the present.

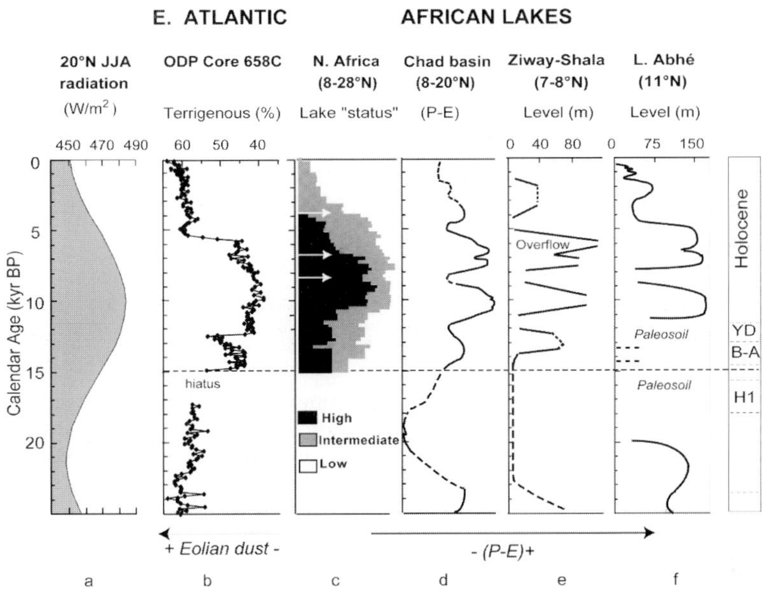

Figure 11-8. Major changes in effective moisture over the past 25,000 years in northern Africa, as inferred from lake records and from eolian dust influx from West Africa to the Atlantic Ocean off Mauritania, compared with changes in summer solar radiation at 20°N. After: (a) Berger and Loutre (1991); (b) deMenocal et al. (2000); (c) Hoelzmann et al. (1998, in press); (d) Servant and Servant-Vildary (1980); (e) Gillespie et al. (1983), Chalié and Gasse (2002); (f) Gasse (2000).
YD: Younger Dryas; B-A: Bölling-Allerod; H1: Heinrich event 1.

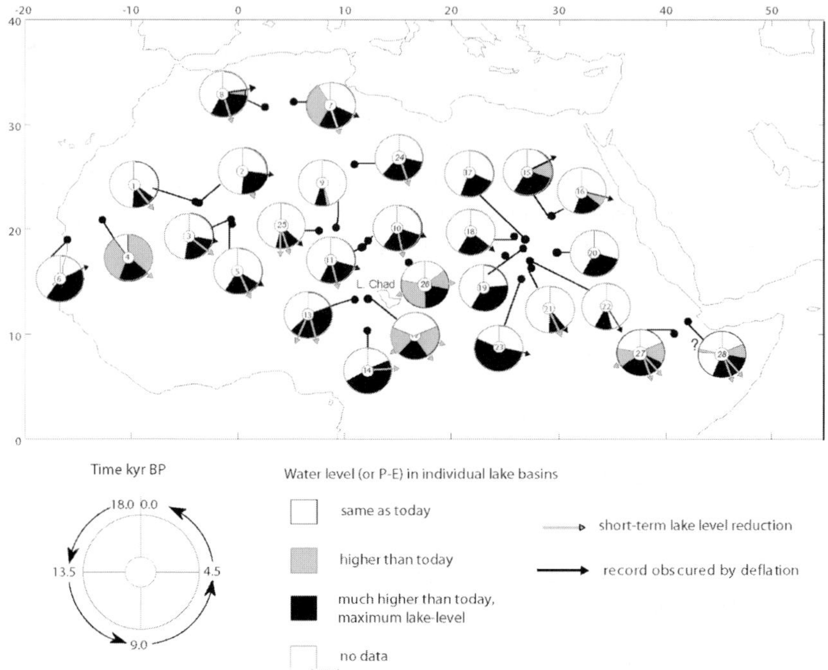

Figure 11-9. Summary results on the spatial distribution and evolution of closed lake basins of northern Africa. Circles represent time from 18,000 BP to the present. Changes in relative water levels reflect the changes in P–E in individual basins. Diagrams are based on the compilation by Hoelzmann et al. (in press) for sites S1 to 23 (see that work for detailed information and references on individual sites). Sites S24 to S28 after Cremashi and di Lernia (1998; S24: southern Libya); Gasse (2002; S25: Izouzadden); Servant and Servant-Vildary (1980; S27: Bahr-el-Ghazal); Gillespie et al. (1983); and Chalié and Gasse (2002; S27: Lake Ziway-Shala); Gasse (2000; S28: Lake Abhé).

The main points arising from lake-level evidence (Figs. 11-8 and 11-9) are as follows.

- Long-term changes in lake-inferred moisture availability are roughly consistent with changes in summer insolation in the northern tropics. However, strongly nonlinear feedback processes with sea surface temperature and vegetation are required to explain the abrupt onset and termination of the AHP observed in lake records and in the eolian dust influx to the ocean (deMenocal et al. 2000).

- LGM conditions were generally at least as dry as today.

- Almost no paleolakes were observed between 24°N–25°N and 30°N, suggesting that the P–E balance has remained low there. Only ephemeral water bodies (sebkhas) were found in the presently waterless core of the northeastern Sahara, east of 24°E / north of 23°N (Hoelzmann et al., in press). Nevertheless, permanent lakes supplied by local rainfall and groundwater flowing from the Atlas Mountains occurred north of 30°N in the western Sahara (e.g., S7, S8, Fig. 11-9).

- All investigated lakes were much higher than they are today at about 9,000 BP, and very low after 3,000–2,000 BP. The AHP is, however, complex. For example, many lakes show two high-stand episodes separated by a marked dry spell at about 8,000 BP (Gasse and Van Campo 1994; Gasse 2000). This large-scale drought may be due to a sudden freshwater discharge in the North Altantic inducing a slowing down of the oceanic thermohaline circulation (Barber et al. 1999), or to an external factor (solar activity; Neff et al. 2001). Schematically, the general period of maximum lake level is longer in the southernmost sites than in the desert core (Fig. 11-9). It is of intermediate length (ca. 10,500 to 5,000 years) in the northernmost sites (S7, S8; Fig. 11-9).

- Changes in mean annual precipitation (ΔP) compared to modern (P) were estimated for extreme climatic conditions from a few lake basins. Using a simple water balance model, Street (1979) suggested ΔP of –9% to –32% (ca. –86 to –300 mm yr^{-1}) during the LGM, and of +28% to +47% (ca. +270 to +450 mm yr^{-1}) at about 10,000–9,000 BP in the Ziway-Shala basin (14,700 km^2) in the Ethiopian Rift (Fig. 11-8; S27, Fig. 11-9). In the presently hyperarid Afar desert, the presently small Lake Abhé (Fig. 8; S28, Fig. 9) extended to over about 5,500 km^2 around 30,000–20,000 BP and 10,000–9,000 BP, and desiccated completely during the LGM. Combined water and salt balance equations suggest a Δ(P–E) of +25% in its basin (81,000 km^2) at 10,000–9,000 BP (Gasse, unpublished). Lake Chad (Fig. 11-8; S26, Fig. 11-9), with its huge catchment area (ca. 5°N–25°N, 10°E–24°E), extends today over about 20,000 km^2. During the AHP, this lake, the "Megachad," covered about 340,000 km^2. ΔP was estimated at ≥ +86% from a combined water and energy balance model (Kutzbach 1980). This result suggests a ΔP ≥ 450–600 mm yr^{-1} in the vicinity of the modern lake.

The most dramatic changes occurred in the driest areas of the Sahara: In northwestern Sudan, at about 18.5°N, the existence of a large West Nubian paleolake (≥1,100 km^2 in area; S21, Fig. 11-9) from 9,500 to 4,000 BP suggests mean annual precipitation of 500–900 mm in a region that today receives less than 15 mm yr^{-1} of rainfall (Hoelzmann et al. 2000).

Lacustrine sediment analyses, especially reconstructed lake water salinity and isotopic composition, also provide information on the moisture availability and sources. Lacustrine $\delta^{18}O_{carbonate}$ values (δ_C) can sometimes be regarded as representative of rainfall or groundwater properties before evaporation processes in the lake (Gasse 2002).

In West Africa, the major points arising from the sediment study of Holocene paleolakes lying along a north-to-south transect (32°N–13°N) are the following (Gasse et al. 1990; Gasse 2002).

- In the northern Sahara, precipitation was abundant enough to sustain permanent lakes at 30°N–32°N (S7, S8, Figs. 11-9 and 11-10a), but their water isotope composition, as that of the regional groundwaters, suggests rainfall events with δ_p values close to modern, when most modern precipitation comes as winter rains of extratropical origin.

- The core of the present-day desert was characterized by heavy rains strongly depleted in heavy isotopes, and by a very low evaporation rate. For example, in the Ténéré dune fields (Fig. 11-4), at ≈20°N in northern Niger (S9, Figs. 11-9 and 11-10b), early Holocene δ_C suggests lake water $\delta^{18}O$ values of –6‰ to –8.5 ‰ (taking T_{water} = 25°C±5°C), while $\delta^{18}O$ values of modern precipitation are positive. The paleolake had the same isotope anomalies as Holocene groundwaters in the region (Fig. 11-6b). The water table, which lies today tens of meters belowground, outcropped in numerous interdunal depressions as permanent freshwater lakes. The low diatom-inferred lake water conductivity is similar to that of the less concentrated rain and river flood waters observed today in the Aïr highlands (Gallaire 1995). The combination of salt and isotope balance equations suggests a mean relative air humidity greater than 50%.

- Such a strong isotopic depletion was not observed southward, in the present-day Sahel. For example, at 13°N (S12, Figs. 11-9 and 11-

10c), the lowest δ_C values, at about 10,000 BP, can be accounted for by a decrease in $\delta_p \leq 2‰$, consistent with an increase in monsoon rainfall amount $\leq 200–400$ mm yr^{-1}. The region probably received more continuous monsoon rains less depleted in heavy isotopes than squall line showers.

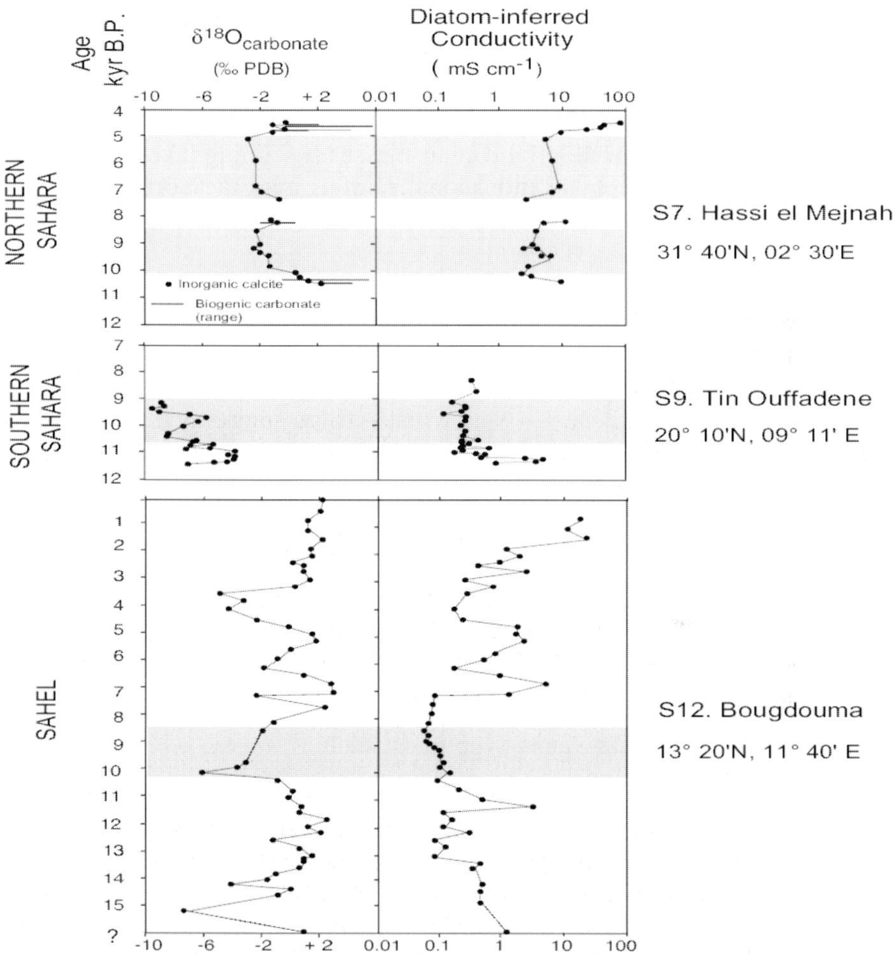

Figure 11-10. Carbonate oxygen isotope and diatom-inferred conductivity profiles in three groundwater-fed paleolakes from West Africa. Shaded zones represent periods of maximum P–E. After Gasse (2002). Sites S7, S9, and S12 are located in Figure 11-9. Besides direct rainfall, S7 was supplied by the Great Western Aquifer (Fig. 11-6a), S9 by the Aïr-Ténéré aquifer (Fig. 11-6b), and S12 by a local aquifer west of Lake Chad.

In the eastern Sahara-Sahel, comparison of stable isotope analyses of lacustrine sediments (17°N–19°N; Abell et al. 1996) with Sahelian and Saharan groundwater (12°N–15°N; Thorweihe 1990) shows extremely depleted $\delta^{18}O$ values that indicate intense tropical summer rainfall. From isotopic studies of lacustrine sediments along a north-to-south transect, the early to mid-Holocene boundary between extratropical and tropical rainfall was placed between 21°N and 23°N (Abell et al. 1996; Abell and Hoelzmann 2000; Hoelzmann et al., in press). This finding is in agreement with a study of eolian landforms that indicates west-to-northwest winds over Egypt's Western Desert (Fig. 11-4) in the early Holocene; these winds steered moist Atlantic/Mediterranean air masses sustaining lakes and playas north of the limit of tropical monsoonal rainfall over the northern Sahara (Brookes 2003).

3.3. Paleovegetation

During the LGM, the limit between the Sahara and the Sahel has been placed at about 12°N–14°N in West Africa, suggesting a southward shift of the ITCZ of 4°–5° (Lézine 1989; Dupont 1993; Hoelzmann et al., in press).

During the early to mid-Holocene, the desert belt was considerably reduced in intensity and extent. Paleovegetation data show a northward migration of vegetation belts requiring tropical rainfall up to 20°N–24°N. Sudanian-Sahelian taxa are present in early Holocene pollen records from the southern Sahara; for example, taxa such as *Hibiscus*, whose pollen is rather poorly dispersed, provide reliable evidence that tropical plants were growing there locally at the time (Ritchie and Haynes 1987). In northwestern Sudan, the occurrence of some savanna-type trees, such as *Acacia albicans* around 7,000 BP at 23.5°N, suggests that conditions were not as intensely hyperarid as they are at present (Kröpelin 1993). Northward, few data are available. Desert-type vegetation with a few tropical elements occurred in southern Libya and in southern Egypt; a semidesert shrub characteristic of the winter rainfall regime prevailed in Egypt north of 25°N (see Hoelzmann et al. [in press] for detail). According to a vegetation reconstruction (Hoelzmann et al. 1998) and a biome model (Prentice et al. 2000) at 6,000 BP, savanna vegetation developed over the southern Sahara and steppe vegetation extended northward. Hoelzmann et al. (1998) suggested that the steppe-savanna boundary lay at 19°N–21°N across most of northern Africa and may have reached 24°N–25°N over southern Algeria and Libya.

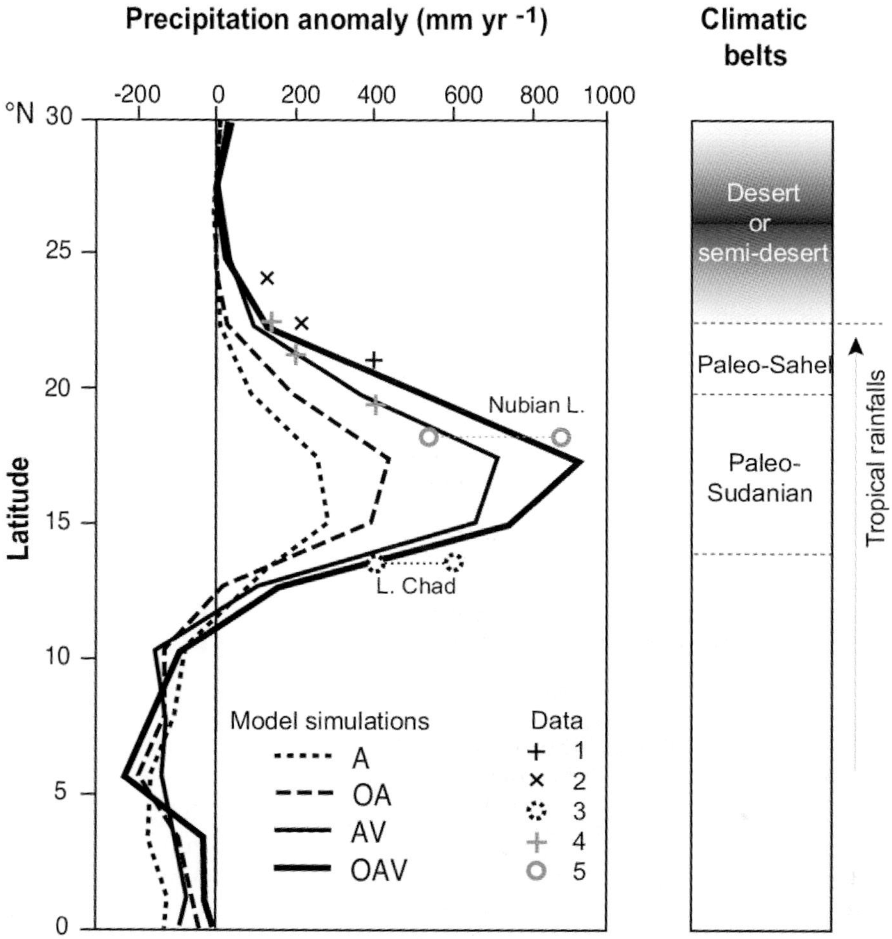

Figure 11-11. Simulations of precipitation change between 6,000 BP and the present day over West Africa, compared with some data-inferred ΔP at the early Holocene. The four simulation curves (redrawn from Braconnot et al. [1999] and IPCC [2001]) derive from a series of model experiments using the same model (IPSL—Institut Pierre et Simon Laplace, Paris—model), and illustrate the importance of astronomical forcing and of ocean and vegetation feedbacks on the monsoon system. A: astronomical forcing alone; AO: interactive ocean; AV: interactive vegetation; OAV: interactive ocean and vegetation. Data-inferred ΔP for the early Holocene in the western (1–3) and the eastern (4–5) Sahara-Sahel. ΔP derived from pollen (1 and 4, after Lézine [1989], Ritchie et al. [1985], Ritchie et al. [1987]), geological data and climate model (2, after Street-Perrott et al. [1991]), paleolake modeling (3 and 5, after Kutzbach [1980], Hoelzmann et al. [2000]). For Lake Chad, ΔP is greater than or equal to values plotted on the graph. The right panel shows the early Holocene climate belts as suggested by data.

The position of the 400 mm isohyet during the early Holocene was estimated at ~19°N in the eastern Sahara (Ritchie et al. 1985) and at ≈21°N in the western Sahara (Lézine 1989). North of this paleo-isohyet, paleovegetation reflects a marked decrease in precipitation over short distances. For example, in the eastern Sahara, pollen-inferred precipitation falls from 400 mm yr^{-1} at 19°N (Ritchie et al. 1985) to 150–200 mm yr^{-1} at 21°N–22°N (e.g. at Selima, S15, Fig.11-9; Ritchie and Haynes 1987). In the western Sahara, a sharp negative south-north paleoprecipitation gradient also emerges across what is today one of the driest parts of the Sahara, from the coupled analysis of geological data and mesoscale climate modeling for northern Mali (Street-Perrott et al. 1991). The simulated actual evapotranspiration (AET) values—and therefore paleoprecipitation values—showed a strong negative paleomoisture gradient between 16°N and 18°N and 22°N and 24°N. Calculated early Holocene precipitation reaches a minimum value of 150–240 mm yr^{-1} at 22°N–24°N. This steep meridional paleoprecipitation gradient (Fig. 11-11) is very similar to that observed today along the Sahel-Sahara fringe (Fig. 11-5).

4. DISCUSSION AND CONCLUSIONS: IMPLICATIONS OF OBSERVED CHANGES FOR ATMOSPHERIC PALEOCIRCULATIONS

Across northern tropical Africa, landscape boundaries that today are closely associated with features of tropical atmospheric circulation shifted up to 1,000 km north or south of their present position during the latter stages of the last glacial period and the early part of the current Holocene interglacial. For example, the southern limit of the Sahara desert lay at 12°N–14°N during the last glacial maximum, and migrated to 19°N–22°N during the early Holocene as a result of summer tropical rains extending further north. The northern Africa's subtropical arid zone contracted (in the early Holocene) or expanded (in the LGM) on both northern and southern margins, and experienced significant meridional shifts in both monsoon-type summer rainfall and winter cyclonic precipitation.

Mean hydrologic conditions, as inferred from different proxies for the Late Pleistocene wet phase, the LGM, and the wet early to mid-Holocene period, are discussed below for their significance in terms of atmospheric circulation and climate forcing factors.

4.1. Glacial Times

During the wet Late Pleistocene phase, groundwaters and the southernmost lakes record evidence for a cold period of increased available moisture. Orbitally induced monsoon strengthening has likely affected part of northern Africa, but groundwaters indicate that heavy rainfall events in the northern Sahara were of extratropical origin. The zonal westerly circulation predominated over this region, reflecting a strengthening and a southward displacement of cyclonic disturbances associated with the mid-latitude westerlies. This suggests an equatorial shift of the subtropical westerly jet and of the Saharan anticyclone. Increased intensity or frequency of westerly depressions originating from the Atlantic is consistent with data from the Mediterranean and the Near East (e.g., Roberts and Wright 1993).

During the LGM, generally dry conditions over northern Africa are inferred from lake records, the hiatus in aquifer recharge, and paleovegetation. Eolian deposits and pollen data indicate reinforced northeast trade winds and a 4°–5° southward shift of the ITCZ, suggesting a stronger Northern Hemisphere Hadley cell circulation. Lower monsoon precipitation is consistent with astronomical forcing, but other climate forcing factors also acted on atmospheric circulation and precipitation over the tropical African continent. Indeed, the southern African tropics, where orbital forcing predicts wetter conditions during the LGM, were at least as dry as today at that time in response to changes in the zonal patterns and gradients of SST in the surrounding oceans and reduced vegetation cover (Barker and Gasse 2003). Last glacial maximum SSTs in the eastern equatorial (e.g., Schneider et al. 1995) and subtropical Atlantic Ocean (e.g., deMenocal et al. 2000) were significantly lower than they are today, due to an enhanced meridian temperature gradient inducing an increased northward ocean heat transport out of the tropics (Bush and Philander 1998). Stronger northeast trade winds over northern Africa and intensified upwelling have further contributed to the ocean tropical cooling.

As a whole, generally dry LGM conditions in northern Africa are in agreement with model simulations. A decreased moisture supply in the ITCZ led to lower WAM precipitation and a reduced northward monsoon flow (e.g., Braconnot et al. 2000). The effect of reduced summer insolation in the northern tropics was enhanced by tropical cooling, a decrease of the global hydrological cycle, and a stronger Northern Hemisphere Hadley cell circulation (Ganopolski et al. 1998; Bush and Philander 1998).

In turn, aridity in northern Africa might have affected large-scale climate processes, especially through dust emission into the atmosphere. Dust affects the climate by reducing the heating of sea surface waters and through cloud microphysical processes that alter rainfall patterns. Today, northern African dust crosses the Atlantic Ocean (e.g., Prospero and Lamb 2003) and sometimes reaches northern Europe (Littmann et al. 1990). The LGM desert expansion and stronger northeast trade winds might have enhanced the cooling of the tropical Atlantic Ocean and aridity in the surrounding continents.

4.2. The Early to Mid-Holocene Period

Data indicate a northward migration of tropical rainfall belts as far as 20°N–24°N, but suggest that moisture in the northern Sahara was of extratropical origin as is the case today.

South of 20°N–24°N, the southern origin of precipitation is mainly inferred from isotopic and paleovegetation evidence. Recycled moisture from the continent may have enhanced the southern Sahara wetting, especially in the presently waterless part of the eastern Sahara. The marked decrease in pollen-inferred paleoprecipitation between ≈18°N and 22°N indicates that the Sahara-Sahel boundary lay 500–600 km northward of its present-day position. This suggests a northward shift of the summer ITCZ, but a large migration of its surface position is unlikely, as data place the limit between the winter and the summer rainfall domains close to the modern one. The slope of the intertropical convergence front might have been less flattened than it is today (Fig. 11-3b), favoring the penetration of the large-scale monsoon wind system inland. High intensity/frequency of squall line rainfall at about 20°N suggests a northward migration of 5°–6° of the African easterly jet and of the tropical easterly jet tracks. Higher energy of these upper-level thermal winds is expected at that time due to the orbitally induced increase in surface heating of the Saharan and of the Tibetan-Indian landmasses. The strengthening of these jets may have reduced their subsidence over the Sahara, which today limits the WAM inland penetration. A northward position of the AEJ core could have favored the mid-level advection of moisture fluxes from the north to the WAM, making a bridge between the tropical and extratropical domains.

All GCM simulations show a strengthening and a northward penetration of the WAM front over northern Africa at 6,000 BP, in response to increased summer insolation over the Northern Hemisphere and an enhanced north-to-south temperature gradient over the Sahara that reinforced the ocean-land pressure gradient. However, atmospheric GCM (AGCM) simulations consistently underestimate the observed northward shift (e.g., Joussaume et al. 1999). Simulations in which vegetation changes were prescribed for an atmospheric model (e.g., Broström et al. 1998; Texier et al. 2000) or computed by using coupled atmosphere-vegetation models (e.g., de

Noblet-Ducoudré et al. 2000) show that vegetation feedbacks amplify and modify the monsoon system's response to orbital forcing. In particular, these feedbacks may explain the abrupt offset of the African Humid Period at 6,000–4,000 BP (Claussen et al. 1999). Soil moisture and wetland extent also produce positive feedbacks (Coe and Bonan 1997). Coupled ocean-atmosphere models show that ocean feedbacks enhance the African monsoon and shift the belt of maximum precipitation further north (e.g., Kutzbach and Liu 1997; Hewitt and Mitchell 1998; Voss and Mikolajewicz 2001; Braconnot et al., in press). The monsoon season is lengthened in coupled simulations. Cold Atlantic SSTs in spring, when the land surface is already beginning to warm, favor an early initialization of the monsoon flow. In late summer, a strong gradient in SSTs at about 10°N, with cooler SSTs than today to the south and warmer SSTs to the north, helps maintain the ITCZ in its northerly position over the Atlantic and western Africa. A similar dipole over the Atlantic occurs today during years when Sahelian rainfall is above average (e.g., Fontaine and Janicot 1996; Giannini et al. 2003). The best consistency between observations and simulations is found with coupled ocean-atmosphere-vegetation models (Fig. 11-11). Such a model (Braconnot et al. 1999, in press; IPCC 2001) shows a maximum tropical anomaly at about 14°N–19°N and a marked decrease northward from 19°N to 22°N–23°N, in rather good agreement with the data summarized above (Fig. 11-11).

Shifts in the WAM intensity and extent seem unlikely to have affected the Sahara north of 20°N–24°N. The moderate wetting of the northern Sahara was associated with west and northwest winds (Brookes 2003) and was in phase with a relative water surplus in the Mediterranean region (e.g., Roberts et al. 2001). The bulk of precipitation in the northern Sahara and the Mediterranean may have been of cyclonic rather than convective origin, and fell during the winter (or transitional seasons) as simulated by climate models (e.g., Hewitt and Mitchell 1996). This conclusion appears to be in agreement with marine core records from the northern Red Sea (Arz et al. 2003) where reconstructed paleosalinity and terrigenous sediment input changes show that substantially higher rainfall and freshwater runoff occurred between about 9,250 and 7,250 BP; this humid interval was attributed to the enhancement and southward extension of rainfall from Mediterranean sources. We agree with Arz et al. (2003) that the monsoonal rains did not cross the subtropical desert of Africa during the early to mid-Holocene. Times of increased winter precipitation might reflect a weakening of the subtropical anticyclone and of northeast trade winds over the Sahara, slower moving of the mid-latitude westerlies originating from the Atlantic, and stronger or more frequent temperate frontal disturbances embedded within them.

In any case, it is clear that the Saharan desert zone, which today is associated with the high-pressure belt of descending air, was both smaller in extent and less intensely arid during the early Holocene. This, and similarly large-scale arid zone shifts at the LGM, imply that significant changes must have taken place in the status of Africa's Hadley cell circulation over Milankovitch time scales.

5. REFERENCES

Abell, P.I., P. Hoelzmann, and H.P. Pachur. 1996. Stable isotope ratios of gastropod shells and carbonate sediments in NW Sudan as paleoclimatic indicators. *Palaeoecology of Africa* 24: 33–52.

Abell, P.I., and P. Hoelzmann. 2000. Holocene paleoclimates in northwestern Sudan: Stable isotope studies on molluscs. *Global and Planetary Change* 26: 1–12.

Arz, H.W., F. Lamy, J. Pätzold, P.J. Müller, and M. Prins. 2003. Mediterranean moisture source for an Early-Holocene humid period in the northern Red Sea. *Science* 300: 118–121.

Barber D.C., A. Dyke, C. Hillaire-Marcel, A.E. Jennings, J.T. Andrews, M.W. Kerwin, G. Bilodeau, R. McNeely, J.R. Southon, M.D. Morehead, and J.M. Gagnon. 1999. Forcing of the cold event of 8,200 years ago by catastrophic drainage of Laurentide lakes. *Nature* 400: 344–348.

Barker, P., and F. Gasse. 2003. New evidence for a reduced water balance in East Africa during the Last Glacial Maximum. Implication for model-data comparison. *Quaternary Science Reviews* 22: 823–837.

Berger, A., and M.F. Loutre. 1991. Insolation values for the climate of the last 10 million years. *Quaternary Science Reviews* 10: 297–317.

Braconnot, P., S.P. Harrison, S. Joussaume, C.D. Hewitt, A. Kitoh, J.E. Kutzbach, Z. Liu, B. Otto-Bliesner, J. Syktus, and N Weber. In press. Evaluation of PMIP coupled ocean-atmosphere simulations of the mid-Holocene. In, Battarbee, R.W., F. Gasse, and C.S. Stickley (eds.). *Past Climate through Europe and Africa. Developments in Paleoenvironmental Research.* Vol. 6. Dordrecht: Kluwer Academic Press.

Braconnot, P., S. Joussaume, N. de Noblet, and G. Ramstein. 2000. Mid-Holocene and Last Glacial Maximum African monsoon changes as simulated within the Paleoclimate Modelling Intercomparison Project. *Global and Planetary Change* 26: 51–66.

Braconnot, P., S. Joussaume, O. Marti, and N. de Noblet. 1999. Synergistic feedbacks from ocean and vegetation on the African monsoon response to mid-Holocene insolation. *Geophysical Research Letters* 26: 2481–2484.

Brookes, I.A. 2003. Geomorphic indicators of Holocene winds in Egypt's Western Desert. *Geomorphology* 56: 155–166.

Broström, A., M.T. Coe, S.P. Harrison, J.E. Kutzbach, J. Foley, I.C. Prentice, and P. Behling. 1998. Land surface feedbacks and paleomonsoons in northern Africa. *Geophysical Research Letters* 25: 3615–3618.

Bush, A.G.B., and S.G.H. Philander. 1998. The role of ocean-atmosphere interactions in tropical cooling during the Last Glacial Maximum. *Science* 279: 1341–1344.

Chalié, F., and F. Gasse. 2002. Late Glacial-Holocene diatom record of water chemistry and lake level change from the tropical East African Rift Lake Abiyata (Ethiopia). *Palaeogeography, Palaeoclimatology, Palaeoecology* 187: 259–283.

Claussen, M., C. Kubatzki, V. Brovkin, A. Ganopolski, P. Hoelzmann, and H.-J. Pachur. 1999. Simulation of an abrupt change in Saharan vegetation in the mid-Holocene. *Geophysical Research Letters* 26(14): 2037–2040.

Coe, M.T., and G.B. Bonan. 1997. Feedbacks between climate and surface water in northern Africa during the middle Holocene. *Journal of Geophysical Research* 102: 11087–11101.

Conrad, G., and J.-Ch. Fontes. 1970. Hydrologie isotopique du Sahara nord-occidental. In, *Isotope Hydrology 1970*. Vienna: International Atomic Energy Agency (IAEA), IAEA-SM-129/24, pp. 405–419.

Cremaschi, M., and S. di Lernia. 1998. *Wadi Teshuinat palaeoenvironment and prehistory in south-western Fezzan (Libyan Sahara)*. All'Insegna del Giglio, CNR, Milano.

deMenocal, P.B., and D. Rind. 1996. Sensitivity of subtropical African and Asian climate to prescribed boundary conditions changes: Model implications for the Plio-Pleistocene evolution of low-latitude climate. In, Johnson, T.C., and E.O. Odada (eds.). *The Limnology, Climatology and Paleoclimatology of the East African Lakes*. Amsterdam: Gordon and Breach, pp. 57–77.

deMenocal, P., J. Ortiz, T. Guilderson, and M. Sarnthein. 2000. Coherent high- and low latitude climate variability during the Holocene warm period. *Science* 288: 2198–2202.

de Noblet-Ducoudré, N., M. Claussen, and I.C. Prentice. 2000. Mid-Holocene greening of the Sahara: First results of the GAIM 6000 year BP Experiment with two asynchronously coupled atmosphere/biome models. *Climate Dynamics* 16: 643–659.

Dodo, A., and G.M. Zuppi. 1997. Groundwater flow study in the Bilma-Djado Basin (Niger) by means of environmental isotopes. *C. R. Acad. Sci,, Paris* 325: 845–852.

Dodo, A., and G.M. Zuppi. 1999. Variabilité climatique durant le Quaternaire dans la nappe de Tarat (Arlit, Niger). *C. R. Acad. Sci., Paris* 328: 371–379.

Dupont, L.M. 1993. Vegetation zones in NW Africa during the Brunhes Chron reconstructed from marine palynological data. *Quaternary Science Reviews* 12: 189–202.

Edmunds, W.M., A. Dodo, D. Djoret, F. Gasse, C.B. Gaye, I.B. Goni, Y. Travi, K. Zouari, and G.M. Zuppi. In press. Groundwater as an archive of climatic and environmental change. In, Battarbee, R.W., F. Gasse, and C.S. Stickley (eds.). *Past Climate through Europe and Africa. Developments in Paleoenvironmental Research*. Vol. 6. Dordrecht: Kluwer Academic Press.

Flohn, H., and S. Nicholson. 1980. Climatic fluctuations in the arid belt of the Old World since the last glacial maximum: Possible causes and future implications. *Palaeohydrology of Africa* 12: 3–21.

Fontaine, B., and S. Janicot. 1996. Sea surface temperature fields associated with West African rainfall anomaly types. *Journal of Climate* 9: 2935–2940.

Fontaine, B., P. Roucou, and S. Trzaska. 2003. Atmospheric water cycle and moisture fluxes in the West African monsoon: Mean annual cycles and relationship using NCEP/NCAR reanalysis. *Geophysical Research Letters* 30(3): 1117–1120.

Fontes, J.-C., F. Gasse, and J.N. Andrews. 1993. Climatic conditions of Holocene groundwater recharge in the Sahel zone of Africa. In, IAEA. *Isotope Techniques in the Study of Past and Current Environmental Changes in the Hydrosphere and the Atmosphere*. Vienna, IAEA-SM-329/59, pp. 231–248.

Gallaire, R. 1995. Hydrologie en milieu subdésertique d'altitude, le cas de l'Aïr (Niger). Thesis, Université Paris-Sud.

Ganopolski, A., S. Rahmstorf, V. Petoukhov, and M. Claussen 1998. Simulation of modern and glacial climates with a coupled global model of intermediate complexity. *Nature* 391: 351–356.

Gasse, F. 2000. Hydrological changes in the African tropics since the Last Glacial Maximum. *Quaternary Science Reviews* 19: 189–211.

Gasse, F. 2002. Diatom-inferred salinity and carbonate oxygen isotopes in Holocene water-bodies of the western Sahara and Sahel (Africa). *Quaternary Science Reviews* 21: 737–767.

Gasse, F., and E. Van Campo. 1994. Abrupt post-glacial climate events in West Asia and North Africa monsoon domains. *Earth and Planetary Science Letters* 126: 435–456.

Gasse, F., R. Téhet, A. Durand, E. Gibert, and J.-C. Fontes. 1990. The arid-humid transition in the Sahara and the Sahel during the last deglaciation. *Nature* 346: 141–156.

Giannini, A., R. Saravanan, and P. Chang. 2003. Oceanic forcing of Sahel rainfall on interannual to interdecadal time scales. *Science* 302: 1027–1030.

Gillespie, R., F.A. Street-Perrott, and R. Switsur. 1983. Post-glacial arid episodes in Ethiopia have implications for climate prediction. *Nature* 306: 680–683.

Gong, C., and E. Eltahir. 1996. Sources of moisture for rainfall in West Africa. *Water Resources Research* 32: 3115–3121.

Gonfiantini, R., G. Conrad, J.-Ch. Fontes, G. Sauzay, and B.R. Payne. 1974. Etude isotopique dez la nappe du Continental Intercalaire et de ses relations avec les autres nappes du Sahara Septentrional. In, IAEA. *Isotope Techniques in Groundwater Hydrology*. Vienna: IAEA, pp. 227–241.

Griffiths, J.F. 1972. Climates of Africa. In, Landsberg, H.E., and J.F. Griffiths (eds). *World Survey of Climatology* Vol. 10, p. 603.

Gröning, M. 1994. Edelgase und Isotopentracer in Grundwasser: Paläo-Klimaänderungen und Dynamik regionaler grundwasserfließSsysteme. Thesis, Universität Heidelberg.

Guendouz, A. Moulla, W.M. Edmunds, P. Shand, J. Poole, K. Zouari, and A. Mamou. 1998. Paleoclimatic information contained in groundwaters of the Grand Erg Oriental, North Africa. In, Isotope Techniques in the Study of Past and Current Environmental Changes in the Hydrosphere and the Atmosphere. Proc. Vienna Symp. 1997. Vienna: IAEA, pp. 555–571.

Hastenrath, S. 1991. *Climate Dynamics of the Tropics*. Dordrecht: Kluwer Academic Press.

Hewitt, C.D., and J.F.B. Mitchell. 1996. GCM simulations of the climate at 6 kyr BP: Mean changes and interdecadal variability. *Journal of Climate* 9: 3505–3529.

Hewitt, C.D., and J.F.B. Mitchell. 1998. A fully coupled GCM simulation of the climate of the mid-Holocene. *Geophysical Research Letters* 25: 361–364.

Hoelzmann, P., F. Gasse, L.M. Dupont, U. Salzmann, M. Staubwasser, D.C. Leuchner, and F. Sirocko. In press. Palaeoenvironmental changes in the arid and subarid belt (Sahara-Sahel-Arabian Peninsula) from 150 ka to present. In Battarbee, R.W., F. Gasse, and C.S. Stickley (eds.). *Past Climate through Europe and Africa.. Developments in Paleoenvironmental Research.* Vol. 6. Dordrecht: Kluwer Academic Press.

Hoelzmann, P., D. Jolly, S.P. Harrison, F. Laarif, R. Bonnefille, and H.-J. Pachur. 1998. Mid-Holocene land-surface conditions in northern Africa and the Arabian peninsula: A data set for the analysis of biogeophysical feedbacks in the climate system. *Global Biochemical Cycles* 12: 35–51.

Hoelzmann, P., H.-J. Kruse, and F. Rottinger. 2000. Precipitation estimates for the eastern Saharan palaeomonsoon based on a water balance model of the West Nubian palaeolake basin. *Global and Planetary Change* 26: 105–120.

Hsu, C.-P.F., and J.M. Wallace. 1976. The global distribution of the annual and semiannual cycles in precipitation. *Monthly Weather Review* 104: 1093–1101.

International Atomic Energy Agency (IAEA). 1992. *Statistical Treatment of Data on Environmental Isotopes in Precipitation.* Vienna: IAEA.

International Atomic Energy Agency and World Meteorological Organization, IAEA/WMO, I. 1998. Global Network for Isotopes in Precipitation. The GNIP database. Release 3, October 1999. URL: http://www.iaea.org/programs/ri/gnip/gnipmain.htm.

Intergovernmental Panel on Climate Change (IPCC). 2001. *Climate Change 2001, The Scientific Basis. Contribution of Working Group I to the Third Assessment Report of the Intergovernmental Panel on Climate Change.* Cambridge: Cambridge University Press, 881 pp.

Janicot, S. 1990. Variabilité des précipitations en Afrique de l'Ouest et circulations quasi stationnaires durant une phase de transition climatique. Thesis, Université Paris-6.

Joussaume, S., and Coauthors 1999. Monsoon changes for 6000 years ago: Results of 18 simulations from the Paleoclimate Modeling Intercomparison Project (PMIP). *Geophysical Research Letters* 26: 859–862.

Kröpelin, S. 1993. Late Quaternary environments at the Lower Wadi Howar (southeastern Sahara/NW Sudan). *Berliner Geographische Abhandlungen* 54: 1–193.

Kutzbach, J.E. 1980. Estimates of past climate at Paleolake Chad, North Africa, based on a hydrological and energy balance model. *Quaternary Research* 14: 210–223.

Kutzbach, J.E., and Z. Liu. 1997. Response of the African monsoon to orbital forcing and ocean feedbacks in the Middle Holocene. *Science* 278: 440–443.

Kutzbach, J.E., and F.A. Street-Perrott. 1985. Milankovitch forcing of fluctuations in the level of tropical lakes from 18 to 0 kyr BP. *Nature* 317: 130–134.

Lamb, P.J., M.A. Bell, and J.D. Finch. 1998. Variability of Sahelian disturbance lines and rainfall during 1951–1987. In, Servat, E., D. Hughes, J.M. Fritsch, and M. Hulme (eds.). *Water Resources Variability in Africa during the XXth Century.* IAHS Publication 252, pp. 19–26.

Leroux, M. 1983. *Le climat de l'Afrique Tropicale.* Paris: Champion.

Lézine, A.M. 1989. Late Quaternary vegetation and climate of the Sahel. *Quaternary Research* 32: 314–334.

Littmann, T., P. Steinrücke, and F. Gasse. 1990. African mineral aerosol deposition in West Germany. *Geödynamik* 11: 163–189.

McGregor, G.R., and S. Nieuwolt. 1998. *Tropical Climatology.* Chichester: John Wiley & Sons.

Neff, U., S.J. Burns, A. Mangini, M. Mudelsee, D. Fleitmann, and A. Matter. 2001. Strong coherence between solar variability and the monsoon in Oman between 9 and 6 kyr ago. *Nature* 411: 290–293.

Nicholson, S.E. 1996. A review of climate dynamics and climate variability in Eastern Africa. In, Johnson, T.C., and E.O. Odada (eds.). *The Limnology, Climatology and Paleoclimatology of the East African Lakes.* Amsterdam: Gordon and Breach, pp. 25–56.

Nicholson, S.E. 2000. The nature of rainfall variability over Africa on time scales of decades to millennia. *Global and Planetary Change* 26: 137–158.

Prentice, I.C., D. Jolly, and BIOME 6000 Participants. 2000. Mid-Holocene and glacial-maximum vegetation geography of the northern continents and Africa. *Journal of Biogeography* 27: 507–519.

Prospero, J.M., and P.J. Lamb. 2003. African droughts and dust transport to the Caribbean: Climate change implications. *Science* 302: 1024–1027.

Ritchie, J.C., and C.V. Haynes. 1987. Holocene vegetation zonation in the eastern Sahara. *Nature* 330:645–647.

Ritchie, J.C., C.H. Eyles, and C.V. Haynes. 1985. Sediment and pollen evidence for an early to mid-Holocene humid period in the eastern Sahara. *Nature* 314: 352–355.

Roberts, N., and H.E. Wright. 1993. Vegetational, lake-level, and climate history of the Near East and Southwest Asia. In, Wright, H.E., J.E. Kutzbach, T. Webb, III, W.F. Ruddiman, F.A. Street-Perrott, and P.J. Bartlein (eds.). *Global Climates since the Last Glacial Maximum*. Minneapolis: University of Minnesota Press, pp. 194–220.

Roberts, N., J.M. Reed, M.J. Leng, C. Kuzucuoglu, M. Fontugne, J. Bertaux, H. Woldring, S. Bottema, S. Black, E. Hunt, and M. Karabiyikoglu. 2001. The tempo of Holocene climatic change in the eastern Mediterranean region: New high resolution crater-lake sediment data from central Turkey. *The Holocene* 11: 721–736.

Rodwell, M.J., and B.J. Hoskins. 1996. Monsoons and the dynamics of deserts. *Quarterly Journal of the Royal Meteorological Society* 122: 1385–1404.

Rozanski, K., L. Araguas-Araguas, and R. Gonfiantini. 1993. Isotopic patterns in modern global precipitation. In, P.K. Swart, K.C. Lohmann, J. McKenzie, and S. Savin (eds.). *Climate Changes in Continental Isotopic Records*. Geophysical Monograph, American Geophysical Union, pp. 1–36.

Schneider, R.R., P.J. Müller, and G. Ruhland. 1995. Late Quaternary surface circulation in the east-equatorial South Atlantic: Evidence from alkenone sea surface temperatures. *Paleoceanography* 10: 197–219.

Servant, M., and S. Servant-Vildary. 1980. L'environnement quaternaire du bassin du Tchad. In, Williams, M.A.J., and H. Faure (eds.). *The Sahara and the Nile*. Rotterdam: Balkema, pp. 133–162.

Sonntag, C., E. Klitzsch, E.P. Lohnert, E.M. El Shazly, K.O. Munnich, C. Junghans, U. Thorweihe, K. Weistroffer, and F.M. Swailem. 1978. Paleoclimatic information from deuterium and oxygen-18 in carbon-14-dated north Saharian groundwaters. In, IAEA (ed.). *Isotope Hydrology*. Vienna: IAEA, pp. 569–580.

Sonntag, C., U. Thornweihe, and J. Rudolph. 1982. Isotopenuntersuchungen zur Bildungsgeschichte saharaisher Paläowasser. *Geomethodica* 7: 55–78.

Sonntag, C., U. Thornweihe, J. Rudolph, E.P. Löhnert, C. Junghans, K.O. Münnich, R. Klitsch, and E.M. El Shazly. 1980. Isotopic identification of Saharian groundwater: Groundwater formation in the past. *Palaeoecology of Africa* 12: 159–171.

Street, F.A. 1979. Late Quaternary precipitation estimates for the Ziway-Shala Basin, Ethiopia. *Palaeoecology of Africa* 11: 135–143.

Street-Perrott, F.A., and R.A. Perrott. 1993. Holocene vegetation, lake levels and climate in Africa: Global climates since the Last Glacial Maximum. In, Wright, H.E., J.E. Kutzbach, T. Webb, III, W.F. Ruddiman, F.A. Street-Perrott, and P.J. Bartlein (eds.). *Global Climates since the Last Glacial Maximum*. Minneapolis: University of Minnesota Press, pp. 318–356.

Street-Perrott, F.A., J.F.B. Mitchell, D.S. Marchand, and J.S. Brunner. 1991 Milankovitch and albedo forcing of the tropical monsoons: A comparison of geological evidence and numerical simulations for 9,000 yr BP. *Transactions of the Royal Society of Edinburgh, Earth Science* 81: 407–427.

Stuiver, M., and P.J. Reimer. 1993. Extended ^{14}C data base and revised CALIB 3.0 ^{14}C age calibration program. *Radiocarbon* 35: 215–230.

Stuiver, M., P.J. Reimer, E. Bard, J. Warrenbeck, G.S. Burr, K.A. Hughen, B. Kromer, G. McCormac, J. Vean der Plicht, and M. Spurk. 1998. INTCAL98 radiocarbon age calibration, 24,000–0 cal BP. *Radiocarbon* 40: 1041–1083.

Sultan, M., N. Sturchio, F.A. Hassan, M.A. Rahman Hamdan, A.M. Mohmood, Z. El Alfy, and T. Stein. 1997. Precipitation source inferred from stable isotopic composition of Pleistocene groundwater and carbonate deposits in the Western Desert of Egypt. *Quaternary Research* 48: 29–37.

Swezey, C. 2001. Eolian sediment responses to late Quaternary climate changes: Temporal and spatial patterns in the Sahara. *Palaeogeography, Palaeoclimatology, Palaeocecology* 167: 119–155.

Taupin, J.-D., A. Coudrain-Ribstein, R. Gallaire, G.M. Zuppi, and A. Filly. 2000. Rainfall characteristics ($\delta^{18}O$, $\delta^{2}H$, ΔT and ΔH_r) in western Africa: Regional scale and influence of irrigated areas. *Journal of Geophysical Research* 105: 11911–11924.

Texier, D., N. de Noblet, and P. Braconnot. 2000. Sensitivity of the African and Asian monsoons to mid-Holocene insolation and data-inferred surface changes. *Journal of Climate* 13: 164–181.

Thorweihe, U. 1990. The Nubian Aquifer. In, Said, R. (ed.). *The Geology of Egypt.* Rotterdam: Balkema, pp. 601–722

Umer, M., D. Legesse, F. Gasse, R. Bonnefille, H. Lamb, M. Leng, and A. Lamb. In press. Late Quaternary climate changes in the Horn of Africa. In, Battarbee, R.W., F. Gasse, and C.S. Stickley (eds.). *Past Climate through Europe and Africa. Developments in Paleoenvironmental Research.* Vol. 6. Dordrecht: Kluwer Academic Press.

Voss, R., and U. Mikolajewicz. 2001. The climate 6000 years BP in near-equilibrium simulations with a coupled AOGCM. *Geophysical Research Letters* 28: 2213–2216.

White, F. 1983. *The Vegetation of Africa.* Paris: UNESCO.

Chapter 12

VARIABILITY OF THE MARINE ITCZ OVER THE EASTERN PACIFIC DURING THE PAST 30,000 YEARS

Regional Perspective and Global Context

Athanasios Koutavas[1] and Jean Lynch-Stieglitz[2]

[1]*Department of Earth Atmospheric and Planetary Sciences, Massachusetts Institute of Technology, Cambridge, Massachusetts 02139, U.S.A.*

[2]*School of Earth and Atmospheric Sciences, Georgia Institute of Technology, Atlanta, Georgia 30332, U.S.A.*

Abstract The Intertropical Convergence Zone (ITCZ) is manifested as a circum-global atmospheric belt of intense, moist convection and rainfall, marking the confluence of the northern and southern trades and the rising branch of the Hadley cell. It regulates the hydrologic cycle over the tropical continents and interacts tightly with the tropical oceans, notably with the seasonal appearance of the equatorial cold tongues of the Atlantic and Pacific. While it undergoes a regular seasonal migration, today the ITCZ maintains a nearly permanent Northern Hemisphere bias. Here we address the question of variability in the mean latitude of the marine ITCZ over the eastern Pacific on time scales of 100–10,000 years, with emphasis on the past 30,000 years. Our strategy relies on reconstructing the intensity of the prominent oceanographic front of the cold tongue–ITCZ complex, using oxygen isotope and magnesium thermometry techniques. We show that a weaker cold tongue–ITCZ front prevailed during the last glacial maximum (LGM), which indicates a more southerly ITCZ at that time. We further show that the Holocene history of sea surface temperature (SST) near the Galapagos Islands is consistent with progressive southward migration of the ITCZ during the last ~7,000 years, in accord with records from South America and the tropical Atlantic. In the more recent past, evidence from eastern Pacific corals supports a northward ITCZ shift since the end of the Little Ice Age (LIA), in agreement with the hydrologic record of the nearby Cariaco Basin. Collectively, the evidence points to coherent behavior of the Pacific, Atlantic, and South American ITCZs over a broad range of time scales. All regions have responded to Northern Hemisphere cooling by southward (equatorward) ITCZ displacements. In the Pacific, such displacements are likely to have been unfavorable to divergent upwelling at the equator, resulting in weaker zonal and meridional SST gradients and more uniform equatorial SSTs, analogous to modern El Niño conditions.

H.F. Diaz and R.S. Bradley (eds.), The Hadley Circulation: Present, Past and Future, 347–369.

1. INTRODUCTION

A peculiar and incompletely understood aspect of the tropical atmospheric circulation concerns the preferred location of the Intertropical Convergence Zone (ITCZ). Over the oceans, particularly over the eastern Pacific and the Atlantic, the ITCZ maintains a nearly permanent Northern Hemisphere bias, rarely crossing the equator or forming spontaneously south of it (Waliser and Gautier 1993). Current theories attribute this northern bias to a variety of mechanisms involving asymmetries in the distribution of land, in the geometry of coastlines, and in stratus cloud cover, or to processes due to upwelling–sea surface temperature (SST) and wind-evaporation-SST feedbacks (Philander et al. 1996; Xie and Saito 2000).

In the modern climate, the ITCZ undergoes a regular seasonal migration toward the summer hemisphere. It reaches its northernmost latitude during late boreal summer (Aug–Sep) and subsequently approaches the equator during boreal winter (Feb–Mar). In the eastern tropical Pacific this annual migration occurs in concert with a large annual cycle in equatorial SST, in excess of 5°C. Due to the presence of a shallow thermocline in this region, equatorial SSTs are efficiently modulated by the changes in winds that accompany the migration of the ITCZ. When the ITCZ is located farthest north, steady cross-equatorial southeast trades drive enhanced upwelling at the equator, promoting surface cooling (e.g., Chelton et al. 2001). Conversely, as the ITCZ migrates south, the equatorial region comes under the influence of weaker winds (the "doldrums"), which cause the upwelling to subside and SST to rise. An example of the seasonal cycle of the ITCZ in relation to SST in the eastern Pacific is illustrated in Figure 12-1.

Given the susceptibility of the ITCZ to seasonal migration during the annual cycle, an obvious question concerns the factors controlling its position over longer time scales, and the physical mechanisms through which it is linked to the global climate system. Through observations made during the instrumental era, it is now understood that at least over interannual time scales associated with the El Niño/Southern Oscillation (ENSO), the position of the Pacific ITCZ varies predictably with the phase of ENSO (e.g., Deser and Wallace 1990). El Niño events are typically marked by equatorward expansion of the ITCZ, whereas La Niña occurrences are marked by a more northerly ITCZ. On the time scale of decades and longer, however, we must rely on paleoclimatic reconstructions to assess whether systematic variations in the ITCZ accompanied global climate changes, and to understand their forcing mechanisms and impacts. Our objective here is to examine coupled ocean-atmosphere interactions between the equatorial cold tongue and the ITCZ (hereafter referred to as the cold tongue–ITCZ complex) of the eastern tropical Pacific, from the last glacial maximum

(LGM) to the present. We primarily utilize measurements of oxygen isotopic composition ($\delta^{18}O$) and magnesium/calcium ratios (Mg/Ca) in planktonic foraminifera from deep-sea cores spanning the last 30,000 years. In parallel, we attempt to integrate our observations in this region with a growing number of records documenting global-scale ITCZ variability over the same period.

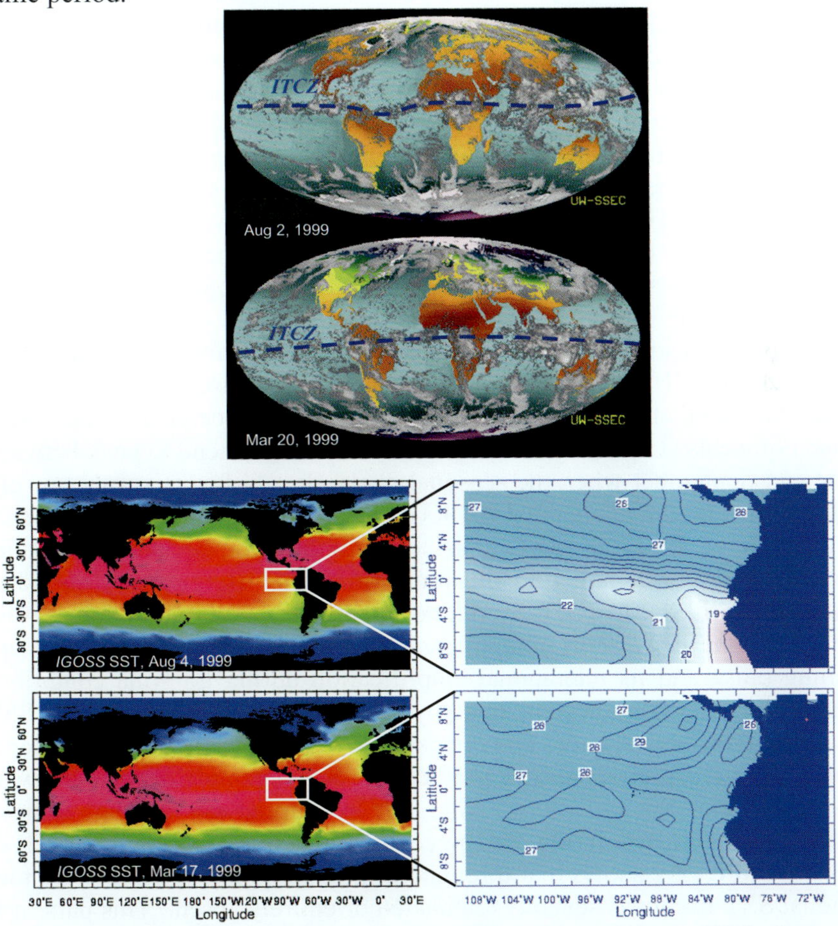

Figure 12-1. Illustration of the annual migration of the ITCZ in relation to SST in the eastern equatorial Pacific. Satellite cloud composites (top, from University of Wisconsin) delineate the approximate ITCZ position on August 2 and March 20, 1999. Corresponding weekly global SSTs from IGOSS are shown at bottom left, and eastern equatorial Pacific SSTs are shown in more detail at bottom right. An intensified equatorial cold tongue and a strong SST gradient to its north accompany the northerly excursion of the ITCZ in boreal summer (Aug–Sep). Conversely, both the cold tongue and SST gradient are suppressed in winter (Feb–Mar), when the ITCZ approaches the equator and the winds diminish.

2. THE SEASONAL CYCLE OF THE ITCZ IN THE EASTERN PACIFIC

The seasonal cycle of the eastern Pacific cold tongue–ITCZ complex is illustrated in Figure 12-1. Using an example from 1999, this figure shows that the seasonal appearance and intensification of the cold tongue occurs as the ITCZ moves north and reaches the northern edge of its latitudinal range in peak boreal summer (Aug–Sep). Strong southeast trades blowing across the equator and into the Northern Hemisphere drive intense upwelling at this time. At the same time, Figure 12-1 demonstrates the prominent SST front that develops just north of the equator, marking the transition from the equatorial upwelling source to the stratified warm pool underneath the ITCZ. In the opposite season (boreal winter, Feb–Mar) the ITCZ approaches the equator, bringing with it diminished winds, which fail to induce vigorous upwelling and to sustain the cold tongue and the frontal zone to its north.

This seasonal succession is repeated annually and is punctuated interannually by ENSO episodes. ENSO is presently phase locked to the seasonal cycle so that a typical El Niño event peaks in December–January (the season of weakest upwelling), whereas La Niña events tend to peak between June and September when the seasonal cycle already favors strong upwelling. As a result, the interannual ENSO signal acts as an amplifier of the seasonal cycle of the ITCZ, SST, and the strength of the equatorial front between 0° and 5°N. Figure 12-2 illustrates this effect using weekly SST data obtained from 1982 to 2003, from a north-south transect of locations across the equator at 90°W. This comparison confirms that both the seasonal upwelling pulse and its interannual amplification during La Niña conditions (e.g., during 1988) are accompanied by amplification of the meridional SST gradient across the equator. The opposite is true during El Niño conditions, which tend to suppress the SST gradient. A clear example of the latter was observed during the 1997–98 El Niño, which matured early enough to have a discernible impact during the preceding upwelling season (summer of 1997). Figure 12-2 shows that throughout the duration of this event the meridional SST front was severely attenuated or entirely absent. This pattern is also borne out clearly in even longer instrumental records spanning the second half of the twentieth century, allowing a more statistically robust analysis of the ENSO impacts in the cold tongue–ITCZ frontal complex (Deser and Wallace 1990).

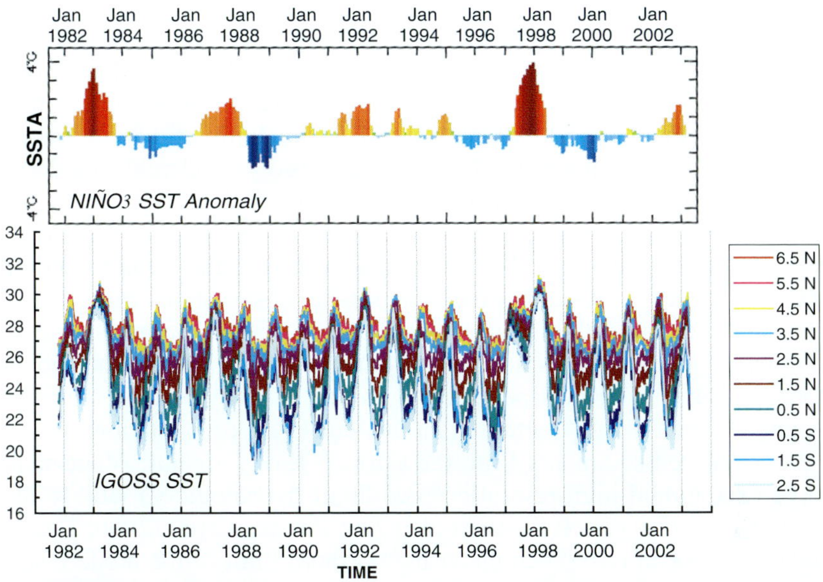

Figure 12-2. Weekly SST time series (IGOSS data accessed at: http://iridl.ldeo.columbia.edu/SOURCES/.IGOSS/.nmc/.Reyn_SmithOIv2/.weekly/sst over the1982–2003 period in ten 1° by 1° grid boxes from 7°N to 3°S at 90°W longitude. These grid boxes were chosen because they traverse the equatorial SST gradient of the eastern Pacific (Fig. 12-1). The "concertina" effect in the data results from the seasonal intensification and collapse of the SST gradient, which accompanies the appearance and disappearance of the equatorial cold tongue in response to shifting ITCZ winds (Fig. 12-1). A weak gradient in the early part of the year occurs under the influence of the "doldrums" while the ITCZ is positioned near the equator. Later in the year, as the ITCZ shifts north, the southeast trades across the equator strengthen and drive enhanced upwelling; hence a strong gradient develops. The top panel shows the monthly NIÑO3 SST anomaly index over the same period. Prominent El Niño anomalies (e.g., 1982–83, 1987, 1997–98) are marked by unusually weak gradients and more southerly ITCZ, whereas La Niñas (e.g., 1988) are marked by an enhanced and longer-persisting SST gradient, and more northerly ITCZ.

Collectively, these observations illustrate the key principle that forms the basis for our strategy; namely, that a reconstruction of the cross-equatorial SST gradient through time can be used as a reliable index of upwelling intensity and relative position of the marine ITCZ with respect to the equator. In the following sections we discuss evidence for past variability in this gradient and place it in the context of evidence for global ITCZ shifts from a tropics-wide network of sites.

3. LGM CONDITIONS IN SOUTH AMERICA

The question of prevailing conditions over tropical South America during the LGM has been subject to prolonged debate and remains contested. This issue is of special relevance here because the continent is flanked by the tropical Atlantic and Pacific, both of which have well-developed "linear" marine ITCZs that undergo considerable modern variability in intensity and position (Mitchell and Wallace 1992; Waliser and Gautier 1993). In particular, while many studies have attributed hydrologic (wet or dry) anomalies in the continent to a shift in the mean latitude of the ITCZ, evidence to this effect requires parallel observations from many locations and can rarely be invoked on the basis of data from a single site. A number of recent studies addressing climatic conditions in and around the Amazon Basin offer new perspective on this issue by placing multiple constraints on the spatial structure of anomalous LGM precipitation.

Varved sediments from the Cariaco Basin offshore Venezuela preserve exceptional high-resolution records of hydrologic variability through the last glacial cycle (Hughen et al. 1996; Peterson et al. 2000; Haug et al. 2001). The basin is situated under the northern limit of the modern seasonal range of the ITCZ and thus experiences a large rainfall cycle, which results in deposition of laminated sediments. Deposition of dark-colored varves occurs during the rainy season (summer) when the ITCZ is positioned overhead and high river runoff supplies dark minerals (Fe and Ti) eroded from the continent. Plankton-rich, light-colored varves are deposited during the dry season (winter) when terrigenous input is reduced and strong northeasterly trades drive coastal upwelling, stimulating the production and sedimentary flux of biogenic carbonate. Thus both the color and mineralogy of the sediments can be used to reconstruct rainfall intensity near Cariaco. The evidence leaves little doubt that significantly cooler and more arid conditions prevailed on the northern margin of the continent during the LGM (Hughen et al. 1996; Peterson et al. 2000; Lea et al. 2003). Glacial aridity in this region and in neighboring Central America is supported by paleoecologic studies in a number of lakes from nearby localities; e.g., Lake Valencia, Venezuela (Leyden 1985; Curtis et al. 1999), Lake Fuquene and the El Abra Valley region, Colombia (van Geel and van der Hammen 1973; Schreve-Brinkman 1978, van der Hammen and Hooghiemstra 2003), Lake El Valle, Panama (Bush 2002), and La Chonta Bog, Costa Rica (Islebe et al. 1995). These studies converge on a consistent picture of enhanced glacial aridity in Central America and northern South America between approximately 10ºN and 5ºN latitude. In addition, in those records where the resolution permits insights on shorter time scales, the deglacial progression bears strong similarity to the Bölling/Allerod–Younger Dryas sequence documented in high latitudes. This is especially true for the Cariaco Basin (Hughen et al. 1996; Lea et al. 2003) but is also evident in the palynologic studies of Islebe et al. (1995) and van der Hammen and Hooghiemstra (1995), from Costa Rica and Colombia, respectively. The rather compelling conclusion from these observations is, therefore, that terrestrial climate in the northern tropics of Central and South America sustained sizable cooling and aridification during the LGM and, moreover, fluctuated in step with the northern high latitudes during deglaciation. Given that the climate of this region is affected prominently by the northward migration of the ITCZ, this observation bears central importance for constraining the northern range of the LGM ITCZ.

Further south, in the Amazon Basin, the evidence is more complex. According to an early paradigm to which some authors still subscribe, widespread Amazonian aridity in glacial times caused the rain forest to shrink and be replaced by savanna, surviving only in isolated patches or "refugia" to later recolonize the basin when the Pleistocene gave way to the more humid Holocene (Haffer 1969; Haffer and Prance 2001). Recent studies, however, refute this hypothesis, and instead indicate biome stability under a persistently wet, albeit cooler, climate (Colinvaux and De Oliveira 2000; Colinvaux et al. 2001). Pollen records from Lake Pata (Colinvaux et al. 1996) and from ODP site 932 in the mouth of the Amazon (Haberle and Maslin 1999) attest to forest stability and imply that moisture availability remained sufficiently high to support tropical rain forest through the LGM. Likewise, offshore Fortaleza in northeastern Brazil (Nordeste), a semiarid region today, Arz et al. (1998) have reported elevated concentrations of Ti and Fe in sediments of the last glacial period. These are especially prominent during the coldest parts of the glacial (corresponding to Heinrich events) and are attributed to increased river runoff and more humid conditions. Wetter LGM conditions in Nordeste are also indicated by travertine and speleothem deposits from local caves (Auler and Smart 2001).

In the Bolivian Altiplano, a wetter LGM climate has been inferred from lake-level fluctuations of Lake Titicaca (Baker et al. 2001), and from the concentration of dust in ice cores from Sajama (Thompson et al. 1998) and Illimani (Ramirez et al. 2003). In Peru, however, the Huascarán ice core has been interpreted to indicate colder and drier conditions, on the basis of depleted $\delta^{18}O$ in LGM ice and elevated dust concentrations at the base of the core (Thompson et al. 1995). Nevertheless, Ramirez et al. (2003) offer an alternative view of the Huascarán data. They attribute the depleted $\delta^{18}O$ values to increased LGM precipitation rather than cooling, and reject the high basal dust content as an artifact of disturbances at the ice-bedrock interface.

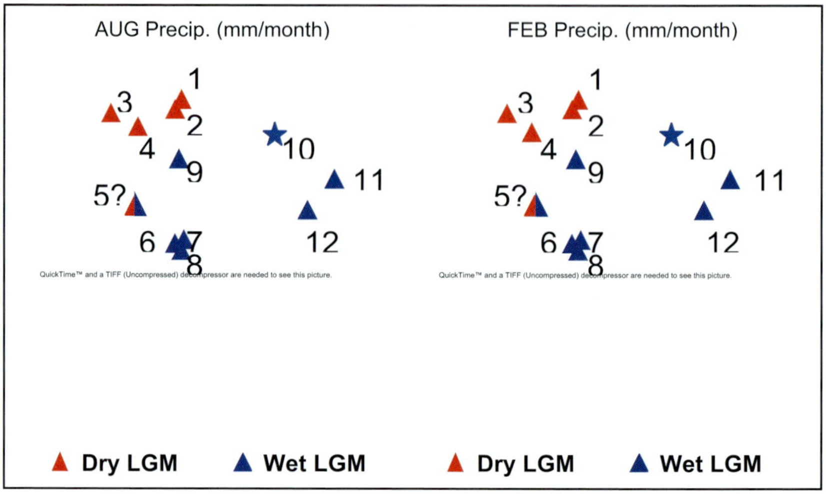

Figure 12-3. Terrestrial records of hydrologic change in South America during the LGM. The background maps show modern monthly precipitation values for August (left) and February (right) in millimeters per month. Vectors indicate prevailing winds at 925 mb. Red symbols indicate more arid conditions during the LGM and blue symbols (more) humid conditions relative to today. The anomalous LGM pattern mimics the pattern arising today from the seasonal southward migration of the ITCZ over the continent. This suggests that to first order the LGM anomalies can be explained by a systematic shift of the ITCZ to the south. Sites are numbered as follows: 1, Cariaco Basin (Peterson et al. 2000); 2, Lake Valencia, Venezuela (Leyden 1985); 3, El Valle Lake, Panama (Bush 2002); 4, Lake Fuquene, Colombia (van Geel and van der Hammen 1973); 5, Huascarán, Peru (Thompson et al. 1995); 6, Sajama, Bolivia (Thompson et al. 1998); 7, Illimani, Bolivia (Ramirez et al. 2003); 8, Lake Titicaca, Bolivia (Baker et al. 2001); 9, Lake Pata, Brazil (Colinvaux et al. 1996); 10, ODP 932, Amazon Fan (Haberle and Maslin 1999); 11, GeoB 3912 (Arz et al. 1998); 12, Northeast Brazil (Auler and Smart 2001). See text for more thorough discussion of these sites. The evidence from site 5 (Huascarán ice core) remains controversial, hence the ambivalent symbol. The star symbol near site 10 (ODP 932) denotes that the pollen record from this marine site is in fact a surrogate for integrated conditions over the inland Amazon Basin.

It is clear from these collective studies that LGM hydrologic changes in South America did not follow a simple pattern or proceed in a uniform direction. Nevertheless, a consistent picture is beginning to emerge which is composed of three main elements: (1) aridification of the northern part of the continent (as well as of Central America), (2) stability of biomes and persistently moist climate in the Amazon Basin, and (3) enhanced rainfall in the southern tropics, including the Altiplano of Bolivia and Nordeste. Figure 12-3 summarizes the evidence discussed above and explores the emerging spatial pattern. Using as a template the present-day seasonal precipitation cycle (which follows the latitudinal migration of the ITCZ), Figure 12-3 shows that the observed hydrologic anomalies during the LGM can be adequately explained by a generalized shift of the locus of precipitation to the south, not unlike its modern southward displacement during austral summer (Feb). As a whole then, these records are entirely consistent with a more southerly position of the ITCZ over South America during the LGM.

The discussion of these data sets may also be convincingly framed in terms of an LGM increase in the strength of the South American summer monsoon, as a consequence of the austral summer (Jan) insolation maximum 20,000 years before present (YBP). Little formal distinction exists today between ITCZ- or monsoon-type convection schemes over tropical South America, hence it is hard to conceive of much substantive difference between a monsoon- versus ITCZ-centered interpretation of the paleorecords (except perhaps in the context of winter versus summer rainfall budgets, a distinction that for the most part defies present reconstruction skills). In either case, the first-order result remains one of a southerly LGM bias in the continental rainfall belts with respect to the present. Of greater relevance here is the interaction between the monsoon circulation over the continent and the narrow oceanic ITCZs of the Atlantic and Pacific, and this is presently not fully understood even on seasonal time scales despite extensive observational data (e.g., Mitchell and Wallace 1992). Present understanding does not preclude the possibility that meridional shifts of rainfall over the continent are independent of their oceanic counterparts, and for this reason it is essential to obtain independent reconstructions of ITCZ variability over the oceans for comparison with the land records. In the following section we present evidence for a southward LGM shift of the eastern Pacific ITCZ, which we assert represents an extension of the observed pattern over South America.

4. SOUTHWARD SHIFT OF THE EASTERN PACIFIC ITCZ DURING THE LGM

As was discussed earlier, shifts in the latitude of the marine ITCZ over the eastern Pacific are accompanied by profound changes in SST structure. More specifically, the frontal zone between 0° and 5°N intensifies when the ITCZ moves north, and weakens or vanishes as the ITCZ shifts south (Figs. 12-1 and 12-2). We therefore expect that any long-term systematic shifts in ITCZ position would leave an imprint in the intensity of this front. The $\delta^{18}O$ composition of planktonic foraminifera is well suited to monitor this front and to reconstruct its past variations. Depleted (more

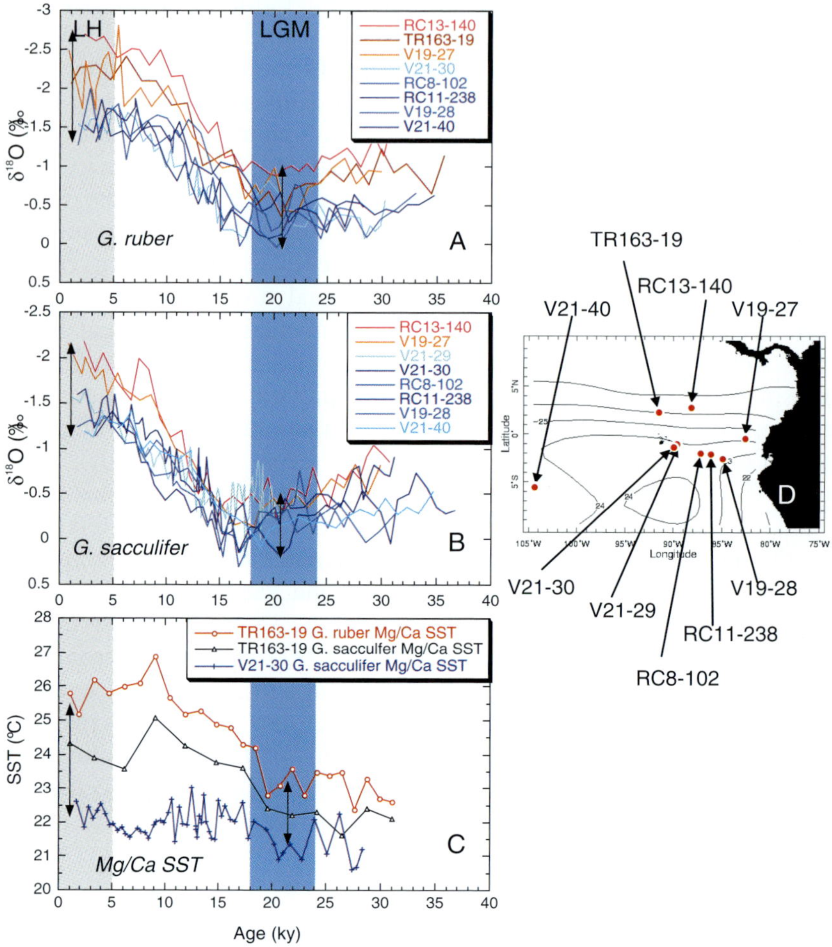

Figure 12-4. $\delta^{18}O$ and Mg/Ca SST reconstructions in a suite of cores from the eastern equatorial Pacific. (A) and (B) show down-core foraminiferal $\delta^{18}O$ profiles of the last 30,000 years from *G. ruber* and *G. sacculifer*, respectively. (C) Mg/Ca reconstructions of SST from cores TR163-19 (Lea et al. 2000; Dekens et al. 2002) and V21-30 (Koutavas et al. 2002). (D) Core locations are relative to annual mean climatologic SST (Levitus and Boyer 1994). Both proxies ($\delta^{18}O$ and Mg/Ca) indicate a reduced meridional SST gradient across the equator during the LGM compared to the Late Holocene (LH) (see vertical arrows in panels A–C). A more thorough discussion of these data is given by Koutavas and Lynch-Stieglitz (2003)

negative) $\delta^{18}O$ values result from either warmer SST (–0.21‰ per ºC) or lower surface salinity (0.27‰ per salinity unit). Thus in the northern edge of the front, where warm SSTs are accompanied by strong ITCZ rainfall and low salinity, planktonic $\delta^{18}O$ values are strongly depleted compared to those on and south of the equator, where lower SSTs and higher salinities prevail (Levitus and Boyer 1994).

We use a suite of cores located between 5°S and 3°N (Fig. 12-4D) to assess the intensity of this front in the LGM in comparison with the Holocene. $\delta^{18}O$ records of the last 30,000 years from these cores, measured in two species of planktonic foraminifera (*G. ruber* and *G. sacculifer*) are shown in Figures 12-4A and 12-4B. Both species capture the modern (late Holocene) $\delta^{18}O$ gradient of ~1‰ across the front quite well. Both species also reflect a significant decrease of the gradient during the LGM, as would be expected from less intense upwelling due to a more southerly ITCZ (Koutavas and Lynch-Stieglitz 2003). Because the interpretation of these data is complicated by the influence of salinity on $\delta^{18}O$, an independent test of this result is provided by direct reconstructions of SST based on foraminiferal Mg/Ca. Mg/Ca has been calibrated versus temperature in a number of studies (Nürnberg et al. 1996; Lea et al. 2000; Dekens et al. 2002) and has been successfully used for down-core SST reconstructions in many tropical locations (Lea et al. 2000; Stott et al. 2002; Koutavas et al. 2002; Rosenthal et al. 2003). Figure 12-4C compares the Mg/Ca SST records from cores V21–30 at 1.2°S (Koutavas et al. 2002) and TR163-19 at 2.2°N (Lea et al. 2000; Dekens et al. 2002). This comparison confirms that a reduced SST gradient between the two cores existed at the LGM compared to the Holocene.

In our view, these data reflect a systematic southward shift of the mean ITCZ latitude during the LGM, in the same manner that a severely reduced meridional SST gradient today accompanies the equatorward retreat of the ITCZ seasonally (during winter) and interannually (during El Niño). We hypothesize that the LGM response reflects a more restricted northward range of the summer ITCZ, as a consequence of cooler Northern Hemisphere summers in the face of permanent land ice fields. In light of the preceding discussion of terrestrial conditions in South America, we regard the southward response of the eastern Pacific ITCZ as an extension of the observed pattern over South America and the tropical Atlantic under common LGM forcing.

5. MIGRATION OF THE ITCZ DURING THE HOLOCENE

Following minimum values during the LGM, northern summer (Jul) insolation gradually increased to a maximum centered at about 10,000 BP, and has been declining ever since. This insolation maximum is linked to the early Holocene climatic optimum observed in high-latitude records from land (for example, the North GRIP ice core from Greenland, Fig. 12-5A) and ocean (Liu et al. 2003, and references therein), and is also believed to have forced the strengthening of the Indian and Asian monsoon systems (Clemens et al. 1991; Gupta et al. 2003; Fleitmann et al. 2003, and Chapter 9, "Holocene Records of Rainfall Variation and Associated ITCZ Migration from Stalagmites from Northern and Southern Oman," this volume). Evidence has recently emerged that this period was also marked by a more northerly mean ITCZ latitude followed by gradual southward migration. Support for this hypothesis is provided by several paleoclimatic archives distributed tropics-wide, some of which are illustrated in Figure 12-5. $\delta^{18}O$ values in cave stalagmites from Oman (Fig. 12-5B) are strongly depleted during the early and middle Holocene, indicating enhanced rainfall under the influence of a more northerly Indian Ocean ITCZ (Fleitmann et al. 2003, and Chapter 9, this volume). The terrigenous record from ODP site 658C off West Africa (Fig. 12-5C) indicates an abrupt transition to a more arid climate at about 5,500 BP, which reflects the southward expansion of North African desert at the expense of grassland (deMenocal et al. 2000). And the Cariaco Basin mineralogy (Fig. 12-5D) documents early Holocene increases in riverine input of Fe and Ti, ascribed to a more northerly ITCZ position (Haug et al. 2001). All three of these records are located in the northern tropics between 10ºN and 20ºN, where rainfall is largely controlled by the northerly summer excursions of the ITCZ. All indicate a common pattern of a wetter climate in the early Holocene up to about 5,000–6,000 BP, and a shift to more arid conditions thereafter. These trends are best explained by a more northerly ITCZ position in the early and middle Holocene followed by a gradual southward retreat.

This interpretation is supported by complementary evidence from the Southern Hemisphere. Lake Titicaca in the Altiplano of Bolivia reached its lowest levels on record between 8,000 and 4,500 BP, then rose again toward the present time, signifying a mid-Holocene transition from dry to wetter conditions (Baker et al. 2000). The vegetation also reflects a transition to a more humid climate. Bolivian rain forest expanded southward at the expense of savanna since at least 3,000 BP (Mayle et al. 2000). And the $\delta^{18}O$ record of the ice core from Nevado Illimani, Bolivia, shows progressive isotopic depletion through the Holocene, consistent with a long-term increase in precipitation (Ramirez et al. 2003). Finally, in Lake Malawi in the southern tropics of Africa (10ºS), mid-Holocene increases in biogenic silica and volcanic dust have been tentatively linked to a southward ITCZ shift there as well (Johnson et al. 2002).

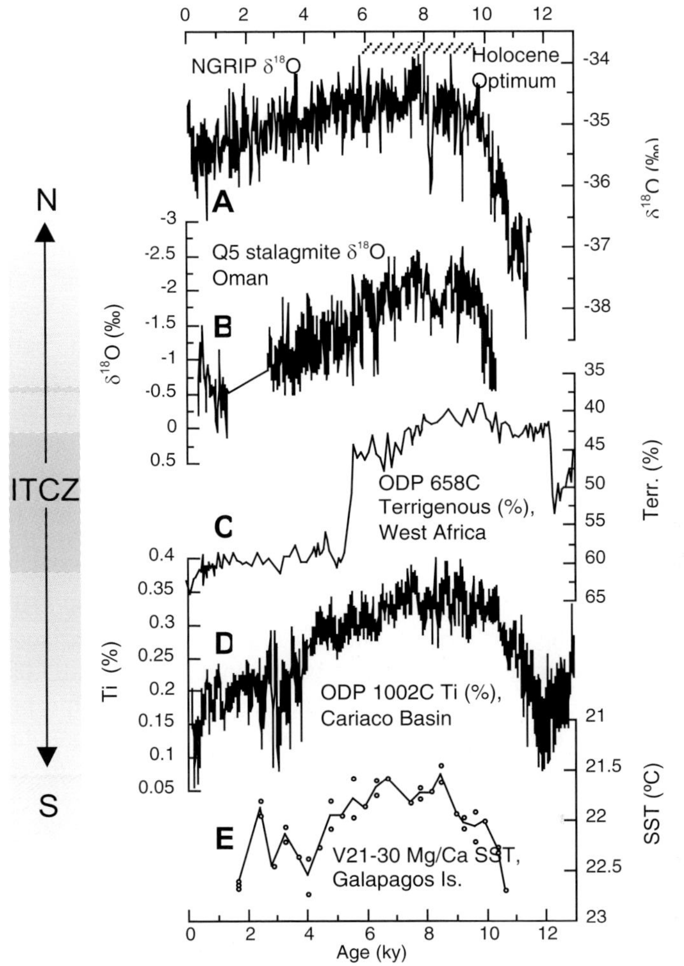

Figure 12-5. Holocene climate variations with relation to the ITCZ. (A) North GRIP (75°N) ice core $\delta^{18}O$ (Johnsen et al. 2001) showing the climatic optimum of the early Holocene, followed by progressive cooling. (B) Q5 stalagmite $\delta^{18}O$ from Qunf Cave, Oman (17°N) (Fleitmann et al. 2003). (C) Terrigenous fraction of ODP 658C (21°N), off West Africa (deMenocal et al. 2000). (D) Cariaco Basin (10°N) titanium concentration (Haug et al. 2001). (E) Mg/Ca SST (inverted) from V21-30, Galapagos Islands (1.2°S) (Koutavas et al. 2002). The records from the northern tropics (B, C, and D) indicate wetter climate in the early to middle Holocene, which is consistent with a more northerly ITCZ in the Indian and Atlantic Oceans. The Galapagos SST record (E) shows early to middle Holocene cooling due to stronger upwelling and supports a more northerly ITCZ in the Pacific as well.

It appears therefore that the combined evidence from these diverse tropical localities in both hemispheres supports a consistent picture of a southward ITCZ movement through the Holocene. Wetter climatic conditions in the northern tropics during the early and middle Holocene gave way to more arid conditions in the late Holocene, whereas the opposite trend occurred in the southern tropics.

In the context of this evidence, we pose the question of whether a coherent pattern of variability is evident in the eastern tropical Pacific, involving an analogous shift of the ITCZ and its impact on equatorial upwelling and SST. The Holocene SST record based on foraminiferal Mg/Ca in core V21-30 in the heart of the equatorial cold tongue (Fig. 12-5E), shows a broad interval of cool SSTs in the early and middle Holocene between 9,000 and 5,000 BP (Koutavas et al. 2002). This is precisely the response expected from a more northerly ITCZ at this time, which would have promoted strong upwelling due to persistent southeast trades blowing across the equator. Likewise, the transition to warmer SSTs after about 5,000 BP in the same record is fully consistent with a more southerly ITCZ in the late Holocene.

One incongruous element in this discussion concerns the timing of the northerly ITCZ extreme, in relation to the summer insolation maximum. While July insolation peaked between about 11,000 and 8,000 BP, enhanced ITCZ rainfall in the northern tropics (for example in the Cariaco Basin) occurred during the broader interval between about 10,000 and 5,000 BP (Fig. 12-5). We see two reasons for this apparent "lag." In the first place, we argue that it is not a real lag: The more appropriate monthly insolation curves to consider with respect to tropical rainfall and the ITCZ are those of August and September. September (not July) is the month during which the ITCZ attains its northernmost position today, and often maintains it well into October (Waliser and Gautier 1993). September insolation peaked between 6,000 and 9,000 BP, which is precisely the time when maximum rainfall is observed in the Cariaco Basin (Fig. 12-5D). This timing also matches the Lake Titicaca low-stand (Baker et al. 2001) and the observed SST minimum in the equatorial cold tongue (Fig. 12-5E). In effect, the 6,000–9,000 BP September insolation maximum may have served to lengthen the northern summer and allowed the ITCZ to linger near its northern limit longer. The second possible reason for the "delayed" ITCZ response may be that the melting of continental glaciers was not completed until about 7,000 BP, as is evident from reconstructions of global sea level (Fleming et al. 1998). More precisely, sea level was 40 m below its present level at 10,000 BP and therefore approximately one-third of the full amount of excess LGM ice remained in place at that time. Land ice restricts the northward migration of the ITCZ by promoting a steep latitudinal temperature gradient, which strengthens the northern trades (Chiang et al. 2003). The final melting of this ice between 10,000 and 7,000 BP combined with maximum September sunshine must have favored the more northerly position of the ITCZ.

An issue related to the ITCZ history concerns the Holocene evolution of ENSO. Evidence indicates less frequent or absent El Niño occurrences in the early and middle Holocene (Tudhope et al. 2001; Moy et al. 2003) and is supported by a simple model of the tropical Pacific forced by insolation (Clement et al. 2000). The onset of El Niño episodes today requires westerly wind anomalies near the equator; that is, weaker easterlies. The easterlies weaken annually during December through April, as part of the seasonal approach of the ITCZ to the equator. El Niño typically matures during the winter months (Dec–Jan), which suggests that it feeds upon and exacerbates the normally encountered conditions of weak easterlies favored by a southerly ITCZ. It is thus likely that a northward-displaced ITCZ in the early to middle Holocene would have a profound effect on the dynamics of ENSO. If the ITCZ were anchored off the equator for a longer season each year, or were forced to oscillate within a range farther removed from the equator, the overall potential for weaker equatorial easterlies capable of triggering El Niño would have been correspondingly reduced. Conversely, it seems plausible that sometime around 5,000 BP, as the ITCZ migrated southward, it attained sufficient proximity to the equator and began to perturb the equatorial easterlies in a manner that favored frequent genesis of El Niño events. In this context it may be significant that some of the largest-amplitude ENSO variability of the last millennium, reconstructed from corals, occurred within the recent Little Ice Age (LIA; Cobb et al. 2003), a period for which there is compelling evidence for a more southerly ITCZ (Haug et al. 2001) (see later discussion). If this reasoning shows validity, one might extrapolate to the future and predict that climatic warming in the next century, whether natural or anthropogenic, may induce a northward excursion of the mean ITCZ, which in the absence of other forcing would tend to inhibit ENSO. It is likely, however, that this effect would require a hemispherically asymmetric warming favoring the Northern Hemisphere whose source remains unclear.

6. ITCZ VARIABILITY ON SUBORBITAL TIME SCALES

Thus far we have examined variability of the ITCZ over the last glacial-interglacial transition and through the Holocene, periods marked by strong orbital influence due to earth's precession. Nevertheless, climate has varied dramatically on shorter time scales as well, particularly in the millennial frequency band, without an apparent external forcing. The observations of distinctly paced Dansgaard-Oeschger (D-O) oscillations (e.g., Rahmstorf

2003) and ice rafting (Heinrich) events in the North Atlantic region (e.g., Bond et al. 1993) have by now been extended by the discovery of correlative events throughout the globe.

In the tropics the evidence for these events has grown steadily during the past decade and in many cases points to hydrologic variability associated with shifts of the ITCZ. In the Cariaco Basin at 10ºN, the cold phases of the D-O oscillations are marked by correlative reductions in Fe and Ti and increases in sediment reflectance, indicating more arid conditions likely due to more southerly ITCZ confinement (Peterson et al. 2000). Conversely, south of the equator off the coast of Brazil (3.5ºS), Heinrich events correlate with increases in Fe and Ti as expected from more humid conditions due to a more southerly ITCZ (Arz et al. 1998). Likewise, the Altiplano of Bolivia and southern Amazonia experienced wetter climate during the Younger Dryas and possibly Heinrich event 1 (Baker et al. 2001). In the eastern equatorial Atlantic, systematic increases in the relative abundance of *Florisphaera profunda*, a deep-living marine alga, correlate with Heinrich events and indicate decreased divergence due to weaker zonal winds (McIntyre and Molfino 1996) best explained by an equatorward shift of the ITCZ "doldrums." In the Indian Ocean isotopic evidence from stalagmites on Socotra Island suggests a weaker monsoon during cold stadials and is basically consistent with a more southerly ITCZ (Burns et al. 2003), but here it is exceedingly difficult to separate the response of the ITCZ from its interaction with the monsoonal circulation. In the western equatorial Pacific, within the bounds of the modern warm pool, stadial conditions correspond with elevated surface salinities (Stott et al. 2002), pointing to a shift of the locus of convection to the south, or perhaps to the east as during modern El Niño conditions.

Unfortunately, available records from the eastern tropical Pacific are as yet unable to resolve millennial climate changes with enough dating precision to constrain the role of the ITCZ in this basin. Low accumulation rates and uncertainties in the surface ocean ^{14}C reservoir correction make the detection and dating of millennial-scale events here a formidable challenge. Nevertheless, given the patterns of millennial variability observed over adjacent South America and the tropical Atlantic, which support southward ITCZ excursions during stadials, we speculate that a similar pattern may eventually be documented in the eastern Pacific as well. This expectation is based on the fact that both ocean basins possess similar equatorial cold tongue–ITCZ complexes governed by similar physics, and today exhibit coherent seasonal variability (Mitchell and Wallace 1992). If such a millennial response of the eastern Pacific ITCZ is confirmed, it would imply that stadial periods were unfavorable to vigorous equatorial upwelling, perhaps promoting instead a shift to more zonally uniform equatorial SSTs, similar to the modern El Niño pattern (e.g., Koutavas et al. 2002). Alternatively, it is possible that the Pacific ITCZ may have been far less sensitive to millennial-scale climate shifts than the observations suggest for the Atlantic, and correspondingly any impact on equatorial Pacific divergence may have been weak. Distinguishing between these scenarios requires advances in identifying high-resolution marine archives from this region, as well as in narrowing the present dating uncertainties.

Millennial-scale oscillations, while dominant during the last glaciation, have persisted with reduced amplitude into the Holocene (Bond et al. 1997; deMenocal et al. 2000). The most recent millennial oscillation involved the transition from the Medieval Warm Period (MWP) to the Little Ice Age, followed by renewed warming in the latter half of the nineteenth century. The detailed mineralogic (Fe and Ti) records from the Cariaco Basin show that the MWP and LIA had correlative hydrologic events in the tropical Atlantic (Haug et al. 2001). The LIA in Cariaco was evidently a drier interval than the preceding MWP and was followed by increases in Fe and Ti resulting from wetter conditions. According to Haug et al. (2001) these variations are diagnostic of fine-scale century-to-decade shifts of the ITCZ during the last millennium, and imply a more southerly ITCZ during the LIA followed by return to a more northerly latitude today.

Did these ITCZ shifts extend over the neighboring tropical Pacific? Coral records spanning the last three to four centuries offer a parallel perspective from that region. Figure 12-6 compares the coral $\delta^{18}O$ records of Linsley et al. (1994) and Dunbar et al. (1994) from Panama and the Galapagos, respectively. The two records straddle the cross-equatorial front between the ITCZ and the cold tongue, and their lengths overlap over most of the last three centuries. As was noted by Dunbar et al. (1996), the opposite $\delta^{18}O$ trends evident in these corals both on centennial and decadal time scales can best be explained by movements of the ITCZ, which, as was discussed earlier, have a large impact on the strength of the cross-equatorial SST/salinity gradient (Figs. 12-1 and 12-2). Thus, the increasing $\delta^{18}O$ gradient between the two locations since about AD 1850 is most likely a manifestation of the same northward ITCZ movement that produced the observed increases in Fe and Ti in the Cariaco Basin at the end of the LIA (Haug et al. 2001). It is likely that the higher-frequency fluctuations evident in the coral $\delta^{18}O$ have correlative signatures in Cariaco as well; however, the dating of the latter record is not precise enough to allow exact matching. These observations confirm that the ITCZ has varied coherently over the Atlantic and Pacific during the last three centuries. Moreover, they add to the growing evidence that the position of the mean ITCZ latitude is linked across both basins (and probably on a global scale) and undergoes systematic shifts over a wide range of time scales, such that climatic cooling in the Northern Hemisphere favors a more southerly ITCZ and vice versa.

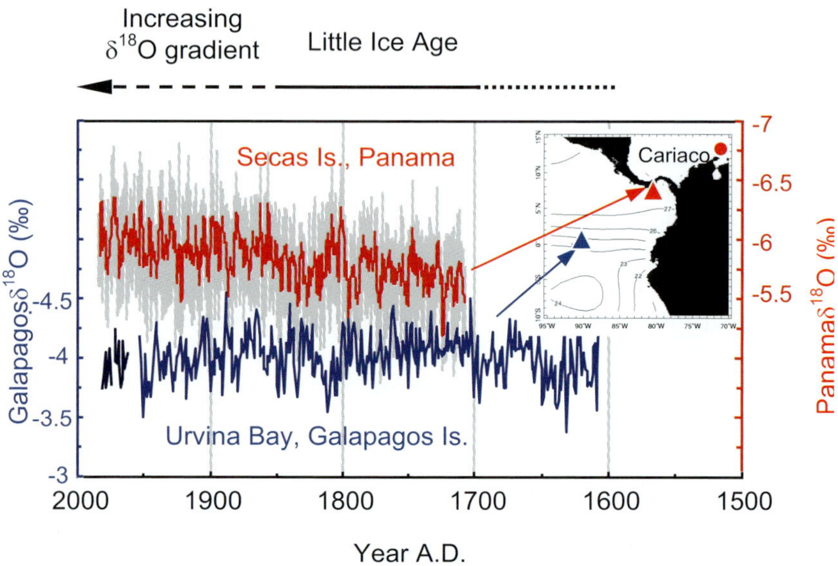

Figure 12-6. Coral $\delta^{18}O$ data of the last ~400 years from two sites straddling the modern SST gradient of the eastern Pacific cold tongue–ITCZ. At the top is the $\delta^{18}O$ record from Secas Island, in the Gulf of Chiriquí, Panama (8°N) (Linsley et al. 1994). The red curve is a 10-point running average of the raw data, in gray. The blue series (bottom) shows the $\delta^{18}O$ of a coral from Urvina Bay, Galapagos Islands (0.4°S) (Dunbar et al. 1994). The inset shows the coral locations relative to the annual mean SST front (Levitus and Boyer 1994), and the Cariaco Basin in the Atlantic. Evidence from Cariaco indicates a northward shift of the ITCZ since the end of the LIA (Haug et al. 2001), which is reflected in the Pacific by the increasing isotopic gradient between the two corals since the mid-nineteenth century.

7. SUMMARY AND CONCLUSIONS

In the present climate, the ITCZ displays a remarkable sensitivity to the annual cycle and to interannual disturbances due to ENSO. The evidence summarized here argues that this sensitivity likely persists over longer time scales in response to a variety of forcing mechanisms, such as the presence and extent of land ice, Earth's orbital geometry, and possibly others operating on shorter time scales. Most important, the patterns of variability across different ocean basins and over the continents appear to share common characteristics: Systematic, coeval ITCZ shifts in the same direction appear to have occurred globally. This finding confronts us with the possibility that variability in the ITCZ will prove to be instrumental in understanding high- to low-latitude climate linkages on a global scale. Nevertheless, the precise dynamical underpinnings through which these and other possible forcing mechanisms (e.g., the strength of the thermohaline circulation or the solar output) operate, or their relative importance, remain elusive. Large gaps remain in the paleoclimate record, both spatial and temporal, which need to be filled in order to gain further insights. Over the eastern tropical Pacific in particular, where the ITCZ is coupled tightly to the dynamics of the equatorial cold tongue and thus controls the distribution of SSTs across the entire basin, it will be exceedingly important to acquire a network of high-quality data sets. The available decadal-scale records from the Cariaco Basin and from Pacific corals leave no doubt that the system has been far from stable in the recent past. Its future evolution is uncertain and so is its potential impact on ENSO, tropical rainfall, and the vast human populations that inhabit the circum-global zone under the influence of the ITCZ.

8. ACKNOWLEDGMENTS

We thank Peter deMenocal for access to his ICP-AES lab and support with Mg/Ca analyses; Ray Bradley and Henry Diaz for their kind invitation to attend the Hadley circulation meeting in Hawaii in November 2002, and for helpful comments on the manuscript; and all of the Hadley circulation conference participants for stimulating presentations and discussions. This work was supported by a grant/cooperative agreement from the National Oceanic and Atmospheric Administration (NOAA). A. Koutavas was supported through a NOAA Climate and Global Change postdoctoral fellowship administered by the University Corporation for Atmospheric Research (UCAR). The views expressed herein are those of the authors and do not necessarily reflect the views of NOAA or any of its subagencies. Sample material used in this study was provided by the Lamont-Doherty Earth Observatory Deep-Sea Sample Repository. Support for the collection and curating facilities of the core collection is provided by the National Science Foundation (NSF) through Grant OCE00-02380 and the Office of Naval Research through Grant N00014-02-1-0073.

9. REFERENCES

Arz, H.W., J. Pätzold, and G. Wefer. 1998. Correlated millennial-scale changes in surface hydrography and terrigenous sediment yield inferred from last-glacial marine deposits off northeastern Brazil. *Quaternary Research* 50: 157–166.

Auler, A.S., and P.L. Smart. 2001. Late Quaternary paleoclimate in semiarid northeastern Brazil from U-series dating of travertine and water-table speleothems. *Quaternary Research* 55: 159–167.

Baker, P.A., G.O, Seltzer, S.C. Fritz, R.B. Dunbar, M.J. Grove, P.M. Tapia, S.L. Cross., H.D. Rowe, and J.P. Broda. 2001. The history of South American tropical precipitation for the past 25,000 years. *Science* 291: 640–643.

Bond, G., W. Broecker, S. Johnsen, J. McManus, L. Labeyrie, J. Jouzel, and G. Bonani. 1993. Correlations between climate records from North Atlantic sediments and Greenland ice. *Nature* 365: 143–147.

Bond, G., W. Showers, M. Cheseby, R. Lotti, P. Almasi, P. deMenocal, P. Priore, H. Cullen, I. Hajdas, and G. Bonani. 1997. A pervasive millennial-scale cycle in North Atlantic Holocene and glacial climates. *Science* 278: 1257–1266.

Burns, S.J., D. Fleitmann, A. Matter, J. Kramers, and A.A. Al-Subbary. 2003. Indian Ocean climate and an absolute chronology over Dansgaard/Oeschger events 9–13. *Science* 301: 1365–1367.

Bush, M.B. 2002. On the interpretation of fossil Poaceae pollen in the lowland humid neotropics. *Palaeogeography, Palaeoclimatology, Palaeoecology* 177: 5–17.

Chelton, D.B., S.K. Esbensen, M.G. Schlax, N. Thum, M.H. Freilich, F.J. Wentz, C.L. Gentemann, M.J. McPhaden, and P.S. Schopf. 2001. Observations of coupling between surface wind stress and sea surface temperature in the eastern tropical Pacific. *Journal of Climate* 14: 1479–1498.

Chiang, J.C.H., M. Biasutti, and D.S. Battisti. 2003. Sensitivity of the Atlantic ITCZ to Last Glacial Maximum boundary conditions. *Paleoceanography* 18: 10.1029/2003PA000916.

Clemens, S., W. Prell, D. Murray, G. Shimmield, and G. Weedon. 1991. Forcing mechanisms of the Indian-Ocean monsoon. *Nature* 353: 720–725.

Clement, A.C., R. Seager, and M.A. Cane. 2000. Suppression of El Niño during the mid-Holocene by changes in the Earth's orbit. *Paleoceanography* 15: 731–737.

Cobb, K.M., C.D. Charles, H. Cheng, and R.L. Edwards. 2003. El Niño/Southern Oscillation and tropical Pacific climate during the last millennium. *Nature* 424: 271–276.

Colinvaux, P.A., and P.E. De Oliveira. 2000. Paleoecology and climate of the Amazon Basin during the last glacial cycle. *Journal of Quaternary Science* 15: 347–356.

Colinvaux, P.A., P.E. De Oliveira, J.E. Moreno, M.C. Miller, and M.B. Bush. 1996. A long pollen record from lowland Amazonia: Forest and cooling in glacial times. *Science* 274: 85–88.

Colinvaux, P.A., G. Irion, M.E. Räsänen, M.B. Bush, and J.A.S. Nunes de Mello. 2001. A paradigm to be discarded: Geological and paleoecological data falsify the HAFFER & PRANCE refuge hypothesis of Amazonian speciation. *Amazoniana* 16(3/4): 609–646.

Curtis, J.H., M. Brenner, and D.A. Hodell. 1999. Climate change in the Lake Valencia Basin, Venezuela, ~12,600 yr BP to present. *The Holocene* 9: 609–619.

Dekens, P.S., D.W. Lea, D.K. Pak, and H.J. Spero. 2002. Core top calibration of Mg/Ca in tropical foraminifera: Refining paleotemperature estimation. *Geochemistry Geophysics Geosystems* 3(4): 1022, 10.1029/2001GC000200.

deMenocal, P.B., J. Ortiz, T. Guilderson, and M. Sarnthein. 2000. Coherent high- and low-latitude climate variability during the Holocene warm period. *Science* 288: 2198–2202.

Deser, C., and J.M. Wallace. 1990. Large-scale atmospheric circulation features of warm and cold episodes in the tropical Pacific. *Journal of Climate* 3: 1254–1281.

Dunbar, R.B., B.K. Linsley, and G.M. Wellington. 1996. Eastern Pacific corals monitor El Niño/Southern Oscillation, precipitation, and sea surface temperature variability over the past 3 centuries. In, Jones, P.D., R.S. Bradley, and J. Jouzel (eds.). *Climatic Variations and Forcing Mechanisms of the Last 2000 Years.* Berlin: Springer-Verlag, pp. 375–407.

Dunbar, R.B., G.M. Wellington, M.W. Colgan, and P.W. Glynn. 1994. Eastern Pacific sea surface temperature since 1600 A.D: The $\delta^{18}O$ record of climate variability in Galapagos corals. *Paleoceanography* 9: 291–315.

Fleitmann, D., S.J. Burns, M. Mudelsee, U. Neff, J. Kramers, A. Mangini, and A. Matter. 2003. Holocene forcing of the Indian monsoon recorded in a stalagmite from Southern Oman. *Science* 300: 1737–1739.

Fleming, K., P. Johnston, D. Zwart, Y. Yokoyama, K. Lambeck, and J. Chappell. 1998. Refining the eustatic sea-level curve since the Last Glacial Maximum using far- and intermediate-field sites. *Earth and Planetary Science Letters* 163: 327–342.

Gupta, A.K., D.M. Anderson, and J.T. Overpeck. 2003. Abrupt changes in the Asian southwest monsoon during the Holocene and their links to the North Atlantic Ocean. *Nature* 421: 354–357.

Haberle, S.G., and M.A. Maslin. 1999. Late Quaternary vegetation and climate change in the Amazon Basin based on a 50,000 year pollen record from the Amazon Fan, ODP Site 932. *Quaternary Research* 51: 27–38.

Haffer, J. 1969. Speciation in Amazonian forest birds. *Science* 165: 131–137.

Haffer, J., and G.T. Prance. 2001.Climatic forcing of evolution in Amazonia during the Cenozoic: On the refuge theory of biotic differentiation. *Amazoniana,* 16(3/4): 579–607.

Haug, G.H., K.A. Hughen, D.M. Sigman, L.C. Peterson, and U. Röhl. 2001. Southward migration of the Intertropical Convergence Zone through the Holocene. *Science* 293: 1304–1308.

Hughen, K.A., J.T. Overpeck, L.C. Peterson, and S. Trumbore. 1996. Rapid climate changes in the tropical Atlantic region during deglaciation. *Nature* 380: 51–54.

Islebe, G.A., H. Hooghiemstra, and K. van der Borg. 1995. A cooling event during the Younger Dryas Chron in Costa Rica. *Palaeogeography, Palaeoclimatology, Palaeoecology* 117: 73–80.

Johnsen, S.J., D. Dahl-Jensen, N. Gundestrup, J.P. Stefensen, H.B. Clausen, H. Miller, V. Masson-Delmotte, A.E. Sveinbjörnsdottir, and J. White. 2001. Oxygen isotope and paleotemperature records from six Greenland ice-core stations: Camp Century, Dye-3, GRIP, GISP2, Renland and NorthGRIP. *Journal of Quaternary Science* 16: 299–307.

Johnson, T.C., E.T. Brown, J. McManus, S. Barry, P. Barker, and F. Gasse. 2002. A high-resolution paleoclimate record spanning the past 25,000 years in southern East Africa. *Science* 296: 113–116.

Koutavas, A., and J. Lynch-Stieglitz. 2003. Glacial-interglacial dynamics of the eastern equatorial–Pacific cold tongue–Intertropical Convergence Zone system reconstructed from oxygen isotope records. *Paleoceanography* 18: 10.1029/2003PA000894.

Koutavas, A., J. Lynch-Stieglitz, T.M. Marchitto, Jr, and J.P. Sachs. 2002. El Niño-like pattern in ice age tropical Pacific sea surface temperature. *Science* 297: 226–230.

Lea, D.W., D.K. Pak, and H.J. Spero. 2000. Climate impact of late Quaternary equatorial Pacific sea surface temperature variations. *Science* 289: 1719–1724.

Lea, D.W., D.K. Pak, L.C. Peterson, and K.A. Hughen. 2003. Synchroneity of tropical and high-latitude Atlantic temperatures over the last glacial termination. *Science* 301: 1362–1364.

Leyden, B. 1985. Late Quaternary aridity and Holocene moisture fluctuations in the Lake Valencia Basin, Venezuela. *Ecology* 66: 1279–1295.

Levitus, S., and T.P. Boyer. 1994. *World Ocean Atlas 1994*, Vol. 4: *Temperature*. NOAA Atlas NESDIS 4, Washington, D.C.: U.S. Department of Commerce.

Linsley, B.K., R.B. Dunbar, G.M. Wellington, and D.A. Mucciarone. 1994. A coral-based reconstruction of Intertropical Convergence Zone variability over Central America since 1707. *Journal of Geophysical Research* 99: 9977–9994.

Liu, Z., E. Brady, and J. Lynch-Stieglitz. 2003. Global ocean response to orbital forcing in the Holocene. *Paleoceanography* 18: 10.1029/2002PA000819.

Mayle, F.E., R. Burbridge, and T.J. Killeen. 2000. Millennial-scale dynamics of southern Amazonian rainforest. *Science* 290: 2291–2294.

McIntyre, A., and B. Molfino. 1996. Forcing of Atlantic equatorial and subpolar millennial cycles by precession. *Science* 274: 1867–1870.

Mitchell, T.P., and J.M. Wallace. 1992. The annual cycle of equatorial convection and sea surface temperature. *Journal of Climate* 5: 1140–1156.

Moy, C.M., G.O. Seltzer, D.T. Rodbell, and D.M. Anderson. 2002. Variability of El Niño/Southern Oscillation activity at millennial time scales during the Holocene epoch. *Nature* 420: 162–165.

Nürnberg, D., J. Bijma, and C. Hemleben. 1996. Assessing the reliability of magnesium in foraminiferal calcite as a proxy for water mass temperatures. *Geochim. Cosmochim. Acta*, 60: 803–814.

Peterson, L.C., G.H. Haug., K.A. Hughen., and U. Röhl. 2000. Rapid changes in the hydrologic cycle of the tropical Atlantic during the last glacial. *Science* 290: 1947–1951.

Philander, S.G.H., D. Gu, D. Halpern, G. Lambert, N.-C. Lau, T. Li, and R.C. Pacanowski. 1996. Why the ITCZ is mostly north of the equator. *Journal of Climate* 9: 2958–2972.

Rahmstorf, S. 2003. Timing of abrupt climate change: A precise clock. *Geophysical Research Letters* 30(10): 1510, doi:10.1029/2003GL017115.

Ramirez, E., G. Hoffmann, J.D. Taupin, B. Francou, P. Ribstein, N. Caillon, F.A. Ferron, A. Landais, J.R. Petit., B. Pouyaud, U. Schotterer, J.C. Simoes, and M. Stievenard. 2003. A new Andean deep ice core from Nevado Illimani, (6350 m), Bolivia. *Earth and Planetary Science Letters* 212: 337–350.

Rosenthal, Y., D.W. Oppo, and B.K. Linsley. 2003. The amplitude and phasing of climate change during the last deglaciation in the Sulu Sea, western equatorial Pacific. *Geophysical Research Letters* 30: 1428, 10.1029/2002GL016612.

Schreve-Brinkman, E.J. 1978. A palynological study of the upper Quaternary sequence in the El Abra corridor and rock shelters (Colombia). *Palaeogeography, Palaeoclimatology, Palaeoecology* 25: 1–109.

Stott, L., C. Poulsen, S. Lund, and R. Thunell. 2002. Super ENSO and global climate oscillations at millennial time scales. *Science* 297: 222–226.

Thompson, L.G., E. Mosley-Thompson, M.E. Davis, P.-N. Lin, K.A. Henderson, J. Cole-Dai, J.F. Bolzan, and K.-b. Liu. 1995. Late glacial stage and Holocene tropical ice core records from Huascarán, Peru. *Science* 269: 46–50.

Thompson, L.G., M.E. Davis, E. Mosley-Thompson, T.A. Sowers, K.A. Henderson, V.S. Zagorodnov, P.-N. Lin, V.N. Mikhalenko, R.K. Campen, J.F. Bolzan, J. Cole-Dai, and B. Francou. 1998. A 25,000-year tropical climate history from Bolivian ice cores. *Science* 282: 1858–1864.

Tudhope, A.W., C.P. Chilcott, M.T. McCulloch, E.R. Cook, J. Chappell, R.M. Ellam, D.W. Lea, J.M. Lough, and G.B. Shimmield. 2001. Variability in the El Niño-Southern Oscillation through a glacial-interglacial cycle. *Science* 291: 1511–1517.

van der Hammen, T., and H. Hooghiemstra. 1995. The El Abra stadial, a Younger Dryas equivalent in Colombia. *Quaternary Science Reviews* 14: 841–851.

van der Hammen, T., and H. Hooghiemstra. 2003. Interglacial-glacial Fuquene-3 pollen record from Colombia: An Eemian to Holocene climate record. *Global and Planetary Change* 36: 181–199.

van Geel, B., and T. van der Hammen. 1973. Upper Quaternary vegetational and climatic sequence of the Fuquene area (Eastern Cordillera, Colombia). *Palaeogeography, Palaeoclimatology, Palaeoecology.* 14: 9–55.

Waliser, D.E., and C. Gautier. 1993. A satellite-derived climatology of the ITCZ. *Journal of Climate* 6: 2162–2174.

Xie, S.-P., and K. Saito. 2000. Formation and variability of a northerly ITCZ in a hybrid coupled AGCM: Continental forcing and oceanic-atmospheric feedback. *Journal of Climate* 14: 1262–1276.

Chapter 13

MOUNT LOGAN ICE CORE EVIDENCE FOR CHANGES IN THE HADLEY AND WALKER CIRCULATIONS FOLLOWING THE END OF THE LITTLE ICE AGE

G.W.K. Moore,[1] Keith Alverson,[2] and Gerald Holdsworth[3]
[1]*Department of Physics, University of Toronto, Toronto, Ontario M5S 1A7, Canada*
[2]*PAGES International Project Office, CH-3011, Bern, Switzerland*
[3]*Arctic Institute of North America, University of Calgary, Calgary, Alberta T2N 1N4, Canada and Institute for the Study of Earth Oceans and Space, University of New Hampshire, Durham, New Hampshire 03824, U.S.A.*

Abstract The Hadley and Walker circulations dominate the climate of the tropics and contribute to extratropical climate variability through the forcing of planetary waves that result in the long-range correlation of atmospheric circulation patterns known as teleconnections. Previous work showed that an annually resolved 301-year ice core record of annual snow accumulation from a high-elevation site on Mount Logan in northwestern North America contains an expression of one such teleconnection, the Pacific–North American (PNA) pattern. Here we show that this record contains a related signal associated with the regional Hadley and Walker circulations in the Pacific. We argue that the positive trend in snow accumulation in the ice core that started in the middle of the nineteenth century is a reflection of changes in the intensities of these circulations that has been ongoing since the end of the Little Ice Age (LIA).

1. INTRODUCTION

The climate of the tropics is to a large extent determined by large-scale overturning motion in the meridional and zonal directions. These mo-

H.F. Diaz and R.S. Bradley (eds.), The Hadley Circulation: Present, Past and Future, 371–395.

tions are known respectively as the Hadley (Oort and Rasmusson 1970) and Walker (Bjerknes 1969) circulations. The Hadley circulation was first described by George Hadley in 1735, who pointed out that constraints arising from the earth's rotation would limit the meridional extent of the convective overturning in the atmosphere proposed earlier by Edmund Halley to be the dominant characteristic of the atmospheric general circulation. Traditionally, the Hadley circulation referred to the zonally averaged meridional overturning circulation in the tropics. In recent years, it has been recognized that there is considerable regional variability in this circulation (Krishnamurthy and Goswami 2000; Trenberth et al. 2000; Wang 2002b). In a similar manner, the Walker circulation was originally defined as the zonal overturning circulation spanning the tropical Pacific (Bjerknes 1969). Today, it refers to the family of zonal overturning circulation cells that span the earth's equatorial regions (Lau and Yang 2002).

Both circulations play a role in the redistribution of heat in the climate system (Trenberth et al. 2000) and contribute to phenomena such as the El Niño/Southern Oscillation (ENSO; Oort and Yienger 1996) and the Asian summer monsoon (Webster et al. 1998). The upper-level convergence associated with the descending branch of the boreal winter Hadley circulation in the Pacific interacts with the planetary wave field in the Northern Hemisphere extratropics, resulting in the Pacific–North American (PNA) teleconnection, which plays an important role in the climate of North America (Wallace and Gutzler 1981; Trenberth et al. 1998).

Recent analyses of satellite observations of the earth's radiation budget indicate that there is considerable decadal-scale variability in the energy budget of the tropics that may be related to variability in the strength of the Hadley and Walker circulations (Chen et al. 2002; Hartmann 2002; Wielicki et al. 2002). In addition to, or perhaps as part of, this decadal variability, some evidence suggests a trend over the past several decades. For example, data from island stations in the western Pacific indicate a trend towards increasing precipitation over the period from 1970 to 1990 (Morrissey and Graham 1996). Meteorological data also indicate a trend towards increasing rainfall over the Amazon that has occurred over the past four decades. This trend is related to an intensification of the moisture transport associated with the global divergent circulation (Chen et al. 2001).

Clearly the study of such trends in the presence of strong decadal variability is severely hampered by the brevity of most instrumental atmospheric data sets (Trenberth et al. 2002), and it is therefore essential to extend our knowledge into the past through the use of paleoclimate data (Alverson et al. 2001). In this chapter, we make use of a 301-year-long annual snow accumulation time series (ATS) of an ice core from a high-elevation site (5,340 m) on Mount Logan (Holdsworth et al. 1992; Moore et al. 2002a) to explore the variability and trends in the Hadley and Walker circulations. Mount Logan, Canada's highest mountain, is situated in the heavily glaciated Saint Elias Mountains just 100 km from the Gulf of Alaska (Fig. 13-1). The synoptic-scale atmospheric moisture transport into the region from the North Pacific is maximum during the winter months (Cohen et al. 2000; Smirnov and Moore 2001). In addition, snowfall on the summit of Mount Logan is highest during the winter months (Waszkiewicz 2003). Consistent with these observations, the ATS exhibits a strong signal related to boreal winter atmospheric circulation in the North Pacific region related to modulation of regional atmospheric moisture transport by the PNA (Moore et al. 2001; Moore et al. 2002b; Moore et al. 2003).

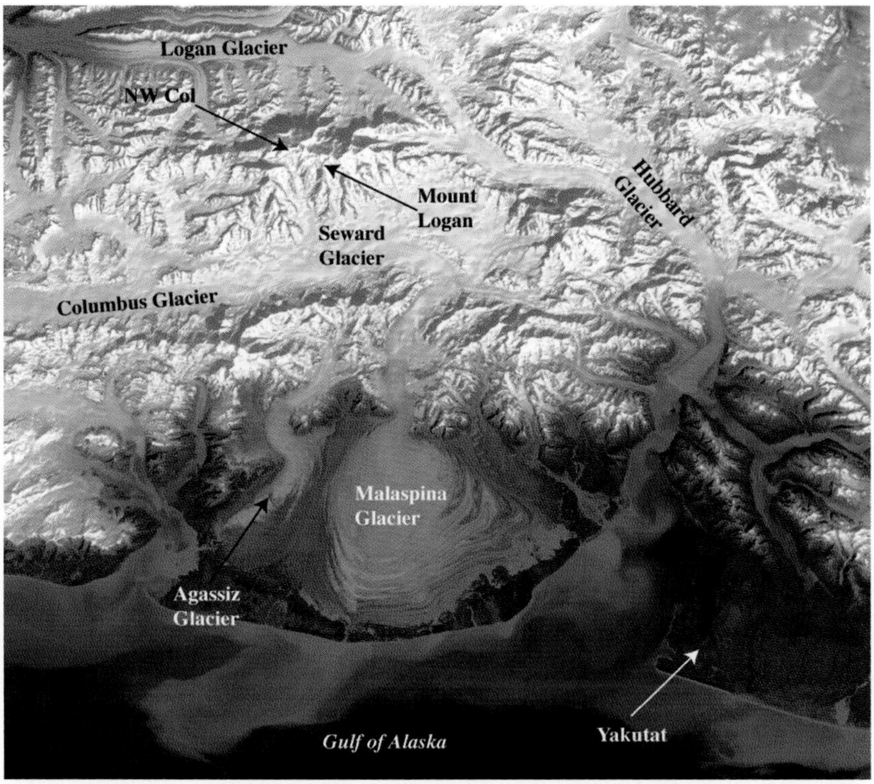

Figure 13-1. Landsat 7 Thematic Mapper false-color image from September 9, 2000, showing the Mount Logan region. The ice core was extracted at the NW Col at an elevation of 5,340 meters above sea level (masl).

Since the end of the Little Ice Age (LIA) in the middle of the nineteenth century, the ATS has exhibited a statistically significant and accelerating positive trend (Moore et al. 2002a). This trend is paralleled by a boreal winter warming at the surface and in the atmosphere over northwestern North America (Moore et al. 2002a). Other studies have indicated that during the Little Ice Age the trade winds across the tropical Pacific were stronger than they are today and that the middle of the nineteenth century was the period when they weakened, attaining more modern values (Linsley et al. 1994; Hendy et al. 2002). It is of great interest to see that this transition was associated with changes in the regional Hadley and Walker circulations in the Pacific. In this chapter, we document the expression of the boreal winter Hadley and Walker circulations in the ATS and in so doing provide information regarding the variability in the tropical atmosphere over the past 300 years.

2. METHODS

As a result of the atmosphere's large aspect ratio, synoptic-scale motion is quasi-two-dimensional. Although small, the divergent flow is nevertheless important in driving vertical motion in tropical overturning cells such as the Hadley and Walker circulations. The divergent wind has traditionally been poorly observed and as a result, our knowledge of its structure and variability has been uncertain (Trenberth and Olson 1988). The recent availability of reanalyses from various meteorological agencies has facilitated the removal of much of this uncertainty. In this chapter, we use the reanalysis produced by the National Centers for Environmental Prediction (NCEP) to generate the divergent and rotational components of the wind field. This reanalysis covers the period from 1948 to the present (Kalnay et al. 1996; Kistler et al. 2001). Recently, a number of studies have documented the ability of this reanalysis to produce realistic divergent winds associated with the Hadley and Walker circulations in the tropics (Trenberth et al. 2000; Wang 2002a, b). This view has been disputed by other studies that have highlighted problems associated with the representation of this field in reanalyses (Waliser et al. 1999; Newman et al. 2000).

The Helmholtz theorem (Arken 1985) allows one to partition the horizontal wind field $\vec{V}$ into rotational $\vec{V}_r$ and divergent $\vec{V}_d$ components:

$$\vec{V} = \vec{V}_r + \vec{V}_d = \hat{\mathrm{k}} \times \vec{\nabla}\psi + \vec{\nabla}\chi, \tag{1}$$

where ψ is the stream function, χ is the velocity potential, and $\hat{k}$ is a unit vector in the vertical direction. In this formulation, the Hadley and Walker circulations are contained in the divergent wind field $\vec{V}_d$, while planetary wave motion is contained in the rotational wind field $\vec{V}_r$. The decomposition was performed on the monthly mean wind fields on each of the 17 pressure levels in the NCEP reanalysis with the SPHEREPACK software package (Adams and Swarztrauber 1999).

Spatially averaged representations of the Hadley and Walker circulations can be obtained by the meridional and zonal averaging of the three-dimensional continuity equation respectively (Trenberth et al. 2000). This process leads to the definition of zonal or meridional mass stream functions whose divergence is the mass flux in that plane. This approach, however, does not allow for information on the regional expression of the Hadley and Walker circulations. One technique that provides this information is to consider the vertical shear of the divergent wind (Webster and Yang 1992; Oort and Yienger 1996; Goswami et al. 1999). For example, the meridional overturning index (MOI), defined as the difference between the 200 mb and 850 mb meridional components of the divergent wind, provides information on the regional Hadley circulation. Positive values of the MOI are associated with overturning motion that has northward transport at upper levels and southward transport at lower levels, with the opposite occurring for negative values. Similarly, the zonal overturning index (ZOI), defined as the difference between the 200 mb and 850 mb zonal components of the divergent wind, provides information on the regional Walker circulation. Positive values of the ZOI are associated with overturning motion in the zonal direction that has eastward transport at upper levels and westward transport at lower levels, with the opposite occurring for negative values.

3. CLIMATOLOGY OF THE HADLEY AND WALKER CIRCULATIONS

Figure 13-2 shows the boreal winter mean climatology of the divergent wind field in the upper, 200 mb, and lower, 850 mb, troposphere over the tropical Pacific from the NCEP reanalysis for the period 1948–2000. In the upper troposphere, centers of divergence over the warm pool in the western Pacific centered on 10°S, 160°E, and over South America are the dominant features. The poleward flow out of these centers is associated with the regional Hadley circulation. The northward branch is more intense, as is to be expected during the boreal winter (Trenberth et al. 2000), and a region of upper-level convergence exists near 30°N, 160°E. A region of upper-level convergence is also situated off the west coast of South America. In the zonal direction, eastward and westward flow emanate from the regions of upper-level divergence. The eastward flow over the Pacific Ocean is associated with the so-called Pacific Walker cell, while the westward flow over the Indian Ocean is associated with another cell known as the transverse Walker cell. The situation at 850 mb is essentially the opposite of that at 200 mb.

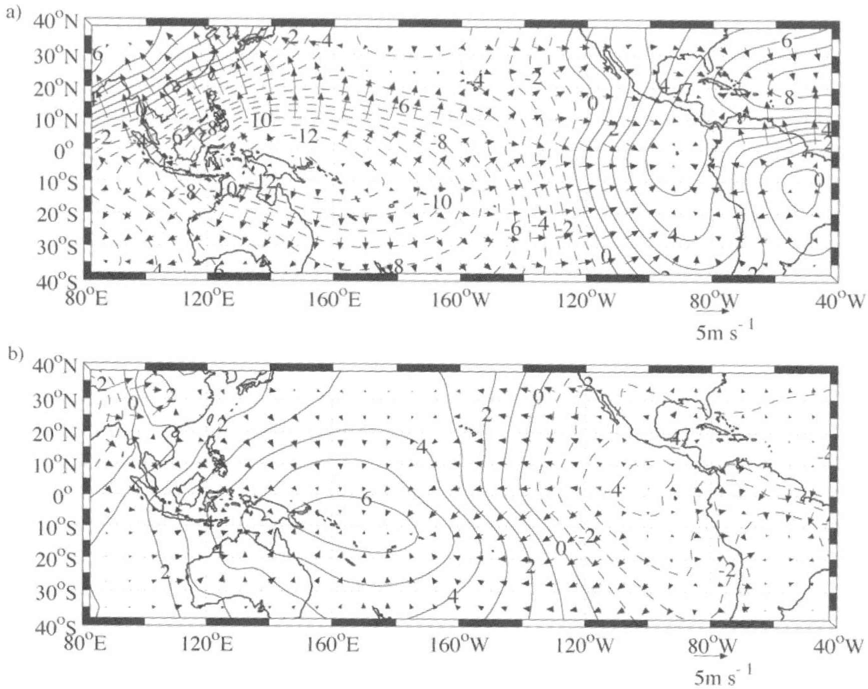

Figure 13-2. Winter mean January–March (JFM) climatology of the divergent wind over the tropical Pacific from the NCEP reanalysis, 1948–2000. The divergent wind field (m s^{-1}) and velocity potential (10^6 m^2s^{-1}) at: (a) 200 mb and (b) 850 mb.

The spatially averaged representations of these boreal winter mean circulations, shown in Figure 13-3, more clearly delineate the various overturning cells. The strength of the northern Hadley cell as compared to the southern cell is evident. The circulation associated with the thermally indirect Ferrel cell can also be seen in Figure 13-3 polewards of the northern Hadley cell (Trenberth et al. 2000). The Pacific and transverse Walker cells can also be clearly seen in Figure 13-3. Comparison of Figures 13-2 and 13-3 indicate that information on the regional variability of the Hadley and Walker circulations is lost through spatial averaging. As can be seen in Figure 13-4, this information can be regained through the use of the meridional and zonal overturning indices MOI and ZOI. In particular, there is a regional intensification of the Hadley circulation in the western Pacific to the north of the maritime continent and northern South America. There is also evidence of weaker regional Hadley cells in the Southern Hemisphere as well as an extension of the northern cell into the Southern Hemisphere along the South Pacific Convergence Zone (SPCZ). The Ferrel circulation is regionally intensified over the eastern Pacific to the north of Hawaii and over the Caribbean Sea. The boreal winter mean ZOI field indicates that the Pacific Walker cell is most intense in the Southern Hemisphere near 10°S. In contrast, the transverse Walker cell in the western Pacific is most intense in the Northern Hemisphere near 10°N.

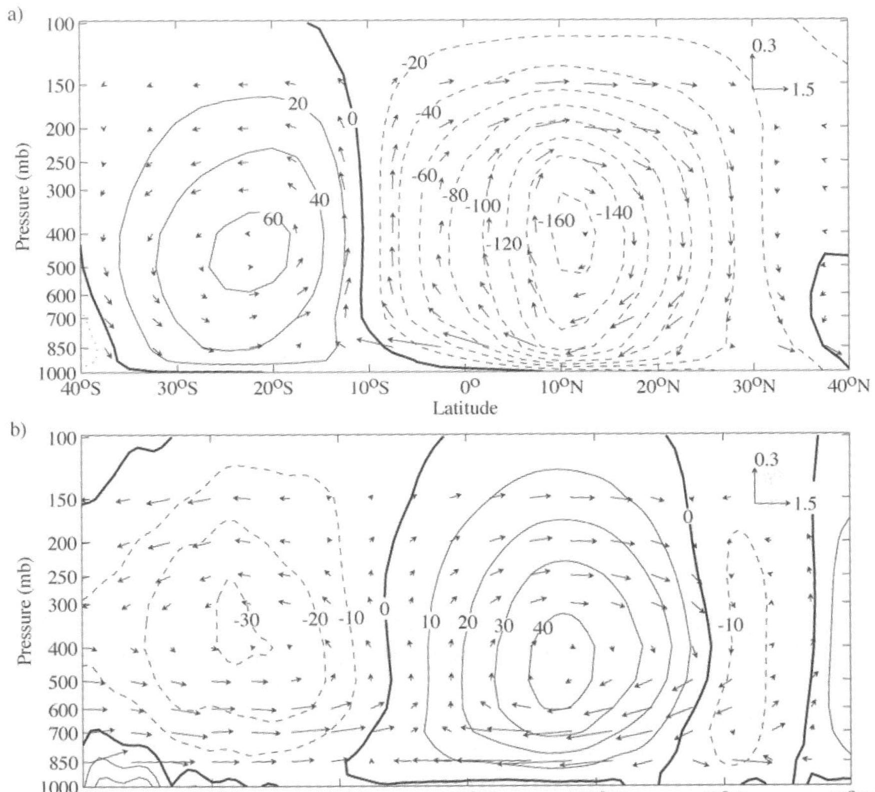

Figure 13-3. Winter mean (JFM) climatology of: (a) the zonally averaged mass flux field (kg m^{-2} s^{-1}) and mass stream function (10^9 kg s^{-1}) over 80°E to 40°W and (b) the meridionally averaged mass flux field (kg m^{-2} s^{-1}) and mass stream function (10^9 kg s^{-1}) over 40°S to 40°N from the NCEP reanalysis, 1948–2000.

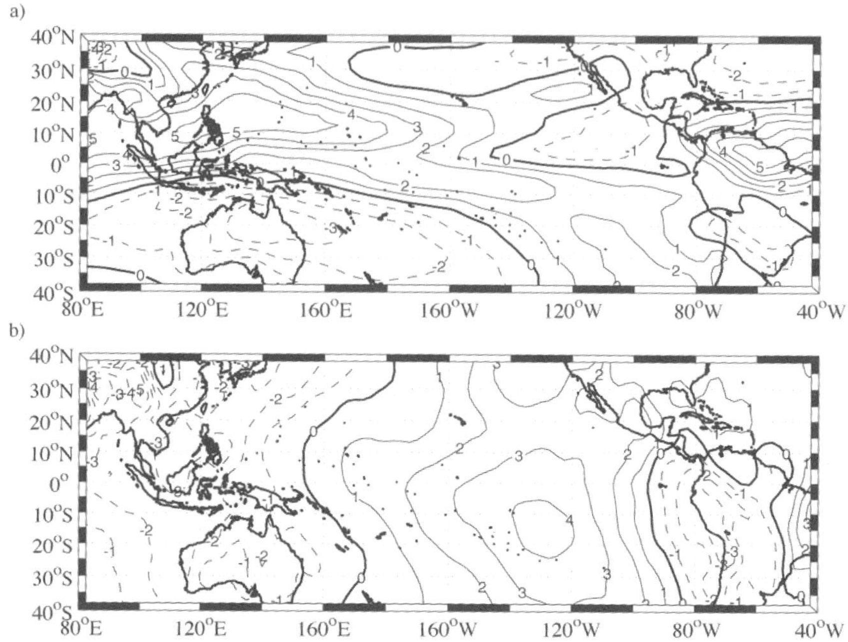

Figure 13-4. Winter mean (JFM) climatology of the: (a) the meridional overturning index, MOI (m s^{-1}), and (b) the zonal overturning index, ZOI (m s^{-1}), from the NCEP reanalysis, 1948–2000.

Information on the global state of the troposphere is unavailable for the period prior to the middle of the twentieth century. For a longer-term perspective, one is restricted to surface fields. Figure 13-5 shows the boreal winter mean sea level pressure (MSLP) field in the tropical Pacific region for the period 1871–1994 as contained in the GMSLP2 (U.K. Meteorological Office Global Mean Sea-Level Pressure) data set (Basnett and Parker 1997). A pressure gradient exists between the equatorial and subtropical region of each hemisphere that is consistent with low-level equatorward flow associated with the Hadley circulation. In addition, sea level pressures are lower in the western Pacific as compared to the eastern Pacific and the Indian Ocean. This is consistent with low-level flow associated with the Pacific and transverse Walker cells.

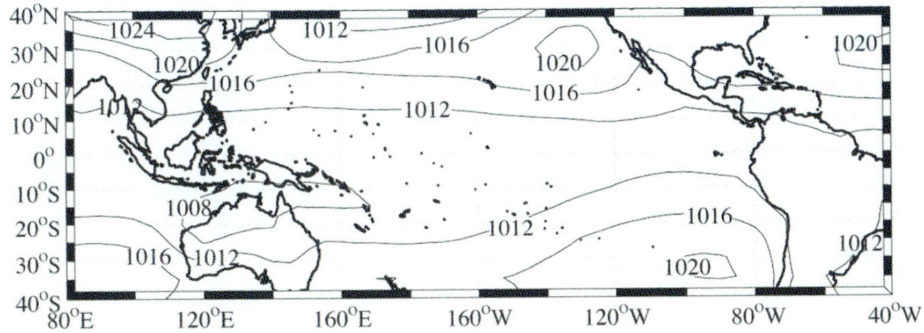

Figure 13-5. Winter mean (JFM) climatology of the sea level pressure (mb) over the tropical Pacific from the GMSLP2 data set, 1871–1994.

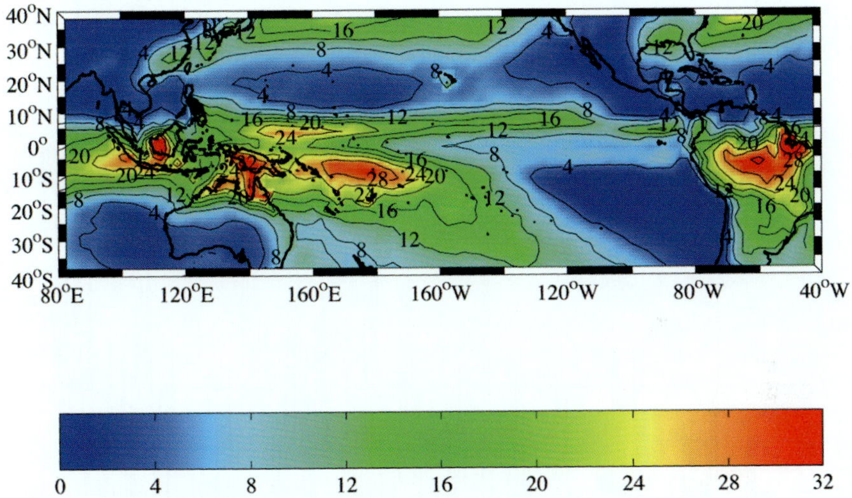

Figure 13-6. Winter mean (JFM) climatology of the precipitation field (cm month^{-1}) over the tropical Pacific from the Global Precipitation Climatology Project data set, 1979–2000.

The precipitation field in the tropics provides useful information on the organization of the various overturning cells (Lau and Yang 2002). Unfortunately, spatially uniform observations of this field over oceanic regions, such as the tropical Pacific, are restricted in length due to the shortness of the satellite record. Figure 13-6 shows the boreal winter mean precipitation field for the period 1979–2000 over the tropical Pacific from the Global Precipitation Climatology Project (GPCP; Huffman et al. 1997). The regions of upper-level divergence over South America and the maritime continent associated with the ascending branches of the Hadley and Walker circulations (Fig. 13-2) are co-located with maxima in the precipitation field. The sub-tropics, in which the descending branches of the Hadley circulation are located, and the equatorial eastern Pacific, in which the descending branch of the Pacific Walker cell is located, are regions where precipitation is at a minimum.

4. EXPRESSION OF THE HADLEY AND WALKER CIRCULATIONS IN THE MOUNT LOGAN SNOW ACCUMULATION TIME SERIES

Figure 13-7 shows the annual snow accumulation time series from the Mount Logan ice core for the period 1700–2000 (Holdsworth et al. 1992; Moore et al. 2002a). The time series has been normalized to have zero

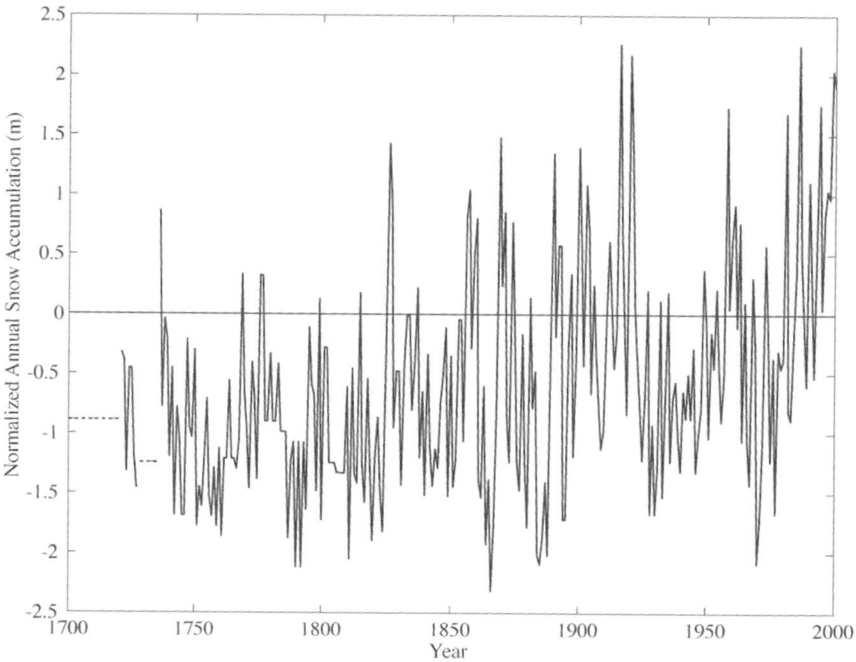

Figure 13-7. Normalized annual snow accumulation time series from the Mount Logan ice core. The time series is normalized to have zero mean, indicated by the thin, straight line, and unit standard deviation over the period 1948–2000. Data for the period prior to 1736 that are not annually resolved, are indicated by dashed lines.

mean and unit variance over the period 1948–2000, which represents the overlap with the NCEP reanalysis. The time series clearly contains variability on the interannual as well as interdecadal time scales as well as the presence of a secular trend over the latter half of the record (Moore et al. 2001, 2002b).

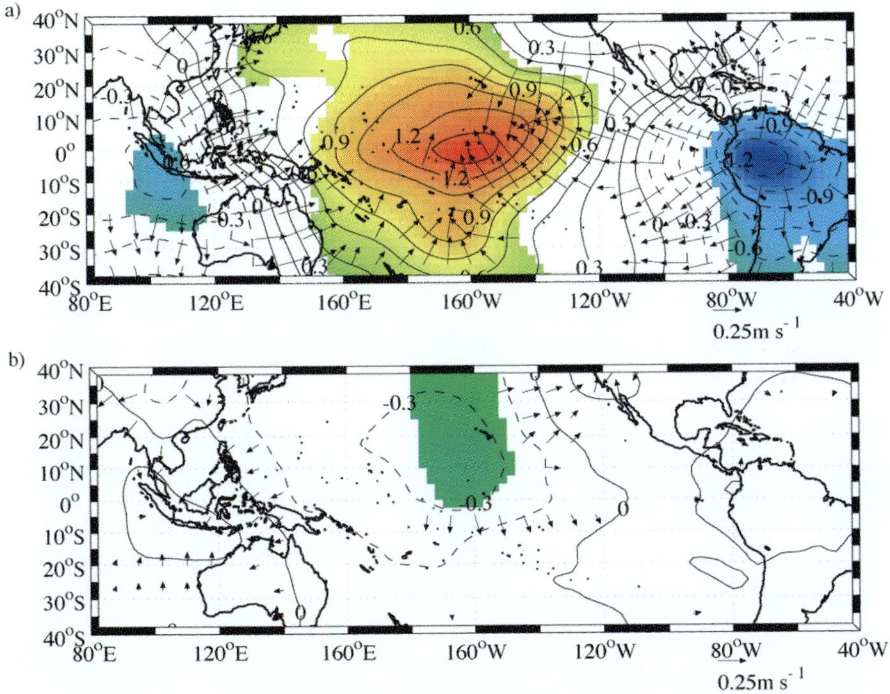

Figure 13-8. Regression of the winter mean (JFM) divergent wind at (a) 200mb and (b) 850 mb over the tropical Pacific from the NCEP reanalysis against the normalized Mount Logan annual snow accumulation time series 1948-2000. The colors indicate regionsthat are statistically significant at or above the 95% level. The vector regressions are shown at those grid points where at least one of the components is statistically significant at or above the 95% level.

In Figure 13-8, we present the regression of the boreal winter mean divergent upper, 200 mb, and lower, 850 mb, tropospheric wind field against the *ATS* for the period 1948–2000. In the upper troposphere, increasing snow accumulation is seen to be associated with an alternating pattern of divergence and convergence that spans the tropics from the Indian Ocean across the Pacific Ocean to South America. In the lower troposphere, the pattern is essentially the opposite, although the statistical significance of the regressions is reduced.

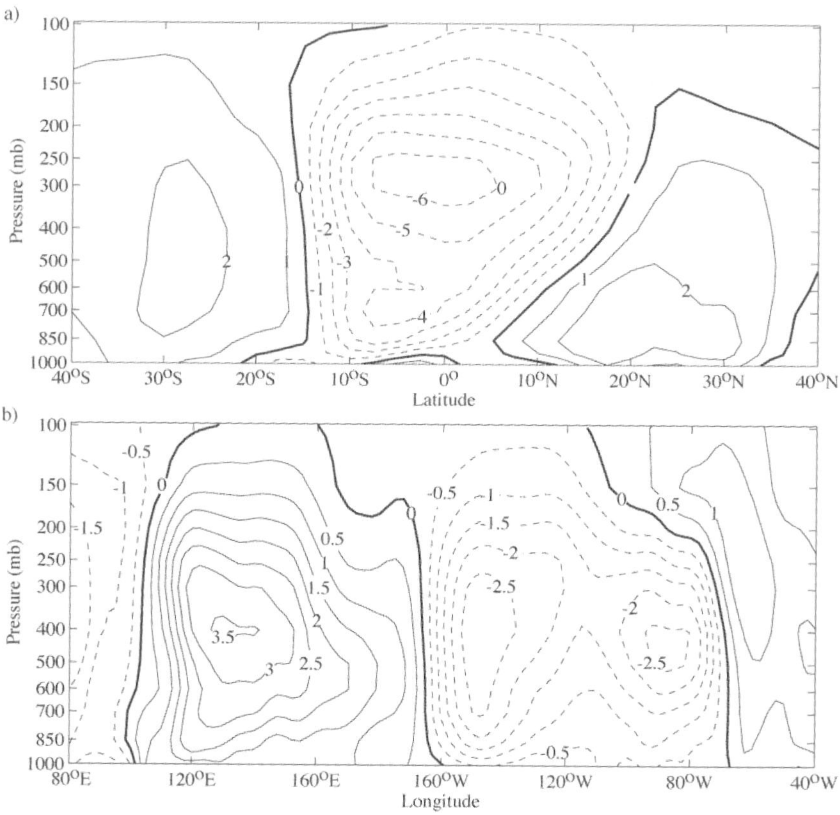

Figure 13-9. Regression of the winter mean (JFM); (a) zonally averaged mass stream function (10^9 kg s^{-1}) over 80°E to 40°W and (b) meridionally averaged mass stream function (10^9 kg s^{-1}) over 40°S to 40°N from the NCEP reanalysis against the normalized Mount Logan annual snow accumulation time series, 1948–2000.

In Figure 13-9, we present the regression of the boreal winter mean zonally and meridionally averaged mass fluxes against the *ATS* for the period 1948–2000. Enhanced snow accumulation at the Mount Logan site is seen to be associated with a zonally averaged circulation that is in the same sense as the climatological Northern Hemisphere Hadley and Ferrel cells (Fig. 13-3a). This implies that snow accumulation at the site is positively correlated with the strength of the Hadley and Ferrel cells. The combination of these two circulation cells results in enhanced upper-level convergence between 10°N and 20°N that is associated with an increase in snow accumulation at the Mount Logan site. High snow accumulation is also associated with a meridionally averaged circulation pattern that is in the opposite sense to the climatological Pacific Walker and transverse cells (Fig. 13-3b). This implies that snow accumulation at the site is negatively correlated with the strength of the Walker cells over the Pacific region.

The regional expressions of the changes in the Hadley and Walker circulations associated with snow accumulation at the Mount Logan site can be seen in Figure 13-10. This figure shows the regression of the *MOI* and *ZOI* against the *ATS* for the period 1948–2000. The *MOI* regression indicates that an increase in snow accumulation at the Mount Logan site is associated with an intensification of the northern and southern Hadley cells over Central and South America and the maritime continent, as well as over the central Pacific. High snow accumulation is also seen to be associated with an intensification of the Ferrel cell over the subtropical northeastern Pacific. The *ZOI* regression indicates that increasing snow accumulation is also associated with a westward overturning circulation over the eastern tropical Pacific and an eastward overturning circulation over Indonesia, the South China Sea, and Australia. As was discussed above, this motion acts against the climatological mean Walker circulation, indicating that increasing snow accumulation at the Mount Logan site is associated with a reduction in the intensity of the Walker circulation over the Pacific and Indian Oceans.

To confirm this connection between snow accumulation at the Mount Logan site and the Hadley and Walker circulations, we present in Figure 13-11 the regression of the boreal winter mean precipitation field from the Global Precipitation Climatology Project against the *ATS* for the period 1979–2000. Enhanced snow accumulation at the Mount Logan site is correlated with an increase in precipitation over South America and the maritime continent and a decrease in precipitation over the central tropical Pacific, as would be expected if it is related to the strength of the Hadley and Walker circulations.

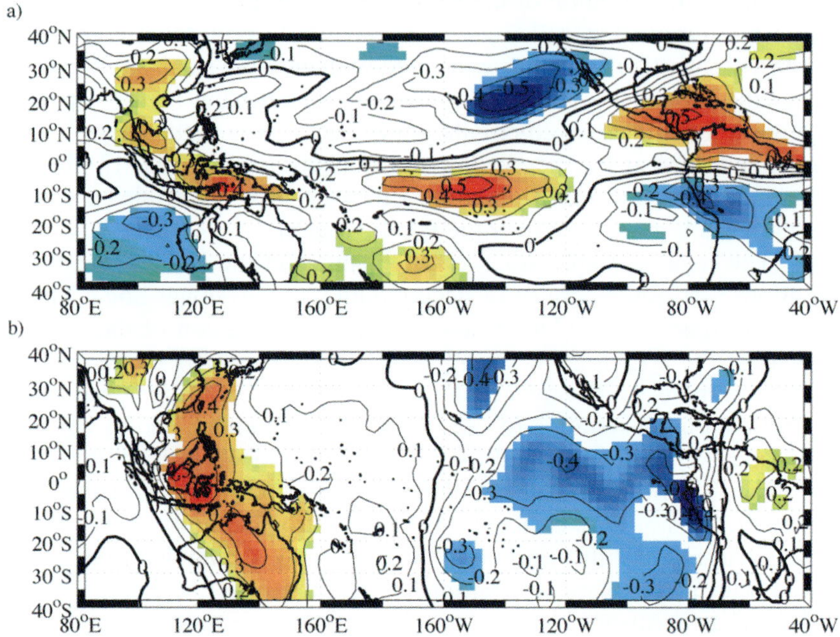

Figure 13-10. Regression of the winter mean (JFM): (a) meridional overturning index, MOI (m s^{-1}), and (b) zonal overturning index, ZOI (m s^{-1}), over the tropical Pacific from the NCEP reanalysis against the normalized Mount Logan annual snow accumulation time series, 1948–2000. The colors indicate regions where the regressions are statistically significant at or above the 95% level.

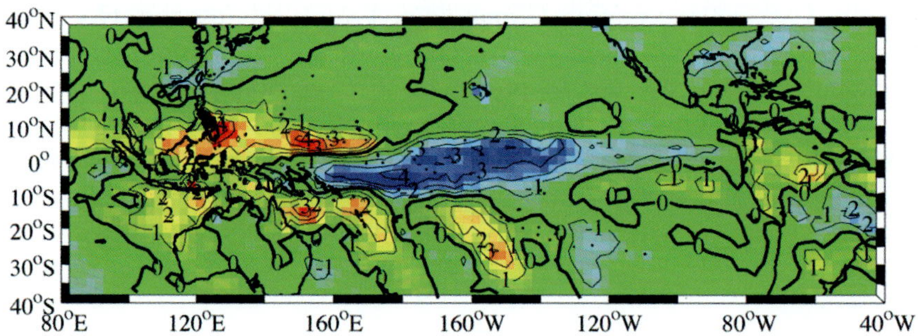

Figure 13-11. Regression of the winter mean (JFM) precipitation from the Global Precipitation Project over the tropical Pacific, 1979–2000, against the normalized Mount Logan annual snow accumulation time series (contours and colors, cm $month^{-1}$).

5. TREND IN THE HADLEY AND WALKER CIRCULATIONS

Given the expression of the Hadley and Walker circulations that exists in the Mount Logan annual snow accumulation time series and the secular trend that is present in the latter, we seek in this section to document whether similar trends exist in the Hadley and Walker circulations. As is the case with most geophysical time series, the assessment of the statistical significance of the trends in these circulations must take into account temporal autocorrelation that reduces the degrees of freedom in the time series. This was accomplished through an expression for the degrees of freedom that is a function of the lag 1 autocorrelation of the *ATS* (Santer et al. 2000; Moore et al. 2002a).

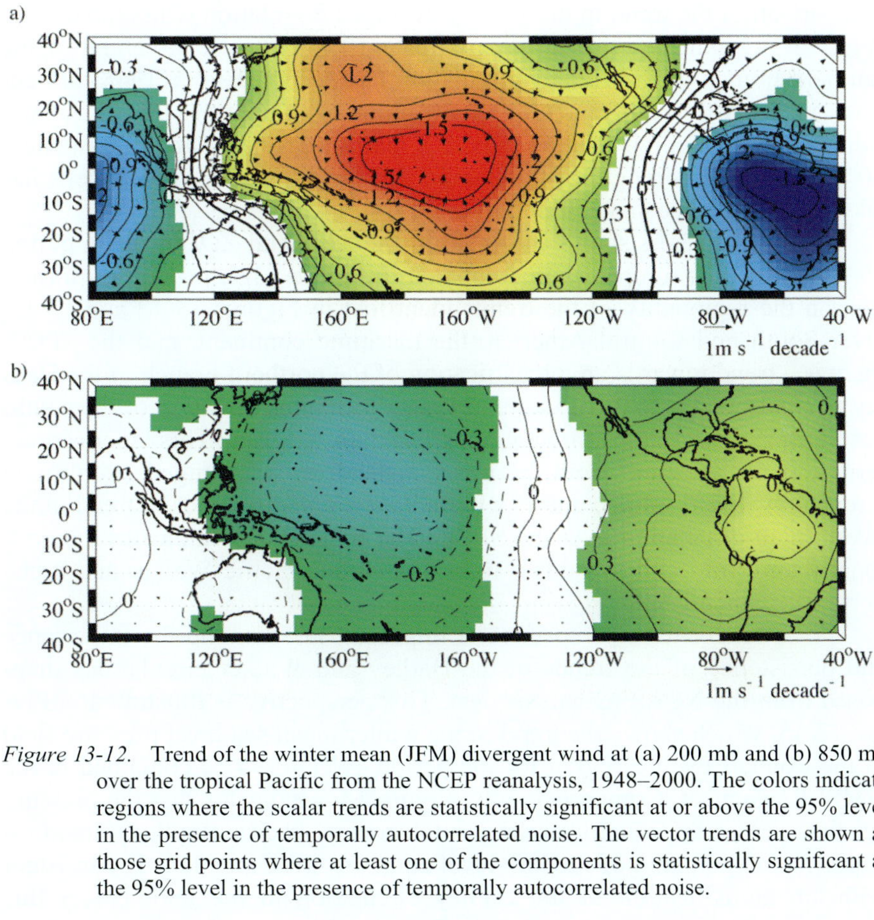

Figure 13-12. Trend of the winter mean (JFM) divergent wind at (a) 200 mb and (b) 850 mb over the tropical Pacific from the NCEP reanalysis, 1948–2000. The colors indicate regions where the scalar trends are statistically significant at or above the 95% level in the presence of temporally autocorrelated noise. The vector trends are shown at those grid points where at least one of the components is statistically significant at the 95% level in the presence of temporally autocorrelated noise.

In Figure 13-12, we present the trend in the boreal winter mean divergent upper, 200 mb, and lower, 850 mb, tropospheric wind field for the period 1948–2000. In the upper troposphere, there exists a trend towards increasing divergence over the Indian Ocean and South America, as well as a trend towards increasing convergence over the central tropical Pacific. There are associated trends towards increasing equatorward flow in the central tropical Pacific and increasing poleward flow over the Indian Ocean and South America. In the zonal direction, there is a trend towards increasing easterly flow over the eastern tropical Pacific and westerly flow over the maritime continent. In the lower troposphere, the trends are generally of the opposite sign and are reduced in magnitude as well as being of more limited statistical significance.

In Figure 13-13, we present the trends of the boreal winter mean zonally and meridionally averaged mass flux for the period 1948–2000. A comparison of the trend in the zonally averaged circulation with climatology (Fig. 13-3a) indicates that an approximate 10%–20% strengthening of the northern branch of the Hadley circulation has occurred since 1948. In contrast, a comparison of the trend with the climatology (Fig. 13-3b) of the meridionally averaged circulation indicates that a weakening of approximately 10%–20% in the strength of the Pacific and the transverse Walker cells has taken place over the same period of time.

Figure 13-14 shows the trend in the *MOI* and *ZOI* over the period 1948–2000. This figure, when compared to Figure 13-4, provides information on the regionality of the trends identified in Figures 13-12 and 13-13. Over South and Central America, the maritime continent, and the SPCZ, there is a trend towards an intensification of the northern branch of the Hadley circulation. There is also a trend towards an intensification of the southern branch of the Hadley circulation over South America. This latter characteristic is not apparent from the trend in the zonally averaged overturning circulation (Fig. 13-13a), due to a cancellation with an opposite trend in the overturning circulation over the SPCZ. The trend in *ZOI* indicates a meridionally uniform trend towards a reduction of the Pacific Walker and transverse cells since 1948.

A longer-term perspective is important in assessing the importance and persistence of the trends in the Hadley and Walker circulations diagnosed from the NCEP reanalysis data. This perspective is illustrated in Figure 13-15, which shows the trend in the winter mean sea level pressure field in the tropical Pacific region for the period 1871–1994 as contained in the GMSLP2 data set. Comparison with Figure 13-5 indicates that the pressure gradient across the equator in the central tropical Pacific implies a trend towards a strengthening of the low-level equatorward wind. This is consistent with an intensification of the Hadley circulation in the region over this longer time scale. In the zonal direction, the trend towards higher pressures in the central Pacific combined with the trend towards lower pressures in the eastern and western Pacific is consistent with a longer-term weakening of the Pacific and transverse Walker cells.

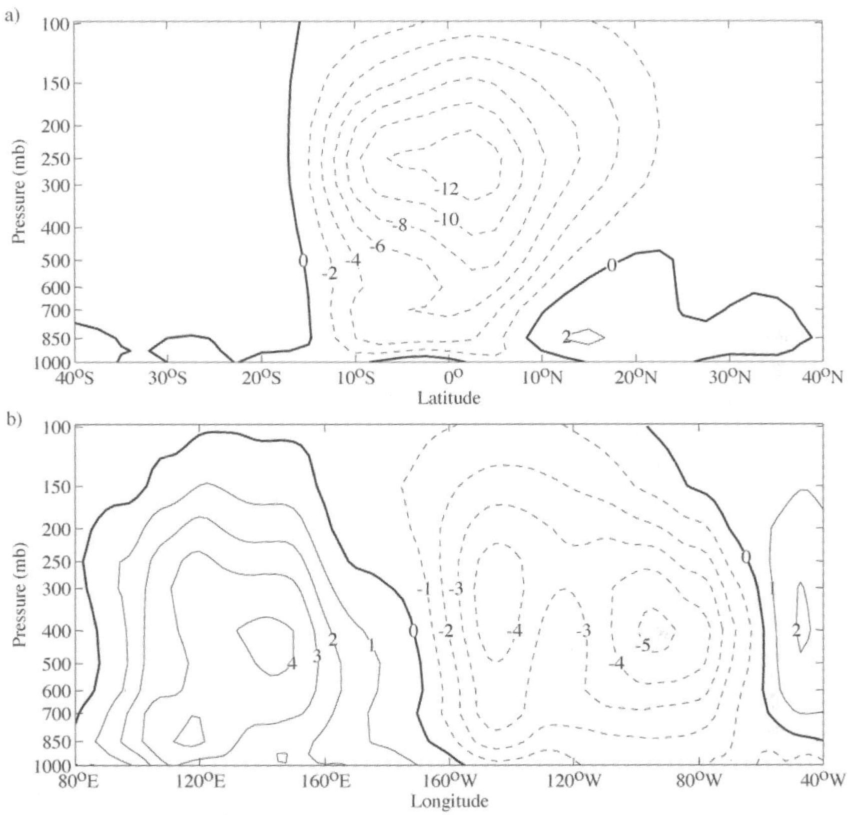

Figure 13-13. Trend in the winter mean (JFM): (a) zonally averaged mass stream function (10^9 kg s^{-1} $decade^{-1}$) over 80°E to 40°W and (b) meridionally averaged mass stream function (10^9 kg s^{-1} $decade^{-1}$) over 40°S to 40°N from the NCEP reanalysis, 1948–2000.

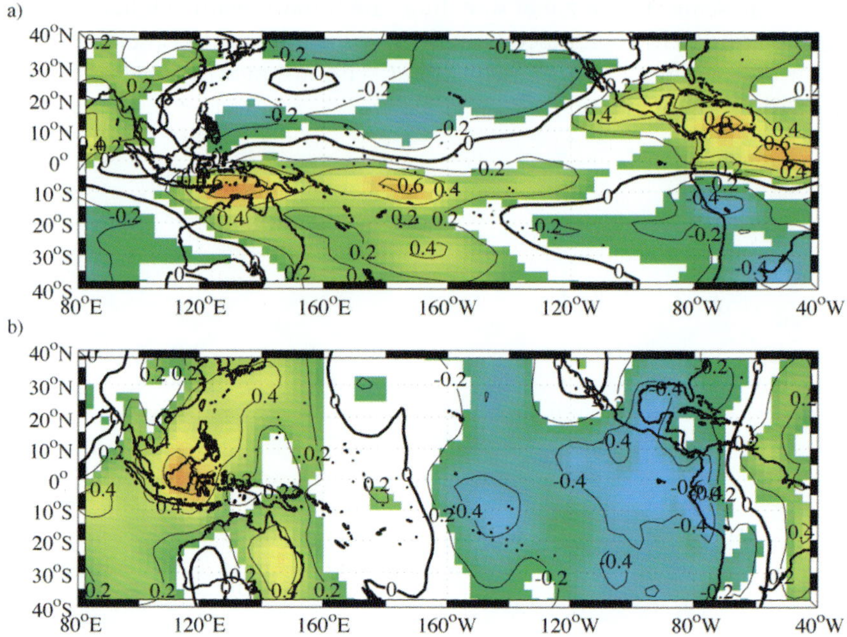

Figure 13-14. Trend in the winter mean (JFM): (a) meridional overturning index (m s^{-1} decade^{-1}) and (b) zonal overturning index (m s^{-1} decade^{-1}) from the NCEP reanalysis over the tropical Pacific, 1948–2000. The colors indicate regions where the trends are statistically significant at or above the 95% level.

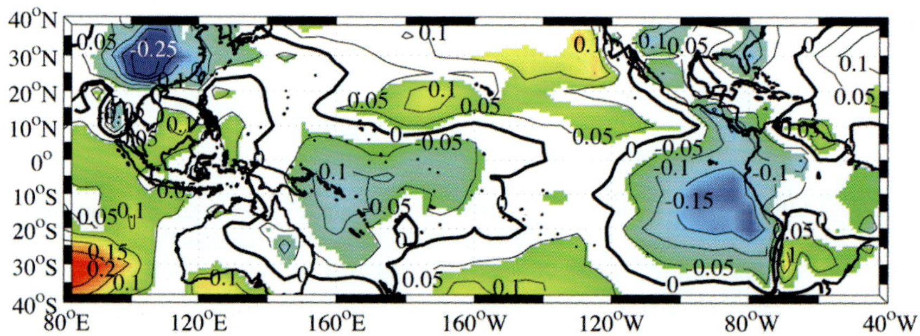

Figure 13-15. Trend in the winter (JFM) mean sea level pressure from the GMSLP2 data set (mb decade^{-1}) over the tropical Pacific, 1871–1994. The colors indicate regions where the trends are statistically significant at or above the 95% level.

To close this section, we illustrate in Figure 13-16 the trend in the boreal winter mean precipitation field from the Global Precipitation Climatology Project for the period 1979–2000. The regions where there is a trend towards an increase in upper-level divergence, the Indian Ocean and South America, are also seen to be regions where there is a trend towards an increase in precipitation. In contrast, the central equatorial Pacific, where there is a trend towards an increase in upper-level convergence, is a region where there is a trend towards a reduction in precipitation. Within the SPCZ, where there is a trend towards an increase in the regional Hadley circulation, there is a corresponding trend towards increased precipitation.

6. ON THE EXISTENCE OF AN EXPRESSION OF THE PNA IN THE MOUNT LOGAN ICE CORE

In previous work (Moore et al. 2001, 2002a, b), we proposed that the climate signal in the *ATS* is the result of a modulation of the regional moisture flux by the PNA. This teleconnection is the result of planetary wave activity forced by the upper-level convergence associated with the descending branch of the regional Hadley circulation in the Pacific (Trenberth et al. 1998). It is this connection between the tropical overturning circulations and the planetary wave activity that we propose results in the strong expression of these circulations in the *ATS*.

Confirmation of this hypothesis is presented in Figure 13-17, which shows the regression of the 200 mb rotational wind field against the *ATS* and its trend for the period 1948–2000. Recall that the rotational wind contains the expression of the planetary wave activity. The regression indicates that increasing snow accumulation is associated with a wave train in the upper-tropospheric stream function extending in an arc from the central subtropical Pacific across North America and southwards over the Caribbean and South America. Mount Logan is situated in a region where enhanced upper-level southerly flow is associated with increasing snow accumulation. The trend shows a remarkably similar picture, with a sequence of regions with alternating positive and negative trends in the stream function over the Pacific Ocean, North America, the Caribbean, and South America. Associated with this trend is an intensification of the southerly flow along the west coast of North America directed towards the Mount Logan region.

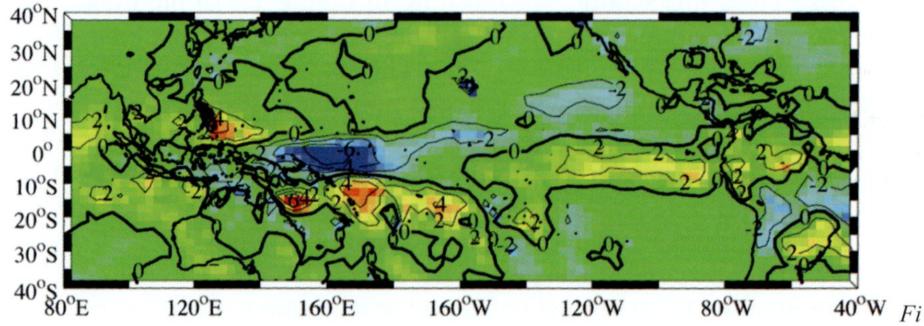

Figure 13-16. Trend in the Global Precipitation Climatology Project winter (JFM) mean precipitation field (cm month^{-1} decade^{-1}) over the tropical Pacific, 1979–2000.

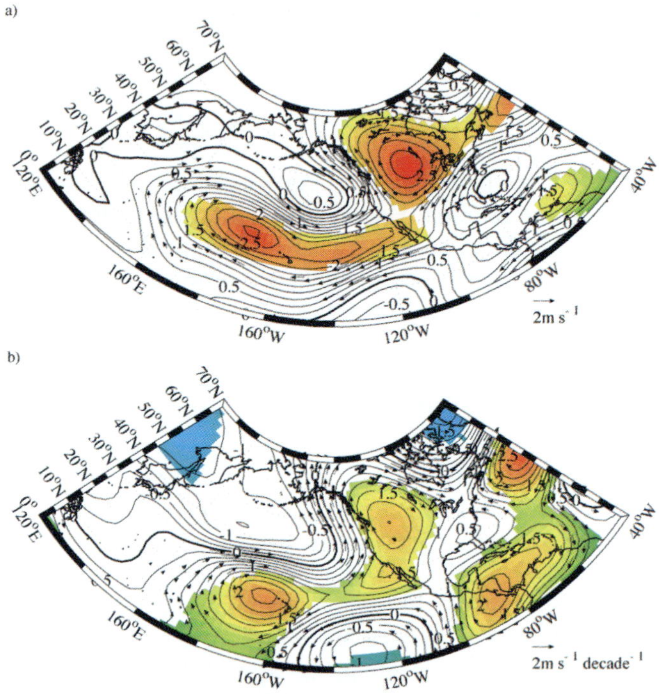

Figure 13-17. (a) Regression of the winter mean (JFM) 200 mb rotational wind (m s^{-1}) and stream function (10^6 m^{2}s^{-1}) from the NCEP reanalysis against the normalized Mount Logan annual snow accumulation time series, 1948–2000. (b) Trend in the winter (JFM) mean 200 mb rotational wind (m s^{-1} decade^{-1}) and stream function (10^6 m^{2}s^{-1} decade^{-1}) from the NCEP reanalysis, 1948–2000. The colors indicate regions where the scalar fields are statistically significant at or above the 95% level. The vector fields are shown at those grid points where at least one of the components is statistically significant at or above the 95% level.

7. DISCUSSION

In this chapter, we have investigated the expression of the Hadley and Walker circulations that is contained in the annual snow accumulation time series from the Mount Logan ice core. Using the divergent component of the wind field, we have shown that increasing snow accumulation at the Mount Logan site is associated with an intensification of the Hadley circulation and a weakening of the Walker circulation over the central Pacific, Indonesia, and South America. This expression has its highest amplitude and statistical significance in the upper troposphere. This expression is consistent with the hypothesis that the signal is the result of the forcing of the planetary wave field by the upper-level convergence associated with the Hadley circulation. Trends in the divergent wind field indicate that since 1948 we have been in a regime in which the Hadley circulation over the Pacific has been intensifying while the Walker circulation in the same region has been weakening.

These characteristics of the Hadley and Walker circulations, which we have identified from the NCEP reanalysis, are consistent with those associated with a long-term sea level pressure record and a shorter-term tropical rainfall record. The trend towards increased upper-level convergence over the tropical Pacific Ocean is situated in a region where a cooling has been observed at the surface in instrumental records since 1900 (Cane et al. 1997). The anticorrelation in the strength of the Hadley and Walker circulations that we have identified is also consistent with what is observed to occur during warm ENSO events (Oort and Yienger 1996). This correspondence is supported by the presence of a strong ENSO signal in the *ATS* (Moore et al. 2001, 2003). Similar conclusions have also been reached with regard to trends in the boreal summer Hadley and Walker circulations across Africa, Asia, and the Pacific (Zhao and Moore, in press) and are in qualitative agreement with those inferred from satellite-based radiation measurements (Chen et al. 2002). There are, however, some quantitative differences that may be related to the difference in time period used or seasonality. In particular, Chen et al. (2002) looked at trends over the period from January 1985 to August 1995, and therefore all seasons were included in their analysis. In contrast, we have focused on the boreal winter season over the period

1948–2000. Therefore, the strong seasonal cycle in the Hadley and Walker circulations (Trenberth et al. 2000) may contribute to the differences between our results and those of Chen et al. (2002).

The Mount Logan ice core record spans the period from the terminal stages of the Little Ice Age to the warmest decade in the last millennium (Esper et al. 2002). Our results suggest that the variability we see in the *ATS* throughout the period under consideration is positively correlated with the intensity of the Hadley circulation and inversely correlated with the intensity of the Walker circulation in the Pacific region. Our results suggest that the secular trend in snow accumulation at Mount Logan that has been ongoing since the end of the Little Ice Age, in middle of the nineteenth century, is a reflection of an intensification of the Hadley circulation and a weakening of the Walker circulation during this period. Previous work has indicated that a shift in the tropical climate associated with a weakening of the trade winds over the Pacific occurred around this time (Linsley et al. 1994; Hendy et al. 2002). Our results suggest that this transition was most likely a manifestation of a weakening of the Walker circulation that occurred as a more vigorous Hadley circulation was established. An important and as yet unanswered question is the role that anthropogenic forcing may have played in these changes. In particular, our results point towards an intensification in the tropical circulation that has been ongoing since the middle of the nineteenth century, before widespread burning of fossil fuels. However, this early transition is not necessarily inconsistent with other anthropogenic forcings, such as land cover change, which some have argued had a substantial impact on greenhouse gases and climate long before industrialization (Holdsworth et al. 1996; Ruddiman 2003). The resolution of the cause of the trends we have diagnosed, and in particular the question of whether they are due to natural variability or anthropogenic influences, is a subject for future work.

8. ACKNOWLEDGMENTS

We acknowledge the efforts of the 2001 field crew to retrieve the updated ice core from Mount Logan. We thank M. Waszkiewicz and D. Fisher for providing access to the Mount Logan weather data. The authors would like to thank Dr. George Kiladis for suggesting the use of the divergent wind field. Funding for this research was provided by the Natural Sciences and Engineering Research Council of Canada, the U.S. International Arctic Research Center (Fairbanks), the Geological Survey of Canada, the United States National Science Foundation (NSF), and the Swiss National Science Foundation. The NCEP reanalysis data were provided by the National Oceanic and Atmospheric Administration Climate Diagnostics Center (NOAA-CDC) in Boulder, Colorado. The Mount Logan annual snow accumulation time series can be accessed via the World Data Center for Paleoclimatology at http://www.ngdc.noaa.gov.

9. REFERENCES

Adams, J.C., and P.N. Swarztrauber. 1999. SPHEREPACK 3.0: A model development facility. *Monthly Weather Review* 127: 1872–1878.

Alverson, K., R. Bradley, K. Briffa, J. Cole, M. Hughes, I. Larocque, T. Pedersen, L. Thompson, and S. Tudhope. 2001. A global paleoclimate observing system. *Science* 293: 47–48.

Arken, G. 1985. *Mathematical Methods for Physicists*, 3rd ed. San Diego, California: Academic Press, 1029 pp.

Basnett, T.A., and D.E. Parker. 1997. *Development of the Global Mean Sea Level Pressure Data Set GMSLP2.* Meteorological Office, Exeter, U.K. Hadley Center Climate Research Technical Note 79.

Bjerknes, J. 1969. Atmospheric teleconnections from equatorial Pacific. *Monthly Weather Review* 97: 163–172.

Cane, M.A., A.C. Clement, A. Kaplan, Y. Kushnir, D. Pozdnyakov, R. Seager, S.E. Zebiak, and R. Murtugudde. 1997. Twentieth-century sea surface temperature trends. *Science* 275: 957–960.

Chen, J.Y., B.E. Carlson, and A.D. Del Genio. 2002. Evidence for strengthening of the tropical general circulation in the 1990s. *Science* 295: 838–841.

Chen, T.C., J.H. Yoon, K.J. St Croix, and E.S. Takle. 2001. Suppressing impacts of the Amazonian deforestation by the global circulation change. *Bulletin of the American Meteorological Society* 82: 2209–2216.

Cohen, J.L., D.A. Salstein, and R.D. Rosen. 2000. Interannual variability in the meridional transport of water vapor. *Journal of Hydrometeorology* 1: 547–553.

Esper, J., E.R. Cook, and F.H. Schweingruber. 2002. Low-frequency signals in long tree-ring chronologies for reconstructing past temperature variability. *Science* 295: 2250–2253.

Goswami, B.N., V. Krishnamurthy, and H. Annamalai. 1999. A broad-scale circulation index for the interannual variability of the Indian summer monsoon. *Quarterly Journal of the Royal Meteorological Society* 125: 611–633.

Hartmann, D.L. 2002: Climate change—Tropical surprises. *Science* 295: 811–812.

Hendy, E.J., M.K. Gagan, C.A. Alibert, M.T. McCulloch, J.M. Lough, and P.J. Isdale. 2002. Abrupt decrease in tropical Pacific sea surface salinity at end of Little Ice Age. *Science* 295: 1511–1514.

Holdsworth, G., H.R. Krouse, and M. Nosal. 1992. Ice core climate signals from Mount Logan, Yukon, 1700–1987. In,. Bradley, R.S., and P.D. Jones (eds.). *Climate since A.D. 1500.* London, U.K.: Routledge, pp. 483–504.

Holdsworth, G., K. Higuchi, G.A. Zielinski, P.A. Mayewski, M. Wahlen, B. Deck, P. Chylek, B. Johnson, and P. Damiano.1996. Historical biomass burning: Late 19th century pioneer agriculture revolution in Northern Hemisphere ice core data and its atmospheric interpretation. *Journal of Geophysical Research—Atmospheres* 101: 23317–23334.

Huffman, G.J., R.F. Adler, P. Arkin, A. Chang, R. Ferraro, A. Gruber, J. Janowiak, A. McNab, B. Rudolf, and U. Schneider. 1997. The Global Precipitation Climatology Project (GPCP) Combined Precipitation Dataset. *Bulletin of the American Meteorological Society* 78: 5–20.

Kalnay, E., M. Kanamitsu, R. Kistler, W. Collins, D. Deaven, L. Gandin, M. Iredell, S. Saha, G. White, J. Woollen, Y. Zhu, M. Chelliah, W. Ebisuzaki, W. Higgins, J. Janowiak, K.C. Mo, C. Ropelewski, J. Wang, A. Leetmaa, R. Reynolds, R. Jenne, and D. Joseph.1996. The NCEP/NCAR 40-year reanalysis project. *Bulletin of the American Meteorological Society* 77: 437–471.

Kistler, R., E. Kalnay, W. Collins, S. Saha, G. White, J. Woollen, M. Chelliah, W. Ebisuzaki, M. Kanamitsu, V. Kousky, H. van den Dool, R. Jenne, and M. Fiorino. 2001. The NCEP-NCAR 50-year reanalysis: Monthly means CD-ROM and documentation. *Bulletin of the American Meteorological Society* 82: 247–267.

Krishnamurthy, V., and B.N. Goswami. 2000. Indian monsoon-ENSO relationship on interdecadal timescale. *Journal of Climate* 13: 579–595.

Lau, K.-M., and S. Yang. 2002. Walker circulation. In, Holton, R., J. Pyle, and J.A. Curry (eds.). *Encyclopedia of Atmospheric Sciences*. San Diego, California: Academic Press.

Linsley, B.K., R.B. Dunbar, G.M. Wellington, and D.A. Mucciarone. 1994. A coral-based reconstruction of Intertropical Convergence Zone variability over Central-America since 1707. *Journal of Geophysical Research—Oceans* 99: 9977–9994.

Moore, G.W.K., G. Holdsworth, and K. Alverson. 2001. Extra-tropical response to ENSO as expressed in an ice core from the Saint Elias Mountain range. *Geophysical Research Letters* 28: 3457–3460.

Moore, G.W.K., G. Holdsworth, and K. Alverson., 2002a. Climate change in the North Pacific region over the past three centuries. *Nature* 420: 401–403.

Moore, G.W.K., K. Alverson, and G. Holdsworth. 2002b. Variability in the climate of the Pacific Ocean and North America as expressed in an ice core from Mount Logan. *Annals of Glaciology* 35: 423–429.

Moore, G.W.K., K. Alverson, and G. Holdsworth. 2003. The impact that elevation has on the ENSO signal in precipitation records from the Gulf of Alaska region. *Climatic Change* 59: 101–121.

Morrissey, M.L., and N.E. Graham. 1996. Recent trends in rain gauge precipitation measurements from the tropical Pacific: Evidence for an enhanced hydrologic cycle. *Bulletin of the American Meteorological Society* 77: 1207–1219.

Newman, M., P.D. Sardeshmukh, and J.W. Bergman. 2000. An assessment of the NCEP, NASA, and ECMWF reanalyses over the tropical West Pacific warm pool. *Bulletin of the American Meteorological Society* 81: 41–48.

Oort, A.H., and E.M. Rasmusson. 1970. On annual variation of monthly mean meridional circulation. *Monthly Weather Review* 98: 423–442.

Oort, A.H., and J.J. Yienger. 1996. Observed interannual variability in the Hadley circulation and its connection to ENSO. *Journal of Climate* 9: 2751–2767.

Ruddiman, W.F. 2003. Orbital insolation, ice volume, and greenhouse gases. *Quaternary Science Reviews* 22: 1597–1629.

Santer, B.D., et al. 2000. Statistical significance of trends and trend differences in layer-average atmospheric temperature time series. *Journal of Geophysical Research—Atmospheres* 105: 7337–7356.

Smirnov, V.V., and G.W.K. Moore. 2001. Short-term and seasonal variability of the atmospheric water vapor transport through the Mackenzie River basin. *Journal of Hydrometeorology* 2: 441–452.

Trenberth, K.E., and J.G. Olson. 1988. An evaluation and intercomparison of global analyses from the National Meteorological Center and the European Centre for Medium-Range Weather Forecasts. *Bulletin of the American Meteorological Society* 69: 1047–1057.

Trenberth, K.E., G.W. Branstator, D. Karoly, A. Kumar, N.-C. Lau, and C. Ropelewski. 1998. Progress during TOGA in understanding and modeling global teleconnections associated with tropical sea surface temperatures. *Journal of Geophysical Research* 103: 14291–14324.

Trenberth, K.E., D.P. Stepaniak, and J.M. Caron. 2000. The global monsoon as seen through the divergent atmospheric circulation. *Journal of Climate* 13: 3969–3993.

Trenberth, K.E., B.A. Wielicki, A.D. Del Genio, T. Wong, J. Chen, B.E. Carlson, R.P. Allan, F. Robertson, H. Jacobowitz, A. Slingo, D.A. Randall, J.T. Kiehl, B.J. Soden, C.T. Gordon, A.J. Miller, S.-K. Yang, and J. Susskind. 2002. Changes in tropical clouds and radiation. *Science* 296: 2095a [DOI: 10.1126/science.296.5576.2095a]

Waliser, D.E., Z. Shi, J.R. Lanzante, and A.H. Oort. 1999. The Hadley circulation: Assessing NCEP/NCAR reanalysis and sparse in situ estimates. *Climate Dynamics* 15: 719–735.

Wallace, J.M., and D.S. Gutzler. 1981. Teleconnections in the geopotential height field during the Northern Hemisphere winter. *Monthly Weather Review* 109: 784–812.

Wang, C.Z. 2002a. Atlantic climate variability and its associated atmospheric circulation cells. *Journal of Climate* 15: 1516–1536.

Wang, C.Z. 2002b. Atmospheric circulation cells associated with the El Nino–Southern Oscillation. *Journal of Climate* 15: 399–419.

Waszkiewicz, M. 2003. Unpublished data.

Webster, P.J., and S. Yang. 1992. Monsoon and ENSO—Selectively interactive systems. *Quarterly Journal of the Royal Meteorological Society* 118: 877–926.

Webster, P.J., V.O. Magana, T.N. Palmer, J. Shukla, R.A. Tomas, M. Yanai, and T. Yasunari.. 1998. Monsoons: Processes, predictability, and the prospects for prediction. *Journal of Geophysical Research—Oceans.* 103: 14451–14510.

Wielicki, B.A., T.M. Wong, R.P. Allan, A. Slingo, J.T. Kiehl, B.J. Soden, C.T. Gordon, A.J. Miller, S.K. Yang, D.A. Randall, F. Robertson, J. Susskind, and H. Jacobowitz. 2002. Evidence for large decadal variability in the tropical mean radiative energy budget. *Science* 295: 841–844.

Zhao, H., and G.W.K. Moore. In press. Trends in the Walker and Hadley circulations as expressed in precipitation records from Asia and Africa during the latter half of the 20th century. *Journal of Climate.*

SECTION C

Causes of Variability in the Hadley Circulation: Past and Future

Chapter 14

THE RESPONSE OF THE HADLEY CIRCULATION TO CLIMATE CHANGES, PAST AND FUTURE

David Rind and Judith Perlwitz
NASA/GISS at Columbia University, New York, New York 10025 U.S.A

Abstract A suite of altered climate experiments for the Paleocene, the last glacial maximum (LGM), and a 2 x CO_2 climate were compared to assess the factors responsible for producing variations in Hadley cell intensity and extent. The climate simulations used best-guess topography and marine surface fields, as well as feasible alternative sea surface temperature (SST) patterns. The individual contributions to the circulations were quantified, and compared among the different simulations. The results show that the Hadley cell intensity is associated with the gradient in latent heat release from the tropics to the subtropics, driven in the model by the gradient in sea surface temperature. It is not related to the absolute warmth of the climate, or of the tropical sea surface temperatures. Eddy forcing, primarily through transient eddy heat transport, amplified the subtropical portion of the cell, as well as the mid-latitude Ferrel cell. The poleward extent of the Hadley cell is affected by numerous processes, including the influence of topography in the extratropics. It also does not vary systematically with global mean temperature. Only the strongest Hadley cell changes are longitudinally homogeneous; there is little relationship between the change in Hadley cell intensity and the change in strength of the Walker cell, and the Pacific Ocean is the most important basin for the zonal average Hadley cell response. Although the latitudinal average precipitation does respond interactively with Hadley cell intensity and extent, the soil moisture variations are less correlated, due to differing seasonal effects and the influence of temperature/evaporation changes. The importance of the Hadley cell variations for assessing past and future water availability changes should not be overestimated, although it is a contributing factor.

H.F. Diaz and R.S. Bradley (eds.), The Hadley Circulation: Present, Past and Future, 399–435.

1. INTRODUCTION

As the dominant zonally averaged circulation feature at low and subtropical latitudes, the Hadley cell represents an organizing principle in which to understand locations of climatological features, such as deserts and rain forests. In the current climate, the distribution of deserts in the descending zone of the Hadley cell at a wide range of longitudes is evidence of its applicability. This then is a lure for viewing paleoclimate and future climate in the same context; how did/will the Hadley cell change, with its implications for past/future water availability, vegetation, etc.

How are we to understand and compare the climate-induced changes in the various forcing functions influencing the Hadley cell? The regions of large-scale meridional and vertical motion that characterize this circulation feature respond to the governing principles of conservation of energy and momentum. Starting from these equations, we can derive the applicable terms. These can be written with respect to the mass stream function ψ. In addition to the definitions given below, the various terms used in Equations (1) and (2) are given in an appendix at the end of the chapter.

Define (1) $\bar{v} = -\frac{\partial \bar{\psi}}{\partial p} \quad \bar{\omega} = \frac{\partial \bar{\psi}}{\partial y}$; as $\frac{\partial \bar{v}}{\partial y} + \frac{\partial \bar{\omega}}{\partial p} = 0$,

$$-\frac{\partial^2 \bar{\psi}}{\partial y \partial p} + \frac{\partial^2 \bar{\psi}}{\partial y \partial p} = 0$$

From the zonally averaged equations of motion, we can derive:

$$(2)\ \frac{\partial^2 \bar{\psi}}{\partial y^2} + \frac{f^2}{S_p}\frac{\partial^2 \bar{\psi}}{\partial p^2} = \overline{\nabla^2 \psi} = \underbrace{\frac{-f}{S_p}\frac{\partial}{\partial p}\left(\frac{\partial \bar{u}}{\partial t} + \bar{v}\frac{\partial \bar{u}}{\partial y} + \bar{\omega}\frac{\partial \bar{u}}{\partial p}\right)}_{NONLINEAR\ MOMENTUM\ FLUX}$$

$$+ \underbrace{\frac{1}{S_p}\frac{\partial}{\partial y}\left(\frac{\partial \bar{T}}{\partial t} + \bar{v}\frac{\partial \bar{T}}{\partial y}\right)}_{NONLINEAR\ THERMAL\ FLUX} \underbrace{- \frac{f}{S_p}\frac{\partial \bar{L}}{\partial p} - \frac{f}{S_p}\frac{\partial \bar{F}_x}{\partial p}}_{MOMENTUM\ FORCING} \underbrace{- \frac{1}{S_p c_p}\frac{\partial \bar{Q}}{\partial y} + \frac{1}{S_p}\frac{\partial \bar{G}}{\partial y}}_{THERMAL\ FORCING}$$

where friction F includes surface friction, mountain torque, and cumulus friction; and heating Q includes short- and long-wave radiation, latent heat release, convective motions, and sensible heat flux.

The eddy forcing terms are

$$(3) \qquad \overline{G} \equiv \frac{\partial \overline{(v'T')}}{\partial y} + \frac{\partial \overline{(\omega' T')}}{\partial p} \quad \text{eddy heat flux divergence}$$

$$\overline{L} \equiv \frac{\partial \overline{(u'v')}}{\partial y} + \frac{\partial \overline{(u'\omega')}}{\partial p} \quad \text{eddy momentum flux divergence}$$

The Laplacian of the stream function that appears on the left-hand side of Equation (2) is negatively related to the stream function, assuming a sinusoidal relationship. So the various terms on the right-hand side force a negative value of $\overline{\psi}$. With the stream function defined as above, the Hadley circulation in December–February (DJF) is characterized by a negative value of the stream function, and positive values of the forcing terms as written intensify the Hadley cell.

Which terms are dominant for Hadley cell intensity and extent? Various researchers have emphasized different components, often in simple models in which the other terms are minimized. For example, Schneider and Lindzen (1977) emphasized the importance of cumulus heating and cumulus friction in driving the intensity of the axisymmetric circulation, while Becker and Schmitz (2001) (for stationary eddies), Trenberth and Stepaniak (2003) (for transient eddies), and Kim and Lee (2001) emphasized the importance of eddy fluxes in driving the Hadley cell. Lindzen and Hou (1988) calculated that the intensity of the circulation increases dramatically as heating moves off the equator, and hence is dominated by near-equatorial displacements of the thermal equator, while Fang and Tung (1999) noted that the time-dependent solution does not show this dominance as clearly, and the results of Dima and Wallace (2003), using reanalysis data, support the view that the seasonal variations are instead driven by the monsoons.

Considering the latitudinal or poleward extent, Held and Hou (1980) emphasized the importance of the latitudinal temperature gradient and tropopause height (in a model assuming conservation of angular momentum in the Hadley cell regime, and no eddy forcing at mid-latitudes), while Taylor (1980), for example, emphasized the importance of eddy momentum transports. As was noted by Kim and Lee (2001), one difficulty is that these sources are not independent of one another; eddies are associated with latent heat release that influences dQ/dy and surface friction affecting dF/dp, as well as heat and momentum transport convergences (dG/dy, dL/dp). The results in the individual calculations also appear to be sensitive to the various parameters used, including the thermal relaxation time (Fang and Tung 1997) or the effective viscosity (Kim and Lee 2001).

Pfeffer (1981) used radiosonde data and estimated atmospheric heating profiles to calculate some of the budget terms shown in Equation

(2). Comparing the relative importance of eddies and diabatic heating for the current climate, he concluded that eddy fluxes of heat and momentum were responsible for 1/3 to 1/4 of the total Hadley circulation intensity, and the latitudinal extent (from his figures) was the result of both eddy forcing and the diabatic circulation. This conclusion is hindered by the approximate nature and sparse sampling of the data.

General circulation model (GCM) studies obviously include all of these terms together, but they can be varied systematically to study the relationships amongst them. Rind and Rossow (1984) performed such experiments, and discussed the complex interactions between the various thermal and momentum forcing terms that arose in the model. For example, solar forcing during Northern Hemisphere winter actually weakened Hadley cell intensity, for it provided out-of-phase forcing with gradients in latent heat release associated with low-latitude sea surface temperature (SST) gradients (driven ultimately by solar heating in the previous seasons). Removing the frictional forcing terms from the model completely destroyed the Hadley circulation even without altering solar radiative forcing. Experiments in such models emphasize the need to incorporate as many terms as possible, even those which from scale analysis would seem to be secondary; they may decide the outcome of competition between the larger forces, and are especially important when the ageostrophic flow field is being considered.

The GISS GCM has also been used to evaluate the dependence of the Hadley cell on systematically varied sea surface temperature gradients in warmer and colder climates (Rind 1998, 2000). An increased (decreased) gradient led to an intensified (weakened) Hadley circulation independent of any change in global mean temperature; the warmer climate by itself actually led to a small decrease in Hadley cell intensity. The latitudinal gradient did not affect the width of the Hadley circulation. We return to these results in the Discussion section.

Finally, GCM studies of other climates have been compared to assess how the Hadley cell changes with climate. In transient future warming experiments (IPCC 2001), tropical precipitation increases, especially over the oceans, with some decreases in the subtropics, implying an increased Hadley cell intensity. In 2 x CO_2 experiments there is little consistency; Rind et al. (2001a) found that even within the GISS suite of models, there were both increases and decreases in Hadley circulation intensity. In ice age experiments, much depends on the magnitude of tropical cooling, a question of considerable uncertainty (e.g., Rind and Peteet 1985). We compare the results given below with other model simulations in the Discussion section, in an effort to assess their generalizability.

Given the potential importance of various forcing terms, and the inconsistency in models, we need to determine how the individual components change in a wide range of climate simulations. That is the focus of this chapter.

2. MODEL SIMULATIONS

In contrast to some of our other studies in which one variable was changed at a time, here we look at a range of modeled paleo/future climate simulations in which many aspects of the climate are different. The sea surface temperatures are uncertain for these periods, especially in the tropics; therefore, each simulation is accompanied by an alternate with a different, plausible sea surface temperature pattern. All the simulations are done with the GISS Global Climate/Middle Atmosphere Model (GISS GCMAM), which extends up to the mesosphere, at 8° x 10° resolution. The reason for including the full stratosphere is to allow the effect of gravity wave drag parameterizations to be apparent; separate sensitivity experiments are used to illustrate this effect on the Hadley cell via the momentum budget. The effect of the coarse resolution and model physics is discussed by reference to higher-resolution runs that were performed for several of these climate simulations. Each of the simulations discussed here has been published previously, albeit without particular emphasis on its Hadley cell response.

How accurately does this model simulate the observed Hadley cell and its variability? To address this question, we ran the model with the observed sea surface temperatures for the time period 1950 through 1997. Results are shown in Figure 14-1, giving the stream function values and their variability. A negative value represents a clockwise mass flux circulation in the frame of the figure, hence the DJF Hadley cell, whereas a positive value gives the opposite circulation sense; i.e., the June–August (JJA) Hadley cell. The control run simulation (top panels) has a magnitude about 10%–20% weaker than in the NCEP analysis (which differs somewhat from the ECMWF analysis [Trenberth et al. 2000]), and also a weaker Ferrel cell, likely due to its coarse resolution. For a similar reason, the poleward extent is about 10° poleward of its observed position. The model's transient eddy energy is about 85% of the observed value, while its standing eddy energy is actually greater than observed (the coarse resolution results in waves moving too slowly). What is important for the budget terms, however, are the eddy heat and momentum transports and their convergences. Hansen et al. (1983) show a comparison of the coarse grid model transports with observations; while the magnitude of both the heat and momentum transports by eddies is appropriate, the momentum transport peaks around 10° poleward of the observed peak, altering the eddy forcing of the Ferrel cell.

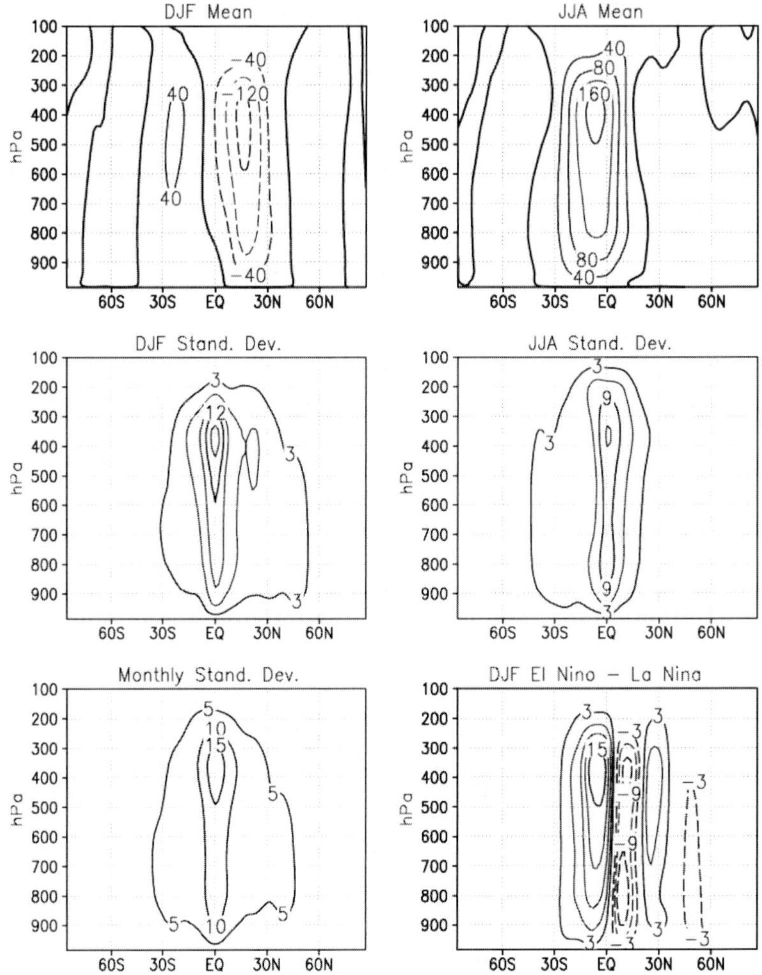

Figure 14-1. Characteristics of the control run meridional circulations derived from a 48 y run with observed SSTs (1950–97). Mean values (top) and interannual variability (middle) are shown for the solstice seasons, as well as monthly variability and the difference between seven El Niño and La Niña events (bottom).

The model's variability for the two solstice seasons is shown in the middle panels. Peak values are on the order of 10% of the seasonal mean. The monthly standard deviations, with the seasonal cycle removed, and the difference between seven El Niño and La Niña time periods are shown in the bottom figures. These can be compared with results given by Waliser et al. (1999) from NCEP/NCAR reanalysis data. The reanalysis results showed maximum variability near the equator of some 15 x 10^9 kg s^{-1}, in excellent agreement with the model values; the chief discrepancy is that the observed variability ("5" contour) extended further poleward into the extratropics. The reanalysis data also showed a stream function difference between El Niño and La Niña conditions of about 10 x 10^9 kg s^{-1} in both hemispheres, again in agreement with model values. Note that variability estimated using in situ radiosonde observations (Oort and Yienger 1996) produces larger-magnitude patterns off the equator, which appear to be due to undersampling of the Pacific Ocean region (Waliser et al. 1999).

The primary simulations used in this study are shown in Table 14-1, along with references that provide greater details on the simulations. In addition to the present climate control run, they include simulations for the Paleocene (58 million years ago), characterized by very low topography and a reduced latitudinal temperature gradient, with extensive high-latitude warming (indicated by PAL); for the last glacial maximum (LGM; 21,000 BP), with increased topography due to ice sheets, and an increased latitudinal temperature gradient (indicated by IA); and for the doubled CO_2 climate (indicated by 2CO2). The standard simulations for the Paleocene and last glacial maximum include the "best guess" sea surface temperature patterns for those periods, while the alternate simulations include prescriptions that provide more of a tropical response, consistent with some data or model studies (see the noted references). Two additional ice age runs were also performed that have reduced topographic gravity wave drag, as is discussed below. The standard doubled CO_2 simulation with the GISS model has a large tropical response, and the alternate, similar to that originally performed with the Geophysical Fluid Dynamics Laboratory (GFDL) model, has less tropical and more high-latitude warming.

Some characteristics of these simulations are provided in Table 14-2, and the changes in DJF sea surface temperatures from the control run are shown for the different experiments in Figure 14-2. The warmer conditions in the tropics and high latitudes for the doubled CO_2 and the Paleocene are clearly visible, with much the reverse conditions for the ice age runs. Also evident is the difference between the standard and alternate simulations, especially in the tropics. In this and several of the following figures, we show the run with the relatively warmer sea surface temperatures for each climate in the left-hand column (hence PAL-A is shown on the lower left).

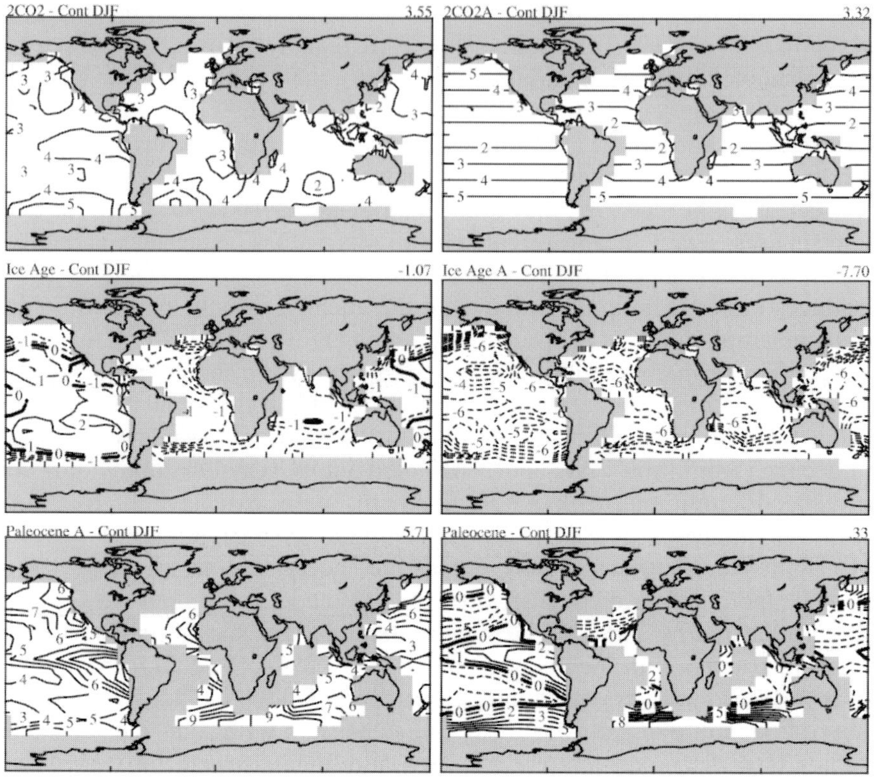

Figure 14-2. Change in sea surface temperatures relative to the control run for the different experiments. Sea ice regions are omitted. The altered land/ocean configuration in the Paleocene reduces the number of grid points that are also oceanic in the current climate.

The zonally averaged surface air temperature changes for the solstice seasons are given in Figure 14-3. The global surface air temperature range of these simulations is some 17°C, the global topography differs by up to 280 m, and the equator-to-pole surface air temperature gradient changes by 30°C. The Paleocene simulations also have somewhat different land-ocean distribution, with a smaller Atlantic Ocean and no ice sheets on Antarctica (thus the large warming in that region in JJA). These simulations represent an extremely wide range of conditions to test hypotheses concerning Hadley cell variations.

Table 14-1. Primary experiments used.

Experiment	Description	Reference
Control	Current sea surface temperatures (SSTs)	Rind et al. (1988)
2CO2	Doubled atmospheric CO_2, GISS calculated SSTs with significant tropical warming	Rind et al. (1990)
2CO2-A	Doubled atmospheric CO_2, GFDL-like SSTs with minimal tropical warming	Rind et al. (1990)
IA	Last glacial maximum with CLIMAP SSTs	CLIMAP (1981); Rind et al. (2001)
IA-A	Last glacial maximum with colder tropical SSTs	Webb et al. (1997); Rind et al. (2001)
PAL	Paleocene (58 million years ago) with doubled atmospheric CO_2, "best guess" SSTs	O'Connell et al. (1996); Rind et al. (2001b)
PAL-A	Paleocene with doubled atmospheric CO_2 with GISS 2 x CO_2 SSTs	Rind et al. (2001b)

Table 14-2. Characteristic changes in the climate experiments.

	ΔHadley Cell Peak (%)		**ΔPoleward Extent (°)**		**ΔGlobal Surface Temperature (°C)**		ΔEquatoria 1 SST (°C)		**ΔTransient EKE (%)**	
	DJF	**JJA**	**DJF**	**JJA**	**DJF**	**JJA**	**DJF**	**JJA**	**DJF**	**JJA**
2CO2	**–2.4**	**– 23.9**	0.5	**–3.9**	**4.36**	**4.19**	**3.50**	**3.58**	**–6.2**	**–6.4**
2CO2-A	**– 18.8**	**– 13.5**	**1.4**	**1.3**	**4.29**	**4.38**	**1.54**	**1.54**	**–12.4**	**–16.1**
IA	**10.1**	**– 23.3**	**4.3**	0.7	**– 4.12**	**– 4.45**	**– 0.50**	**–1.95**	**55.1**	**30.6**
IA-A	**9.4**	**67.5**	**5.6**	0	**– 7.98**	**– 10.3**	**– 7.30**	**–7.70**	**52.9**	**59.4**
PAL	**49.2**	**16. 0**	**–7**	**–9**	**4.09**	**2.60**	**1.50**	**0.45**	**52.2**	**21.1**
PAL-A	**– 11.7**	**– 25.8**	–0.1	**–2**	**8.60**	**6.58**	**5.55**	**5.42**	**18.8**	**9.0**

All experiments are run with the specified sea surface temperatures/sea ice, obtained either through paleodata analysis or model simulations. The runs are for 20 years following a spin-up year; with the specified ocean conditions the simulations are stable throughout the course of the integration. The differences among the experiments are evaluated for their significance from the interannual variability in the control run.

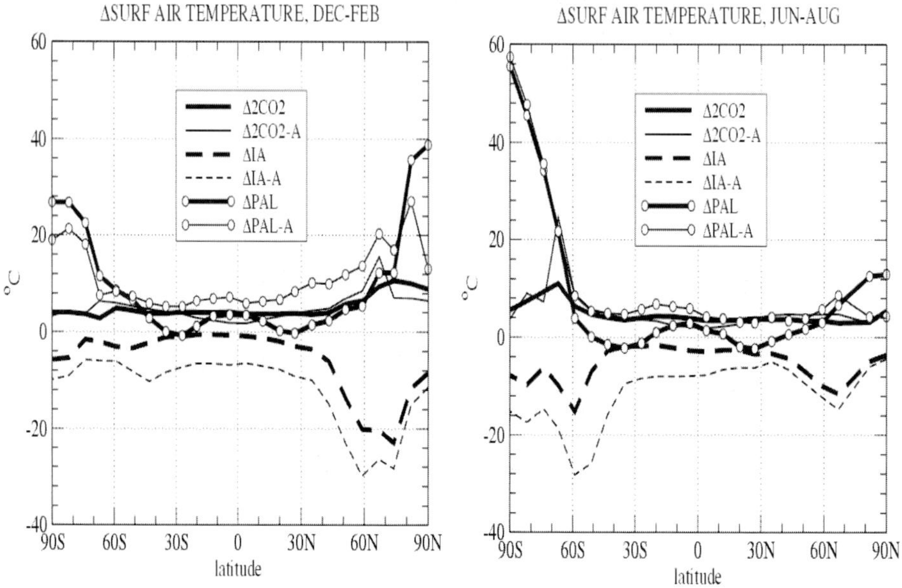

Figure 14-3. Zonal average surface air temperature changes in the two solstice seasons.

3. RESULTS

The changes in the stream function for the two solstice seasons are given in Figure 14-4(a, b), with the significance of the change indicated by the shading. Light (dark) shading means a significant increase at the 95% level in the negative (positive) value of the stream function, hence an increase in DJF (JJA) Hadley cell value.

It can be seen from these presentations that the Hadley cell changes involve alterations in both the peak magnitude and latitudinal/altitude distribution. The percentage change in the peak value is listed in Table 14-2, with the significant changes indicated in boldface (for the 20 y simulations in these extreme climates; for seasonal averages, all the peak changes are significant, as are most of the changes shown in Fig. 14-4; all the temperature and energy changes in Table 14-2 are highly significant).

The results show that across the suite of experiments the Hadley cell intensity is not simply related to global or equatorial temperatures.

In DJF, the Hadley cell was strongest in the Paleocene and weakest in 2CO2-A despite their similar tropical SST magnitudes; in JJA, the Hadley cell was strongest in IA-A and weakest in PAL-A despite the ice age tropical SSTs being some 13°C colder. Hence, in the Southern Hemisphere for these particular experiments, the intensity is actually inversely related to tropical SSTs.

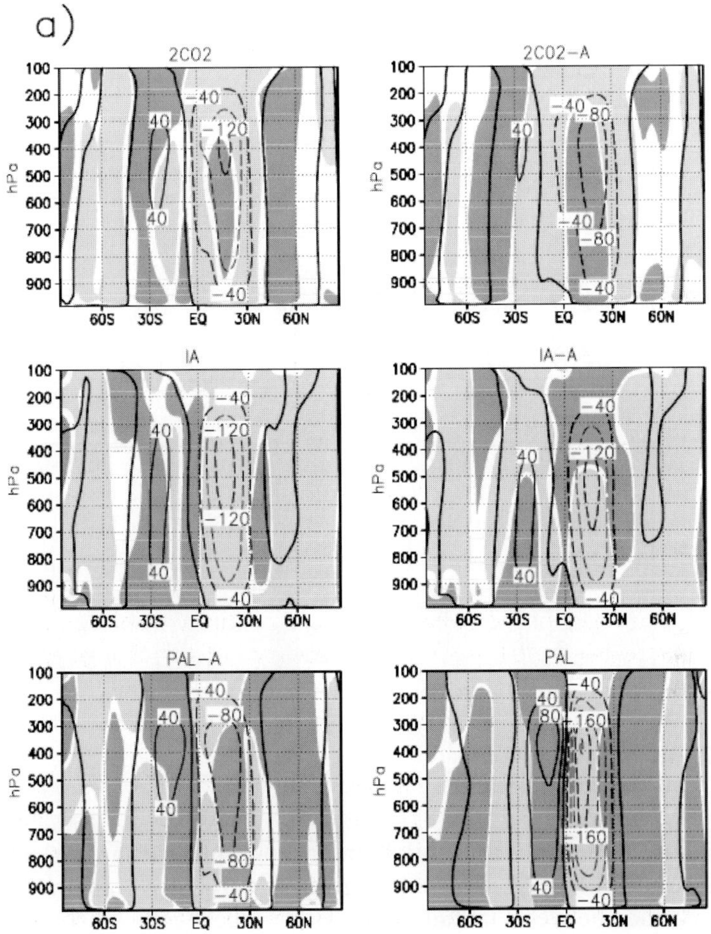

Figure 14-4. Stream functions (10^9 kg s^{-1}) for the two solstice seasons (a) Dec–Feb and (b) Jun–Aug. Light (dark) shading indicates where the negative (positive) changes between the experiment and the control are significant at the 95% level.

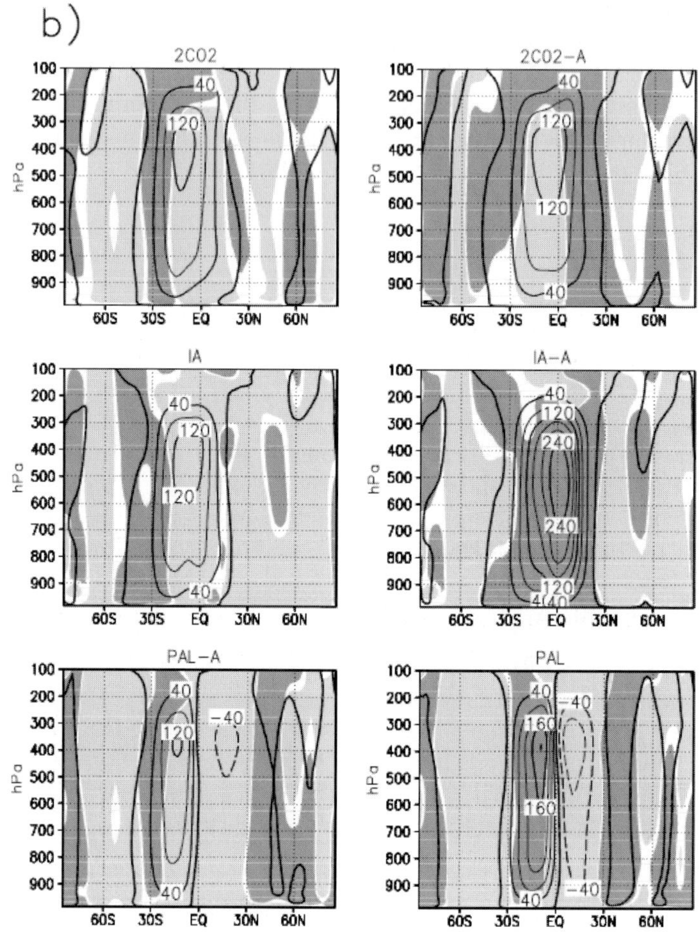

Figure 14-4. Continued.

Also given in the table is the change in poleward extent, which we produce by linearly interpolating the vertically integrated stream function value between latitudes; the Hadley cell ends where the stream function changes sign. When this technique was used with an 8° x 10° average of results from our 2° x 2.5° model, it produced generally similar latitudinal changes to those defined at the finer resolution. Nevertheless, it is at best only a crude approximation. The interannual variation of this value from the control run is 1.33° latitude for DJF, and 2.04° latitude for JJA. The significant changes are again shown in boldface in Table

14-2, and are in effect also indicated in Figure 14-3 by significant changes in the poleward extension of the stream function.

Again, there is little relationship between the warmth of the climate and the poleward extent of the circulation. The Hadley cell has its greatest poleward expansion in the coldest (ice age) climate, and is most contracted in one of the warmest climates (Paleocene). To understand these results, we refer to the budget terms.

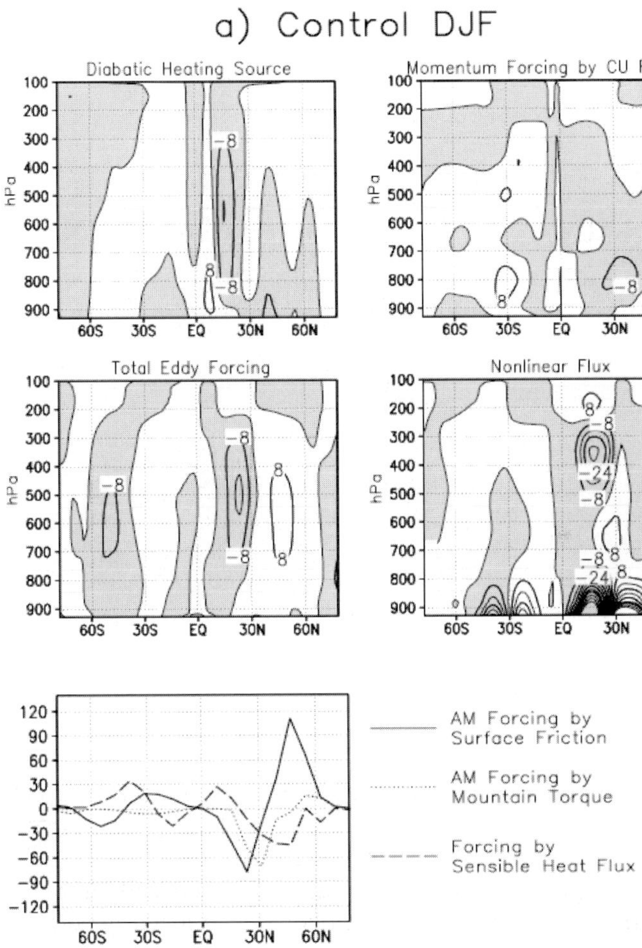

Figure 14-5. Contributions of individual terms (W m^{-2}/σ) to the Hadley cell for the control run for (a) Dec–Feb and (b) Jun–Aug. Negative values are shaded.

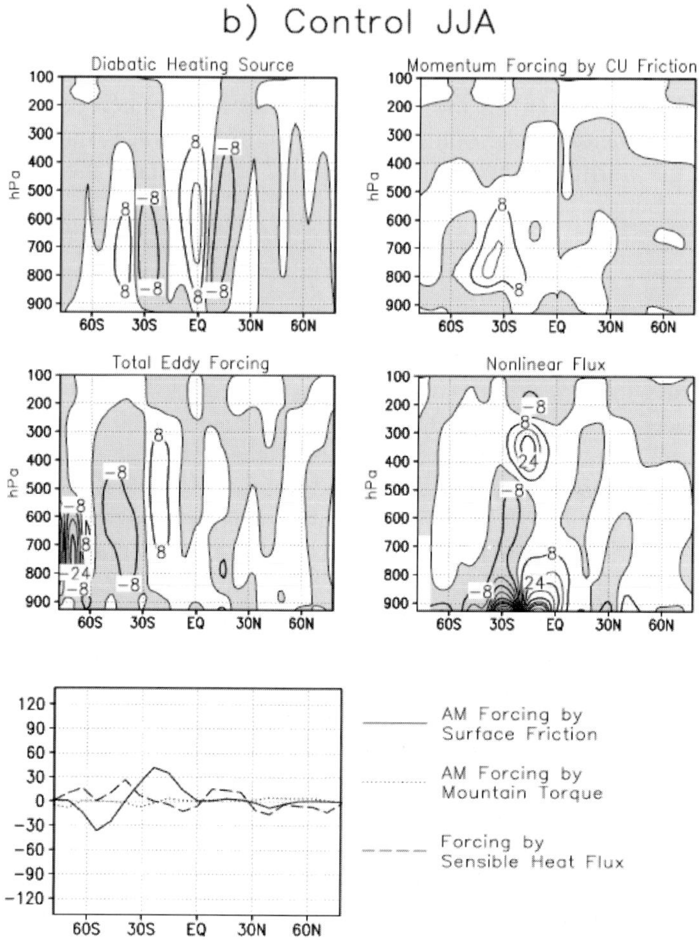

Figure 14-5. Continued.

We calculate the contributions of the different terms proportional to the negative of the stream function, given in Equation (2) above. While these are generated in association with the stream function and its changes, to some extent they represent "forcing terms," as commented on further in the Discussion section. The results are shown for the control run solstice seasons in Figures 14-5a and b (in units of W m^{-2} normalized by the static stability). They are divided into the contribution associated with the latitudinal gradient in diabatic heating (dQ/dy), the mixing of momentum by convection (cumulus friction), the total eddy forcing ($G + L$), and the nonlinear momentum and heat flux terms that act once the circulation is set up. On the bottom is shown the contribution from the various surface forcings that act in the lower atmosphere due to sensible heat flux, surface friction, and mountain torque.

As is shown in the figure, a negative value helps generate a clockwise circulation in the frame of the figure, thereby enhancing the Hadley cell during DJF, while a positive value enhances the Hadley cell during JJA (or the Ferrel cell in Northern Hemisphere winter). In the model, the Hadley cell is forced at low latitudes primarily by the diabatic circulation; and in the subtropics by eddy forcing, cumulus friction, and surface friction/mountain torque at low levels; while the nonlinear terms amplify the circulation in the upper and lower troposphere. The mid-latitude Ferrel cell is forced primarily by eddies, with a contribution due to the nonlinear flux and surface friction. Compared to the selected results shown by Pfeffer (1981), the model's diabatic heating term is properly positioned but somewhat weaker, and the eddy contribution to the Hadley cell is somewhat greater and located further poleward; the eddy contribution to the Ferrel cell is properly positioned but too weak. The latter effect is likely due to the coarse model resolution, while the other discrepancies may be due to a mixture of model inaccuracies and sampling/heating estimate uncertainties in the observations.

Each of these forcing terms was calculated for the different model simulations noted above. The change in the diabatic heating source for June–August is shown in Figure 14-6, along with the control run value for reference. Comparison with Figure 14-4b indicates a general correspondence between the variation of this term and the peak low-latitude intensity in the different runs. The change in the total eddy forcing (the sum of the eddy contributions to the momentum and thermal forcing as given in the introduction) is depicted in Figure 14-7 for DJF. Comparison with Figure 14-4a (and Table 14-2) indicates that this term is important for the subtropical Hadley cell strength (along with the diabatic heat source).

Shown in Figures 14-8 and 14-9 are the contributions of the different terms in the runs with the biggest Hadley cell increases, the Paleocene during DJF and the alternate ice age simulation during June–August (Table 14-2). The dominance of the diabatic heating source change is evident in these figures, along with the subtropical/mid-latitude eddy forcing changes. In addition, the surface forcing terms can be quite important at specific latitudes.

What produced these responses? We correlated the intensity and poleward extent changes in these seven simulations (six experiments plus control) with various aspects of the modeled climate. The results are given in Table 14-3. A significant response (indicated in bold) requires a correlation greater than 0.7. With respect to the general idea that a warmer climate has increased Hadley cell intensity, the model simulations show no such effect; what correlations do exist are if anything the reverse. However, the Hadley cell intensity is strongly correlated with the sea surface temperature gradient—with a stronger gradient from low latitudes to the subtropics, the intensity increases. The correlation is almost as high with the precipitation gradient, the latent heat release being a prime component of the diabatic heating term.

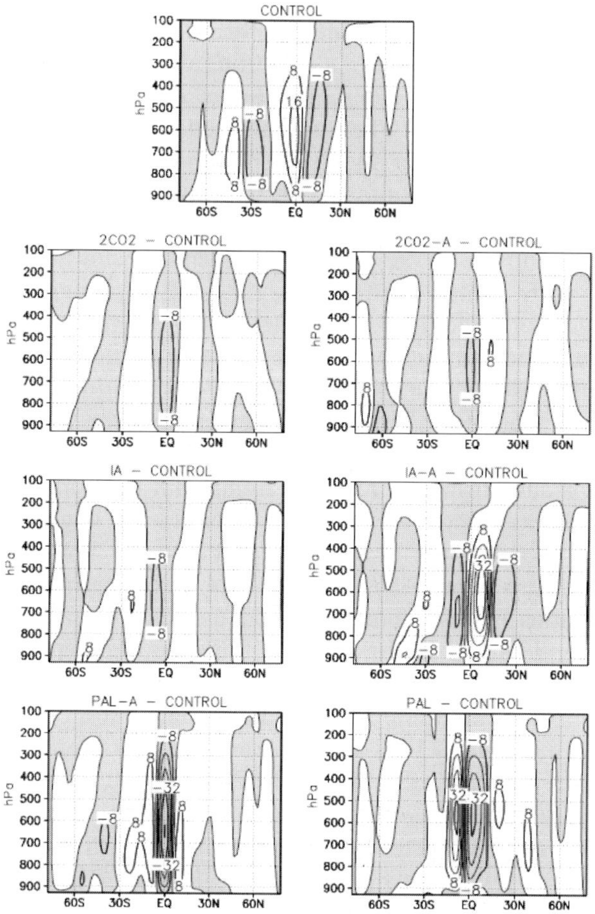

Figure 14-6. Change in diabatic heating source (W m^{-2}/σ) for Jun–Aug. Negative values are shaded.

This effect is less important in the Southern Hemisphere, as monsoon precipitation over land at higher northern latitudes also influences Hadley cell intensity. Even here the relationship is strong at times; in June–August, the large Hadley cell change in IA-A is associated with a

68% increase in the SST gradient between 27°N and 27°S, and in PAL, the change south of the equator is related to a 40% increase in the SST gradient between 4°S and 27°S, while the decrease north of the equator is due to an SST gradient reduction between 27°N and 4°N. There is no significant correlation between Hadley cell intensities in the two hemispheres for their respective solstice seasons, nor between the intensity and poleward extent.

Table 14-3. Correlation of Hadley cell characteristics and climate parameters.

	Intensity		**Poleward Extent**	
	NH, DJF	**SH, JJA**	**NH, DJF**	**SH, JJA**
Surface Temperature	–0.20	**–0.73**	–0.65	–0.35
Equatorial SST	–0.23	**–0.83**	–0.54	–0.28
SST Gradient 4°N–27°N (D–F) 12°N–27°S (J–A)	**0.99**	**0.81**	–0.59	–0.44
Precipitation Gradient 4°N–27°N (D–F) 12°N–27°S(J–A)	**0.90**	0.62	**–0.80**	**–0.73**
NH/SH	0.41	0.41	**0.86**	**0.86**
Intensity/ Extent	–0.56	–0.05	0.56	0.05
Eddy Kinetic Energy (EKE)	0.46	**0.80**	0.21	0.43
Transient EKE	**0.87**	**0.74**	0.28	0.01
Standing EKE	–0.55	–0.07	0.47	0.54
Northward Transient Static Energy by Eddies	0.50	0.64	0.20	0.11
Gravity Wave Drag	0.14	–0.67	**0.82**	–0.47
Surface friction	0.28	–0.25	–0.54	0.63

The poleward extent is inversely proportional to the precipitation gradient, because when the subsidence previously in the subtropics moves poleward, it allows for more subtropical precipitation. The poleward extent is highly correlated between the two hemispheres, but it is not simply related to either the SST gradient or eddy kinetic energy. Held and Hou (1980) calculated that the latitudinal extent should be related to the equator-to-pole temperature gradient, all else (e.g., the height of the tropopause) being assumed equal. A general correspondence can be seen in that the poleward extent and gradient decrease in the doubled CO_2 climate, and increase in the ice age climate, although the latitudinal change is not closely related to the magnitude of the gradient (e.g.,

2CO2-A has greater contraction than 2CO2, although its latitudinal gradient decrease is smaller). It does not hold for the Paleocene simulations, or IA in June–August. The theory relates only to the symmetric circulation driven in the absence of mid-latitude eddies, and assumes conservation of angular momentum at low latitudes (i.e., no cumulus friction), both at variance with the model simulations and the real world.

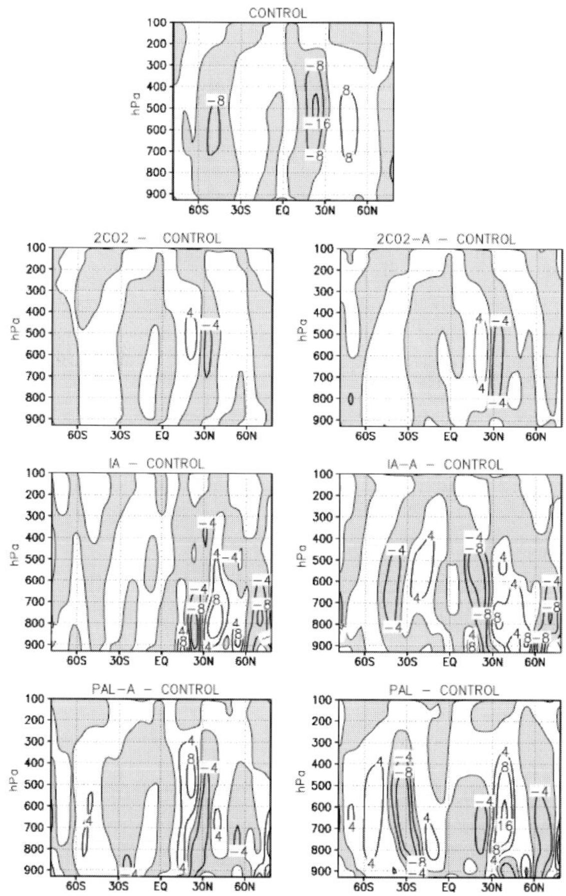

Figure 14-7. Change in total eddy forcing (W m$^{-2}/\sigma$) for Dec–Feb. Negative values are shaded.

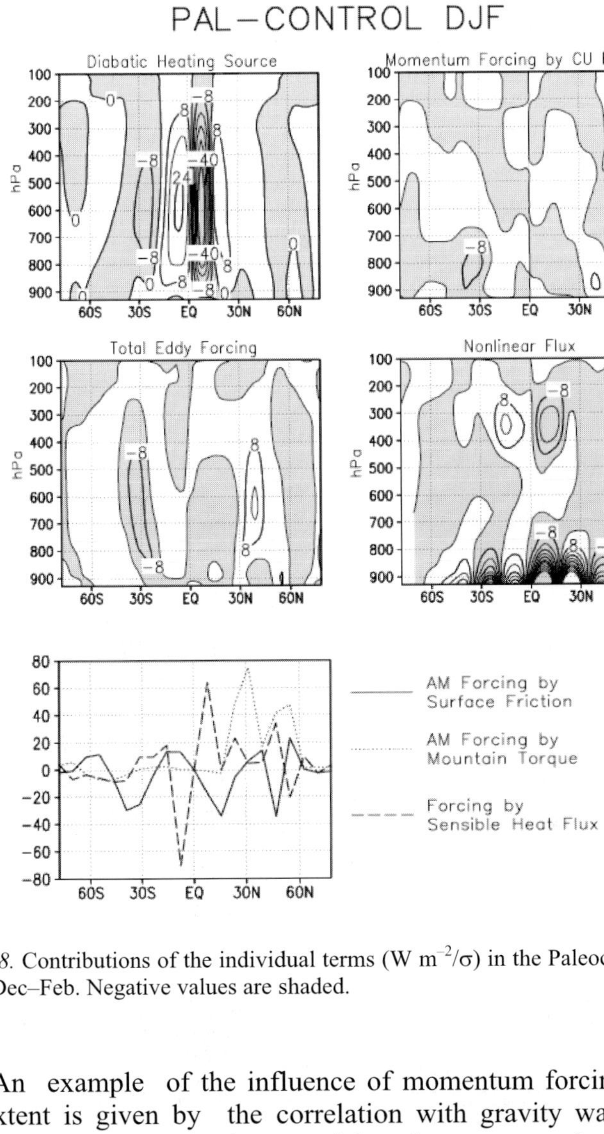

Figure 14-8. Contributions of the individual terms (W m$^{-2}/\sigma$) in the Paleocene simulation for Dec–Feb. Negative values are shaded.

An example of the influence of momentum forcing on the poleward extent is given by the correlation with gravity wave drag in the Northern Hemisphere. The increased ice sheet topography at upper mid-latitudes in the last glacial maximum runs also gave rise to increased parameterized mountain wave drag in the lower stratosphere. This parameterization is a function of the roughness (standard deviation) of the topography. Because that is not well known for the big ice sheets (e.g., Laurentide), the ice age experiments were run in two different ways. The standard simulations assumed that this standard deviation is similar to that of current-day ice sheets of equivalent height (even though those ice sheets, on Greenland and Antarctica, are in regions of actual mountains); this assumption provided a large increase in mountain wave drag. Alternatively, they were assumed to be plateau-like, and thus relatively smooth, with little increase in mountain drag.

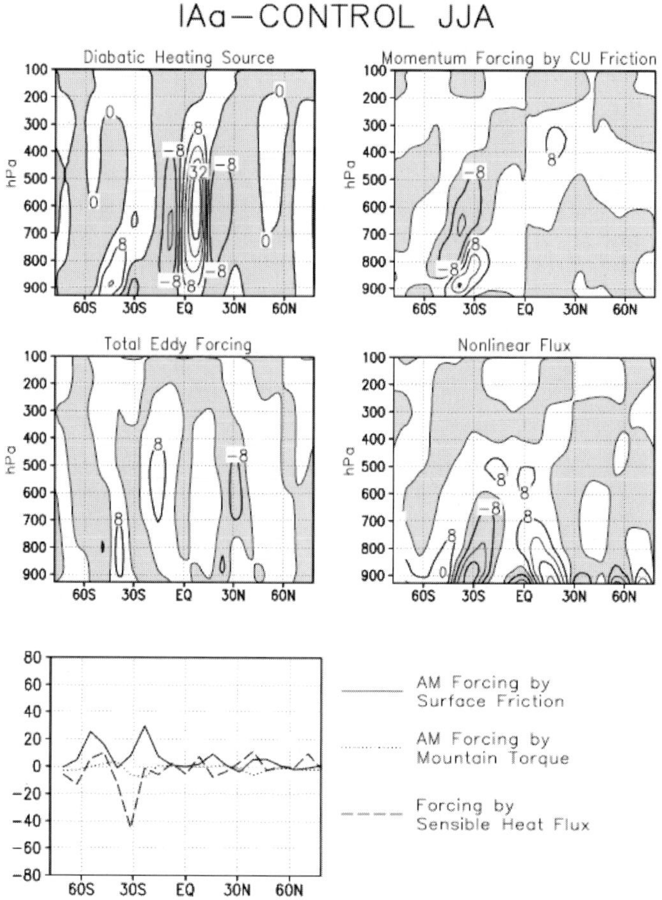

Figure 14-9. Contributions of the individual terms (W m^{-2}/σ) during the alternate ice age for Jun–Aug. Negative values are shaded.

The difference in results between these two formulations is shown in Figure 14-10 for both the standard and alternate ice age sea surface temperature simulations. Figure 14-10 (top left) shows the change during DJF of the gravity wave drag between the rough and smooth simulations, with much larger reduction of momentum in the lower stratosphere in both cases of rough topography. Since the model attempts to conserve angular momentum, there is a corresponding increase in angular momentum input by the surface friction at the corresponding extratropical latitudes (Fig. 14-10, top right). Figure 14-10 (bottom left) gives the sea level pressure (SLP) differences responsible for this surface friction change. In order for the angular momentum to be increased by surface friction, there have to be relatively greater east winds at the surface, produced by having higher pressure to the north, and lower pressure to the south, as occurred. The vertical motion field change that would produce this pressure response consists of subsidence at higher latitudes and rising air at lower latitudes. This more direct circulation is illustrated by the vertically integrated stream function change (Fig. 14-10, bottom right), with the more negative values poleward of 40°N implying an extension of the Hadley cell direct circulation to much higher latitudes.

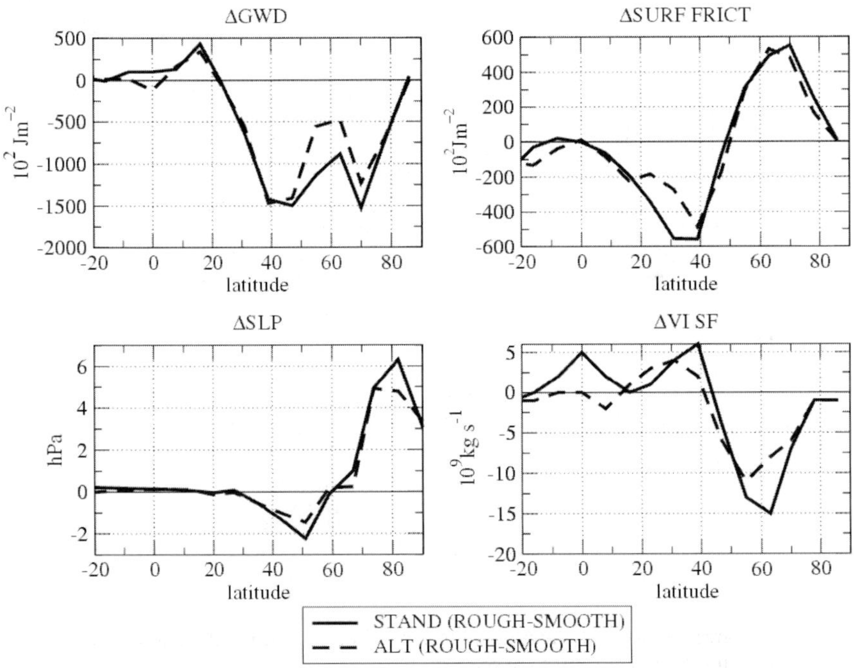

Figure 14-10. Differences between the rough ice sheet and smooth ice sheet ice age experiments for: the change of angular momentum by gravity wave drag (top left); change of angular momentum by surface friction (top right); sea level pressure (bottom left); vertically integrated stream function (bottom right).

4. DISCUSSION

4.1. The Importance of Sea Surface Temperature Gradient—Model Dependency

The model experiments over this wide range of paleo/future climates show no simple relationship between the Hadley cell intensity and the global or equatorial temperature magnitudes, responding instead to the latitudinal temperature gradient from the tropics through the subtropics. Some theories of this region of the globe assume that the gradient is negligible, perhaps minimized by radiation feedback (e.g., Pierrehumbert 1995). The so-called "weak temperature gradient" approximation has been proposed as a balance model for the tropics (Sobel and Bretherton 2000), and Polvani and Sobel (2002) utilize it while calculating the Hadley cell response to heating. Although this gradient is smaller than that across the mid-latitudes, it still controls the model's precipitation gradient, and thus to a good extent the heating gradient, involving interactive feedbacks with the vertical motion field, convection, and Hadley cell intensity.

The effect is so strong that the Hadley cell differences between the alternate SST depictions for the individual climates often exceed the differences between the different climate regimes. Similarly, the change in one solstice season bears little relationship to the change in the other, given how the gradients can differ seasonally. During Northern Hemisphere summer, the temperature (and precipitation) response over the subtropical land areas also affects the intensity, lowering the correlations with SST gradients (and emphasizing the importance of monsoon circulations for that season, as in Dima and Wallace [2003]).

It is not the latitudinal gradient per se that is driving the Hadley circulation, but the precipitation feedback response to the induced circulation. To the extent that the analyses by Schneider and Lindzen (1977) and Schneider (1977) using an axisymmetric model are relevant, the latitudinal gradient itself initially drives a weak low-level circulation via sensible heating. The convective feedback with latent heat release, and to some extent cumulus friction, amplifies the gradient effects and the circulation in both intensity and depth. Schneider (1977) calculated that the low-level circulation driven by the gradient was sufficient to provide for enough moisture convergence to generate a significant Intertropical Convergence Zone (ITCZ).

How general are these results? In idealized experiments performed with a finer resolution (and newer) version of the GISS GCM, sea surface temperature gradients were increased (decreased) in a linear fashion as a function of latitude without altering the mean temperature, and also warmed (cooled) while maintaining the same (altered) temperature gradient (Rind 1998). The January and July Hadley cell intensities were then compared. Qualitatively similar to the results shown here, in that study, with the increased gradient in January, the Hadley cell increased by 15%; with the decreased gradient, it decreased by 12%. Simply warming the climate decreased the intensity by about 9% (with either gradient). In July the sensitivities were reduced: +3% for the increased gradient, –1% for the decreased gradient, and ±10% with warming. As in the results shown here, the July situation is further complicated by the upward motion generated in the northern subtropics due to heating over land. The poleward extent was basically unaffected by the latitudinal gradient change, but it was increased by a few degrees of latitude in the warmer climate, an effect that is apparently overridden by the other changes in the warm (cold) experiments presented here.

The magnitude of the Hadley cell response to SST gradients does appear to be model dependent. In several recent publications (Rind et al. 2001a, 2002a), we have noted that the sensitivity of a model to sea surface temperature variations varies with its control run characteristics. If the model's boundary layer scheme does not easily transfer these surface changes into the free atmosphere, perhaps due to weak surface winds, and if the convection scheme does not translate lower atmosphere changes into the middle and upper troposphere, perhaps due to reduced mass fluxes or lack of penetrating convection, the atmospheric results including the Hadley circulation can be much less sensitive to sea surface temperature variations. With respect to all these features, the model used for these experiments appears to be quite sensitive, although as indicated in the discussion of Figure 14-1, its sensitivity to SST variations is similar to that in the NCEP/NCAR reanalysis data.

In particular, the same altered SST gradients imposed here in 2CO2 were also utilized in a newer version of the GISS GCMAM (Rind et al. 2002a). Its Hadley cell response had certain similarities and some differences with the results presented here, a result as discussed in that paper that was influenced by the differences in physics parameterizations. Nevertheless, in comparison to a run with weaker SST gradients, its Hadley cell was still amplified during DJF.

Doubled CO_2 simulations by other modeling groups often show an increase in Hadley cell intensity, although the specific characteristics of the changes differ greatly. For example, Watterson et al. (1995), using the CSIRO GCM, found that a Hadley cell shift upwards was associated with increased latent heat release (similar to the results shown for the doubled

CO_2 experiments in Fig. 14-4), and there was also a poleward shift in the regions of tropical ascent. In contrast, Dettinger et al.[1], using the NCAR PCM model, found an equatorward shift in the region of ascent, with poleward shifts in the region of descent. Douville et al (2002) with the CNRM GCM and Thompson and Pollard (1995) with the GENESIS model found an intensified Hadley circulation, although in the latter study that occurred only with a penetrating convection scheme. Ramstein et al. (1998) found that the LMD model simulated a 10% reduction in intensity for doubled CO_2. None of these authors clearly relate the results to the pattern of sea surface temperature changes, in particular to the magnitude of tropical warming, a feature that differs strongly from model to model (Rind 1987) and so may be an influencing factor in model differences.

A more direct comparison can be made with the results of Magnusdottir (2001). With the NCAR CCM3 model, the SST gradient was changed so that it was increased from 10°S to about 12°N, and then decreased to 30°N. The Hadley cell response in DJF mimicked the gradient change: increased intensity from 5°S to 10°N, and decreased intensity from 10°N to 30°N. The results were reversed when these gradient changes were reversed. The gradient changes used were those estimated from annual average changes between 1903 and 1994; they have the same pattern over the tropical region as the 2CO2 results (see Rind et al. 2002b, their Fig. 1) and produced a Hadley cell response qualitatively similar to that shown in Figure 14-4a. Our GCM experiments show that the precise patterns of SST change, including the small (latitudinal average) peaks in SST, have a dominant influence on the (modeled) Hadley cell, especially in DJF, that is qualitatively similar to the response of axisymmetric models (Schneider and Lindzen 1977; Schneider 1977; Lindzen and Hou 1988).

4.2. Budget Implications

As was discussed in the introduction, Pfeffer (1981) and others have used the budget terms to imply causality; however, Equation (2), whether for the current climate or a climate change situation, simply implies equivalence. The results for the different climates represent the equilibrium response of the model (or the observations) consistent with the climatological stream function, arrived at theoretically via interaction with the "forcing" terms. To the extent that any stream function change does not in itself alter the forcing terms, the distinction would be minimal. For example, the equilibrium radiative budget for a 2% solar irradiance increase

---

[1] Paper presented at the Hadley Circulation Workshop, Honolulu, Hawaii, November 2002. See Diaz and Bradley (2003) *EOS Newsletter 84(16)* 22 April 2003 for a summary.

would show no net imbalance, as the outgoing long-wave radiation would have adjusted to the temperature increase, but it would still indicate a 2% increase in solar irradiance impinging upon the upper atmosphere. How important, then, are the Hadley cell changes for the individual budget terms?

Hadley cell changes can influence the magnitude of dQ/dy by altering the vertical motion field, affecting convective heat release, although in these experiments they cannot alter the sea surface temperature gradients, which are specified. The sea surface temperature gradients are strongly associated with the resulting Hadley cell changes, implying that they are forcing the Hadley cell response in these experiments. In the real world, for both current and paleoclimates, the boundary forcing terms would be interactive with the Hadley cell, especially at low and subtropical latitudes.

Similarly, subsidence in the subtropics can warm the atmosphere in that region, amplifying the latitudinal temperature gradient across mid-latitudes and influencing baroclinic instability and eddy forcing. While the Hadley cell change can therefore affect the eddies, in these experiments there are very strong changes in the extratropical latitudinal temperature gradient (driven by the specified SSTs and radiative forcing) and topography from the different climate regimes that will likely have a much stronger impact on the eddies. In this sense, the eddy changes are more like the +2% solar irradiance forcing in the example given above, relatively independent of the stream function response, and so more closely represent true "eddy forcing." Hence the different budget terms will have varying degrees of interaction with the stream function changes.

As is implied by the correlations, the diabatic forcing term is driven by the gradient in precipitation, largely via the latent heat of condensation associated primarily with moist convection. Radiation (short-wave plus long-wave) aids the circulation (i.e., dQ/dy) but is much less important in the equilibrium calculation. However, as was noted by Betts and Ridgway (1988), long-wave radiative cooling drives subsidence, which brings dry air down to the boundary layer and is associated with subtropical evaporation, the moisture source for the latent heat release that drives the circulation. A crude calculation, using the approach of Betts and Ridgway (1988), shows that the increased outgoing long-wave radiation in the subtropics in the standard Paleocene simulation (due to the warmer temperatures) results in increased subsidence and hence increased latent heat flux, on the order of 70% of the observed change. For the ice age, with decreased outgoing long-wave radiation, the reduced subsidence would have led to about 50% of the observed latent heat flux decrease. Given that the outgoing radiation change is due to the altered temperatures, which are themselves influenced by the change in condensation and thus latent heat fluxes, this is another example of the interdependence of various forcing terms. Ultimately, of course, radiation differences drive the SST gradients and their changes in the different climate regimes, and the Hadley cell differences.

The sensible heat flux term is proportional to the gradient between the ground and surface air temperatures. Processes that act to warm the atmosphere minimize this flux; hence the sensible heating term in equilibrium works against the condensation term (note in Fig. 14-5 bottom, the contribution to reducing the Hadley cell from the sensible heating term near the equator).

What about the suggested importance of eddy heat fluxes to the intensity of the Hadley cell? The total eddy forcing does affect the model's Hadley cell intensity (Fig. 14-7). This is composed of four terms as is apparent from the definitions of L and G (Eq. 3), involving the convergences of the horizontal and vertical transports of heat and momentum. In both hemispheres, the modeled eddy horizontal heat transport is somewhat more important than the eddy horizontal momentum transport to the total eddy forcing of the Hadley cell, while both vertical transport terms are even less important. This is true both for the control run itself, and for the change between climate experiments. Note that the effect of the model's coarse resolution might influence this result. The contribution of the stationary eddies to the horizontal heat transport varies with climate; in the subtropics during the last glacial maximum, the stationary eddy forcing was more than 50% of the eddy heat transport effect, while the contribution was negligible in the Paleocene (with little topography and stationary wave energy). As a result of this eddy forcing, the Hadley cell intensity does correlate significantly with transient eddy energy (Table 14-3) (in general agreement with the analysis of Trenberth and Stepaniak [2003]; see footnote 1). In both solstice seasons, the eddy forcing of the Hadley cell maximizes poleward of the diabatic source, and is more important for the off-equatorial intensity, a result also seen, to a lesser extent, in the analysis of Pfeffer (1981).

The nonlinear terms augment the Hadley circulation intensity in the lower and upper troposphere, once the circulation is developed. This too is the result of two terms, with the advective thermal flux amplifying the Hadley cell at low levels, and the advection of momentum providing amplification in the lower and especially the upper troposphere. This result occurs in the climate change experiments as well, and in all cases the terms appear to act as positive feedbacks, strengthening the circulations. Similarly, Schneider (1977) calculated that inclusion of the nonlinear advection of momentum by the mean circulation amplified its components in the upper and lower troposphere.

The diabatic forcing, the nonlinear flux response, and the eddy forcing help in generating the Ferrel circulation (Figs. 14-5 through 14-9).

Offhand, one would suppose that with a stronger Ferrel cell, the Hadley cell extent would be reduced. Nevertheless, there is no correlation between transient eddy energy and Hadley cell poleward extent. One reason is that the eddies are generating a stronger subtropical portion of the Hadley cell at the same time that they generate the Ferrel cell, as can be seen in the figures; the effects then nullify one another.

The different experiments have greatly different topography, which could influence the Hadley cell extent through its effects on both standing wave energy and the mountain torque. However, the correlation between standing wave energy and extent is not significant, due to the existence of other forcing factors. The mountain torque in the control run acts to extend the Hadley cell (Fig. 14-5a, bottom), and its lack in the Paleocene therefore helps limit the extent (Fig. 14-8, bottom). As is noted in Table 14-2, the Paleocene has the most equatorward extent, and as can be seen from Figure 14-8, all of the terms contribute to this result.

During the ice ages, the increase in topography is at fairly high latitude, and as big high-pressure systems build over the ice sheets, it is largely within an east wind regime. The increased mountain torque thus acts to limit the Hadley cell extent at lower mid-latitudes, while inducing a negative contribution to the stream function further poleward. As a secondary effect to the mountain wave drag, this action helps to extend the vertically integrated stream function poleward.

4.3. Hadley Cell Changes and Soil Moisture

A prime reason for understanding Hadley cell sensitivity is to gauge how soil moisture changes with climate, given that the current regions of Hadley cell descent are arid, while Hadley cell ascent regions are moist. As is shown in Table 14-3, both the Hadley cell intensity and extent are related to precipitation gradients. However, Hadley cell changes are not a perfect predictor of soil moisture, which is the variable whose effect is most likely observed in paleoclimate studies, and is of most importance for understanding societal impacts. We correlated the precipitation changes during the two solstice seasons with the actual soil moisture changes among the simulations; the results are given in Table 14-4. In about half the cases the correlation was significant at the 95% level. Overall, the latitudinal average precipitation change accounts for about 40% of the variance in soil moisture in the two solstice seasons in the different experiments. Soil moisture is affected by evaporation changes as well, a function of temperature and humidity, and so during the ice age, a precipitation decrease does not necessarily mean a soil moisture reduction. Also, the soil moisture values in individual seasons are affected by precipitation/soil moisture changes in other seasons.

Given that the Hadley cell changes themselves do not account for all the precipitation differences, the results suggest that the importance of Hadley cell changes for paleomoisture variations, while significant, do not account for a majority of the soil moisture variability in the model.

Table 14-4. Correlation between latitudinal precipitation and soil moisture changes in each experiment. Significance level is in parentheses.

Experiment	DJF	JJA
2CO2	0.79 (.001)	0.75 (.005)
2CO2-A	0.59 (.04)	0.33 (0.30)
IA	0.50 (.10)	–0.36 (.25)
IA-A	0.54 (.06)	0.41 (.18)
PAL	0.74 (.006)	0.80 (.001)
PAL-A	0.86 (.0003)	0.65 (.02)

4.4. Longitudinal Variations

In this section, we discuss a number of issues related to longitudinal variations, which can affect the Hadley cell and the interpretation of its change in the paleodata. An overall question concerns the relative importance of zonally averaged circulation changes like the Hadley cell to the actual site-specific results found by paleoclimate researchers. How much of what is seen locally in climate change scenarios is the result of an organized, latitudinal response to large-scale patterns, and how much is determined locally by particulars of geography, nearby ocean currents, etc.? This is an unanswerable question in general; Lonnie Thompson (see footnote 1) showed some ice age results from six geographically spaced tropical ice cores that indicated a general concurrence in response, therefore appearing to represent a true Hadley cell change. The reality is obviously that any particular observational site will represent a mixture of large- and small-scale phenomena, and only multiple sites covering a wide range of longitudes will allow for generalization related to the Hadley cell.

Conversely, when the Hadley cell changes, how representative is that of the change at any particular longitude? We can assess the longitudinal variance in meridional cell response by comparing the changes in upwelling in the tropics. Figure 14-11 shows the results for the December–February time series of the different sets of runs. When the Hadley cell changes are strong (as in PAL, or 2CO2-A), most of the longitudes have a similar response. When the changes are smaller (e.g., the ice age runs), there is much greater inconsistency in longitudinal response. Thus even when the Hadley cell change is statistically significant, it may not have much bearing on the change in the meridional cell or its expected consequences at any particular longitude (location).

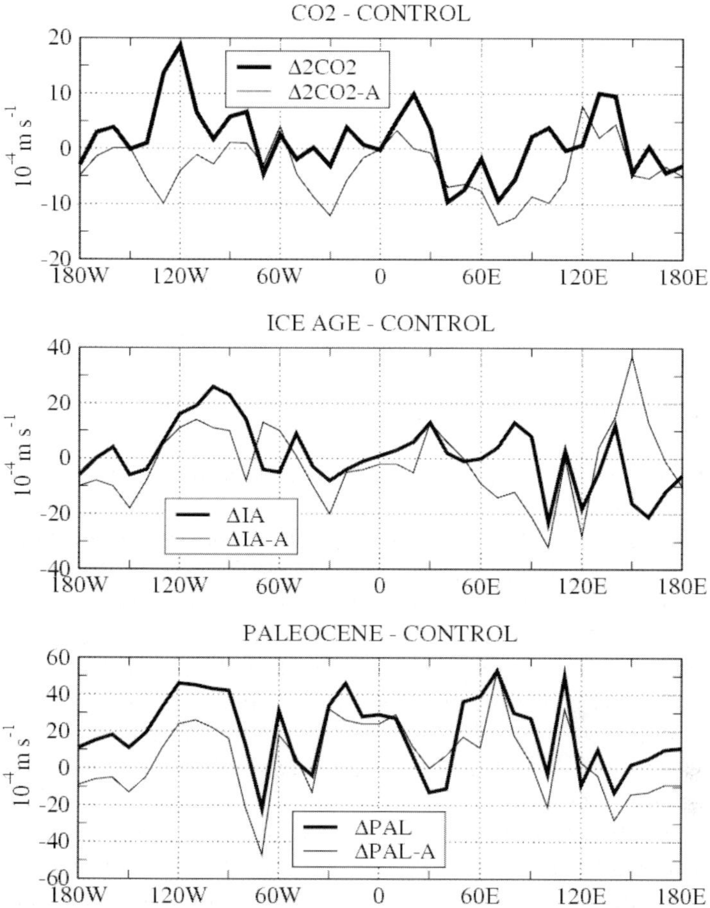

Figure 14-11. Change in tropical vertical velocity (8°N–16°S) in Dec–Feb in the different experiments relative to the control run. Shown are the runs for the two doubled CO_2 experiments (top), ice age experiments (middle), and Paleocene runs (bottom).

It can be seen in Figure 14-11 that there is considerable similarity in the longitudinal response in the two simulations for each of the different climates, and to the extent this is true, one cannot easily use longitudinal variations in paleodata to "back out" the appropriate latitudinal variation in SST. Between the two doubled CO_2 simulations, the longitudinal changes in tropical vertical motion correlate at 0.4 (98% significance), while for the two runs in the Paleocene, the correlation is 0.81 (>99%). Therefore, even when the magnitude of the Hadley cell change is very different due primarily to different latitudinal gradients in SSTs, the longitudinal variation in response has certain similarities, most likely driven by continental position, topography, and perhaps a warming/cooling climate. Cook (2003) noted from GCM experiments that the presence of land with its increased surface friction intensified the Hadley circulation, presumably with preferred longitudinal locations for the direct cell. Nakamura (1978) noted that mountains shift the subtropical high and subtropical subsidence northward via their blocking effect on west winds, requiring geostrophic adjustment of the pressure field. The various angular momentum effects of the continental ice sheets during the LGM were discussed above. The different ice age and Paleocene experiments each have in common the differing land/topography variations from the current climate.

How important are longitudinal variations in SSTs to what happens to the Hadley cell? Longitudinal circulation cells responding to such SST gradients could theoretically divert energy from meridional cells (e.g., during El Niños, both the Hadley cell and Walker circulations are altered). All of the runs discussed differed in their longitudinal gradients across the ocean basins, in somewhat random ways (see Fig. 14-2). The impact on these circulations can be estimated from the tropical vertical motion changes in Figure 14-11; the Walker circulation changes, for example, can be estimated by comparing the results at 90°W–120°W with 140°E–170°E. With this as an index, compared to the control run for DJF, the two Paleocene simulations showed a decrease in the Walker circulation by 70%–80%, IA had a decrease of 58%, and 2CO2 a decrease of 15%. IA-A showed a small increase (9%) with little change in 2CO2-A. When these changes are compared to the Hadley cell peak changes (Table 14-2), there is no apparent relationship.

A complicating factor in the real world is that the relationship is complex: Currently, the Hadley cell intensity is increased during El Niño time periods (e.g., Fig. 14-1). But as is discussed by Gagan and Thompson (Chapter 10, "Evolution of the Indo-Pacific Warm Pool and Hadley-Walker Circulation since the Last Deglaciation," this volume), the response of the Hadley cell to other forcings (such as the precession cycle) can even force the ocean response in the opposite fashion (i.e., greater Hadley cell due to the orbital forcing resulting in more La Niñas).

Finally, what about the longitudinal gradient changes between ocean basins, as might happen with a change in the frequency of El Niños, or reduction in North Atlantic Deep Water (NADW) production and associated ocean heat transports in the Atlantic? To better address this issue, we can use idealized experiments in which the latitudinal gradients in different ocean basins are altered one at a time, or increased in one ocean basin and decreased in another, setting up a change in the longitudinal gradients between the basins (Rind et al. 2001b). The response, compared to altering the gradients similarly in each basin, was that there is now no relationship with the zonally averaged SST gradient. As was discussed in Rind et al. (2001b), when tropical temperatures warm in one ocean basin (i.e., Atlantic or Pacific) relative to the other, a longitudinal circulation cell is set up with relative rising air in the warmer basin and relative descent in the other. In the basin with relative descent, the Hadley cell intensity also diminishes, as the tropical descent mitigates the normally rising air. The net effect is to make the zonally averaged circulation practically independent of the zonally averaged SST gradient. Given the width of the Pacific relative to the Atlantic, the strongest (zonally averaged) Hadley cell occurred with a decreased gradient (i.e., cooler tropics) in the Atlantic, which led to rising air in the Pacific and an overall amplification of the zonally averaged meridional circulation. The weakest Hadley cell response came with an increased gradient in the Atlantic and decreased gradient in the Pacific, as both effects minimized the direct meridional circulation cell in the Pacific. With respect to a particular past or future climate scenario, it is clearly important to know how each of the major ocean basins is responding.

4.5. Comparison with Paleodata

The discussions given above obviously bear on any comparisons of the model Hadley cell response with paleodata. If the variation in soil moisture in any given season is primarily affected by processes other than the latitudinal average precipitation, then Hadley cell changes for individual seasons are not necessarily indicative of what appears in the paleorecord. With this major caveat, we note that for the LGM, in the model the Hadley cell during December–February intensified and expanded with both sets of SSTs, while in June–August it weakened (strengthened) and expanded (little change) in IA (IA-A). Thompson et al. (see footnote 1) showed results from six different tropical/semitropical ice core locations, which implied a weakening of the Hadley cell, a result not really obtained with either SST pattern used (except for Jun–Aug in IA with the CLIMAP data). Ramstein et al. (1998), with the LMD model and CLIMAP (1981) SSTs (hence like IA), also found the Hadley cell intensified, only in their case it occurred especially in June–August (with a 20% increase).

Chylek et al. (2001) suggested that enhancement of dust source areas during past glacial periods implied a contraction of the Hadley cell, inconsistent with IA and to some extent IA-A, although as noted in the Results section, the model effect was largely due to the inclusion of added mountain drag. Without the added drag, there was little change in latitudinal extent with either set of SSTs in DJF. Ramstein et al. (1998) also found that the Hadley cell poleward limit expanded, by about 3° latitude, roughly similar to the values estimated here. To emphasize the model dependency of such specific results, we note that in a finer-resolution GISS model (with a top at 10 mb and hence no explicit gravity wave drag), the ice age Hadley cell was slightly contracted (Rind 1988). Koutavas and Lynch-Stieglitz (Chapter 12, "Variability of the Marine ITCZ over the Eastern Pacific during the Past 30,000 Years: Regional Perspective and Global Context," this volume) concluded that the equatorward extent of the Hadley cell as indicated by the ITCZ location shifted southward during the ice ages, especially during JJA. This result occurred in the model only with the warmer tropical SSTs (IA) (Fig. 14-4b); during December–February it occurred with the cooler values (IA-A).

While there is little direct evidence of Hadley cell changes in the Paleocene, Farrel (1990) has suggested that the more "equable" climates may well have had a more expanded Hadley cell, allowing for the symmetric circulation to transfer energy to high latitudes. With both sets of SSTs, the model's Hadley cell for this time period contracted its poleward extent (Table 14-3 and Fig. 14-4).

5. CONCLUSIONS

There are many suppositions concerning how the Hadley cell responds to climate changes, and the implications for the paleo/future climate of such responses. We explore the general rules governing Hadley cell changes in the context of the various thermal and momentum forcing terms contributing to the mass stream function. We compare general circulation model experiments for the Paleocene (58 million years ago), the last glacial maximum (21,000 BP), and the doubled CO_2 climate with respect to the current climate. In addition to the standard experiments, each simulation is also run with an altered prescription of sea surface temperatures incorporating a greater tropical sensitivity. Finally, we comment on what the Hadley cell changes actually mean with respect to soil moisture variations that might appear in the paleorecord.

The primary results are given below.

With respect to Hadley cell intensity:

- There is little relationship between the intensity and the magnitude of the global or equatorial temperatures.

- The peak intensity is most strongly related to the gradients of SST between the tropics and subtropics, which affect the gradient of latent heat release.

- When the SST latitudinal gradient change differs strongly between ocean basins, the Hadley cell is much less related to the overall gradient.

- Eddy forcing, particularly transient eddy transports of heat, affects the off-equatorial Hadley cell intensity.

- There is no significant correlation between the Hadley cell intensity in the two winter hemispheres.

With respect to Hadley cell extent:

- The Hadley cell poleward extent is also not simply related to the mean temperature, being affected by numerous processes, including those associated with the altered topography in these experiments.

- In particular, the extent is not strongly correlated with the equator-to-pole temperature gradient, as has been postulated from simple axisymmetric models.

With respect to paleoclimate interpretations:

- When the Hadley cell changes are strong, there is good coherence between the sense of the local meridional circulation change (at any particular longitude) and the overall response.

- In experiments with weaker, though still significant changes, there is much variation in longitudinal response, making the Hadley cell a poor interpreter of local paleodata.

- The Hadley cell change, while significantly correlated with seasonal soil moisture changes over land, does not account for a majority of the model's percentage changes due to conflicting influences of precipitation and temperature variations, and the influence of soil moisture changes in preceding seasons.

- The model results do not agree particularly well with the [rather?] uncertain paleoclimate interpretations of Hadley cell change.

The various budget terms indicate all the processes that must be appropriately modeled for the Hadley cell and its change to be successfully simulated. A number of these processes are affected by SST boundary conditions and topography; thus uncertainties in these input fields will affect the results. As is shown here, uncertainties especially in tropical SSTs can produce larger Hadley cell differences for a particular climate than occur between climate regimes (e.g., the ice ages during June–August, the Paleocene during December–February). The relevant processes are also affected by model parameterizations and resolution, so the different results obtained with different models are not surprising, although there may well be several common denominators in model simulations: the relationship of Hadley cell intensity to the SST gradient between the tropics and subtropics, especially in December–February, and the influence of the monsoon, especially in June–August. Comparison with paleodata interpretations of Hadley cell response is complicated by the uncertain influence of Hadley cell changes on local field data. Understanding the Hadley cell response in past climates, and using it to "back-out" the tropical sensitivity remains a challenging problem, as is predicting the future response of Hadley cell changes and their climatic implications.

6. ACKNOWLEDGMENTS

We thank Patrick Lonergan for help in obtaining climate model diagnostics and figure preparation. Climate modeling at GISS is supported by the NASA Climate Program Office, while the development and use of the stratospheric model for climate change experiments is funded by the NASA ACMAP program.

7. REFERENCES

Becker, E., and G. Schmitz. 2001. Interaction between extratropical stationary waves and the zonal mean circulation. *Journal of the Atmospheric Sciences* 58: 462–480.

Betts, A.K., and W. Ridgway. 1988. Coupling of the radiative, convective and surface fluxes over the equatorial Pacific. *Journal of the Atmospheric Sciences* 45: 522–536.

Chylek, P., L. Lseins, and U. Lohmann. 2001. Enhancement of dust source area during past glacial periods due to changes of the Hadley circulation. *Journal of Geophysical Research* 106: 18477–18485.

CLIMAP Project Members. 1981. Seasonal reconstructions of the Earth's surface at the last glacial maximum. Geological Society of America, Map and Chart Series, MC-36.

Cook, K.H. 2003. Role of continents in driving the Hadley Cells. *Journal of the Atmospheric Sciences* 60: 957–976.

Dima, I.M., and J.M. Wallace. 2003. On the seasonality of the Hadley Cell. *Journal of the Atmospheric Sciences* 60: 1522–1527.

Douville, H., F. Chauvin, S. Planton, J.-F. Royer, D. Salas-Melia, and S. Tyteca. 2002. Sensitivity of the hydrological cycle to increasing amounts of greenhouse gases and aerosols. *Climate Dynamics* 20: 45–68.

Fang, M., and K.K. Tung. 1997. The dependence of the Hadley circulation on the thermal relaxation time. *Journal of the Atmospheric Sciences* 54: 1379–1384.

Fang, M., and K.K. Tung. 1999. Time-dependent non-linear Hadley circulation. *Journal of Atmospheric Sciences* 56: 1797–1807.

Farrell, B.F. 1990. Equable climate dynamics. *Journal of the Atmospheric Sciences* 47: 2986–2995.

Hansen, J., G. Russell, D. Rind, P. Stone, A. Lacis, S. Lebedeff, R. Ruedy and L. Travis. 1983. Efficient three dimensional global models for climate studies: Models I and II. *Monthly Weather Review* 111: 609–662.

Held, I.M., and A.Y. Hou. 1980. Nonlinear axially symmetric circulations in a nearly inviscid atmosphere. *Journal of the Atmospheric Sciences* 37: 515–533.

IPCC (Intergovernmental Panel on Climate Change). 2001. *Climate Change 2001. The Scientific Basis*. Summary for Policymakers and Technical Summary of the Working Group I Report. Cambridge: Cambridge University Press, 98 pp.

Kim, H.K., and S. Lee. 2001. Hadley cell dynamics in a primitive equation model. Part II: Nonaxisymmetric flow. *Journal of the Atmospheric Sciences* 58: 2859–2871.

Lindzen, R.S., and A.Y. Hou. 1988. Hadley circulations for zonally-averaged heating centered off the equator. *Journal of the Atmospheric Sciences* 45: 2416–2427.

Magnusdottir, G. 2001. The modeled response of the mean winter circulation to zonally averaged SST trends. *Journal of Climate* 14: 4166–4190.

Nakamura, M. 1978. Dynamic effects of mountains on the general circulation of the atmosphere. IV. Effects on the general circulation of the baroclinic atmosphere. *Journal of the Meteorological Society of Japan* 56: 353–366.

O'Connell, S., M.A. Chandler, and R. Ruedy. 1996. Implications for the creation of warm saline deep water: Late Paleocene reconstructions and global climate model simulations. *GSA Bulletin* 108: 270–284.

Oort, A.H., and J.J. Yienger. 1996. Observed interannual variability in the Hadley circulation and its connection to ENSO. *Journal of Climate* 9: 2751–2767.

Pfeffer, R.L. 1981. Wave-mean flow interactions in the atmosphere. *Journal of the Atmospheric Sciences* 38: 1340–1359.

Pierrehumbert, R.T. 1995. Thermostats, radiator fins, and the local runaway greenhouse. *Journal of Atmospheric Science* 52: 1784–1806.

Polvani, L.M., and A.H. Sobel. 2002. The Hadley circulation and the weak temperature gradient approximation. *Journal of the Atmospheric Sciences* 59: 1744–1752.

Ramstein, G., Y. Serafini-Le Treut, J. Le Treut, M. Forichon, and S. Joussaume. 1998. Cloud processes associated with past and future climate changes. *Climate Dynamics* 14: 233–247.

Rind, D. 1987. The doubled CO_2 climate: Impact of the sea surface temperature gradient. *Journal of the Atmospheric Sciences* 44: 3235–3268.

Rind, D. 1988. Dependence of warm and cold climate depiction on climate model resolution. *Journal of Climate* 10: 965–997.

Rind, D. 1998. Latitudinal temperature gradient and climate change. *Journal of Geophysical Research* 103: 5943–5971.

Rind, D. 2000. Relating paleoclimate data and past temperature gradients: Some suggestive rules. *Quaternary Science Reviews* 19: 382–390.

Rind, D., and D. Peteet. 1985. LGM terrestrial evidence and CLIMAP SSTs: Are they consistent? *Quaternary Research* 24: 1–22.

Rind, D., and W. Rossow. 1984. The effects of physical processes on the Hadley circulation. *Journal of the Atmospheric Sciences* 41: 479–507.

Rind, D., J. Lerner, and C. McLinden. 2001a. Changes of tracer distributions in the doubled CO_2 climate. *Journal of Geophysical Research* 106: 28061–28079.

Rind, D., P. Lonergan, N.K. Balachandran, and D. Shindell. 2002b. 2xCO_2 and solar variability influences on the troposphere through wave-mean flow interactions. *Journal of the Meteorological Society of Japan* 80: 863–876.

Rind, D., P. Lonergan, J. Lerner, and M. Chandler. 2001b. Climate change in the middle atmosphere. Part V: The paleostratosphere in warm and cold climates. *Journal of Geophysical Research* 106: 20195–20212.

Rind, D., J. Perlwitz, J. Lerner, C. McLinden, and M. Prather. 2002a. The sensitivity of tracer transports and stratospheric ozone to sea surface temperature patterns in the doubled CO_2 climate. *Journal of Geophysical Research* 107, doi10.1029/2002JD002483.

Rind, D., R. Suozzo, N.K. Balachandran, and M. Prather. 1990. Climate change and the middle atmosphere. Part 1. The doubled CO_2 climate. *Journal of the Atmospheric Sciences* 47: 475–494.

Rind, D., R. Suozzo, N.K. Balachandran, A. Lacis, and G.L. Russell. 1988. The GISS Global Climate/Middle Atmosphere Model Part I: Model structure and climatology. *Journal of the Atmospheric Sciences* 45: 329–370.

Schneider, E.K. 1977. Axially symmetric steady-state models of the basic state for instability and climate studies. Part II: Nonlinear calculations. *Journal of the Atmospheric Sciences* 34: 280–296.

Schneider, E.K., and R.S. Lindzen. 1977. Axially symmetric steady-state models of the basic state for instability and climate studies. Part I: Linearized calculations. *Journal of the Atmospheric Sciences* 34: 263–279.

Sobel, A.H., and C.S. Bretherton. 2000. Modeling tropical precipitation in a single column. *Journal of Climate* 13: 4378–4392.

Taylor, K.E. 1980. The roles of mean meridional motions and large-scale eddies in zonally averaged circulations. *Journal of the Atmospheric Sciences* 37: 1–19.

Thompson, S.L., and D. Pollard. 1995. A global climate model (GENESIS) with a land-surface transfer scheme (LSX). Part II: CO_2 sensitivity. *Journal of Climate* 8: 1104–1121.

Trenberth, K.E., and D.P. Stepaniak. 2003. Seamless poleward atmospheric energy transports and implications for the Hadley circulation. *Journal of Climate* 16: 3705–3721.

Trenberth, K.E., D.P. Stepaniak, and J.M. Caron. 2000. The global monsoon as seen through the divergent atmospheric circulation. *Journal of Climate* 13: 3969–3993.

Waliser, D.E., Z. Shi, J.R. Lanzante, and A.H. Oort. 1999. The Hadley circulation: Assessing NCEP/NCAR reanalysis and sparse in-situ estimates. *Climate Dynamics* 15: 719–735.

Watterson, I.G., M.R. Dix, H.B. Gordon, and J.L. McGregor. 1995. The CSIRO nine-level atmospheric general circulation model and its equilibrium present and doubled CO_2 climates. *Australian Meteorological Magazine* 44: 111–125.

Webb, R.S., S.J. Lehman, D. Rind, R. Healy, and D. Sigman. 1997. Influence of ocean heat transport on the climate of the Last Glacial Maximum. *Nature* 385: 695–699.

APPENDIX

$\bar{u}$ latitudinal average zonal wind

$\bar{\mathrm{v}}$ latitudinal average meridional wind

$\bar{\omega}$ latitudinal average vertical wind in pressure coordinates

$\bar{\psi}$ mass stream function in pressure coordinates

f coriolis parameter

S_p, σ static stability parameters in pressure coordinates

$\bar{L}$ zonal average eddy momentum flux divergence

$\bar{G}$ zonal average eddy heat flux divergence

$\bar{F}$ zonal average friction

$\bar{Q}$ zonal average heating

Chapter 15

THE SENSITIVITY OF THE HADLEY CIRCULATION TO PAST AND FUTURE FORCINGS IN TWO CLIMATE MODELS

Bette L. Otto-Bliesner[1] and Amy Clement[2]

[1]*Climate and Global Dynamics Division, National Center for Atmospheric Research, Boulder, Colorado 80307, U.S.A.*

[2]*Rosenstiel School of Marine and Atmospheric Sciences, University of Miami, Miami, Florida 33149, U.S.A.*

Abstract

A comparison is made of the Hadley cell response to altered external forcing and climatic boundary conditions in two different climate models. The experiments performed cover past and future climatic forcings that range from the last glacial maximum (LGM) to a doubling in atmospheric CO_2. Both models have a consistent response to all forcings in the depth and meridional extent of the Hadley cell in both solstitial seasons. The strength of the December–January–February (DJF) cell varies consistently between the two models, increasing in colder climates and weakening in warmer climates. For June–July–August (JJA), however, the sign of the response is opposite between models for the range of forcing. It is suggested that the origin of this different response in the JJA cell is related to the type of ocean model used. Nevertheless, the results from the LGM and doubled CO_2 experiments show a consistent correlation between the surface temperature gradient between the winter and summer hemisphere tropics and Hadley cell strength: When the temperature contrast is larger (smaller), the Hadley cell is stronger (weaker). The Holocene, with the response to the forcing peaking over the subtropical and mid-latitude continents, has temperature gradients and Hadley cell responses that deviate from this relationship. The implications of these results to paleoclimate studies are discussed.

H.F. Diaz and R.S. Bradley (eds.), The Hadley Circulation: Present, Past and Future, 437–464.

1. INTRODUCTION

The Hadley circulation, first proposed in 1735 by George Hadley to explain the surface trade winds, is the primary agent of poleward atmospheric heat transport in the tropics and subtropics. Thermally direct cells with rising motion at the Intertropical Convergence Zone (ITCZ) and sinking near 30° latitude are driven by latent heating in the tropics and cooling in the subtropics. The rising motion associated with the Hadley circulations migrates across the equator with the mean ITCZ position in the summer hemisphere. Seasonally, there is a strong intensification and predominance of the winter Hadley cells and only weak summer Hadley cells.

Currently, there is no complete theory for what determines the strength and meridional extent of the Hadley cells. Early studies using axisymmetric models (Held and Hou 1980; Lindzen and Hou 1988) have advanced our understanding of the dynamics of the Hadley cell, but do not account for the full interaction between the atmosphere and ocean circulation, atmospheric physics, and radiation. In the absence of theory, one way to assess the controls on the Hadley cell is to consider situations in which the climate is significantly altered. Conceptual models have been advanced that suggest the Hadley cells might expand in warmer climate regimes (Budyko 1974), and that on planets with differing rotation rates such as Venus and Jupiter, the meridional extent of the tropical cells may be significantly different (Hunt 1979; Williams and Holloway 1982).

For past and future climates, the intensity, depth, and poleward extent of the Hadley cells might be expected to change as the mean surface temperature and latitudinal temperature and heating gradients change. These ideas can be tested by using general circulation models (GCMs). Prior model simulations suggest that the Hadley cell intensity is governed by the latitudinal temperature gradient and the Hadley cell extent is controlled by global mean temperature (Rind 1998). The degree to which the response of the Hadley cell extent and intensity is model dependent is still an open question. Here, this question is addressed by comparing the response of the Hadley cell in two different models to past and future climate forcings.

2. MODELS

Two models are used to perform simulations in which the external forcing and climatic boundary conditions of the earth are altered. The models each contain a comprehensive atmospheric model with parameterizations for radiation and convection but are coupled to ocean models of different complexity. One model is a non-flux-corrected, fully coupled ocean-atmosphere general circulation model, the National Center for Atmospheric Research (NCAR) Climate System Model (CSM) (Otto-Bliesner and Brady 2001; Blackmon et al. 2001). CSM includes atmospheric and ocean general circulation models, a land surface biophysics and hydrology model, and a sea ice dynamics and thermodynamics model. The atmosphere is run at T31 resolution, which corresponds to a grid spacing of approximately 3.75° latitude × 3.75° longitude, with 18 vertical levels. The ocean is run with 25 vertical levels, 3.6° longitude grid spacing, and latitude spacing of 1.8° poleward of 30° latitude smoothly decreasing to 0.9° within 10° of the equator. See Kiehl et al. (1998), Bonan (1998), Gent et al. (1998), and Weatherly et al. (1998) for descriptions of the individual component models of CSM.

The second model is the Geophysical Fluid Dynamics Laboratory (GFDL) atmospheric general circulation model coupled to a mixed-layer ocean model with a heat and water balance model over the continents. The atmosphere is run at R30 resolution, which corresponds to a grid spacing of approximately 2.25° latitude × 3.75° longitude, with 20 vertical levels. The ocean mixed layer is treated as a static layer 50 m deep having a uniform temperature with depth. To represent the horizontal heat transport by ocean currents, an additional heat flux term is applied in the ocean that varies geographically and seasonally, but not from year to year. See Broccoli (2000) for additional model details.

3. EXPERIMENTS

Both models were forced with past and future climatic forcings that range from the last glacial maximum (LGM, approximately 21,000 BP) to a doubling of atmospheric CO_2. The control simulation for both models, termed "Modern" in this chapter, represents preindustrial climate conditions with atmospheric trace gases set to values appropriate for ~AD 1800. For atmospheric CO_2, the NCAR model uses 280 parts per million by volume (ppmv) and the GFDL model uses 300 ppmv for Modern conditions. For the 2 x CO_2 simulations, these values were doubled. For the LGM, both models adopted the reduced levels of atmospheric trace gases sampled in ice cores (Raynaud et al. 1993); lowered sea level by 105 m, exposing the continental shelves; and prescribed continental ice sheets covering North America and Eurasia (Peltier 1994). The small changes to the incoming solar radiation (Fig. 15-1) associated with the Milankovitch cycles of the earth's orbit are also included (Berger 1978)

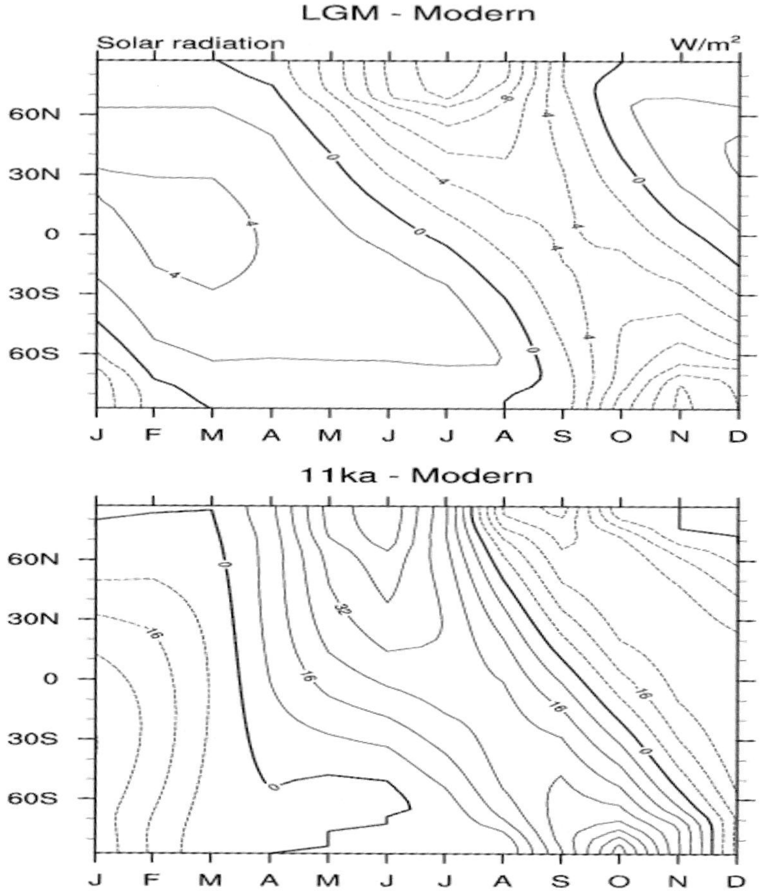

Figure 15-1. Anomalies (W m^{-2}) of solar radiation incoming at the top of the atmosphere. Top: LGM minus Modern. Bottom: 11,000 BP minus Modern.

The primary forcing during the early Holocene (11,000 BP) relative to Modern is the large solar insolation changes at the top of the atmosphere associated with the Milankovitch cycles. At 11,000 BP, obliquity (tilt) was 24.20° (the modern value is 23.45°), perihelion was July 1 (the modern value is January 3), and eccentricity was 0.0195 (the modern value is 0.0167). In the GFDL model, only changes in solar insolation associated with the change in perihelion date are included in the 11,000 BP run; exclusion of the effect of obliquity has only a minor impact on the solar forcing.

Annual mean changes in solar insolation at the top of the atmosphere are small, ranging from –1 W m^{-2} in the tropics to +5 Wm^{-2} at the poles. Seasonal changes are much larger (Fig. 15-1). In the Northern Hemisphere, positive anomalies exceed 16 W m^{-2} from April through July, with comparable negative anomalies in boreal winter. In the Southern Hemisphere, large positive anomalies occur in austral late winter and spring. In both models, atmospheric trace gases and land ice are prescribed to be the same as for Modern, and the residual ice sheet that remained over North America at 11,000 BP is not considered in these simulations.

In this chapter, the wide range of simulations in two different models allows an assessment of the extent to which results may be model dependent, but also to what extent the sensitivity of the Hadley cell may depend on the particular type of climatic forcing. Other aspects of the simulations are documented in Broccoli (2000), Otto-Bliesner et al. (2003), Liu et al. (2003), and Shin et al. (2003).

4. MODERN CLIMATE

4.1. NCAR Model

The zonal mean meridional stream functions simulated by the NCAR model for Modern show the predominance of the strong winter Hadley cell in each hemisphere with mass fluxes over 19 x 10^{10} kg s^{-1} (Fig. 15-2; Table 15-1). The December–January–February (DJF) winter cell has a maximum of 23.6 x 10^{10} kg s^{-1} located at 5°N latitude in the zonal average. The ascending branch of the tropical Hadley cell occurs in the southern tropics in association with a maximum of precipitation associated with the Intertropical Convergence Zone precipitation predicted by the NCAR model (Fig. 15-3). The June–July–August (JJA) winter cell is somewhat weaker with a maximum of 20.2 x 10^{10} kg s^{-1} located at 5°S latitude in the zonal average. The ascending branch of the tropical Hadley cell shifts to the northern tropics in association with a shift in the ITCZ precipitation to north of the equator. The summer Hadley cells are significantly weaker, with mass fluxes of about 5 x 10^{10} kg s^{-1}. The centers of the circulation are near 700 mb in both seasons, somewhat lower than indicated by observational estimates. The widths of the winter cells are 40°–45° latitude, comparable to those observed.

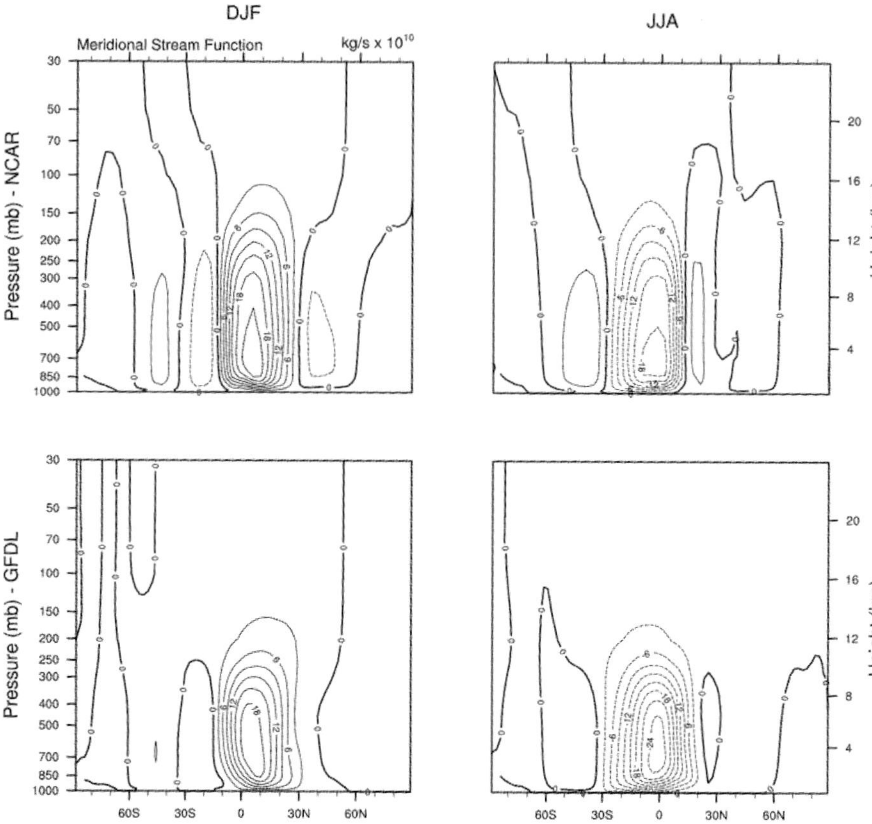

Figure 15-2. Modern meridional stream function (in 10^{10} kg/s) for DJF and JJA as simulated by the NCAR and GFDL models. Positive values indicate clockwise circulations.

4.2. GFDL Model

To first order, the simulated Modern seasonal Hadley circulations and precipitation patterns in the GFDL model are similar to those for the NCAR model (Fig. 15-2; Table 15-2). There are, however, some interesting differences. First, in the GFDL model, the DJF cell is weaker than the JJA cell while the opposite is true for the NCAR model. Also, the JJA cell has a significantly wider meridional extent in the GFDL model than in the NCAR model. Both of these differences are consistent with the different simulations of the Northern Hemisphere monsoons and JJA precipitation in the two models (Fig. 15-3). In the GFDL model, there is a broader maximum of precipitation over the western tropical Pacific and extending into the Asian con-

Table 15-1. Meridional stream function (Ψ, 10^{10} kg s^{-1}) and surface temperature (T_s, K) for LGM, 11,000 BP, Modern, and 2 x CO_2 (top) and corresponding changes from Modern (bottom) as simulated by the NCAR model.
All meridional stream function values are for the winter Hadley cell in the tropics. Maximum (10^{10} kg/s) represents the absolute value of the maximum strength of this cell. Width (degrees of latitude) of this cell is defined by latitudes where the mid-tropospheric stream function at 500 mb goes to zero. Depth (mb) of this cell is defined as the pressure level where the meridional stream function of this cell is 10% of its maximum.
Surface air temperature means are calculated as tropical (20°S to 20°N), southern tropical (20°S to equator), northern tropical (equator to 20°N), northern extratropical (20°N to 90°N), and southern extratropical (20°S to 90°S).

	LGM	11,000 BP	Modern	2 x CO_2
Ψ_{DJF} max	26.3	25.3	23.6	22.3
Ψ_{DJF} width	42	42	42	42
Ψ_{DJF} depth	110	107	105	98
Ψ_{JJA} max	22.5	22.1	20.2	19.2
Ψ_{JJA} width	41	43	41	41
Ψ_{JJA} depth	126	120	118	113
T_s annual global mean	282.9	287.0	287.3	289.0
T_s annual tropical mean	298.2	299.6	300.1	301.3
T_s DJF tropical mean	298.0	298.8	299.9	301.1
T_s DJF northern extratropical mean	264.0	271.7	272.8	276.0
T_s JJA tropical mean	297.9	299.8	299.8	301.0
T_s JJA southern extratropical mean	272.4	277.5	277.5	279.2
T_s DJF tropical minus northern extratropical mean	34.0	27.1	27.1	25.1
T_s JJA tropical minus southern extratropical mean	25.5	22.3	22.3	21.8
T_s DJF southern tropical minus northern tropical mean	2.25	2.32	1.79	1.63
T_s JJA northern tropical minus southern tropical mean	1.94	1.84	1.87	1.80
Differences (percent change or difference [Δ]) from modern				
Percent change Ψ_{DJF}	+11%	+7%		–6%
Percent change Ψ_{JJA}	+11%	+9%		–5%
ΔT_s DJF tropical mean	–1.9	–1.1		+1.2
ΔT_s DJF northern extratropical mean	–8.8	–1.1		+3.2
ΔT_s JJA tropical mean	–1.9	0		+1.2
ΔT_s JJA southern extratropical mean	–5.1	0		+1.7

tinent. In JJA, the zonal mean Hadley cell is dominated by the Asian monsoon, as this is by far the largest heat source for the atmosphere in that season (Trenberth et al. 2000).

Table 15-2. Meridional stream function (Ψ, 10^{10} kg s^{-1}) and surface temperature (T_s, K) for LGM, 11,000 BP, Modern, and 2 x CO_2 (top) and corresponding changes from Modern (bottom) as simulated by the GFDL model. See Table 15-1 for legend.

	LGM	11,000 BP	Modern	2 x CO_2
Ψ_{DJF} max	23.3	21.1	20.1	17.1
Ψ_{DJF} width	43	42	42	42
Ψ_{DJF} depth	124	114	94	65
Ψ_{JJA} max	24.2	25.6	25.7	27.8
Ψ_{JJA} width	47	48	49	49
Ψ_{JJA} depth	158	152	155	153
T_s annual global mean	284.3	288.2	288.3	291.5
T_s annual tropical mean	298.9	300.6	300.7	302.7
T_s DJF tropical mean	298.5	299.9	300.4	302.4
T_s DJF northern extratropical mean	262.2	271.3	271.7	277.0
T_s JJA tropical mean	298.6	300.5	300.3	302.4
T_s JJA southern extratropical mean	274.6	276.8	277.3	280.8
T_s DJF tropical minus northern extratropical mean	36.3	28.6	28.7	25.4
T_s JJA tropical minus southern extratropical mean	24.0	23.7	23.0	21.6
T_s DJF southern tropical minus northern tropical mean	1.0	0.73	0.52	0.01
T_s JJA northern tropical minus southern tropical mean	2.0	2.0	2.3	2.5
Differences (percent or Δ) from modern				
Percent change Ψ_{DJF}	+16%	+5%		–15%
Percent change Ψ_{JJA}	–6%	–1%		+8%
ΔT_s DJF tropical mean	–1.9	–0.5		+2.0
ΔT_s DJF northern extratropical mean	–9.5	–0.4		+5.3
ΔT_s JJA tropical mean	–1.7	+0.2		+2.1
ΔT_s JJA southern extratropical mean	–2.7	–0.5		+3.5

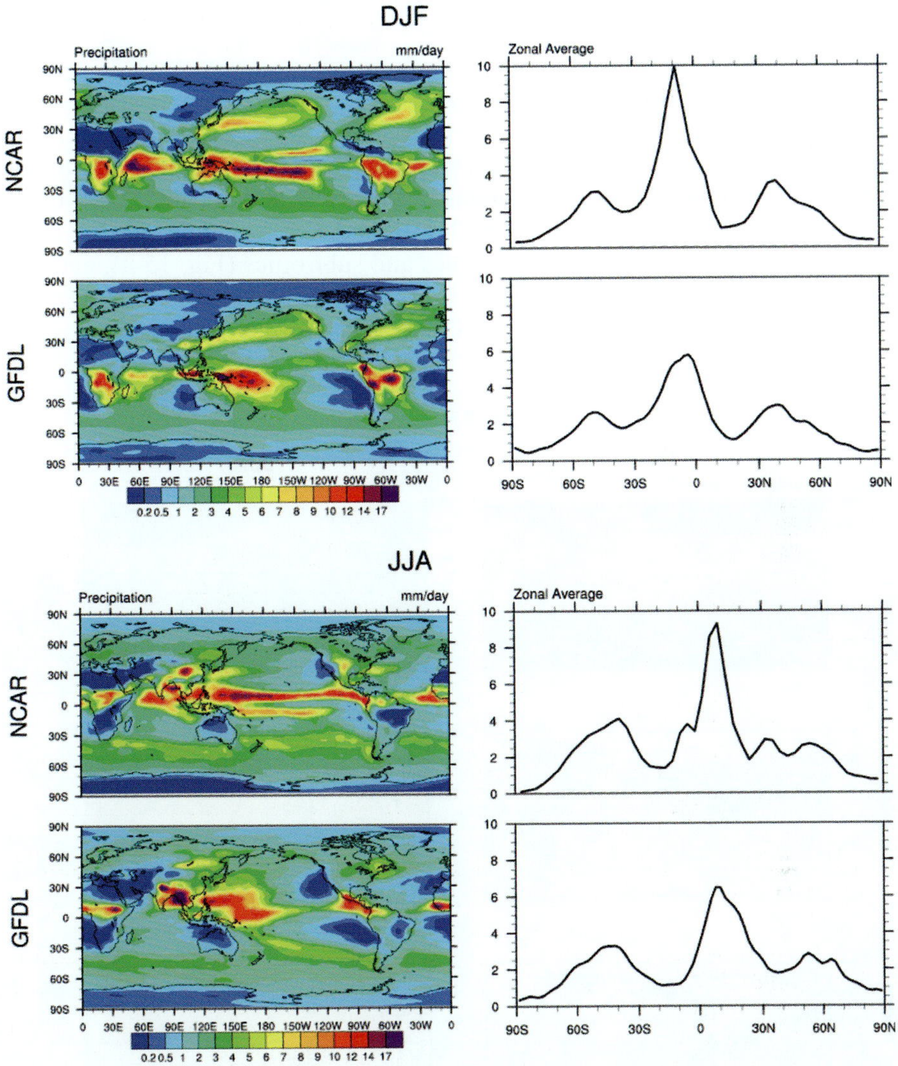

Figure 15-3. Modern precipitation (mm/d) for DJF and JJA as simulated by NCAR and GFDL models.

5. SURFACE TEMPERATURE RESPONSE TO CHANGED FORCING

5.1. NCAR Model

The patterns of surface temperature change for the past and future scenario simulations result in changes in the gradients. For the LGM, cooling occurs at all latitudes and for both DJF and JJA, with enhanced cooling at high latitudes compared to the tropics and subtropics (Fig. 15-4), and larger latitudinal temperature gradients (Fig. 15-5; Table 15-1). Greatest cooling occurs in the winter hemisphere, DJF for the Northern Hemisphere and

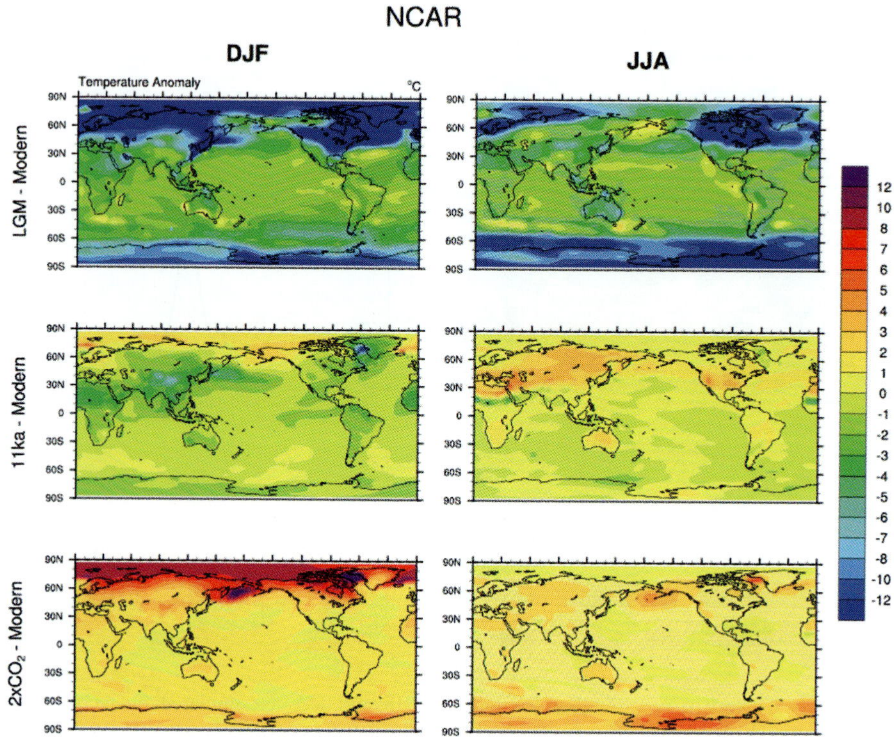

Figure 15-4. Surface temperature anomalies (°C) as compared to Modern for DJF and JJA as simulated by the NCAR model for LGM, 11,000 BP, and 2 x CO_2.

JJA for the Southern Hemisphere, exceeding 15°C cooling at the respective poles. Significant sea ice expansion occurs for both hemispheres. The large continental ice sheets over North America and Eurasia force significant

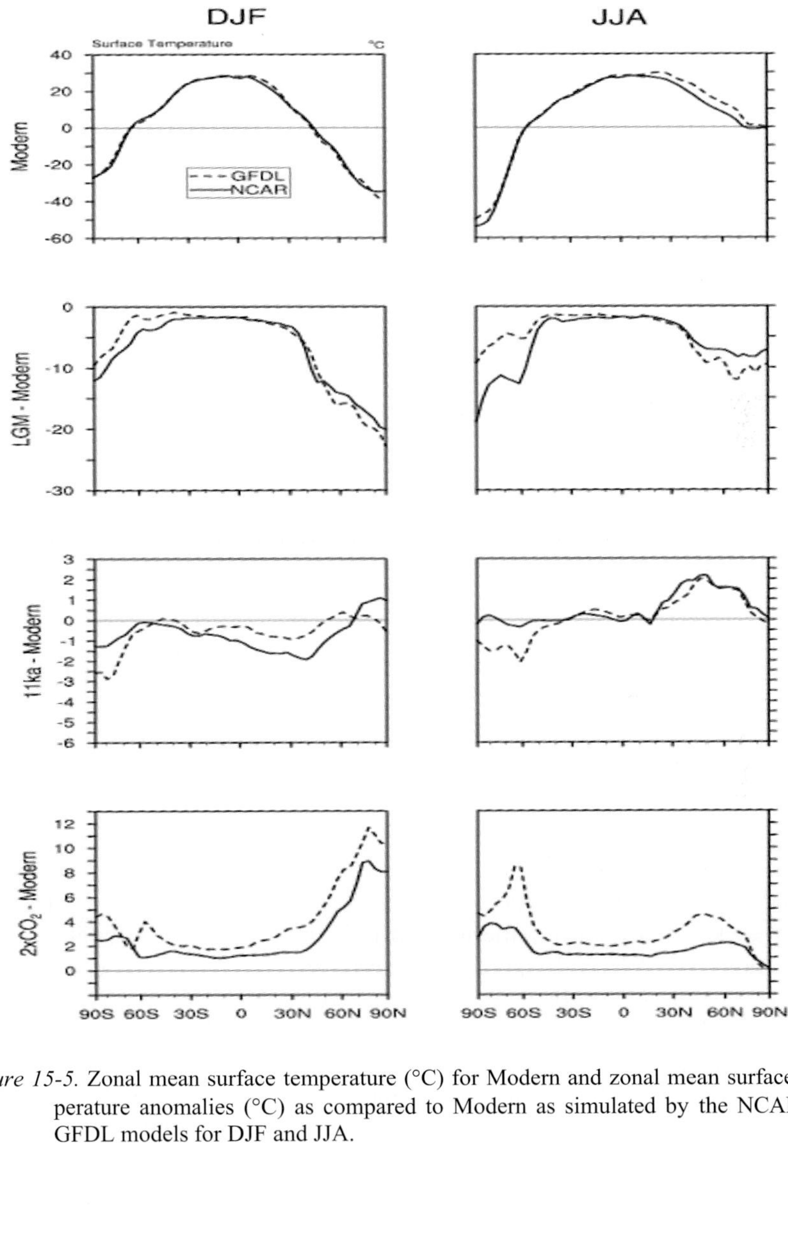

Figure 15-5. Zonal mean surface temperature (°C) for Modern and zonal mean surface temperature anomalies (°C) as compared to Modern as simulated by the NCAR and GFDL models for DJF and JJA.

cooling year-round. The tropics cool on average 1.9°C. Annual global cooling is 4.4°C compared to Modern. For doubled CO_2, surface temperatures warm more at high latitudes than in the tropics and subtropics, with the greatest warming during DJF in the Northern Hemisphere and JJA in the Southern Hemisphere. Hence, latitudinal temperature gradients are reduced. The heating of the large continental landmasses and reduced snow cover and sea ice in DJF results in greater warming at high northern latitudes than corresponding southern latitudes in JJA. High-latitude regions in the Southern Hemisphere show reduced response to the forcing because of the large oceans and prescribed Antarctic ice cap, which is not allowed to melt in these simulations. Annual global warming is 1.7°C compared to Modern.

At 11,000 BP, hemispheric differences in the seasonal surface temperature change exist and are tied to the seasonal nature of the solar forcing and the predominance of oceans in the Southern Hemisphere versus large continents in the Northern Hemisphere. In JJA, the Northern Hemisphere surface temperatures warm in response to the positive anomalies of incoming solar insolation. Northern Hemisphere land areas warm more than nearby oceans, enhancing the land-ocean temperature contrast. The Southern Hemisphere response is minimal except for small warming over Australia, southern Africa, and subtropical latitudes of South America. In DJF, surface temperatures at 11,000 BP cool in response to negative anomalies of incoming solar insolation. The Southern Hemisphere cools by less than the Northern Hemisphere despite larger insolation anomalies. Arctic latitudes remain warmer than for Modern in association with seasonal memory of sea ice. Annual global temperatures are similar at 11,000 BP and Modern.

5.2. GFDL Model

Similar to the NCAR model, the GFDL model responds to LGM forcing with cooling at all latitudes and for both DJF and JJA, with enhanced cooling at high latitudes compared to the tropics and over the continental ice sheets of North America and Eurasia (Fig. 15-6). Global temperatures cool by 4.0°C and tropical temperatures cool by 1.8°C compared to Modern (Table 15-2). In contrast to the NCAR model, the GFDL model shows comparable amounts of cooling in JJA in both hemispheres (Fig. 15-5). For doubled CO_2, surface temperatures warm more at high latitudes than in the tropics, with the greatest warming during DJF in the Northern Hemisphere and JJA in the Southern Hemisphere. The GFDL model shows greater sensitivity to doubled CO_2 than the NCAR model at all latitudes, with the tropics warming 2°C and the globe warming by 3.2°C compared to

Modern. In both seasons, the GFDL model shows less cooling/heating for the more oceanic southern tropics than for the more continental northern tropics in contrast to the more uniform cooling/heating of the NCAR model throughout the tropics.

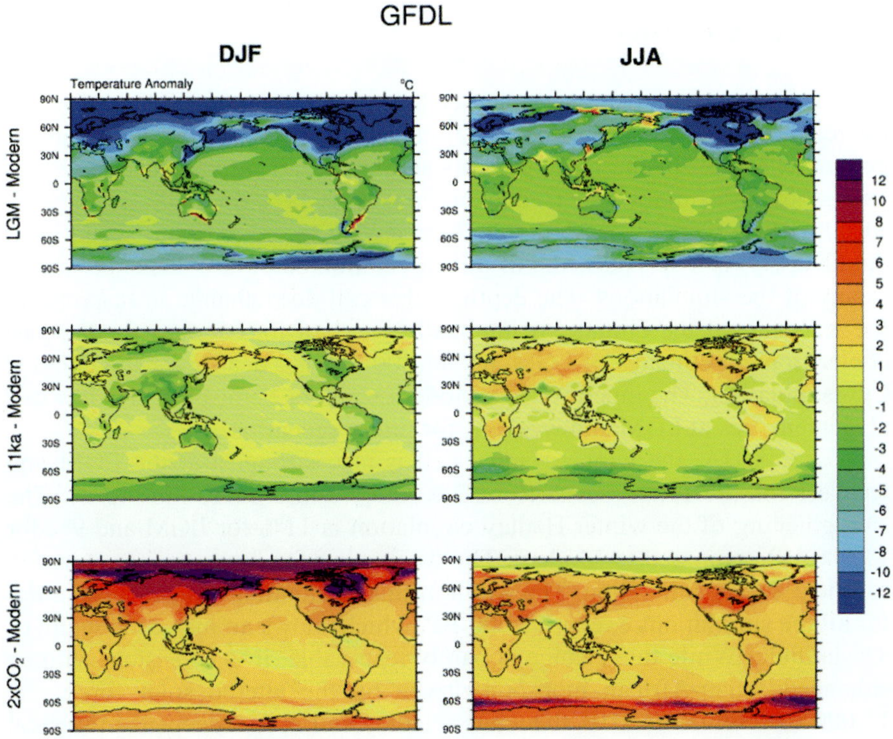

Figure 15-6. Surface temperature anomalies (°C) as compared to Modern for DJF and JJA as simulated by the GFDL model for LGM, 11,000 BP, and 2 x CO_2.

At 11,000 BP, the large positive anomalies in JJA result in warming over North America, Eurasia, and northern Africa similar to that in the NCAR model, but greater cooling occurs in the Southern Hemisphere over Antarctica and the adjacent southern oceans. In DJF the greatest cooling occurs over the continents, similar to the NCAR model. Annual global temperatures are similar between 11,000 BP and Modern in both models.

6. SENSITIVITY OF HADLEY CIRCULATION TO FORCING

6.1. NCAR Model

For DJF, the winter Hadley cell is strongest at the LGM and weakest for the future scenario of 2 x CO_2 (Table 15-1; Fig. 15-7), although the general character of the Hadley cell is similar for all forcing scenarios. The strengthening of the winter Hadley circulation is 11% for LGM and 7% for 11,000 BP compared to Modern. The weakening for a future scenario of doubled CO_2 is 6%. The width of this circulation cell (42°) does not change in any of the simulations. The depth of this cell does change in response to the warming of the climate, being shallower than for Modern for the past cooler tropical climates of LGM and 11,000 BP and deeper for the doubled CO_2 simulation. The much weaker Modern summer hemisphere Hadley cell also intensifies at LGM and weakens for future scenarios.

For JJA, the winter Hadley cell is also strongest at the LGM and weakest for the future scenario of 2 x CO_2 (Table 15-1; Fig. 15-8). The strengthening of the winter Hadley circulation is 11% for LGM and 9% for 11,000 BP compared to Modern. The weakening for future scenarios is 5% for the 2 x CO_2 simulation. The width of this circulation cell (41°) is similar for all the simulations except 11,000 BP, which shows an increased width of 43° latitude. While this change is barely resolvable, it is consistent with an enhanced Asian-African monsoon driven by the altered solar forcing at 11,000 BP. Similar to DJF, the winter Hadley cell in JJA deepens in vertical extent for the warmer future climate simulations.

6.2. GFDL Model

The DJF Hadley cell response to the different forcings is similar between the two models. As in the NCAR model, the width of the circulation cell is similar for all the GFDL simulations (Fig. 15-9; Table 15-2). The depth changes of the DJF cell are also consistent across models, with the cell being shallower in a colder climate and deeper in a warmer climate. The cell is strongest in the LGM simulation and weakest for the 2 x CO_2 simulation. However, the GFDL model Hadley cell has a significantly stronger response to the 2 x CO_2 than the NCAR model. This result is consistent with a different response in the meridional temperature gradient. Overall, the GFDL model has a much stronger sensitivity of surface temperature to CO_2

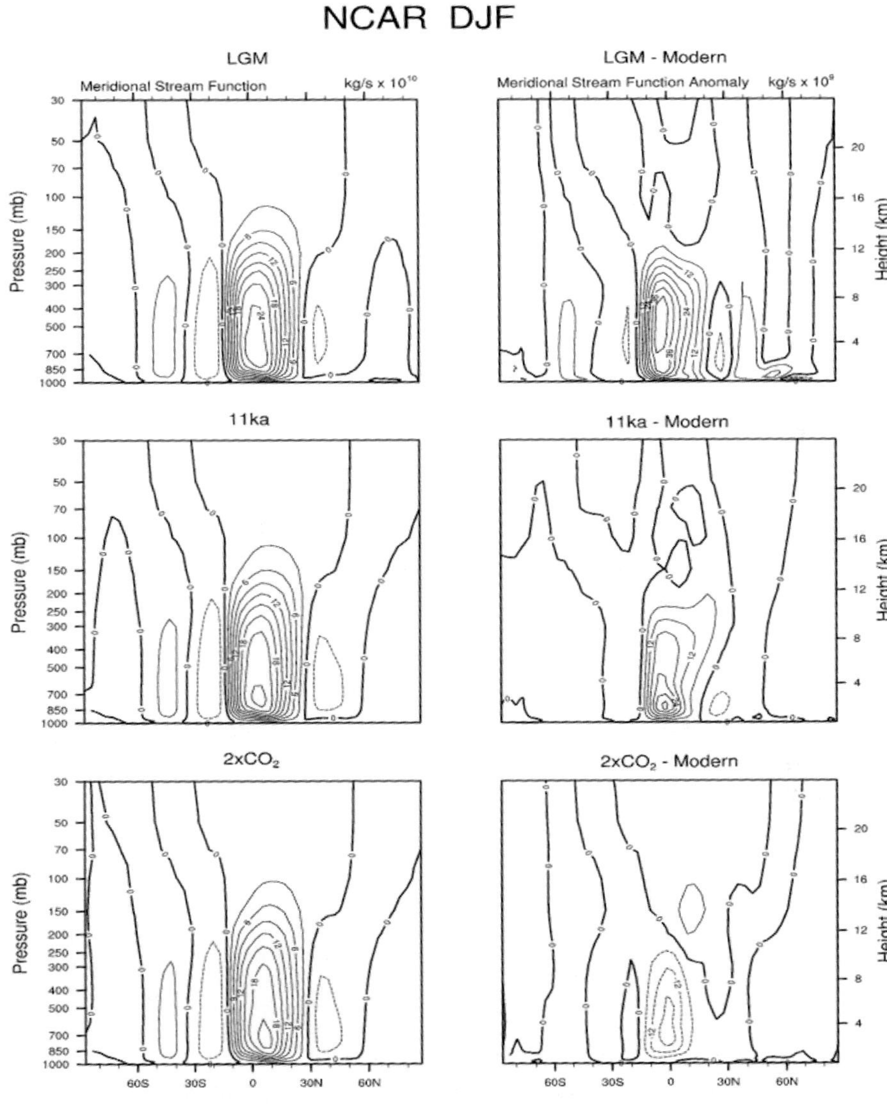

Figure 15-7. Meridional stream function (in 10^{10} kg/s) for LGM, 11,000 BP, and 2 x CO_2 (left) and meridional stream function anomalies (in 10^9 kg/s) compared to Modern (right) as simulated in the NCAR model for DJF. Positive values indicate clockwise circulations.

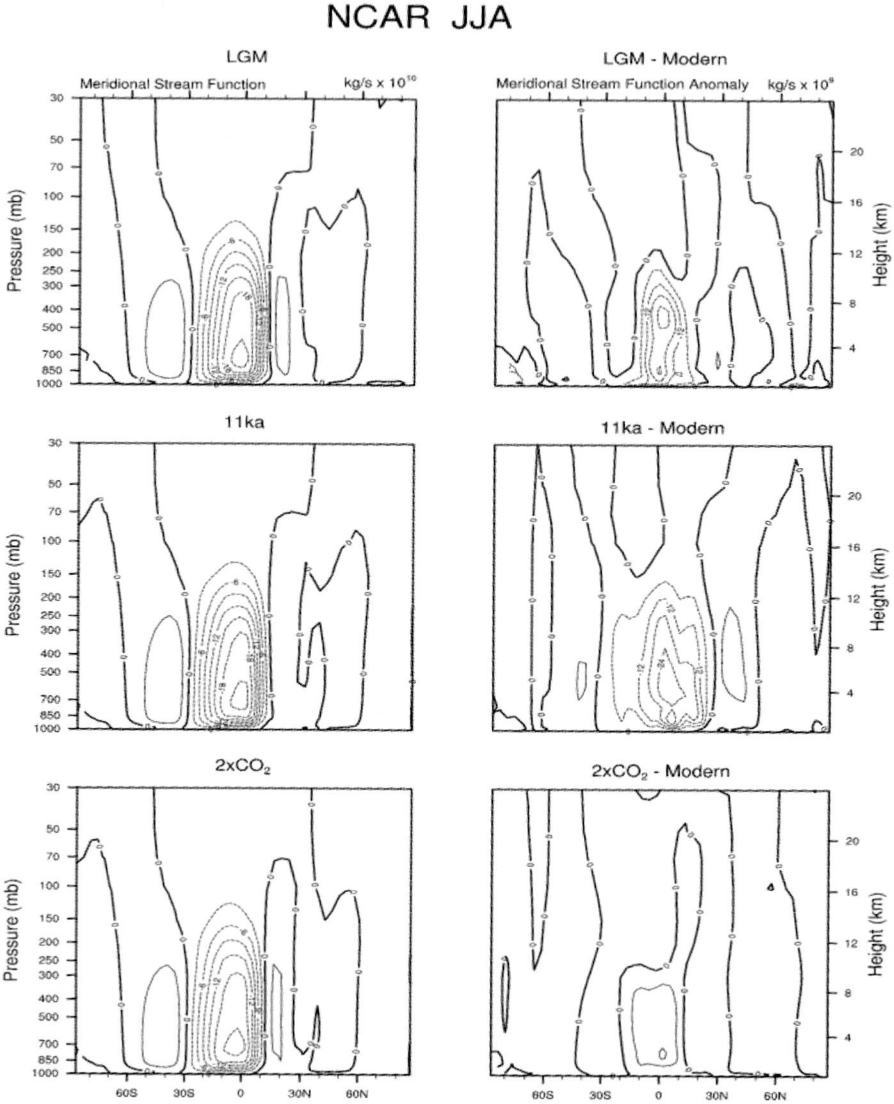

Figure 15-8. Meridional stream function (in 10^{10} kg/s) for LGM, 11,000 BP, and 2 x CO_2 (left) and meridional stream function anomalies (in 10^9 kg/s) compared to Modern (right) as simulated in the NCAR model for JJA. Positive values indicate clockwise circulations.

(Fig. 15-5), and the meridional gradient response is similarly stronger. The GFDL atmosphere model has been shown to have greater surface temperature sensitivity than the CSM atmosphere model to doubled CO_2 with a slab ocean (Cubasch et al. 2001).

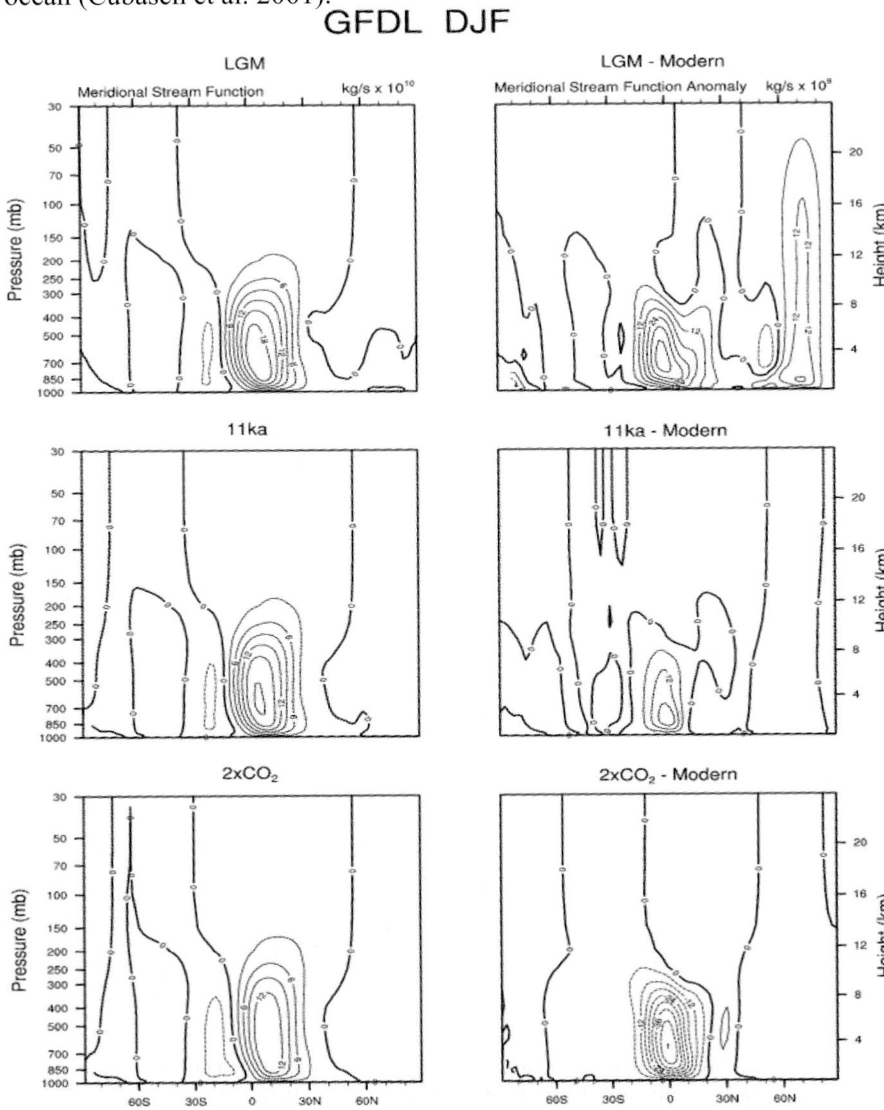

Figure 15-9. Meridional stream function (in 10^{10} kg/s) for LGM, 11,000 BP, and 2 x CO_2 (left) and meridional stream function anomalies (in 10^9 kg/s) compared to Modern (right) as simulated in the GFDL model for DJF. Positive values indicate clockwise circulations.

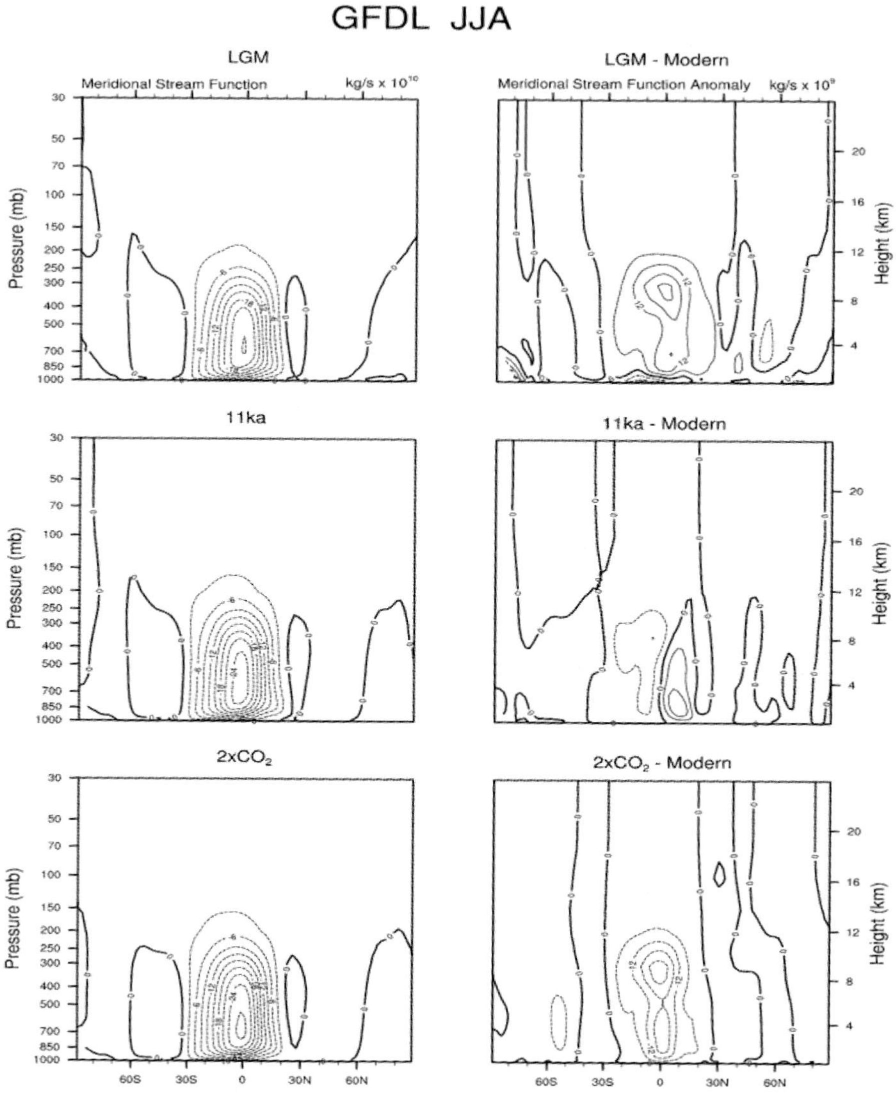

Figure 15-10. Meridional stream function (in 10^{10} kg/s) for LGM, 11,000 BP, and 2 x CO_2 (left) and meridional stream function anomalies (in 10^9 kg/s) compared to Modern (right) as simulated in the GFDL model for JJA. Positive values indicate clockwise circulations.

In JJA, the Hadley cell response to all forcings is different in sign between the GFDL and NCAR models. In the GFDL model, the JJA Hadley cell is weakest for the LGM and strongest in the 2 x CO_2 simulation (Fig. 15-10; Table 15-2). Also, the JJA cell in the GFDL model is much less sensitive than the DJF cell, which is not the case in the NCAR model, where the cells have approximately the same magnitude change in both seasons. The width of the cell appears to scale with the strength of the cell, though variations are slight. In JJA, the depth of the cell is less sensitive than in DJF.

7. DISCUSSION

The Hadley cell vertical extent is not well correlated to seasonal changes (JJA-DJF) in the mean tropical surface temperature within a simulation, but is correlated to the mean global and tropical surface temperature between the past-present-future simulations (Tables 15-1 and 15-2; Fig. 15-11). That is, the winter Hadley cells in DJF and JJA increase in vertical extent as the mean surface temperatures increase in the NCAR and GFDL simulations from LGM to Modern to 2 x CO_2. Other quantities related to Hadley cell depth have been shown to vary with the mean climate. Using an earlier version of the GFDL model, Wetherald and Manabe (1988) showed that cloud-top heights increase as the climate warms via an increase in cloudiness around the tropopause and a decrease in cloudiness in the upper troposphere. Santer et al. (2003) have used the National Centers for Environmental Prediction (NCEP) reanalysis and Goddard Institute for Space Studies (GISS) models to show that there has been an overall increase in tropopause height over the period 1979–97, which they relate to anthropogenic warming. This collection of results suggests that there is a consistent connection between Hadley cell depth/cloud-top height/tropopause height and climate change, which does not appear to be model dependent.

The width of the winter Hadley cells in DJF is similar in the NCAR and GFDL models, but larger in JJA in the GFDL model compared to the NCAR model (Tables 15-1 and 15-2), consistent with the precipitation patterns (Fig. 15-3). The NCAR model with ocean dynamics has a narrower ITCZ and less northward expansion of JJA monsoonal precipitation in Asia and Central America. Changes in the width of the winter Hadley cell show a correlation to global mean surface temperature in the GFDL model, but only in JJA, increasing slightly in poleward extent for warmer climate, similar to the previous GISS model results (Rind 1998).

In both models, the LGM and doubled CO_2 experiments show the DJF Hadley cell strength to be positively correlated with the latitudinal surface temperature gradient both between the tropics and the winter extratropics, and within the tropical belt between the summer tropical latitudes and the winter tropical latitudes (Tables 15-1 and 15-2; Fig. 15-11). Model simulations with idealized latitudinal temperature gradients in the GISS atmospheric model found a similar relationship between Hadley cell intensity and the latitudinal surface temperature gradients.

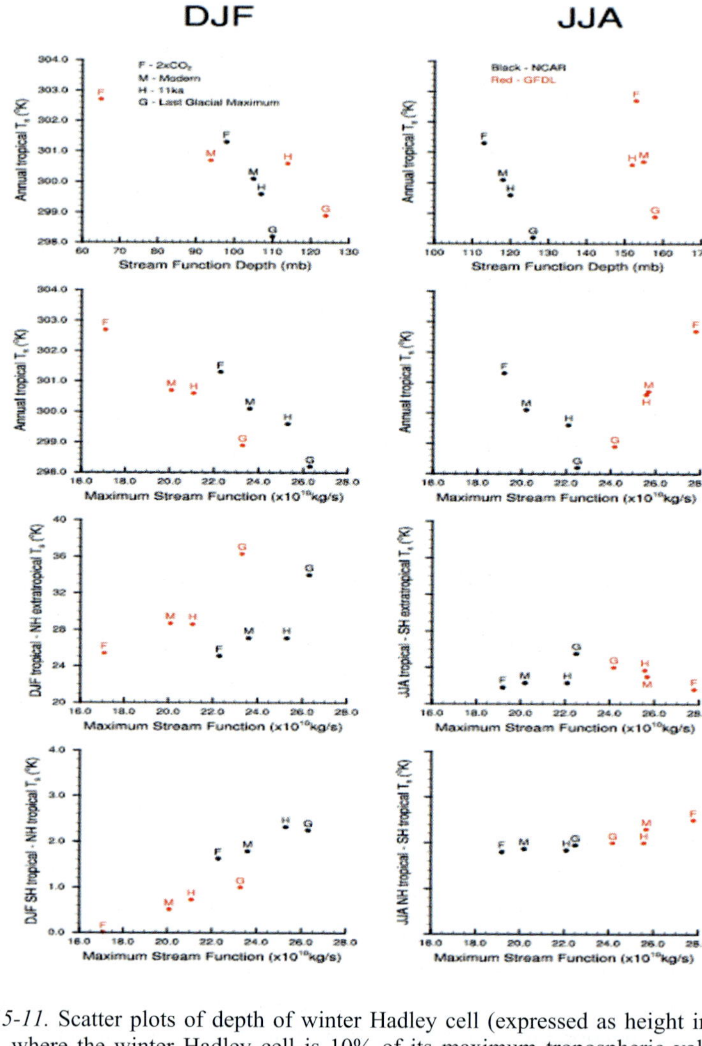

Figure 15-11. Scatter plots of depth of winter Hadley cell (expressed as height in millibars where the winter Hadley cell is 10% of its maximum tropospheric value) versus tropical annual mean surface temperature (°K), and maximum strength of winter Hadley cell (10^{10} kg/s) versus tropical annual mean surface temperature, extratropical-tropical surface temperature gradient (°K), and tropical surface temperature gradient (°K), as simulated by the NCAR and GFDL models for LGM, 11,000 BP, Modern, and 2 x CO_2. See Table 15-1 for definitions of terms.

The doubled CO_2 experiments for both models show a weakened DJF Hadley cell compared to Modern, consistent with greater warming at high latitudes than in the tropics. This result is in contrast to those of Quan et al. (Chapter 3, "Change in the Tropical Hadley Cell since 1950," this volume) using NCEP/NCAR reanalysis data for winds. Their analysis indicates a strengthening of the DJF Hadley cell since 1950. Observed sea surface temperatures (SSTs) have warmed more in the tropics than at high latitudes, particularly in the Indo-Pacific warm pool. In addition, the amplitude and frequency of El Niño events have increased since 1976. Recent studies for the instrumental period using the NCAR coupled climate model suggest that only with a large ensemble of simulations can trends associated with changed forcings be separated from internal climate variations.

For the JJA Hadley cell, the NCAR and GFDL models respond differently. First, the JJA Hadley cell strength in the GFDL model is not related to the tropical-extratropical gradient. Rather, it is the intratropical gradient that has a consistent relationship with the Hadley cell strength between the two models. Second, the response in JJA between the two models is different in sign for each forcing. In the GFDL model, the surface temperatures at 30°S cool by less at LGM and warm by less for 2 x CO_2 than at 30ºN in all seasons (Fig. 15-5). The net result is that the Hadley cell strength increases in JJA as the climate system warms from LGM to Modern to 2 x CO_2, opposite to what happens in the NCAR model.

We hypothesize that the differing responses of the NCAR and GFDL models in JJA is controlled primarily by the formulation of the ocean components of the models and the resulting atmospheric response to changes in these oceans in response to the forcings. This hypothesis is investigated by analyzing LGM and present-day simulations of the NCAR CCSM2 model coupled to a mixed-layer ocean model (instead of the three-dimensional ocean general circulation model). For the mixed-layer ocean model, ocean heat transports are specified geographically and seasonally from a present-day simulation. The primary changes in the atmosphere model from CSM to CCSM2 are the incorporation of a prognostic cloud water parameterization, a new radiation package, and 26 vertical levels.

The NCAR mixed-layer version of CCSM2 has a similar response to LGM forcing in DJF as the NCAR CSM model with a full ocean and the GFDL mixed-layer model. That is, the DJF Hadley cell is stronger at LGM compared to the present in response to stronger surface temperature gradients between the tropics and Northern Hemisphere extratropics and within the tropics. In JJA, on the other hand, the response is consistent between the NCAR and GFDL models with mixed-layer oceans with a slightly weaker Hadley cell, while the NCAR model with a full ocean shows an increase in the cell strength. While the temperature gradient between the tropics and winter extratropics increases in JJA in all models, the Hadley cell in the mixed-layer models appears to be more sensitive to the gradient within the tropics, which decreases in those models. Further studies with atmosphere models forced by idealized SSTs would help to elucidate these issues.

Finally, these experiments reveal that the application of the conceptual model of the Hadley cell to climate change has limitations. For example, consider the response to 11,000 BP orbital forcing. Because of the small heat capacity of the land surface, there is a stronger temperature response to the solar forcing over land than in adjacent ocean regions, resulting in a significant change in the land-sea temperature contrast, which is largest in the subtropics (Figs. 15-4, 15-5). These zonal gradients in surface temperature drive zonally asymmetric circulations, which dominate the atmospheric response to orbital forcing (Clement et al. 2004). There is some residual change in the intratropical temperature gradient in the GFDL model (Fig. 15-5; note the change in scale of the vertical axis), which can be related to the Hadley circulation strength (Table 15-2). However, in the NCAR model, there is no change in the intratropical gradient, but a significant increase of the Hadley cell strength. This response is no doubt related to the vertical motion associated with the Asian and African monsoons (land-ocean temperature contrast). In contrast to the orbitally forced cases, the glacial and 2 x CO_2 forcings produce large changes in the meridional temperature gradient in both models (Fig. 15-5), and thus a significant change in the Hadley circulation strength. While the climate response to these forcings is not perfectly zonally symmetric, the zonal mean captures a significant fraction of the response. Thus, depending on the characteristics of the forcing, Hadley cell changes may or may not be a very good descriptor of the large-scale characteristics of the climate response. For year-round forcings such as the glacial and 2 x CO_2, climate feedbacks are activated in different latitude belts (e.g., ocean circulation, sea ice) and impact the meridional gradients. However, for orbital forcing, which is primarily seasonal and has little annual mean, these feedbacks are less effective, and there is less impact on meridional gradients.

8. IMPLICATIONS FOR PALEOCLIMATE OBSERVATIONS

The model results shown here suggest that, regardless of the details of the model physics, the seasonal zonal mean meridional surface temperature gradients within the tropics can be a good predictor of the strength of the seasonal Hadley circulation. This result is encouraging for paleoclimate studies, because surface temperature is a quantity that is commonly derived from the proxy records. However, there are some potential limitations to inferring past changes in the strength of the Hadley cell from paleoclimate records. In particular, a practical issue arises due to sampling. The Hadley cell can truly be defined only in a zonal mean sense. Even in the modern climate, errors in the estimate of the strength of the Hadley cell of up to 20% can arise due to incomplete sampling of meridional winds (Waliser et al. 1999). Such errors are more problematic for the past climate, for which data are considerably sparser and the meridional winds are not recorded directly. The model results presented here indicate that if the zonal mean temperature gradients at different times in the past are properly sampled, they may provide some indication of the strength of the Hadley cell. The question is, can this sampling be achieved?

We address this question by using the network of available paleotemperature estimates for the LGM. This era is perhaps one of the best sampled periods of the past due in part to the extensive effort of the CLIMAP mapping project, and is therefore an optimistic estimate of the potential for data coverage for a given period. Broccoli (2000) has quantitatively compared the LGM temperature change from the present simulated with the GFDL model with the available paleotemperature data from the LGM, which include oceanographic records (based on faunal assemblages, modern analog, and alkenone methods) and terrestrial records (based on noble gases in groundwater and pollen). The model temperature changes are based on the same set of LGM and control experiments with the GDFL model that are presented here.

Here we are interested in exploring the question of whether this particular network of data is sufficient to represent the zonal mean meridional temperature gradient. Following on work done by Waliser et al. (1999), who addressed this issue for the modern data, we calculate the zonal mean temperature change in three ways: (1) using model temperature change from all model grid points, (2) using model temperature change only from the points where there are paleotemperature estimates, and (3) using all the paleotemperature data. A comparison between methods (2) and (3) gives an estimate of the model's ability to reproduce the data. There are some significant differences between the model and CLIMAP SST change over the tropical oceans (Fig. 15-12). More recent data have opened up a debate as to what the pattern, in particular the zonal gradients, looked like at the LGM in the tropics (Hostetler and Mix 1999; Lea et al. 2000; Koutavas et al. 2002), though there is currently no consensus. Broccoli (2000) showed that the model also significantly underestimated the cooling over land. This discrepancy could be partly due to the fact that the estimates of the "modern" temperature over land and ocean are made in fundamentally different ways. Core top temperatures are used to estimate the modern surface temperature over the ocean, and these temperatures are integrated over a couple of centuries by the sedimentation process. On the other hand, over land, the proxy-based estimates of the modern temperature are generally of sufficient resolution to capture the second half of the twentieth century, when the Northern Hemisphere continents (and most regions of the globe) were warmer. The fact that land- and ocean-based proxies have a different inherent time scale may be of relevance to the ability to represent the zonal mean. A careful re-analysis of the modern proxy temperature estimates may be able to correct for this factor.

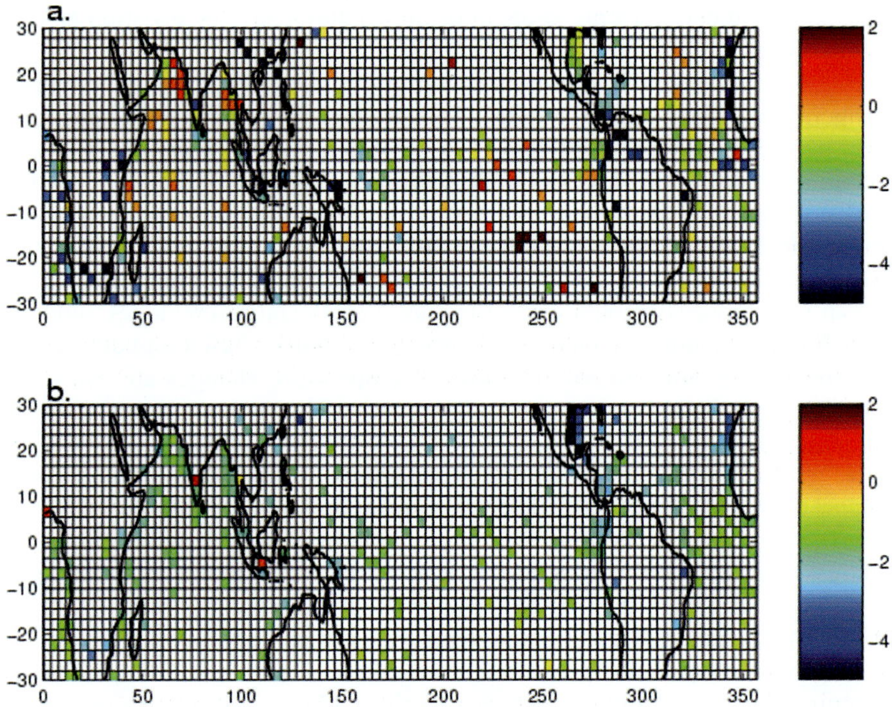

Figure 15-12. (a) Paleotemperature estimates from oceanographic records (based on faunal assemblages, modern analog, and alkenone methods) and terrestrial records (based on noble gases in groundwater and pollen) gridded on the GFDL model grid as described in Broccoli (2000). (b) GFDL model temperature change between the Modern and LGM runs shown only at the grid points where paleotemperature estimates exist. (c) Same as (b) but for NCAR model LGM minus Modern.

While the source of the discrepancy between model and data estimates of LGM cooling is an important area of current and future research, what we are particularly interested in here is a comparison of the zonal mean calculated by methods (1) and (2), which gives an estimate of the errors arising from the sampling network. In Figure 15-12, it is clear the data have considerably more zonal variations than the model does. In general, model temperature changes in response to LGM forcing are fairly uniform in the tropics. This is true whether ocean dynamics, which is an important source for longitudinal variations in the tropics, is included (as in CSM) or not (as in GFDL). There are strong dynamical constraints that limit how large horizontal temperature gradients in the tropics can get (e.g. Lindzen and Nigam 1987; Wallace 1992). As such, there is very good agreement between the full model estimate of the zonal mean and the subsampled estimate (Fig. 15-13). Results are shown only for the GFDL model; CSM results are similar. Interestingly, although there are significant longitudinal variations in the data-model agreement (Fig. 15-12), the zonal means of both the data and the model do show the increased temperature gradient between the northern and southern tropics, which is an indication of the strengthening of the Hadley cell. The anomalous zonal mean temperature gradient between the northern tropics and the southern tropics is –0.3 in the data, –0.37 in the model, and –0.27 in the subsampled model. Thus, it appears that a realistic (if somewhat optimistic) sampling of temperature change in the past is sufficient to represent the zonal mean, and in spite of longitudinally dependent model-data discrepancies, there is good agreement in the zonal mean between the models and the data. We note, however, that in the case of the LGM, mean temperature changes, as well as the meridional gradient changes, are large. For other climate forcings (e.g., precessional forcing), however, it may be difficult to decipher the zonal mean temperature changes unless they are large.

Finally, it is important to point out that all methods of inferring past changes in the Hadley circulation have some errors associated with them. These errors can be related to model parameterizations, assumptions in the proxy calibrations, or sampling issues. In order to get the most complete view of how the Hadley circulation changes in response to climate forcing of either the past or future, both models and data are required along with a full discussion of both the technical and conceptual limitations of the different approaches.

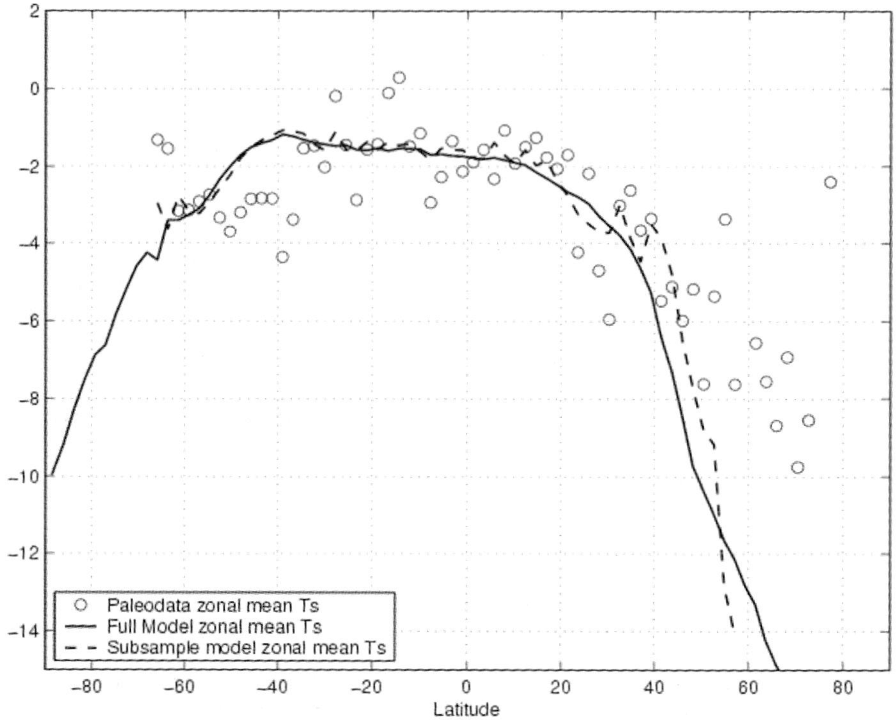

Figure 15-13. Three different estimates of the zonal annual mean surface temperature change between the LGM and Modern: (1) using GFDL model surface temperature change from all model grid points (solid line), (2) using GFDL model surface temperature change only from the points where there are paleotemperature estimates (dashed), and (3) using all the paleotemperature data (circles). Results for the NCAR model are similar and therefore are not shown.

9. ACKNOWLEDGMENTS

We thank Tony Broccoli for contributing the GFDL model results and for his generous and helpful insights. We thank Christine Shields and Robert Tomas for generating the figures and tables. We are also grateful for the comments of two anonymous reviewers. Simulations with the NCAR model were carried out on the Climate Simulation Laboratory (CSL) and Partnership in Modeling Earth System History (PMESH) computing systems funded by the National Science Foundation (NSF). The National Center for Atmospheric Research is sponsored by the National Science Foundation. A. Clement was funded by NSF grant #ATM-0134742 from the NSF Paleoclimate Program.

10. REFERENCES

Berger, A.L. 1978. Long-term variations of caloric insolation resulting from the earth's orbital elements. *Quaternary Research* 9: 139–167.

Blackmon, M., et al. 2001. The Community Climate System Model. *Bulletin of the American Meteorological Society* 82: 2357–2376.

Bonan, G.B. 1998. The land surface climatology of the NCAR Land Surface Model coupled to the NCAR Community Climate Model. *Journal of Climate* 11: 1307–1326.

Broccoli, A.J. 2000. Tropical cooling at the last glacial maximum: An atmosphere–mixed layer ocean model simulation. *Journal of Climate* 13: 951–976.

Budyko, M. 1974. Climate modification techniques. *Meteorology and Hydrology* 2: 110–122.

Clement, A.C., R. Seager, and M.A. Cane. 1999. Orbital controls on the El Niño/Southern Oscillation and the tropical climate. *Paleoceanography* 14: 441–456.

Clement, A.C., A. Hall, and A.J. Broccoli. 2004. The importance of precessional signals in the tropical climate. *Climate Dynamics* 22: 327–341.

Cubasch, U., et al. 2001. Projections of future climate change. *Climate Change 2001. The Scientific Basis*. New York, New York: Cambridge University Press.

Gent, P.R., F.O. Bryan, G. Danabasoglu, S.C. Doney, W.R. Holland, W.G. Large, and J.C. McWilliams. 1998. The NCAR Climate System Model global ocean component. *Journal of Climate* 11: 1287–1306.

Held, I.M., and A.Y. Hou. 1980. Nonlinear axially symmetric circulations in a nearly inviscid atmosphere. *Journal of the Atmospheric Sciences* 37: 515–533.

Hostetler, S.W., and A.C. Mix. 1999. Reassessment of ice-age cooling of the tropical ocean and atmosphere. *Nature* 399: 673–676.

Hunt, B.G. 1979. The influence of the Earth's rotation rate on the general circulation of the atmosphere. *Journal of the Atmospheric Sciences* 36: 1392–1408.

Kiehl, J.T., J.J. Hack, G.B. Bonan, B.A. Boville, D.L. Williamson, and P.J. Rasch. 1998. The National Center for Atmospheric Research Community Climate Model: CCM3. *Journal of Climate* 11: 1131–1149.

Koutavas, A., J. Lynch-Stieglitz, T.N.J. Marchitto, and J. P. Sachs. 2002. El Nino–like pattern in ice age tropical Pacific sea surface temperature. *Science* 297: 226–230.

Lea, D.W., D.K. Pak, and H.J. Spero. 2000. Climate impact of late Quaternary equatorial Pacific sea surface temperature variations. *Science* 289: 1719–1724.

Lindzen, R.S., and A.Y. Hou. 1988. Hadley circulations for zonally averaged heating centered off the equator. *Journal of the Atmospheric Sciences* 45: 2416–2427.

Lindzen, R.S., and S. Nigam. 1987. On the role of sea surface temperature gradients in forcing low-level winds and convergence in the Tropics. *Journal of the Atmospheric Sciences* 44: 2418–2436.

Liu, Z., E.C. Brady, and J. Lynch-Stieglitz. 2003. Global ocean response to orbital forcing. *Paleoceanography* 18, doi: 10.1029/2002PA000819.

Otto-Bliesner, B.L., and E.C. Brady. 2001. Tropical Pacific variability in the NCAR Climate System Model. *Journal of Climate* 14: 3587–3607.

Otto-Bliesner, B.L., E.C. Brady, S. Shin, Z. Liu, and C. Shields. 2003. Modeling El Niño and its teleconnections during the last glacial-interglacial cycle. *Geophysical Research Letters* 30: 2198, doi 10.1029/2003GL018553.

Peltier, W.R. 1994. Ice age paleotopography. *Science* 265: 195–201.

Raynaud, D., J. Jouzel, J.M. Barnola, J. Chappellaz, R.J. Delmas, and C. Lorius. 1993. The ice record of greenhouse gases. *Science* 259: 926–934.

Rind, D. 1998. Latitudinal temperature gradients and climate change. *Journal of Geophysical Research* 103: 5943–5971.

Santer, B.D., R. Sausen, T.M.L. Wigley, J.S. Boyle, K. AchutaRao, C. Doutriaux, J.E. Hansen, G.A Meehl, E. Roeckner, R. Ruedy, G. Schmidt, and K.E. Taylor. 2003. Behavior of tropopause height and atmospheric temperature in models, reanalyses, and observations: Decadal changes. *Journal of Geophysical Research* 108, 10.1029/2002JD002258.

Shin, S.I., Z. Liu, B.L. Otto-Bliesner, E.C. Brady, J.E. Kutzbach, and S.P. Harrison. 2003. A simulation of the Last Glacial Maximum climate using the NCAR CSM. *Climate Dynamics* 20: 127–151.

Trenberth, K.E., D.P. Stepaniak, and J.M. Caron. 2000. The global monsoon as seen through the divergent atmospheric circulation. *Journal of Climate* 13: 3969–3993.

Waliser, D.E., Z. Shi, J.R. Lanzante, and A.H. Oort. 1999. The Hadley circulation: Assessing NCEP/NCAR reanalysis and sparse in situ estimates. *Climate Dynamics* 15: 719–735.

Wallace, J. 1992. Effect of deep convection on the regulation of tropical sea surface temperature. *Nature* 357: 230–231.

Weatherly, J.W., B.P. Briegleb, W.G. Large, and J.A. Maslanik. 1998. Sea ice and polar climate in the NCAR CSM. *Journal of Climate* 11: 1472–1486.

Wetherald, R.T., and S. Manabe. 1988. Cloud feedback processes in a general circulation model. *Journal of the Atmospheric Sciences* 45: 1397–1415.

Williams, G.P., and J.L. Holloway. 1982. The range and unity of planetary circulations. *Nature* 297: 295–299.

Chapter 16

PRESENT-DAY CLIMATE VARIABILITY IN THE TROPICAL ATLANTIC

A Model for Paleoclimate Changes?

John C.H. Chiang

Department of Geography and Berkeley Atmospheric Sciences Center, University of California, Berkeley, California 94720-4740, U.S.A.

Abstract Paleoproxy records of the Holocene and last glacial period suggest that the meridional position of the Atlantic Intertropical Convergence Zone (ITCZ) and hence the thermally direct circulation in that region changed significantly in the past, a behavior similar to that of a leading mode of interannual-decadal climate variability in the present-day tropical Atlantic. This chapter explores how knowledge of this mode of variability may be usefully employed to advance hypotheses for understanding tropical Atlantic paleoclimate change. A review of past coupled general circulation model (CGCM) studies reveals that change to the Atlantic ITCZ meridional position is pervasive in two situations of paleoclimate interest; namely, modification of the Atlantic thermohaline circulation, and adjustment to last glacial maximum (LGM) boundary conditions. Comparison of atmosphere-ocean general circulation model LGM simulations from the Paleoclimate Modeling Intercomparison Project (PMIP) shows, however, that the magnitude of the latter adjustment is model dependent. Accurate paleoproxy records of tropical Atlantic climate may therefore be able to provide crucial constraints on acceptable coupled model behavior.

H.F. Diaz and R.S. Bradley (eds.), The Hadley Circulation: Present, Past and Future, 465–488.

1. INTRODUCTION

Recent sediment core records from the Cariaco Basin region (just off the northern coast of South America, approximately 10.5°N and 65°W) show that, over the Holocene and last glacial period, the climate of the tropical Atlantic experienced significant changes over a wide range of time scales from decadal to glacial-interglacial (Hughen et al. 1996; Peterson et al. 2000; Black et al. 1999; Haug et al. 2001). Furthermore, these variations appear to be linked to the climate of the North Atlantic over virtually all time scales resolved by these records. The results of Peterson et al. (2000) illustrate this point nicely.

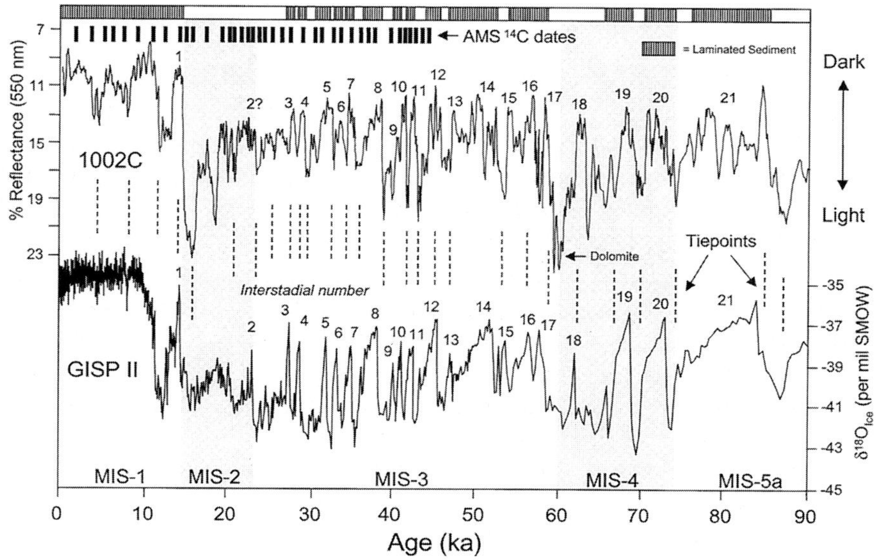

Figure 16-1. From Peterson et al. (2000), showing the linkage between the tropical Atlantic and North Atlantic climate over the last 90,000 years. It is a comparison of measured color reflectance (550 nm) of Cariaco Basin sediments from ODP Hole 1002C to $\delta^{18}O$ from the GISP II ice core. The reader is referred to the original reference for details on dating and tiepoints. Peterson et al. comment on this figure: "Deposition of dark, generally laminated sediments preferentially occurs during warm interglacial or interstadial times (numbered events), whereas deposition of light-colored bioturbated sediments was restricted to colder stadial intervals of the last glacial. Sediment color variations in the Cariaco Basin are driven by changing surface productivity, with increased organic rain leading to darker sediments and, through remineralization reactions, periods of anoxic or near-anoxic conditions in the deep basin."

Figure 16-1 compares color reflectance (550 nm) of a Cariaco Basin sediment core (Ocean Drilling Program [ODP] Hole 1002C)—a measure of surface productivity in that basin—to the $\delta^{18}O$ record from the Greenland Ice Sheet Project (GISP), representative of temperature changes over the North Atlantic. Other studies show similar linkages during the Younger Dryas (Haug et al. 2001; Hughen et al. 1996) and also during the last eight centuries (Black et al. 1999). These records beg the following two questions: (1) What pattern of climate variations are they representing, and (2) why is there such a strong linkage between the tropical and the North Atlantic?

All the studies cited above suggest that the Cariaco Basin is recording changes in the north-south position of the Intertropical Convergence Zone (ITCZ) and the band of maximum rainfall embedded in it. A shift in the ITCZ can impact sediment records either through changes in the runoff that feeds into the basin (Peterson et al. 2000), or through changes in the northeasterly trades, upwelling, and therefore nutrient supply and biological productivity (Hughen et al. 1996). The inferred climate changes resemble the behavior of a leading statistical mode—hereafter referred to as the meridional mode, following Servain et al. (1999)—of present-day tropical Atlantic interannual-to-decadal climate variability, also characterized by meridional displacement in the ITCZ and associated with trade wind variations (e.g., Nobre and Shukla 1996; Chang et al. 1997). Furthermore, this variability is also strongly tied to the North Atlantic variability, particularly on longer (decadal) time scales (e.g. Marshall et al. 2001); they also lead to substantial changes to the thermally direct circulation in the tropical Atlantic (Wang 2002). These facts suggest that the dynamics governing the meridional mode may be usefully applied to understand climate and circulation changes in the past, in much the same way as the El Niño/Southern Oscillation (ENSO; e.g., Cane 1998) and the monsoon have been similarly invoked.

This chapter explores the possible linkage between the tropical Atlantic meridional mode and Atlantic paleoclimate change from a model viewpoint. A companion paper (Chiang et al. 2003) investigates this problem in the context of paleoclimate model experiments using a specific general circulation model (GCM; the Community Climate model version 3, hereafter CCM3 [Kiehl et al. 1998]), whereas here, we sample results from a range of published paleoclimate model simulations, and also model results from the Paleoclimate Modeling Intercomparison Project (PMIP; Joussaume and Taylor 2000). We first review (Section 2) the current knowledge of the phenomenology and dynamics of this mode, and discuss how they may be relevant for understanding paleoclimate change (Section 3). We then review (Section 4) several model papers that demonstrate the sensitivity of the Atlantic ITCZ to two particular situations of paleoclimate interest; namely, change to the Atlantic thermohaline circulation (THC), and the climate of the last glacial maximum (LGM). We contrast the tropical Atlantic model response to LGM boundary conditions in the PMIP simulations (Section 5), which show that the magnitude of the ITCZ response is highly model dependent. We close (Section 6) with a discussion of how paleoproxy records may help with our understanding of tropical Atlantic paleoclimate.

2. THE TROPICAL ATLANTIC "MERIDIONAL MODE"

The meridional mode can be extracted from the data by using various objective statistical techniques; for example, Ruiz-Barradas et al. (1999) used rotated empirical orthogonal function (EOF) analysis on sea surface temperature (SST), ocean heat content, wind stress, and atmospheric dibatic heating. Here, we use maximum covariance analysis[1] (MCA; also known as singular value decomposition [SVD]; Bretherton et al. [1992]) of tropical Atlantic monthly mean sea surface temperature (left field) and surface wind (right field) anomalies from 30°N to 20°S across the entire Atlantic basin, using data over all months from January 1950 to December 2001. Similar MCA calculations have been done before (e.g., Chang et al. 1997, 2003). Prior to analysis, the data were interpolated to a 7.5° x 3.75° longitude/latitude grid, monthly mean anomalies were computed, and the resulting fields were detrended. In this particular analysis, the meridional mode comes out as the leading pattern, explaining the majority (56%) of the squared covariance fraction (SCF). By comparison, the second mode explains 22% of the SCF.

The result for the leading SVD mode (meridional mode) is shown in Figure 16-2. The SST and surface wind pattern[2] (Fig. 16-2a) show an anomalous meridional SST gradient just north of the equator (5°N–10°N) that is also the position of the mean ITCZ; and cross-equatorial flow towards the warmer hemisphere indicates a northward displacement of the ITCZ.[3] The precipitation pattern (Fig. 16-2b) shows a dipole-like structure indicative of the northward ITCZ displacement, though the southern lobe has a significantly larger magnitude compared to the northern lobe. The physical interpretation of this pattern is that the meridional SST gradient drives the cross-equatorial flow by creating a surface pressure gradient through hydrostatic adjustment of the atmospheric boundary layer

[1] MCA extracts patterns that maximize the temporal covariance between the two fields of interest.

[2] The spatial patterns are derived from linear regression on the normalized SST expansion coefficients of the MCA mode

[3] Note that while the ITCZ commonly refers to the convective structure, it is defined in terms of the low-pressure zone where the trades converge.

(Hastenrath and Greischar 1993; Lindzen and Nigam 1987), which then drives a cross-equatorial boundary layer flow, changing the meridional position of maximum surface wind convergence and therefore the ITCZ.

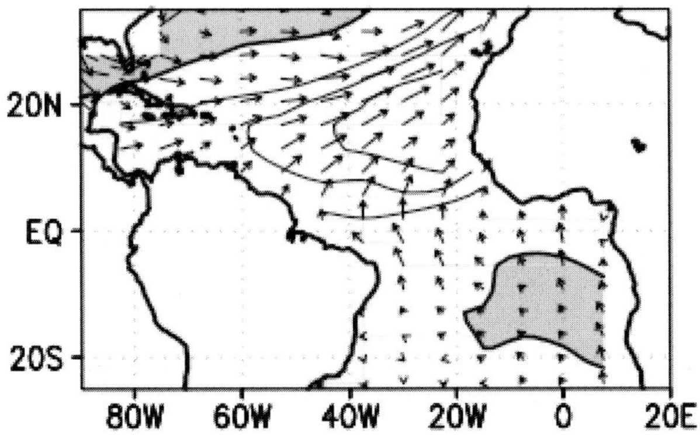

(a)

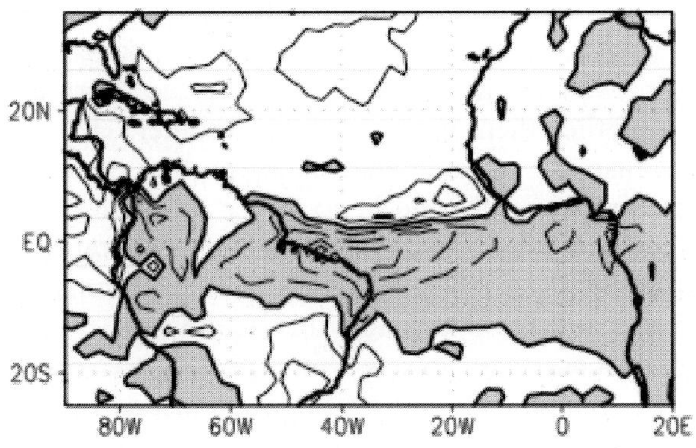

(b)

Figure 16-2. Mode 1 of an MCA analysis on tropical Atlantic SST and surface wind anomalies, showing the structure of the tropical Atlantic "meridional mode." (a) Spatial pattern of SST and surface wind anomalies. Contour interval is 0.1 K, and the reference vector is 1 m/s. (b) Spatial pattern of precipitation anomalies associated with MCA mode 1. Contour interval is 0.2 mm/d. The spatial patterns are derived through linear regression on the normalized SST expansion coefficients of MCA 1. In both (a) and (b), the negative values are shaded.

The upper-level circulation features associated with the meridional mode have been comprehensively explored by Wang (2002). In particular, the displacement in the ITCZ implies changes to the thermally direct circulation: anomalous ascent in regions of increased convection over the anomalously warmer SST, and descent in regions of decreased convection.

The regions of anomalous descent are associated with anomalous divergent outflow (inflow) at lower tropospheric levels and divergent inflow (outflow) at upper tropospheric levels. This divergent flow associated with the meridional mode is concentrated especially over the Amazon region just south of the equator (Wang 2002, his Fig. 7). Also, substantial upper-level zonal wind anomalies occur with the mode: for a warm north–cool south gradient, there are bands of ~20°-wide zonal wind anomalies of alternating polarity over the tropical Atlantic basin, from a positive anomaly concentrated at the 250 mb level between 20°S to just north of the equator, then a negative band also concentrated at the 250 mb level between the equator and 20°N, then another positive band also focused at the 250 mb level between ~20°N and ~40°N. Another negative band exists between 40°N and 65°N, though this one has a barotropic structure. Thus, changes to the meridional mode are associated with substantial changes to the tropical Atlantic basin-wide tropospheric circulation.

How does the meridional SST gradient come about? There are two trains of thought. The "active" interpretation (e.g. Chang et al. 1997) suggests that it arises out of positive ocean-atmosphere feedback between the SST and surface winds: Because the mean trades flow equatorward in both hemispheres, the anomalous winds in Figure 16-2a, induced by the meridional SST gradient, weaken the mean trades in the Northern (warmer) Hemisphere and strengthen them in the Southern (colder) Hemisphere. Their net effect is to decrease the surface latent heat flux out of the ocean in the north, increase them in the south, and in so doing act to strengthen the SST gradient. This feedback is often called wind-evaporation, or wind-evaporation-SST (WES; Xie 1999) feedback. Thus, a small perturbation could conceivably be amplified through this feedback, to be damped by some (unspecified) process. The "passive" interpretation (Nobre and Shukla 1996) states that the anomalous subtropical trade winds are driven by an external source, which then creates the meridional SST gradient across the equator, and that this pattern damps away once the forcing disappears. More recent observational and model analyses (e.g., Chiang et al. 2000, 2002; Czaja et al. 2002; Sutton et al. 2000) suggest that the real picture is weighted towards the "passive" interpretation: that the meridional mode is largely externally driven, most prominently by ENSO and the North Atlantic Oscillation (NAO). Both are thought to act through modulating the strength of the northern tropical Atlantic (NTA) trades, which in turn creates SST anomalies there, and hence a meridional SST gradient across the position of the mean ITCZ. In the deep tropics, however, the ocean and atmosphere are coupled, in the sense that cross-equatorial flow is driven once a meridional SST gradient is established (Chiang et al. 2001). These relationships mean that the meridional mode has a forced component (subtropics) and a coupled component (deep tropics). The coupled deep tropical component does produce a WES feedback near the equator that acts to prolong the anomalous meridional mode event. However, by all indications this feedback is insufficient to sustain the event beyond the boreal summer (Chiang et al. 2002; Czaja et al. 2002).

3. POSSIBLE LINKAGES BETWEEN THE MERIDIONAL MODE AND PALEOCLIMATE DYNAMICS

Several aspects of the origins and dynamics of the meridional mode make it intriguing from a paleoclimate perspective. One such property is the meridional mode sensitivity to external forcing. The teleconnection influence from the tropical Pacific and North Atlantic suggests that, should the climate of either of those regions have changed during the past (as they surely have throughout the Holocene and last glacial), the tropical Atlantic would be required to adjust, in order to at least be consistent with the altered source regions. The observed decadal linkage of the tropical Atlantic to the North Atlantic may be an example of this "adjustment." Observational (e.g. Xie and Tanimoto 1998) and modeling (e.g. Seager et al. 2000) studies have shown that the decadal variability in the Atlantic is characterized by a broad "pan-Atlantic" SST pattern stretching from the North Atlantic through to the south tropical Atlantic (STA), with a well-known "tripole" band of alternating SST anomaly polarities in the North Atlantic, and a further lobe just south of the equator (Fig. 16-3). The tropical part of this pattern is the meridional mode pattern. The suggestion is that at these relatively long time scales, the entire Atlantic from the northern high latitudes through to the southern tropics acts as a unit.

What drives this system? Modeling studies show that the SST variability can largely be explained by the ocean response to observed Atlantic surface winds (Seager et al. 2001), with the surface ocean thermodynamics playing the primary role and the ocean dynamics playing an important role in the mid-latitudes. The results imply that the origins of decadal SST variability are "blowing in the winds," though the origins of the anomalous winds remain unclear. One possibility is that they arise from intrinsic mid-latitude wintertime variability associated with the North Atlantic Oscillation and amplified by thermodynamic interaction with the surface ocean (Barsugli and Battisti 1998) and possibly also by ocean dynamics. Another possibility is that the wind anomalies are at least partly driven from the tropics (Okumura et al. 2001) through displacement of the Atlantic convection center and consequent impact on the tropical and extratropical (in particular the NAO) atmospheric circulation.

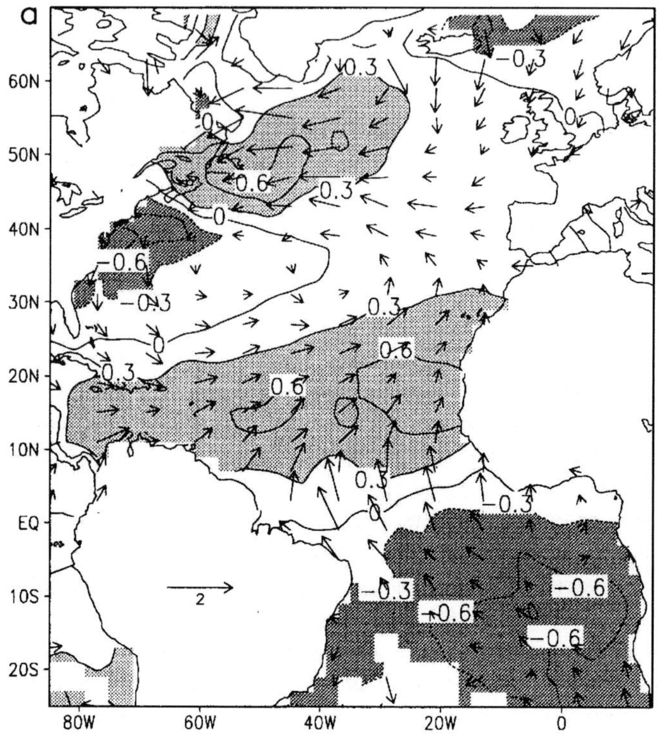

Figure 16-3. The "Pan-Atlantic Decadal Oscillation" pattern from Xie and Tanimoto (1998), showing the association of the climate variability of the tropical Atlantic to that of the North Atlantic. It is a composite of annual-averaged SST (K) and surface wind velocity (m/s) taken from extreme high (1969, 1970, 1978, 1980, 1981, 1992) and low (1972, 1973, 1974, 1984, 1985, 1986) years based on an index of the meridional gradient of SST across the equator (the dipole index defined by Servain [1991]).

The sensitivity of the Atlantic ITCZ position to small SST gradients is another intriguing aspect of the tropical Atlantic climate. The extreme SST variations in the NTA regions associated with the present-day meridional mode are only about ~1 K at most, and yet they are able to displace the position of maximum ITCZ convection thousands of kilometers (Chiang et al. 2002). This means that the marine ITCZ climate is exquisitely sensitive to perturbations. The sensitivity to SST gradients also implies that the ITCZ is open to influence from forcings, other than the well-known NTA trade wind influence, that act to change the meridional SST gradient near the ITCZ mean latitude. On the paleoclimate time scales there exist at least three possible forcings:

(1) Changes to the insolation resulting from precessional changes to the earth's orbit are known to produce anomalous gradients in insolation across the equator that may force meridional SST gradients that in turn displace the ITCZ. Indeed, Haug et al. (2001) argue based on analysis of Cariaco Basin paleoproxy data that ITCZ position changed in response to precessional changes in insolation during the Holocene.

(2) Changes in the Atlantic thermohaline circulation, as proposed by Yang (1999) and also by Johnson and Marshall (2002). The THC can change the meridional SST gradient near the equator through changes to ocean heat transport.

(3) Changes to hemispheric mean temperature (through increases in land or sea ice in the hemisphere, for example) that in turn change the interhemispheric gradient in temperature (e.g., Broccoli 2000).

To what extent are these proposed mechanisms for Atlantic ITCZ change actually realized in paleoclimate; and is the "meridional mode" response a preferred way for the Atlantic climate to respond to paleoclimate changes? We address this question in the following two sections.

4. SENSITIVITY OF THE ATLANTIC ITCZ IN PALEOCLIMATE SIMULATIONS

A review of several paleoclimate modeling studies suggests that meridional ITCZ displacement, along with the attendant meridional SST gradient and cross-equatorial flow, may indeed be a preferred way for the Atlantic to respond to paleoclimate forcing. Two paleoclimate situations are of particular interest: changes to the thermohaline circulation, and conditions during the last glacial maximum.

An Atlantic ITCZ response appears to be a pervasive feature of THC perturbation experiments in coupled general circulation models

(CGCMs). Manabe and Stouffer (1988) have shown in Geophysical Fluid Dynamics Laboratory (GFDL) coupled model simulations that exhibit two stable equilibrium ("on" and "off") Atlantic THC states, that under the "off" condition the north tropical Atlantic is relatively cooler and the ITCZ shifts southwards. This shift is more pronounced in the transient response to a THC shutdown caused by a sudden input of freshwater into the North Atlantic (Vellinga and Wood 2002). In the study by Vellinga and Wood, the model used is the Hadley Center coupled model, HadCM3. A marked southward shift of the Atlantic ITCZ was observed in this model 20–30 years after the collapse of the THC, associated with a strong (2 K) cooling of the NTA trades, but with weaker warming of the south tropical Atlantic region. Dong and Sutton (2002), studying the adjustment process of a similar THC shutdown using the same HadCM3 model, showed that the ITCZ displacement occurs as soon as 4–6 years after the initial freshwater perturbation. An interesting result of this study was the initiation of an El Niño event 7 years after the initial freshwater perturbation in the Atlantic that they attribute to the THC shutdown and more immediately to the ITCZ displacement in the Atlantic. They argue from the basis of this experiment that the climate impact of a THC change could spread globally through changes of the tropical climate.

It is unclear exactly how the THC communicates its influence to the tropical Atlantic in these coupled simulations. While change to the tropical Atlantic ocean heat transport (especially the meridional advection of temperature, according to Yang [1999]) is the mechanism implied in these studies, another plausible candidate is through increasing Northern Hemisphere sea ice extent and communication of that influence via the atmosphere. The idea is that additional sea ice in the Northern Hemisphere, created as a consequence of reduced northward ocean heat transport by the THC shutdown, cools Northern Hemisphere air temperature, introducing an interhemispheric temperature gradient that then drives ITCZ displacement. A simple model experiment highlights this possibility. We ran an atmospheric general circulation model (the Community Climate Model version 3; Kiehl et al. [1998]) coupled to a fixed 50 m slab ocean with flux adjustment applied to bring the slab ocean SST to match the observed present-day SST under present-day model boundary conditions. The model resolution and configuration are similar to those used by Chiang et al. (2003). Two simulations were done: one with sea ice extent prescribed to observed present-day extent ("control"), and another with additional sea ice imposed in the Northern Hemisphere (Fig. 16-4a). The difference in the annual mean precipitation and surface winds between these two runs (Fig. 16-4b) shows a pronounced

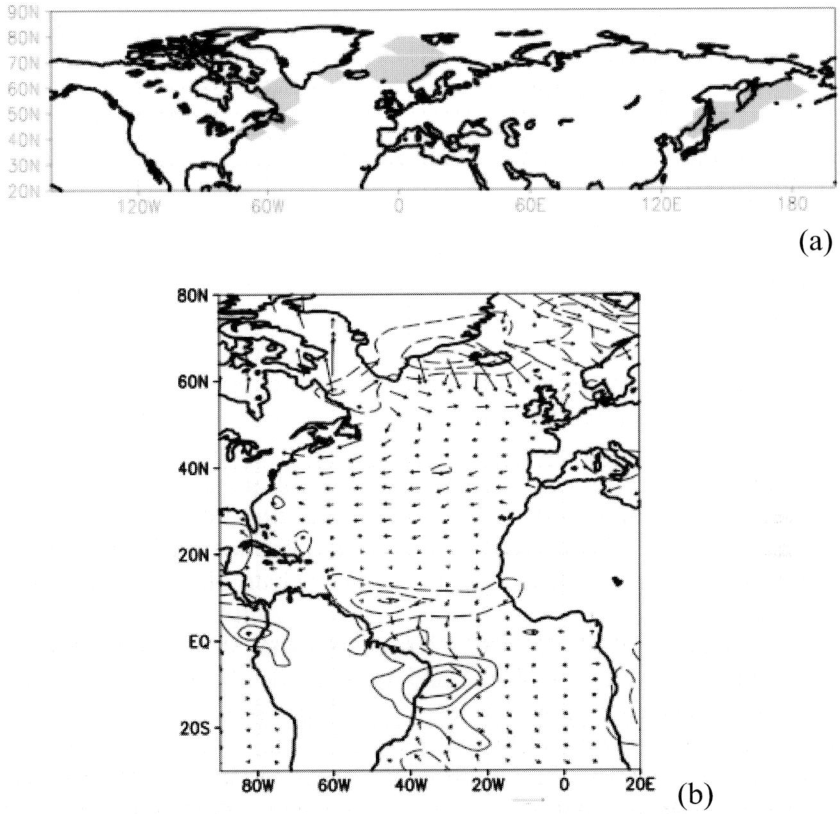

Figure 16-4. Annual mean difference between a present-day "control" simulation, and a simulation with increased sea ice extent in the Northern Hemisphere. (a) Gray shading indicates additional sea ice imposed for the perturbed experiment, for January. Note that in the model used, each grid point is either 100% sea ice or no sea ice. The imposed additional sea ice differs from month to month—in boreal summer, no additional sea ice is imposed. (b) Surface wind (reference vector is 3 m/s) and precipitation (contour interval 0.5 mm/d, negative values are dashed, and the zero contour is not shown) anomalies. A southward shift of the ITCZ is evident from both the surface wind and precipitation anomalies with increased sea ice. Since the ocean component of the model used is only a thermodynamic slab, the method of communication between the sea ice region and the tropics is atmospheric. However, the lack of northern subtropical trade wind anomalies implies that the communication is not through NTA trade wind variations.

southward shift in the model Atlantic ITCZ in the added sea ice experiment compared to the control. So, the influence of additional North Atlantic sea ice works to displace the ITCZ southwards, and apparently in the same direction as the tropical Atlantic ocean heat transport changes due to reduction of THC intensity (Yang 1999). The combination of the two influences may be the reason why the Atlantic ITCZ responds strongly to THC perturbations.

The LGM simulation is another instance in which many models simulate features in the tropical Atlantic resembling the meridional mode spatial signature. Manabe and Broccoli (1985) imposed the LGM land ice sheet on the GFDL model coupled to a fixed-depth ocean mixed layer. The simulation showed substantial cooling in the North Atlantic and very little SST response in the South Atlantic, and a clear signature of north-to-south anomalous cross-equatorial flow over the equatorial Atlantic in the December–February season, indicating a southward displaced ITCZ. A more recent LGM simulation (Broccoli 2000) incorporating LGM forcing as specified by the Paleoclimate Modeling Intercomparison Project (see Section 5), and using an updated and higher-resolution GFDL AGCM slab ocean model, also displays the preferred North Atlantic SST cooling and strengthened NTA trades. Chiang et al. (2003) show (using a similar AGCM slab ocean model configuration in the CCM3) that land ice is the dominant LGM boundary condition that determines the "meridional mode" response in the tropical Atlantic. They show that land ice communicates its influence to the tropics through perturbation of the stationary wave circulation in the North Atlantic due to the presence of the land ice sheet.

A southward-displaced Atlantic ITCZ is also suggested in fully coupled model simulations of the LGM. Kitoh and Murakami (2002) show, in their LGM simulation using the Japanese Meteorological Research Institute (MRI) coupled GCM, increased (2–3 m/s) trades in their NTA region combined with a 3–4 K SST cooling, while the south tropical Atlantic has little surface wind response and a smaller (2–3 K) SST cooling. However, the precipitation response in their model was not well defined in the tropical Atlantic, possibly due to the fact that their Atlantic ITCZ was not well defined in their mean climate. A meridional mode–type response is also suggested in an LGM simulation by Bush and Philander (1999) using the GFDL coupled model. In this case the ITCZ displacement (as indicated in their study by the anomalous cross-equatorial winds) was better defined.

5. THE LGM TROPICAL ATLANTIC IN PMIP SIMULATIONS

How dependent is the simulation of tropical Atlantic LGM climate on the GCM used? The Paleoclimate Modeling Intercomparison Project coordinated and made available simulations of present and LGM climate for several GCMs (Joussaume and Taylor 2000). While they all have virtually the same specifications of LGM boundary conditions, the physics incorporated into the models differ. We extracted all available modern and 21,000 BP (years before present) simulations for which the SST was predicted rather than imposed (see Table 16-1 for the list of models and their labels). Differences in the annual mean climatological SST and 10 m wind vectors in the tropical Atlantic between 21,000 BP and modern values in eight PMIP runs are shown in Figure 16-5, and surface wind speed differences are shown in Figure 16-6. Despite apparent differences in both the wind and SST response, two general features appear common to all models:

- The strengthening of the northern tropical and subtropical trades[4] during 21,000 BP relative to modern (this is more readily visible when wind speed rather than wind vector is plotted [not shown]); and

- The north-south gradient in the SST difference, with the north cooling more than the south during 21,000 BP.

What differs is the relative magnitude of the response: The meridional gradient in SST across the equator ranges from barely existing (as in the CCC2, GFDL, and UGAM models) to very strong (GEN1, UKMO, and MRI2 models). The cross-equatorial flow strength is closely tied to the strength of the anomalous SST gradient—the stronger the gradient, the larger the cross-equatorial flow. We interpret the PMIP results to mean that the fundamental dynamics causing the increase in the northeasterly trades and cooling of the NTA SST is robust. Likewise, so is the response of the tropical cross-equatorial flow (and hence ITCZ displacement) given the meridional SST gradient. What appears to vary significantly is the sensitivity of the tropical Atlantic to the northern subtropical trade wind forcing.

---

[4] The CCM1 model surface wind response is relatively weak; however, a similar LGM simulation using the updated CCM3 model shows clearly strengthened northern subtropical trades and a meridional SST gradient across the equator (Chiang et al. 2003).

Table 16-1. The list of model data from the PaleoclimateModeling Intercomparison Project used in our analysis; and a brief description for each model. Other PMIP models were excluded either because the requisite run was not available, or because the model dynamics were insufficient to simulate Atlantic ITCZ behavior.

Label	**Model and Resolution**	**Description of Ocean**
CCC2	Canadian Center for Climate modeling and analysis model CCMA version 2 (T32 L10)	Mixed-layer model
CCM1	National Center for Atmospheric Research (NCAR) CCM1 (R15 L12)	50 m mixed-layer, fixed annual mean, zonally symmetric ocean heat transport (OHT)
GEN1	National Center for Atmospheric Research/ Pennsylvania State University (NCAR/PSU) GENESIS 1.02A (R15 L12)	50 m mixed-layer, fixed annual mean, zonally symmetric OHT
GEN2	NCAR/PSU GENESIS 2 (T31 L18)	50 m mixed-layer, diffusive OHT
GFDL	Geophysical Fluid Dynamics Laboratory model CDG (R30 L20)	50 m mixed-layer, prescribed seasonally and spatially varying OHT
MRI2	Meteorological Research Institute model MRI GCM-IIb (4 x 5 L15)	50 m mixed-layer, prescribed OHT restores model SST and sea ice amount to observations
UGAMP	UK Universities' Global Atmospheric Modeling Programme model UGAMP UGCM version 2 (T42 L19)	Mixed-layer, prescribed seasonally and spatially varying OHT restores to present-day SST
UKMO	United Kingdom Meteorological Office model UKMO HADAM2 (2.5 x 3.75 L19)	Bryan-Cox primitive equation ocean model

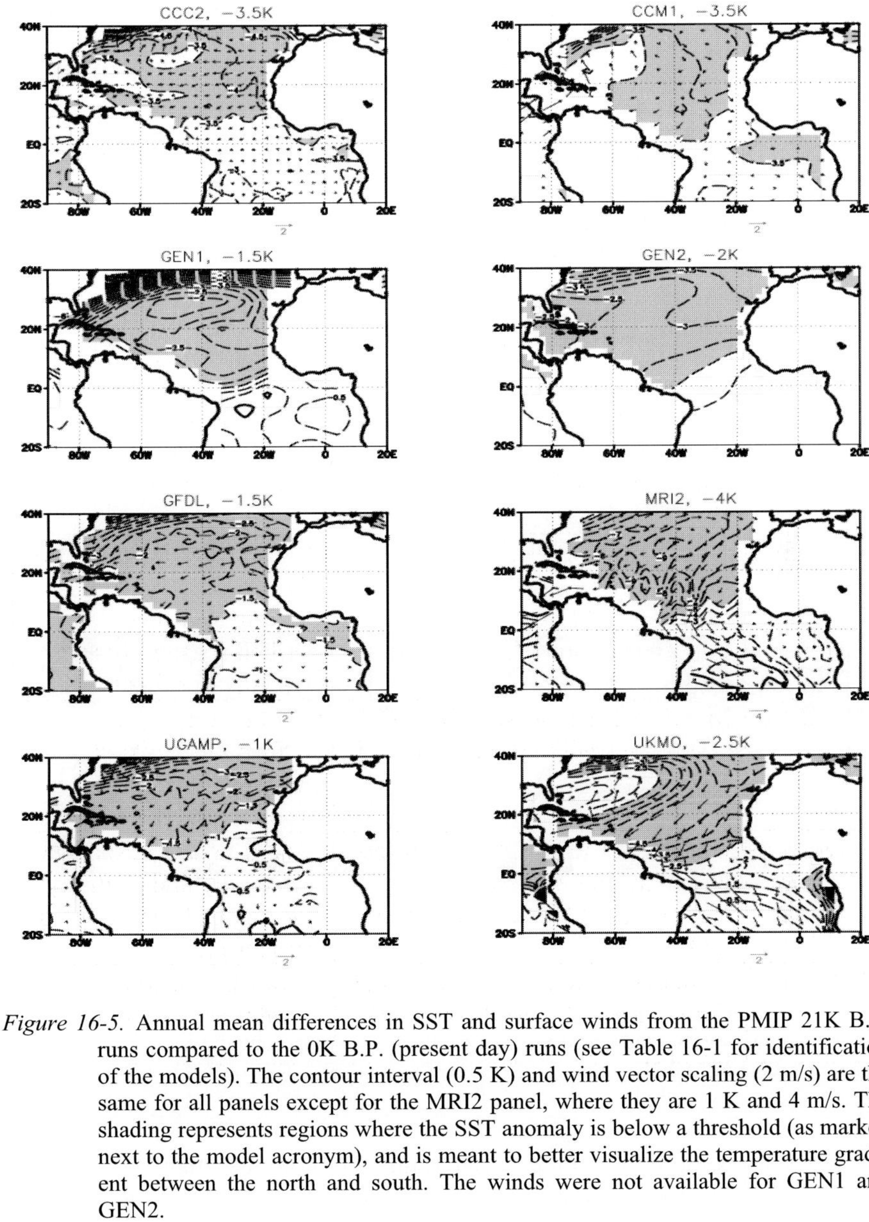

Figure 16-5. Annual mean differences in SST and surface winds from the PMIP 21K B.P. runs compared to the 0K B.P. (present day) runs (see Table 16-1 for identification of the models). The contour interval (0.5 K) and wind vector scaling (2 m/s) are the same for all panels except for the MRI2 panel, where they are 1 K and 4 m/s. The shading represents regions where the SST anomaly is below a threshold (as marked next to the model acronym), and is meant to better visualize the temperature gradient between the north and south. The winds were not available for GEN1 and GEN2.

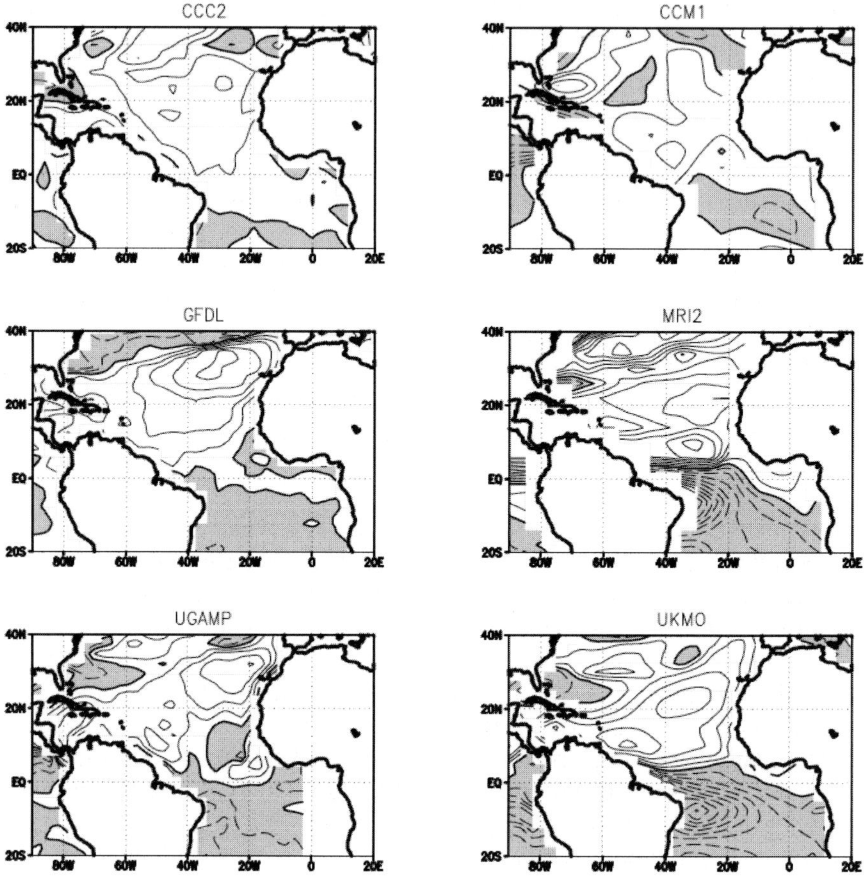

Figure 16-6. Same as Fig. 16-5, but for annual mean differences in surface wind speed between the PMIP 21K B.P. and 0K B.P. runs. The contour interval is 0.4 m/s, negative regions are shaded, and the thick contour denotes zero. Surface wind speeds were estimated from the annual mean zonal and meridional wind components, as annual mean wind speed was not available. The wind output was not available for GEN1 and GEN2.

Why is there this range of behavior? We note that many (though not all) of the PMIP present-day simulations had mean-state SST and surface winds that were significantly different from those observed, and from each other (not shown). These differences may account in part for the range in tropical sensitivity to the NTA trade wind forcing, since it has been shown that the meridional mode response is sensitive to the mean-state climate

(Biasutti 2000). Chiang et al. (2003) note that in their LGM simulation using the CCM3 slab ocean model, while it is the land ice sheet that causes the northern subtropical trade winds to increase, it is the tropical atmosphere–thermodynamic ocean feedback that brings the signal to the deep tropics in their model. The PMIP results suggest that the strength of this tropical feedback varies among models. Whatever the correct reason, it appears that correct sensitivity of tropical processes is prerequisite for an accurate simulation for the tropical Atlantic climate.

Finally, we note that the LGM tropical Atlantic "meridional mode" response in the only fully coupled model (UKMO) in this comparison qualitatively resembles the "meridional mode" response of the other PMIP models that possess only slab oceans. This *might* indicate that ocean mixed-layer thermodynamics, rather than full ocean dynamics, is the more important process determining LGM response. This conclusion would be consistent with the primary role assigned to ocean surface thermodynamics in determining present-day meridional mode variability (e.g. Seager et al. 2001). However, it is difficult to pin down the relative roles of ocean thermodynamics and dynamics in the LGM tropical Atlantic response without a more comprehensive examination of the model response. (Ideally, one would like two simulations, one with and one without the ocean dynamical response, and also comparison with other coupled model LGM simulations to see how robust the response was.) Also, since the LGM perturbation is fairly substantial, one might expect significant feedback from coupled or ocean dynamics, like the Bjerknes-type equatorial coupled dynamics (see the end of the next section for a discussion), or coastal upwelling. Indeed, one region where the UKMO model response qualitatively differs from the mixed-layer-only simulations is the cold-SST region off central and southwest Africa, where upwelling is known to play a role in determining SST.

6. SUMMARY AND DISCUSSION

Paleoclimate change in the tropical Atlantic as inferred from various paleoproxy records has been interpreted as related to displacements in the mean position of the Atlantic ITCZ, analogous to the dominant "meridional" mode of interannual-decadal climate variability in the region. This interpretation raises the question as to what extent the present-day tropical Atlantic climate variability can be used as a model for past changes to the climate and the circulation in the tropical Atlantic. Current understanding suggests that the ITCZ is sensitive to the local meridional gradient in SST, and that this SST gradient in turn is sensitive to forcing from outside the tropical Atlantic region. These features make the ITCZ highly sensitive to climate changes, either locally forced through insolation changes, or remotely induced through changes to the climate in other regions. The displacement of the ITCZ may lead to substantial changes to the tropospheric thermally direct circulation, as suggested by the modern analog (Wang 2002). A review of past model results suggests that meridional ITCZ displacement may indeed be the preferred way for the tropical Atlantic climate to respond to paleoclimate changes. In particular, the ITCZ is sensitive to two paleoclimate situations: a change to the Atlantic thermohaline circulation (though it is uncertain whether it is the ocean heat transport or the Northern Hemisphere sea ice change associated with THC change that is responsible), and to boundary conditions associated with LGM climate. However, as has been demonstrated by PMIP simulations of the LGM, the degree to which the model tropical Atlantic responds with an ITCZ displacement and associated marine conditions is sensitive to the model used.

The demonstrated sensitivity of the Atlantic ITCZ to THC and sea ice changes suggests that the Atlantic ITCZ may also be sensitive to increased atmospheric CO_2 concentrations, since several coupled model studies (e.g. Stouffer et al. 1989; Vinnikov et al. 1999) show changes to polar sea ice concentrations and the THC. The direct influence of CO_2 may not impact the ITCZ position, because of its uniform spatial distribution: Chiang et al. (2003) show from their CCM3 slab ocean model simulations that lowered CO_2 levels (to LGM levels), by itself and without sea ice feedbacks, do not change the Atlantic ITCZ position. The precipitation shows little response, and the tropical Atlantic sea surface temperature cools on the order of 1 K in almost uniform fashion. On the other hand, the THC weakens in the transient phase of most (though not all) coupled model simulations with increased CO_2 (though the equilibrium response may be the same or even strengthened [Stouffer and Manabe 2003]), and many, though not all, coupled models show weakened and shallower quasi-equilibrium THC for CO_2 concentrations approximating LGM levels. Recent coupled model simulations of sea ice extent (Vinnikov et al. 1999), forced by observed greenhouse gases and sulfate aerosols, show decreases to Northern Hemisphere sea ice extent over the last few decades that are consistent with observed trends. So how might the Atlantic ITCZ respond to these feedbacks under enhanced CO_2 concentrations? We can only speculate what might happen, but based on the scenarios presented, increased CO_2 weakens the THC and *reduces* Northern Hemisphere sea ice extent, in contrast to the THC perturbation scenarios described in Section 4, where THC weakens and Northern Hemisphere sea ice extent *increases*. In this case, therefore, the THC and sea ice appear to play opposing roles, and the response of the Atlantic ITCZ is not obvious. There is also additional uncertainty in that changing CO_2 levels may also change Southern Hemisphere sea ice, which may augment (if coverage increases) or counter (if coverage decreases) the effects of decreased Northern Hemisphere sea ice on the ITCZ.

The wide range in magnitude of the PMIP model responses to essentially the same LGM forcing suggests that the climate dynamics in this region have yet to be correctly pinned down by coupled GCMs. This may be due to incorrect sensitivities in the physical parameterizations incorporated into the models (like convection and clouds, and the relative roles of land vs. ocean in the tropical climate), or even missing physics. It is also telling that most current coupled GCMs do a relatively poor job of simulating the tropical Atlantic climate (Davey et al. 2002), and in particular the ITCZ. At the time of this writing, major efforts are under way to improve the ITCZ representation in coupled model simulations. Accurate tropical Atlantic ITCZ paleoproxy records can play an important role in solving this vexing problem, by providing crucial constraints on the acceptable range of model tropical Atlantic behavior given paleoclimate forcing.

As far as paleoproxy research is concerned, the northeast (NE) Brazil (around 4°S and 39°W) region is an ideal location for monitoring Atlantic ITCZ behavior. Because of its strategic position relative to the Atlantic ITCZ and its migration,[5] the variability of rainfall there during its boreal spring rainy season is largely determined by the behavior of the Atlantic ITCZ, which in turn is determined by atmosphere-ocean conditions over the entire tropical Atlantic and also the tropical Pacific (Uvo et al. 1998). Furthermore, boreal spring is also the time when the Atlantic ITCZ is most sensitive to perturbations (Chiang et al. 2002). Northeastern Brazil rainfall is therefore indicative of large-scale tropical marine climate conditions at a time when they are particularly susceptible to change. Also, rainfall changes over NE Brazil are quite dramatic—for example, variations in February–May averaged rainfall over Fortaleza (3.7°S, 38.5°W) over 1950–2002 range from just over 5 cm/month to about 50 cm/month—an order of magnitude difference (Fig. 16-7). Northeast Brazil is therefore a superb amplifier of otherwise relatively subtle changes to the tropical marine SST climate.

While the focus here has been on the tropical Atlantic meridional mode, since it appears to be the mode most relevant to paleoclimate, it should be noted that unlike the tropical Pacific with its ENSO mode, the tropical Atlantic is not dominated by any single mode of climate variability (Sutton et al. 2000). In particular, an El Niño–like mode, the Atlantic "Niño," also operates and has sizable control over the climate over the equatorial Atlantic

[5] Figure 16-2 shows nicely the overlap between the rainfall anomaly associated with the meridional mode and the NE Brazil region.

and neighboring land regions, especially along the Gulf of Guinea and also along the equatorial Amazon regions (Zebiak 1993). This mode is characterized by anomalous SSTs centered over the Atlantic cold tongue, analogous to El Niño. The second mode of the MCA analysis in Section 2 has characteristics resembling the Atlantic Niño (Fig. 16-8). It is likely that this mode will also respond to paleoclimate forcing, potentially complicating the tropical Atlantic response.[6] Because of the potential degrees of freedom of the tropical Atlantic response, and also the potential ambiguity in interpreting single-location paleoproxy records, ideally a network of paleoproxy sites should be used to determine the past climate history of the tropical Atlantic. For optimizing the locations of these sites, it would be useful to take advantage of our knowledge of modern-day tropical Atlantic climate variability.

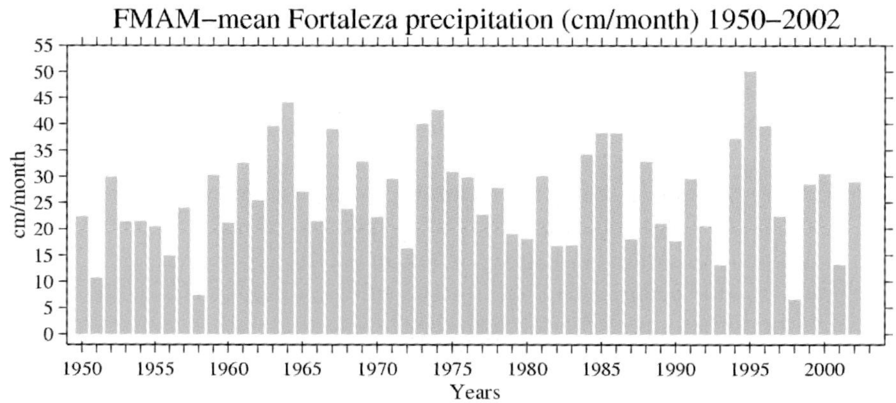

Figure 16-7. February–May averaged rainfall in Fortaleza (3.7°S, 38.5°W) over 1950–2002 (courtesy of Todd Mitchell, University of Washington), showing the extreme variability in the rainfall there associated with Atlantic ITCZ variability. The 1998 value of this time series is not reliable, as it is based on the rainfall for February 1998 only. This figure can be found online at http://tao.atmos.washington.edu/data_sets/brazil.

[6] In fact, there may be a relationship between the meridional mode and the Atlantic Niño, as postulated by Servain et al. (1999)

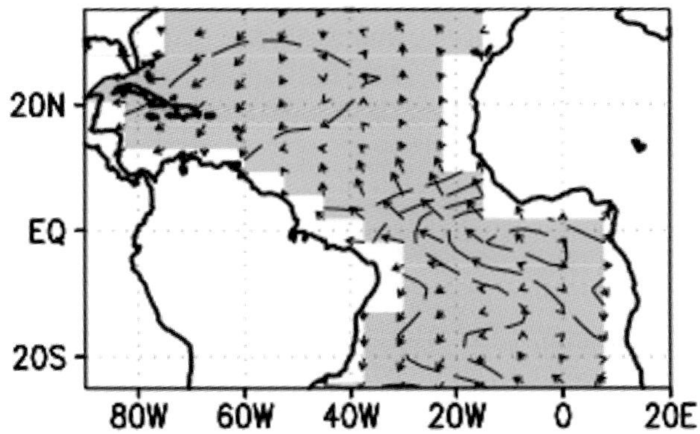

(a)

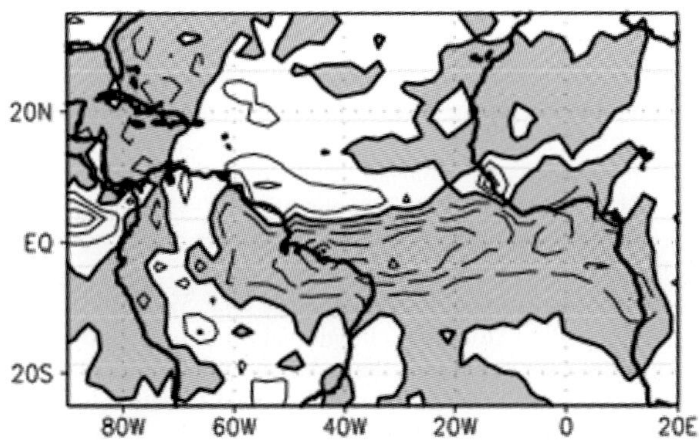

(b)

Figure 16-8. The tropical Atlantic possesses other modes of climate variability. This figure is similar to Fig. 16-2, but for the second MCA mode, showing a pattern resembling the Atlantic "Niño."

7. ACKNOWLEDGMENTS

J.C.H. Chiang was funded in part through a National Oceanic and Atmospheric Administration (NOAA) Climate and Global Change postdoctoral fellowship, administered by the University Corporation for Atmospheric Research (UCAR) visiting scientist programs. The National Science Foundation (NSF) provided travel funds for the author to attend the Hadley circulation conference. The PMIP data were obtained online: http://www-lsce.cea.fr/pmip. The CCM3 simulations were done on the IBM-SP supercomputers managed by the Scientific Computing Division of the

National Center for Atmospheric Research (NCAR). Thanks to T. Mitchell (Fig. 16-7), L. Peterson (Fig. 16-1), and S.-P. Xie (Fig. 16-3) for permission to reproduce their figures, and to D.S. Battisti, M. Biasutti, M. Evans, C. Farmer, A. Koutavas, and G. Seltzer for stimulating discussions on tropical Atlantic paleoclimate.

8. REFERENCES

Barsugli, J.J., and D.S. Battisti. 1998. The basic effects of atmosphere-ocean thermal coupling on midlatitude variability. *Journal of the Atmospheric Sciences* 55: 477–493.

Biasutti, M. 2000. Decadal variability in the Tropical Atlantic as simulated by the Climate System Model and the CCM3 coupled to a slab ocean model, Dept. of Atmospheric Sciences, University of Washington, 43 pp.

Black, D.E., L.C. Peterson, J.T. Overpeck, A. Kaplan, M.N. Evans, and M. Kashgarian. 1999. Eight centuries of North Atlantic Ocean atmosphere variability. *Science* 286: 1709–1713.

Bretherton, C.S., C. Smith, and J.M. Wallace. 1992. An intercomparison of methods for finding coupled patterns in climate data. *Journal of Climate* 5: 541–560.

Broccoli, A.J. 2000. Tropical cooling at the last glacial maximum: An atmosphere–mixed layer ocean model simulation. *Journal of Climate* 13: 951–976.

Bush, A.B.G., and S.G.H. Philander. 1999. The climate of the Last Glacial Maximum: Results from a coupled atmosphere-ocean general circulation model. *Journal of Geophysical Research—Atmospheres* 104: 24509–24525.

Cane, M.A. 1998. A role for the Tropical Pacific. *Science* 282: 59–61.

Chang, P., L. Ji, and H. Li. 1997. A decadal climate variation in the tropical Atlantic Ocean from thermodynamic air-sea interactions. *Nature* 385: 516–518.

Chang, P., R. Saravanan, L. Ji, and G.C. Hegerl. 2000. The effect of local sea surface temperatures on the atmospheric circulation over the tropical Atlantic sector. *Journal of Climate* 13: 2195–2216.

Chiang, J.C.H., M. Biasutti, and D.S. Battisti. 2003. Sensitivity of the Atlantic Intertropical Convergence Zone to Last Glacial Maximum boundary conditions. *Paleoceanography* 18: 10.1029/2003PA000916.

Chiang, J.C.H., Y. Kushnir, and A. Giannini. 2002. Deconstructing Atlantic ITCZ variability: Influence of the local cross-equatorial SST gradient, and remote forcing from the eastern equatorial Pacific. *Journal of Geophysical Research* 107, 10.1029/2000JD000307.

Chiang, J.C.H., S.E. Zebiak, and M.A. Cane. 2001. Relative roles of elevated heating and surface temperature gradients in driving anomalous surface winds over tropical oceans. *Journal of the Atmospheric Sciences* 58: 1371–1394.

Czaja, A., P. van der Vaart, and J. Marshall. 2002. A diagnostic study of the role of remote forcing in tropical Atlantic variability. *Journal of Climate* 15: 3280–3290.

Davey, M.K., M. Huddleston, K.R. Sperber, P. Braconnot, F. Bryan, D. Chen, R.A. Colman, C. Cooper, U. Cubasch, P. Delecluse, D. DeWitt, L. Fairhead, G. Flato, C. Gordon, T. Hogan, M. Ji, M. Kimoto, A. Kitoh, T.R. Knutson, M. Latif, H. Le Treut, T. Li, S. Manabe, C.R. Mechoso, G.A. Meehl, S.B. Power, E. Roeckner, L. Terray, A. Vintzileos, R. Voss, B. Wang, W.M. Washington, I. Yoshikawa, J.Y. Yu, S. Yukimoto, and S.E. Zebiak. 2002. STOIC: A study of coupled model climatology and variability in tropical ocean regions. *Climate Dynamics* 18: 403–420.

Dong, B.W., and R.T. Sutton. 2002. Adjustment of the coupled ocean-atmosphere system to a sudden change in the Thermohaline Circulation. *Geophysical Research Letters*, 29, art. no. 1728.

Hastenrath, S., and L. Greischar. 1993. Circulation mechanisms related to Northeast Brazil rainfall anomalies. *Journal of Geophysical Research—Atmospheres* 98: 5093–5102.

Haug, G.H., K.A. Hughen, D.M. Sigman, L.C. Peterson, and U. Rohl. 2001. Southward migration of the intertropical convergence zone through the Holocene. *Science* 293: 1304–1308.

Hughen, K.A., J.T. Overpeck, L.C. Peterson, and S. Trumbore. 1996. Rapid climate changes in the tropical Atlantic region during the last deglaciation. *Nature* 380: 51–54.

Johnson, H.L., and D.P. Marshall. 2002. Localization of abrupt change in the North Atlantic thermohaline circulation. *Geophysical Research Letters*, 29, art. no. 1083.

Joussaume, S., and K.E. Taylor. 2000. The Paleoclimate Modeling Intercomparison Project. *Proceedings of the Third PMIP Workshop, WCRP report*, 111, 271.

Kiehl, J.T., J.J. Hack, G.B. Bonan, B.A. Boville, D.L. Williamson, and P.J. Rasch. 1998. The National Center for Atmospheric Research Community Climate Model: CCM3. *Journal of Climate* 11: 1131–1149.

Kitoh, A., and S. Murakami. 2002. Tropical Pacific climate at the mid-Holocene and the Last Glacial Maximum simulated by a coupled ocean-atmosphere general circulation model. *Paleoceanography* 17, art. no. 1047.

Lindzen, R.S., and S. Nigam. 1987. On the role of sea-surface temperature gradients in forcing low-level winds and convergence in the tropics. *Journal of the Atmospheric Sciences* 44: 2440–2458.

Manabe, S., and A.J. Broccoli. 1985. The influence of continental ice sheets on the climate of an ice-age. *Journal of Geophysical Research—Atmospheres* 90: 2167–2190.

Manabe, S., and R.J. Stouffer. 1988. Two stable equilibria of a coupled ocean-atmosphere model. *Journal of Climate* 1: 841–866.

Marshall, J., Y. Kushnir, D. Battisti, P. Chang, A. Czaja, R. Dickson, J. Hurrell, M. McCartney, R. Saravanan, and M. Visbeck. 2001. North Atlantic climate variability: Phenomena, impacts and mechanisms. *International Journal of Climatology* 21: 1863–1898.

Nobre, P., and J. Shukla. 1996. Variations of sea surface temperature, wind stress, and rainfall over the tropical Atlantic and South America. *Journal of Climate* 9: 2464–2479.

Okumura, Y., S.-P. Xie, A. Numaguti, and Y. Tanimoto. 2001. Tropical Atlantic air-sea interaction and its influence on the NAO. *Geophysical Research Letters* 28: 1507–1510.

Peterson, L.C., G.H. Haug, K.A. Hughen, and U. Rohl. 2000. Rapid changes in the hydrologic cycle of the tropical Atlantic during the last glacial. *Science* 290: 1947–1951.

Ruiz-Barradas, A., J.A. Carton, and S. Nigam. 1999. Structure of interannual-to-decadal climate variability in the tropical Atlantic sector. *Journal of Climate* 12: 1–43.

Seager, R., Y. Kushnir, P. Chang, N.H. Naik, J. Miller, and W. Hazeleger. 2001. Looking for the role of the ocean in tropical Atlantic decadal climate variability. *Journal of Climate* 14: 638–655.

Seager, R., Y. Kushnir, M. Visbeck, N. Naik, J. Miller, G. Krahmann, and H. Cullen. 2000. Causes of Atlantic Ocean climate variability between 1958 and 1998. *Journal of Climate* 13: 2845–2862.

Servain, J. 1991. Simple climatic indexes for the tropical Atlantic Ocean and some applications. *Journal of Geophysical Research—Oceans* 96: 15137–15146.

Servain, J., I. Wainer, J.P. McCreary, and A. Dessier. 1999. Relationship between the equatorial and meridional modes of climatic variability in the tropical Atlantic. *Geophysical Research Letters* 26: 485–488.

Stouffer, R.J., and S. Manabe. 2003. Equilibrium response of thermohaline circulation to large changes in atmospheric CO_2 concentration. *Climate Dynamics* 20: 759–773.

Stouffer, R.J., S. Manabe, and K. Bryan. 1989. Interhemispheric asymmetry in climate response to a gradual increase of atmospheric CO_2. *Nature* 342: 660–662.

Sutton, R.T., S.P. Jewson, and D.P. Rowell. 2000. The elements of climate variability in the tropical Atlantic region. *Journal of Climate* 13: 3261–3284.

Uvo, C.B., C.A. Repelli, S.E. Zebiak, and Y. Kushnir. 1998. The relationships between tropical Pacific and Atlantic SST and northeast Brazil monthly precipitation. *Journal of Climate* 11: 551–562.

Vellinga, M., and R.A. Wood. 2002. Global climatic impacts of a collapse of the Atlantic thermohaline circulation. *Climatic Change* 54: 251–267.

Vinnikov, K.Y., A. Robock, R.J. Stouffer, J.E. Walsh, C.L. Parkinson, D.J. Cavalieri, J.F.B. Mitchell, D. Garrett, and V.F. Zakharov. 1999. Global warming and Northern Hemisphere sea ice extent. *Science* 286: 1934–1937.

Wang, C.Z.. 2002. Atlantic climate variability and its associated atmospheric circulation cells. *Journal of Climate* 15: 1516–1536.

Xie, S.-P. 1999. A dynamic ocean-atmosphere model of the tropical Atlantic decadal variability. *Journal of Climate* 12: 64–70.

Xie, S.-P., and Y. Tanimoto. 1998. A pan-Atlantic decadal climate oscillation. *Geophysical Research Letters* 25: 2185–2188.

Yang, J.Y. 1999. A linkage between decadal climate variations in the Labrador Sea and the tropical Atlantic Ocean. *Geophysical Research Letters* 26: 1023–1026.

Zebiak, S.E. 1993. Air-sea interaction in the equatorial Atlantic region. *Journal of Climate* 6: 1567–1568.

Chapter 17

MECHANISMS OF AN INTENSIFIED HADLEY CIRCULATION IN RESPONSE TO SOLAR FORCING IN THE TWENTIETH CENTURY

Gerald A. Meehl, Warren M. Washington, T.M.L. Wigley, Julie M. Arblaster, and Aiguo Dai

National Center for Atmospheric Research,[**] *P.O. Box 3000, Boulder, Colorado 80307-3000, U.S.A.*

Abstract Ensemble experiments with a global coupled climate model for the twentieth century with time-evolving solar, greenhouse gas (GHG), sulfate aerosol (direct effect), and ozone (tropospheric and stratospheric) forcing are analyzed to show that solar forcing produces coupled dynamical interactions in the tropics that strengthen regional Hadley and Walker circulation regimes over the first half of the century. Solar forcing produces feedbacks involving temperature gradient–driven atmospheric circulations that can alter clouds. Over relatively cloud-free oceanic regions in the subtropics, greater solar forcing in mid-century compared to the early century produces greater evaporation, more moisture transport into the precipitation convergence zones, intensified regional Hadley and Walker circulations, less clouds over the subtropical ocean regions, and even more solar input. Coupled dynamical interactions produce upper-ocean heat content anomalies in concert with positive sea surface temperature (SST) anomalies that intensify precipitation over the South Indian, South Pacific, and South Atlantic Convergence Zones, as well as the South Asian and West African monsoons. Coupled regional responses are most evident when the solar forcing occurs in concert with increased greenhouse gas forcing of about the same magnitude over the first half of the century. The latter is also altered by interaction with solar forcing, and the base-state tropical SSTs are increased in the relatively cloud-free subtropical regions of low-level moisture divergence to fuel the regional feedbacks induced by the spatially differentiated solar forcing. Consequently, the greater solar forcing acting in concert with increased greenhouse gases during the early

[**] The National Center for Atmospheric Research (NCAR) is sponsored by the National Science Foundation (NSF).

H.F. Diaz and R.S. Bradley (eds.), The Hadley Circulation: Present, Past and Future, 489–511.

twentieth century produces larger increases of tropical precipitation, calculated as a residual for the solar forcing, than for early-century solar-only forcing, even though the size of the solar forcing is the same.

1. INTRODUCTION

Previous climate model experiments have suggested that globally averaged warming during the early twentieth-century warming could have been primarily due to "natural" forcings (mainly solar with some contribution from reduced volcanic activity), while the late twentieth-century warming was likely mainly due to increases of greenhouse gases (GHGs; Cubasch et al. 1997; Stott et al. 2000, 2001; Meehl et al. 2003). However, natural variability also plays a role (Delworth and Knutson 2000). Thus, a combination of internal variability and external forcing could have contributed to the actual early twentieth-century warming. The geographical distributions of the response to external forcing could also play a role (Meehl et al. 2003).

Studies in the paleoclimate context have indicated that during past periods of enhanced solar forcing, there is relatively greater summertime warming of tropical areas, land in particular, which produces stronger tropical monsoon regimes (e.g., Kutzbach et al. 1998; Joussaume et al. 1999). However, these changes are due to orbital shifts that produce enhanced solar radiation in the northern tropics in spring, not to an increase of total solar irradiance. In addition, observations have shown that the signature of enhanced solar forcing is evident in a warming of the tropical troposphere (van Loon and Shea 2000) and an intensified Hadley circulation, implying stronger tropical precipitation (van Loon and Labitzke 1994; Labitzke and van Loon 1995).

Model experiments involving changes of stratospheric ozone associated with the 11-year solar cycle (related mostly to ultraviolet [UV] radiation) with solar forcing changes have been able to reproduce some aspects of these observations (Rind and Balachandran 1995; Haigh 1996; Balachandran et al. 1999; Rind et al. 1999; Shindell et al. 1999, 2001). Those studies usually involve a hypothesized change of stratospheric ozone associated with the 11-year solar cycle (related mostly to UV radiation). This change in ozone levels results in the enhanced UV radiation causing changes in vertical gradients of temperature and wind that lead to altered planetary wave propagation and changes in circulation patterns in the lower stratosphere and troposphere. Though some investigators note either direct (e.g., Balachandran et al. 1999) or implied (e.g., Rind and Balachandran 1995; Haigh 1996) changes in tropical precipitation with fluctuations of solar forcing, sea surface temperatures (SSTs) are usually fixed and there is no capability for coupled feedbacks at the surface to contribute to the response. Rind et al. (1999) use a slab nondynamic ocean in experiments comparing solar and greenhouse forcing. They conclude there is little change in tropical precipitation due to solar forcing, in spite of the Balachandran et al. (1999) result that in a different model version there were small increases in tropical precipitation in response to increases in solar forcing.

Infrared radiation (IR) forcing is relatively spatially uniform over land and ocean in all seasons, with land warming faster than ocean, and with high-latitude amplification of warming due to ice-albedo feedback (e.g., Cubasch et al. 2001). The influence on the strength of monsoon circulations is scenario dependent (Cubasch et al. 2001). In contrast, solar forcing is greater in the tropics in cloud-free areas during daytime. This result suggests that the response of the climate system to solar versus GHG forcing could involve coupled feedbacks at the surface and be different in transient climate change experiments. Indeed, some of the simulated changes in climate response to these different forcings have been previously noted (e.g., Shindell et al. 2001).

In this chapter, we examine the lower-frequency (multi-decadal) transient climate system response of a global coupled atmosphere–ocean–sea ice climate model to study the response of tropical precipitation regimes associated with the regional Hadley and Walker circulations. We first analyze multimember ensembles forced with increases in GHGs and tropospheric ozone, decreases in stratospheric ozone, and increasing sulfate aerosol (direct effect). We then contrast those results with a multimember ensemble of those forcings in combination with solar forcing. In addition, the linearity issue is addressed by comparison of a multimember ensemble of model integrations performed with only solar forcing variability included to compare to the response with solar plus the other forcings. Our goal is to identify differences in the nature of the early and late twentieth-century forcings and responses simulated by the model. We hypothesize that due to the different nature of the forcing (shortwave for solar, and IR for GHGs), the transient response of the climate system could be different, with solar forcing producing relatively warmer tropical areas and stronger tropical regional precipitation regimes (and thus stronger Hadley and Walker circulations), while the response to increased GHGs will be more spatially uniform.

A significant part of solar forcing of the climate system is related to the UV part of the solar spectrum, and this influence could be greatest in the stratosphere, especially with regards to effects involving ozone (e.g., Rind et al. 1995). A notable caveat to the analyses here is that the solar forcing is manifested as changes in total solar irradiance, with no wavelength dependence of the solar forcing in the model (changes in solar forcing are simply total solar irradiance anomalies added to the solar constant). In addition, there is no response of stratospheric ozone to fluctuations of incoming solar radiation, and the atmospheric model does not have a well-resolved stratosphere. We will show that the tropical precipitation responses considered in this chapter occur mainly due to coupled land-atmosphere-ocean interactions.

Results are shown for the average of the ensemble members and differences for early (1936–45 minus 1900–09) and late (1990–99 minus 1961–70) century periods. These differences are assessed in two different ways. First, as a measure of consistency of response across the ensemble members, we calculate a signal-to-noise measure (the ensemble mean difference divided by the intra-ensemble standard deviation). Where this value is greater than 1.0, the ensemble mean response is judged to be consistent across ensemble members (e.g., Cubasch et al. 2001). As a measure of the relative magnitude of the differences in comparison to internal model variability, we calculate a t statistic using a 50-year period from the control integration (assuming equality of variances) to determine if the differences exceed model internal variability.

2. THE MODEL AND EXPERIMENTS

The Parallel Climate Model (PCM) analyzed here is a fully coupled global ocean–atmosphere–sea ice–land surface model without flux adjustment (Washington et al. 2000). The atmospheric component is the National Center for Atmospheric Research Community Climate Model version 3 (NCAR CCM3) with T42 resolution (an equivalent grid spacing of roughly 2.8° by 2.8°) and 18 levels (hybrid coordinates; see details and references in Kiehl et al. [1998]). The land surface model component is the LSM, with specified vegetation types and a comprehensive treatment of surface processes (see Bonan 1998). The CCM3 has an equilibrium climate sensitivity for a doubling of CO_2 of 2.1°C when coupled to a slab ocean with implied ocean heat transports, sometimes called a "q-flux" ocean (Meehl et al. 2000). This sensitivity is on the lower end of other model responses to increasing CO_2 (Cubasch et al. 2001). This is thought to have to do with reduced ice-albedo feedback in this model related to cloud changes, as well as ocean heat transport changes when the model is fully coupled (Meehl et al. 2000).

The PCM ocean component is a version of the Parallel Ocean Processor (POP) with 32 levels in the vertical and 2/3° nominal latitude-longitude resolution reducing to 0.5° in latitude in the tropics (actual longitudinal grid spacing in the equatorial Pacific ranges from about 0.6° in the east to 0.9° in the west due to the effects of the rotated pole over northern North America). El Niño amplitude is comparable to observations (Meehl et al. 2001). Sea ice is thermodynamic with elastic-viscous-plastic (EVP) dynamics as described by Washington et al. (2000).

After spin-up to 1870 conditions (see details in Washington et al. [2000]), an ensemble of five experiments, each starting from a different initial state in the control run at least 20 years apart, is run with time-evolving greenhouse gases, sulfate aerosol (direct effect), and stratospheric and tropospheric ozone changes. This experiment is referred to as "GHG + sulfates." The forcing is shown in Figure 17-1a. Details regarding the forcing are described in Dai et al. (2001a), and results from those experiments are shown in Dai et al. (2001b, c). Further discussion of these experiments is given in Meehl et al. (2003).

A second set of four experiments with varying initial conditions is run with the addition of the time-evolving solar forcing from Hoyt and Schatten (1993) to the GHG plus sulfate forcing. The ensemble mean of the GHG + sulfates is subtracted from these experiments to obtain the solar response as a residual, and is referred to as "solar-residual." A third set of four experiments is run with just the time-varying changes of solar forcing from Hoyt and Schatten (1993), and is referred to as "solar-only." This solar forcing data set is derived from a composite solar irradiance model based on five solar indices plus an added activity component. The five indices are solar cycle length, cycle decay rate, mean level of solar activity, solar rotation, and fraction of penumbral sunspots. For 1979–92, the irradiances are scaled to the mean Nimbus 7 measurements.

A second solar forcing data set is also used in climate model simulations of twentieth-century climate (Lean et al. 1995). The differences between the solar forcing data sets, important for a detection/attribution study, are less crucial in the present chapter since we are focusing on forcing/response aspects. In other words, for a given amount of solar forcing, we are interested in the consequent response. The mechanisms of the response are the focus of this chapter, and should be comparable with either solar forcing data set.

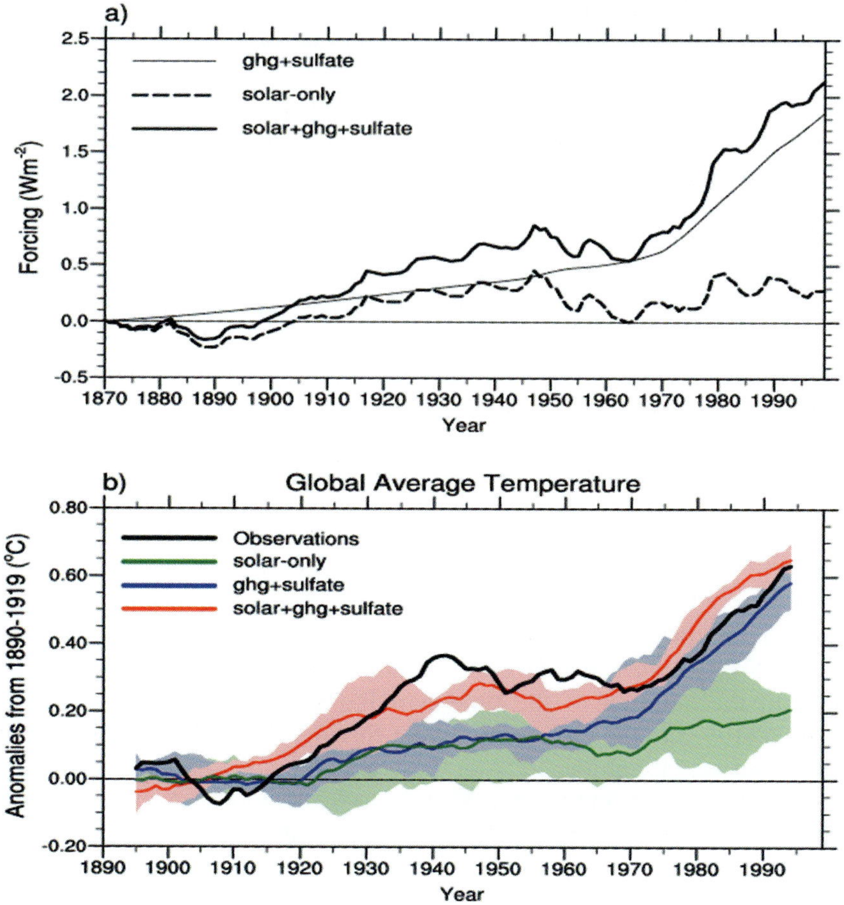

Figure 17-1. (a) Time series of top of troposphere radiative forcing from the coupled model for GHG + sulfates (thin solid line), solar forcing (dashed line) from Hoyt and Schatten (1993), and the sum of the thin solid and dashed lines, the thick solid line, is solar + GHG + sulfates. (b) Global annual mean surface air temperature, 11 y running mean (central value plotted, so the average of, for example, January 1, 1895, to December 31, 1905, is plotted at July 1, 1900), for observations (black line; Jones et al. 1999), ensemble mean (blue line) and range (defined as the maximum and minimum for any member of the ensemble for a given year, shading) for GHG + sulfates experiments from coupled model; ensemble mean (red line) and range (shading) for solar + GHG + sulfates experiments from coupled model; ensemble mean (green line) and range (shading) for solar-only experiments.

The total top-of-troposphere forcing for solar plus GHG plus sulfate plus ozone, as well as the "GHG + sulfates" and the "solar-only" are shown in Figure 17-1a. For GHG + sulfate, positive radiative forcing (relative to the year 1870) increases steadily to a value of about +0.6 W m^{-2} around 1970, and then increases more rapidly to about +1.8 W m^{-2} at the year 2000. The solar forcing (relative to its 1870 value; anomalies divided by 4 to account for spherical geometry, and multiplied by 0.7 to account for planetary albedo) increases to a relative maximum in the mid-1940s of nearly +0.5 W m^{-2}, levels off, and then increases again to late-century maximum values of about +0.6 W m^{-2}.

The early-century solar forcing is about a factor of 3 less than the late-century GHG + sulfate forcing. In comparing the two periods and if the system was linear, the early-century response due to solar forcing should be about a third the size of the late-century response due to increased GHGs. However, nonlinearities could affect the system in several ways. For example, the late century has more forcing history or "warm start" compared to the early century, which has more of a "cold start." The rates of change of the forcing are different in the early and the late century. In addition, single forcings may not produce the same response as additive forcings. The first two are difficult to evaluate, but the last one regarding additive effects will be addressed explicitly in this chapter. We will compare the responses during the two periods to assess the characteristics of the transient climate system response to the two different kinds of forcing (single and additive). We will show results for the GHG + sulfate experiments to illustrate the late-century warming response. The difference of solar + GHG + sulfate minus GHG + sulfate will show the early-century solar response as a residual, and will be compared to the solar forcing–alone ensemble and the GHG + sulfates for the early century.

3. SURFACE TEMPERATURE

Figure 17-1b shows the globally annually averaged surface air temperature from observations (solid black line, 10 y running mean, used by the Intergovernmental Panel on Climate Change [IPCC] and described, for example, by Jones et al. [1999]), along with the model responses (10 y running mean) to the forcings in Figure 17-1b. The range of the ensemble members is depicted by the shading. The range is defined here as the maximum and minimum of any member of the ensemble for a given 10-year mean. Results from the GHG + sulfate experiments in Figure 17-1a are described more fully in Dai et al. (2001b).

The GHG + sulfate ensemble mean is colder than the solar + GHG + sulfate ensemble mean due to less positive radiative forcing in the former compared to the latter (Fig. 17-1a). The GHG + sulfate ensemble mean warms somewhat in early century (0.08°C from a 10 y ensemble mean centered on 1905 to one centered on 1940) but does not show evidence of the large early-century warming seen in the observations. The addition of solar forcing produces warming (0.24°C) closer to the observed during that time period, consistent with previous studies mentioned above. The range of responses from the solar + GHG + sulfate ensembles is about 0.05°C to 0.10°C.

The solar-only ensemble mean follows the GHG + sulfate ensemble for the early century, both with warming around a tenth of a degree. Thus, when the forcings are added, they produce a larger early-century warming closer to the observed as was noted above. The late-century (from the decade centered on 1965 to the one centered on 1995) solar-only ensemble shows warming of about 0.1°C. But to get the rapid rate of warming seen in the observations requires the forcing from the GHG + sulfates (1.1°C). The time series from GHG + sulfates + solar shows some agreement with the observations during the course of the twentieth century, but warms too little in the early century and too much in the late century. Clearly, many issues are involved with such a comparison in a climate change detection/attribution sense, such as relative model sensitivity, uncertainties in the magnitudes of the forcings, and other forcings not included in these experiments. For our purposes here, we will not consider the detection/attribution issue, but will focus on the nature and mechanisms of the coupled climate response to these different forcings.

4. GEOGRAPHICAL CHANGES

To examine the geographical distribution of the changes, Figures 17-2 and 17-3 show precipitation changes for two of the major Northern Hemisphere monsoon systems, the West African and South Asian monsoons, respectively, for early-century solar-only, greenhouse gas plus sulfate, early-century solar-residual, and late-century GHG + sulfate. Stippling on the figures indicates areas where the ensemble mean difference divided by the intra-ensemble standard deviation is greater than 1.0. This result is a measure of consistency of the response across ensemble members (e.g., Cubasch et al. 2001). The contour indicates areas where the ensemble mean difference is significant at the 10% level taken from the measure of variance in the control run over 50 years and assuming that to a first order the variance in each period is similar. Thus there is an ensemble consistency measure and an indication of the relative size of the ensemble mean signal compared to noise in the control run.

In Figures 17-2 and 17-3 for the early-century solar-only and GHG + sulfate, there are few areas of either consistent or significant model response. One area, in the tropical Indian Ocean in the early-century solar-only, is significant in Figure 17-3a, but the area-averaged precipitation index, averaged over the entire monsoon region in Figure 17-3, shows values near zero. However, for the early-century solar-residual in Figure 17-3c, there are a number of areas that are both consistent and significant in the tropical Indian Ocean and in northeastern India and South China, with anomalies of greater than 1 mm d^{-1}. Many of the precipitation increases over the West African monsoon land areas in Figure 17-2c are nearly that size and are both consistent and significant. There is also a large positive increase of area-averaged precipitation of +0.18 mm d^{-1} for both the West African monsoon and the South Asian monsoon for early-century solar-residual. In contrast, although there are increases of precipitation over many areas of the West African and South Asian monsoons for late-century GHG + sulfates, most are neither consistent across ensemble members nor significant, with area-averaged precipitation increases near zero. The solar + GHG + sulfate precipitation differences for the early century (not shown) are characterized mainly by the solar forcing response shown in Figures 17-2c and 17-3c.

One of the most consistent long-term indices of monsoon rainfall is the all-India monsoon rainfall index (Parthasarathy et al. 1991). The linear trend for early-century (1900–45) monsoon rainfall is +1.48 mm y^{-1}. The mid-century linear trend (1945–60) is –0.80 mm y^{-1}, and the late-century linear trend (1960–2000) is –0.10 mm y^{-1}. Thus, the early-century increase is greater than the late-century trend, which is slightly negative. This result is qualitatively consistent with the results in Figure 17-3, though does not provide attribution (note that the solar + GHG + sulfate simulation results, not shown, are dominated by the monsoon changes shown in Fig. 17-3c). Nevertheless, our results do indicate that, for this index, the observed early-century increase of monsoon precipitation compared to late-century might be associated, at least in part, with enhanced solar forcing in the early century. However, we do not include black carbon (soot) over South Asia, which may be an important forcing in that region (Podgorny et al. 2000). Analysis of observed precipitation trends in the IPCC Third Assessment report indicates that early-century precipitation trends over the monsoon regions were mostly positive (Folland et al. 2001), a result that is also consistent with the Parthasarathy record and the results in Figure 17-3, though low-frequency changes in the El Niño/Southern Oscillation (ENSO) effects on the monsoon could also be playing a role.

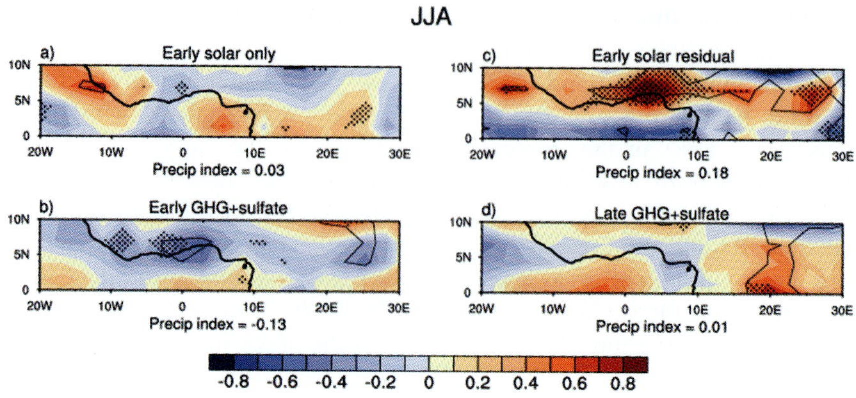

Figure 17-2. (a) June–August precipitation differences (mm d^{-1}) for early-century solar-only for West African monsoon region; precipitation index is area-averaged precipitation for 10°W–10°E, 5°N–10°N; (b) same as (a) except for early-century GHG + sulfate; (c) same as (a) except for early-century solar-residual; (d) same as (a) except for late-century GHG + sulfates. Stippling indicates regions where the ensemble mean divided by the intra-ensemble standard deviation exceeds 1.0; solid black contour shows areas where significance of differences exceeds the 10% level.

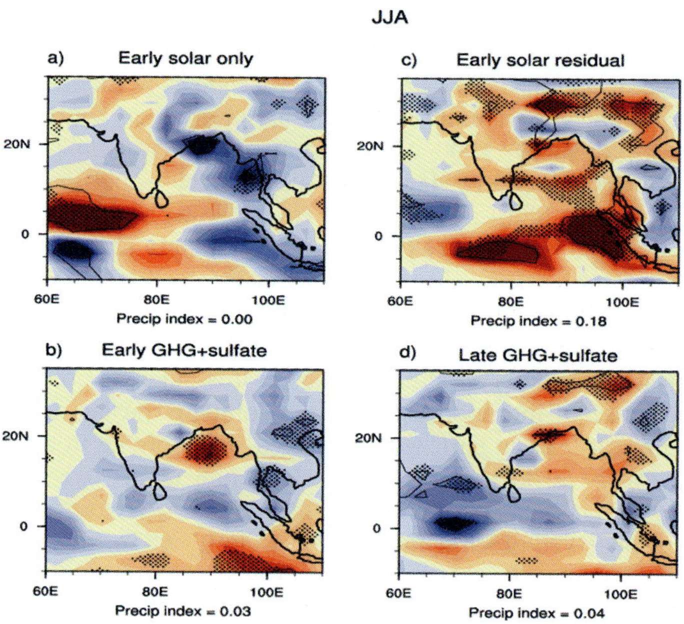

Figure 17-3. Same as Fig. 17-2 except for the South Asian monsoon region; precipitation index is area-averaged precipitation for 10°S–35°N, 60°E–110°E.

For the South Asian monsoon, the change in surface temperature difference for June–August (JJA) from central North Asia to the tropical Indian Ocean is about +0.80°C for the early-century solar-residual forcing ensemble, while for the late-century GHG + sulfates the difference is almost half that, at +0.45°C (not shown). However, this relatively small meridional temperature gradient enhancement for the early-century solar forcing ensemble is during the height of the stronger monsoon. The gradient that sets the stage for monsoon precipitation anomalies appears the previous winter (e.g. Meehl 1997). For early-century solar-residual, the December–February (DJF) warming ranges from values greater than +1.75°C in central Asia to near zero in the tropical Indian Ocean for a meridional temperature gradient enhancement of about +1.75°C (not shown). For late-century GHG + sulfate, the central Asian warming is mostly less than +1°C while the Indian Ocean warms on the order of +0.3°C, a gradient enhancement of about +0.7°C. Thus there is a much greater early-century enhancement in the gradient for solar-residual. There is little gradient enhancement for early-century solar-only or GHG + sulfates, consistent with the weak changes in monsoon precipitation in Figure 17-3.

Analysis of observations shows that a meridional surface temperature gradient increase between the northern Indian Ocean and central Asia of about 1.5°C in seasons prior to the monsoon can contribute to enhanced South Asian summer monsoon rainfall (Meehl and Arblaster 2002a, b). Such a strengthened meridional temperature gradient here contributes to enhancements in South Asian monsoon rainfall in the early-century solar-residual compared to the late-century GHG + sulfates ensembles. There is a factor of 2 or more increase in net solar radiation in the northern tropics during DJF for the early-century solar-residual ensemble mean compared to the late-century ensembles, with the largest values locally over Asia (not shown). This net radiation increase contributes to the enhanced meridional temperature gradients and greater South Asian monsoon rainfall.

There are several possible competing mechanisms for the West African monsoon in addition to the surface energy balance and coupled feedback processes to be described shortly for DJF precipitation changes. In the early-century solar-residual ensembles, there is a slightly enhanced meridional temperature gradient in JJA (not shown), with warming over northern Africa of about +0.3°C and near-zero warming in the tropical Atlantic. This would contribute to enhanced West African monsoon rainfall. In addition, in the late-century temperature differences for GHG + sulfates, there is almost uniform surface temperature warming from North Africa to the tropical Atlantic of about +0.30°C to +0.40°C (not shown), thus providing little enhancement of the meridional temperature gradient. The warmer SSTs in the southern tropical Atlantic in the late century would also contribute to a weaker monsoon over land (e.g., as noted in observations and a modeling study by, for example, Folland et al. [1986]), with positive precipitation anomalies of up to +0.4 mm d^{-1} near 10°S over the south tropical Atlantic Ocean. Taken together, the greater relative warming over land areas of Africa associated with enhanced solar radiation contributes to an enhanced meridional temperature gradient and a stronger West African monsoon in the early-century period compared to the late-century period.

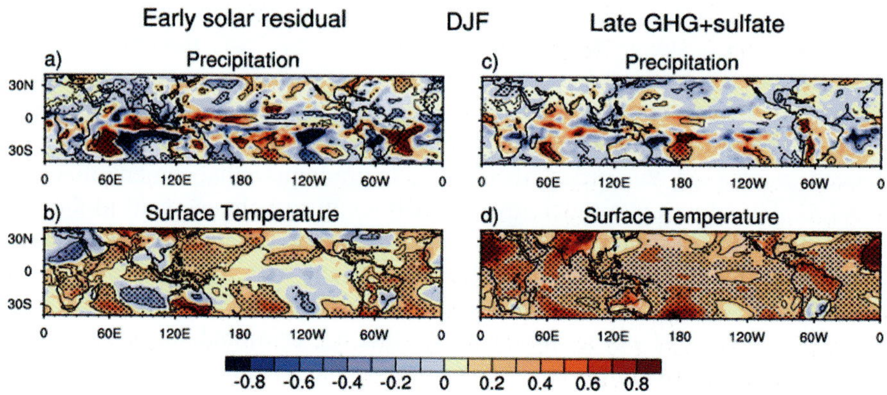

Figure 17-4. (a) December–February precipitation anomalies (mm d^{-1}), for early twentieth-century solar-residual; (b) same as (a) except for surface temperature (C); (c) same as (a) except for late-century GHG + sulfates; (d) same as (b) except for late-century GHG + sulfates. Stippling indicates regions where the ensemble mean divided by the intra-ensemble standard deviation exceeds 1.0; solid black contour shows areas where significance of differences exceeds the 10% level.

For DJF precipitation changes, the regions that have the largest increases for early-century solar-residual (Fig. 17-4a) are mainly in the tropical eastern Indian, Indonesian/northern Australian, and western Pacific Ocean areas, with consistent and significant increases locally of about +1.5 mm d^{-1}. There are also areas of enhanced precipitation in the South Indian Convergence Zone near Madagascar, the South Pacific Convergence Zone (SPCZ) around and southeast of Fiji, and the South Atlantic Convergence Zone (SACZ) southeast of South America for the early-century solar-residual (Fig. 17-4a) compared to late-century GHG + sulfate (Fig. 17-4c). The precipitation differences are about a factor of 2 greater in Figure 17-4a for early-century solar-residual in those regions (about +1 mm d^{-1}, both consistent and significant) compared to late-century GHG + sulfate in Figure 17-4c. These areas of tropical precipitation increase in early-century solar-residual generally correspond to greatest relative increases of SST that are both consistent and significant in Figure 17-4b, with corresponding negative surface temperature anomalies also consistent and significant in the areas of the subtropical high-pressure areas, particularly in the southern Indian and Pacific Oceans. However, for late-century GHG + sulfate, the large, consistent, and significant surface temperature increases in Figure 17-4d have no such spatial differentiation, with more or less uniform spatial increases.

5. MECHANISM FOR STRENGTHENED HADLEY AND WALKER CIRCULATIONS

The late-century GHG + sulfates precipitation changes are a factor of 2 smaller in the tropics compared to the early-century solar-residual changes in Figure 17-4 during DJF. However, the precipitation changes in the tropical convergence zones are a factor of 2 larger for early-century solar-residual compared to solar-only, and both were greater than early-century GHG + sulfates (not shown). The reason for these relative regional precipitation changes in DJF can be traced to changes in surface energy balance in areas upstream of the tropical convergence zones associated with low-level moisture divergence (not shown). Here we choose for analysis these areas, indicated by patterns of low-level moisture divergence, which are low-level moisture sources for the ocean convergence zones where there is enhanced precipitation for the solar forcing ensembles in the early century compared to GHG + sulfates in the late century. These areas include the north and south tropical Indian Ocean, where surface winds collect moisture to converge into the anomalous precipitation enhancement areas in the eastern equatorial Indian Ocean and South Indian Convergence Zone; a region south and west of Hawaii where moisture converges into the enhanced precipitation area in the western tropical Pacific called the western Pacific area; an area in the South Pacific that supplies moisture for the South Pacific Convergence Zone called the South Pacific area; and an area in the tropical Atlantic upstream of the South Atlantic Convergence Zone called the south tropical Atlantic (STA) area (actual latitude-longitude area boundaries are given in Fig. 17-5).

For net solar from the early-century period for solar-residual (Fig. 17-5a), all areas show positive anomalies, though solar-residual values are 30% to a factor of 4 larger than the solar-only values (not shown). The solar-only range from about +0.2 to +0.8 W m^{-2}, all smaller than 1.0 for the ratio of ensemble mean to intra-ensemble standard deviation and smaller than natural variability standard deviations typically of several W m^{-2}. Solar-residual in Figure 17-5a are +0.6 to +2.3 W m^{-2}, with the South Indian and

South Pacific Ocean areas exceeding 1.0 for the ratio of ensemble mean to intra-ensemble standard deviation, and the former about one standard deviation of the natural variability value.

The increase of solar energy at the surface could be expected to increase latent heat flux to maintain surface energy balance. Early-century solar-residual latent heat flux differences are negative for all five areas, indicating more energy removed from the surface by evaporation. The north tropical Indian Ocean and south tropical Indian Ocean areas exceed one intra-ensemble standard deviation, with the latter greater than one standard deviation of natural variability.

The increased low-level moisture from the enhanced evaporation and latent heat flux is then advected into the areas of low-level moisture convergence, thus contributing to the positive precipitation anomalies shown in Figure 17-4a, with the larger and more consistent negative latent heat flux anomalies in the solar-residual anomalies corresponding to the larger positive precipitation anomalies in the oceanic convergence zones, many of the latter consistent and significant in Figure 17-4a. In Figure 17-4b, the enhanced latent heat removal from the surface (Fig. 17-5b) and increased low-level moisture divergence over those areas are associated with consistent and significant SST decreases in areas of enhanced latent heat flux, and increased SST in areas of precipitation increase, indicating a coupling between atmosphere and ocean.

As precipitation increases in the oceanic convergence zones, regional-scale circulations are associated with increased upward vertical motion over the positive precipitation anomalies, and enhanced subsidence over areas of low-level moisture divergence (not shown). This pattern produces decreases of low clouds in four out of five areas in Figure 17-5d for early-century solar-residual. The low cloud changes among the ensembles are fairly noisy, with the south tropical Atlantic area the only one above one intra-ensemble standard deviation. The two areas with negative low cloud changes for early-century solar-only (south tropical Indian Ocean and western Pacific) are smaller than for solar-residual by about 50% (not shown).

For late-century GHG + sulfates in Figure 17-5, the greater amount of CO_2 and other GHGs could be expected to increase downward infrared radiation, which is the case in Figure 17-5c. Late-century values are larger and more consistently significant, and are associated with the greater forcing and more infrared radiation into the surface, with all area averages exceeding one intra-ensemble standard deviation. Since GHG IR forcing is more or less spatially uniform, there are increases in downward IR at most locations (not shown) that produce more spatially uniform, consistent, and significant increases of SST for the late-century GHG + sulfates in Figure 17-4d. This SST increase contributes to enhanced low-level moisture and increases of

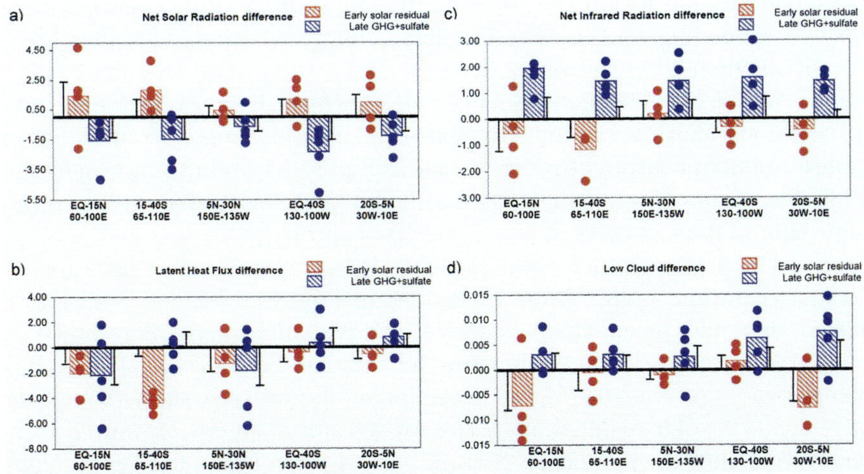

Figure 17-5. (a) Area-averaged net solar radiation (W m^{-2}) for five areas denoting ocean low-level moisture divergence maxima that serve as collection areas for low-level moisture carried to the tropical convergence zones, early-century solar-residual (red hatching), and late-century GHG + sulfates (blue hatching); (b) same as (a) except for latent heat flux (W m^{-2}); (c) same as (a) except for net infrared radiation (values in a, b, and c that are positive denote energy input to the surface); (d) same as (a) except for low cloud fraction. Bars show ensemble means, dots indicate individual ensemble member values, and thin black lines show intra-ensemble standard deviations. The latitude-longitude definitions for each area are given at the bottom of each panel and are referred to in the text as, from left to right, north tropical Indian Ocean, south tropical Indian Ocean, western Pacific, South Pacific, and south tropical Atlantic.

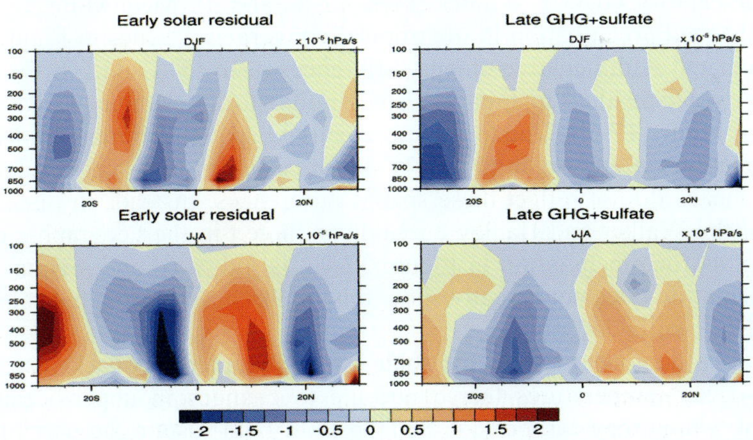

Figure 17-6. (a, upper left) Vertical velocity (omega) anomalies, contour interval is 0.25 hPa s^{-1} x 10^{-5}, negative values (blue) represent anomalous upward motion, DJF, early twentieth-century solar-residual; (b, upper right) same as (a) except for the late twentieth-century GHG + sulfates; (c, lower left) same as (a) except for JJA; (d, lower right) same as (b) except for JJA.

low cloud amount for all five areas in the late century (Fig. 17-5d), with all but two exceeding one intra-ensemble standard deviation. This increase in low clouds then contributes to a decrease in net solar in all five areas in Figure 17-5a for the late century. Four of these meet or exceed one intra-ensemble standard deviation. The end result is a weak and inconsistent response of latent heat flux for most areas in Figure 17-5b in the area averages (though spatially localized increases of latent heat removed from the surface contribute to the increases in low clouds [not shown]).

The early-century solar-residual JJA precipitation values for the West African and South Asian monsoons in Figures 17-2 and 17-3 are reflected in zonal mean changes in vertical velocity (here represented by omega) in Figure 17-6c, compared to the late-century GHG + sulfate changes in Figure 17-6d. Zonal mean precipitation (not shown) indicates early-century solar-residual increases of greater than +0.25 mm d^{-1} and nearly +0.2 mm d^{-1} around 10°S and 20°N as reflected in the geographical differences in Figures 17-2 and 17-3. These precipitation changes are associated with enhanced vertical velocity at those latitudes of about –2 hPa s^{-1} x 10^{-5} and –1 hPa s^{-1} x 10^{-5}, respectively (see Fig. 17-6c). The largest zonal mean increase in tropical precipitation in the late-century GHG + sulfate ensemble is less than +0.15 mm d^{-1} near 10°S (not shown). However, in both early- and late-century ensembles, the increases of precipitation in the tropics are associated with decreases around 30°S of about –0.1 mm d^{-1} due to the resulting intensified Hadley circulation and increased subsidence of over +2 hPa s^{-1} x 10^{-5} for early-century solar-residual and +1 hPa s^{-1}x 10^{-5} for late-century GHG + sulfates (Figs. 17-6c and d). Meanwhile, for DJF, the enhanced precipitation in the tropical convergence zones in Figure 17-4 is associated with negative omega differences near 25°S and 5°S for early-century solar-residual in Figure 17-6a, similar to late-century GHG + sulfates in Figure 17-6b. Due to the more marked spatial differentiation in the precipitation differences in Figure 17-4a for early-century solar-residual, the zonal means do not reflect these spatial differences. Instead, an intensification of the Walker and Hadley circulations, noted in the geographic difference plots for precipitation in Figure 17-4a, is indicated and a zonal mean representation does not actually show this important aspect of the solar versus GHG + sulfate response.

To demonstrate how the solar forcing coupled feedbacks can affect upper-ocean temperatures that contribute to changes in upper-ocean heat content, which then can provide SST "memory" to sustain the precipitation anomalies, Figure 17-7 shows quantities that contribute to the coupled feedbacks among ocean and atmosphere for DJF, early-century solar-residual. At the top is a cross section of omega anomalies averaged from 10°S to 30°S; moisture convergence anomalies in the lowest model layer are on the horizontal plot along with surface wind anomalies; and the section at the bottom shows upper-ocean temperature anomalies at 15°S from the surface to 275 m. Blue indicates enhanced upward vertical motion in the omega section at the top; red in the horizontal plot indicates anomalous positive surface moisture convergence at the surface and corresponds closely to the positive precipitation anomalies in Figure 17-4; vectors are surface wind anomalies with the scaling vector at the upper right; and red in the lower section indicates positive upper-ocean temperature anomalies.

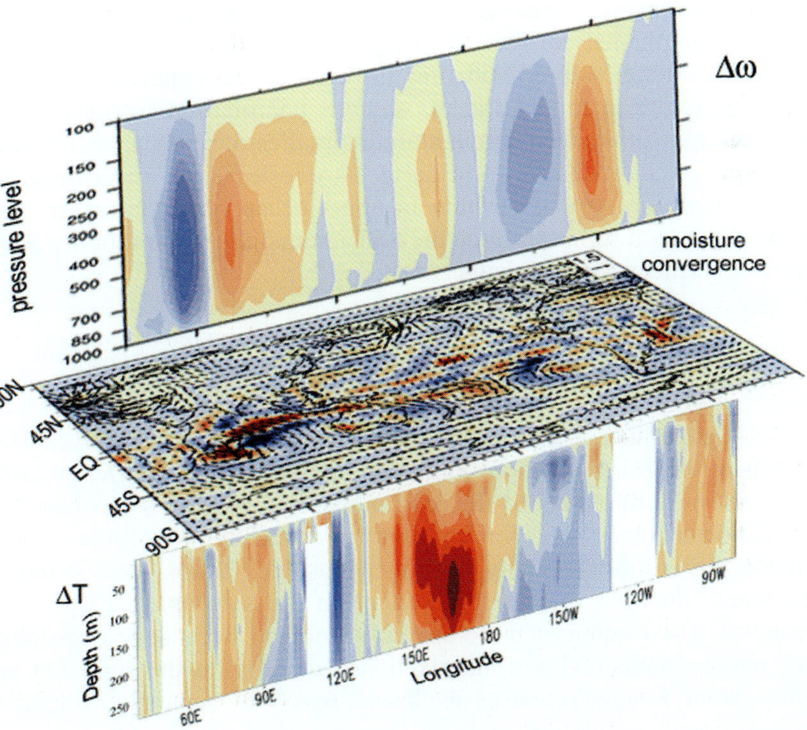

Figure 17-7. Differences for early-century solar-residual; top vertical panel is omega (indicative of vertical velocity) averaged from 10°S to 30°S with a contour interval of 0.25 hPa s^{-1} x 10^{-5}, negative values (blue) represent anomalous upward motion; middle horizontal panel is moisture convergence at the surface, with a contour interval of 0.1 g kg^{-1} s^{-1} x 10^{-5}, orange and red indicate anomalous positive moisture convergence, vectors are surface wind anomalies, scaling arrow on figure; bottom vertical panel is ocean temperature anomalies (C) at 15°S, contour interval 0.1°C, orange and red indicate positive anomalies.

Corresponding to the precipitation anomalies in Figure 17-4 and the surface moisture convergence anomalies from the horizontal plot at the center, there is enhanced upward vertical motion (negative omega anomalies) in the Indian Ocean near 60°E with values greater than –10 hPa s^{-1} x 10^{-5}, in the western Pacific between the date line and around 130°W with differences around –5 hPa s^{-1} x 10^{-5}, and in the eastern Atlantic east of 80°W with differences of roughly –2 hPa s^{-1} x 10^{-5}. There are corresponding areas of anomalous descending motion (the reds and oranges in the omega plot at the top) of comparable magnitude but opposite sign in the eastern Indian Ocean, far western Pacific, and the Pacific east of the date line. Surface wind anomalies roughly converge into areas of anomalous surface moisture convergence as expected. However, of greatest interest in this plot are the upper-ocean temperature anomalies. The positive SST anomalies under the areas of strengthened convection in Figure 17-4 are seen to be associated with positive temperature anomalies throughout the upper ocean, particularly in the western Indian Ocean, western Pacific, and western Atlantic, with maximum subsurface temperature anomalies of up to almost a degree. Thus the coupled feedbacks at the surface are reflected by upper-ocean temperature anomalies that contribute to upper-ocean heat content anomalies that persist, and in turn, reinforce the SST anomalies on the roughly 50-year time scale of the early twentieth century.

6. CONCLUSIONS

Regionally coupled feedbacks during DJF (also working in JJA to supplement the enhanced monsoon circulations) produce a different climate system response to enhanced IR from increased GHGs versus enhanced solar radiation from the solar forcing runs. The fundamental difference is that solar forcing is more spatially heterogeneous (i.e., acting most strongly in areas where sunlight reaches the surface in the subtropical highs), while greenhouse gas forcing is more spatially uniform because the greenhouse gases are assumed to be well mixed horizontally in the model. Consequently, solar forcing produces feedbacks involving temperature gradient–driven regional circulation regimes that can alter clouds. Over relatively cloud-free oceanic regions in the subtropics in areas of low-level moisture divergence (the moisture collection areas for the ocean precipitation convergence zones), the enhanced solar heating produces greater evaporation. This increased moisture then converges into the precipitation zones, intensifying the upward vertical motions of the regional Hadley and Walker circulations. The subsidence associated with this enhanced regional vertical motion results in fewer low clouds over the subtropical ocean regions, allowing even more solar energy to reach the surface, and so on.

Since the greenhouse gases are more spatially uniform, such regional circulation feedbacks are not as strong. SST increases are not as spatially differentiated, SST gradient-driven circulations do not set up, and the oceanic convergence zone precipitation regimes are not particularly intensified compared to the solar-forcing cases. The response for GHG + sulfates for the early and late century are proportional to the strength of the forcing, with a larger response for the greater late-century forcing.

This response is not the case for the early-century solar-only versus solar-residual. Though not shown here, the latter consistently shows a larger and more significant response, even though the magnitude of the solar forcing is the same. This result suggests a nonlinear interaction involving solar and GHG + sulfates forcing for the early-century period, and can be traced to the increase of SST of +0.2°C to +0.3°C in the regions of low-level moisture divergence produced by the GHG + sulfates forcing. Increases of similar magnitude in the tropical Atlantic, Indian, and Pacific Oceans during JJA contribute to intensification of the monsoon regimes (and western Pacific precipitation maximum) through increasing the base-state SSTs and low-level moisture source. When the solar forcing is added to these base-state SST changes from the increased GHGs, the regional feedbacks are more intense and produce a greater response in early-century solar-residual compared to solar-only.

Note that there is a much greater base-state warming in the late twentieth century (Fig. 17-1b), but the low-frequency changes of solar forcing over that time period (Fig. 17-1a) are small, with a dominant contribution from mainly the 11-year solar cycle. This aspect of the relationship between base-state surface temperature and climate system response to the 11-year solar cycle is currently under investigation.

Several different solar forcing reconstructions are currently available, and we have shown results from only one of those products. However, since the present study is a forcing/response analysis, it does not depend critically on the details of the time series of solar forcing. Those details are more important for detection/attribution studies. We qualitatively point to consistency with low-frequency changes in an observed monsoon index over the twentieth century, but that is only a qualitative result and far from a detection/attribution result. The key point we make is, given a certain amount of solar forcing acting over a given time period, the climate system exhibits a certain response that involves the regional tropical precipitation regimes with implications for the Hadley circulation.

The very nature of the low-frequency character of the solar forcing evident in Figure 17-1a recently has been called into question (e.g., Foukal and Milano 2001; Lean et al. 2002). The uncertainty involved with the solar reconstructions is beyond the scope of this chapter. But the present study does show that in a coupled model, a certain amount of solar forcing applied in combination with other forcings can produce a response in the tropics that is different from that to GHGs.

The implications from this study for paleoclimate are that zonally averaged tropical precipitation indicators may not be indicative of the complicated regional coupled processes that can occur to change the response of the tropical precipitation regimes. Thus, a zonally averaged view of the Hadley circulation and its changes must also take into account the east-west, or Walker, component. The regional precipitation convergence zones are connected in three dimensions with dynamical coupled processes that can affect tropical precipitation.

What is apparent from the model simulations is that greater solar forcing produces intensified climatological precipitation maxima, while stronger descent, as part of the Hadley and Walker circulations, produces greater drying. Thus, for greater solar forcing, to a first order in the tropics, climatologically wet areas get wetter, and dry areas get drier. Therefore, each region must be examined in that context when one interprets paleoclimate indicators of precipitation before a general and possibly erroneous conclusion concerning "Hadley circulation" changes is reached.

7. ACKNOWLEDGMENTS

The authors thank Harry van Loon for invaluable discussions and comments, and Henry Diaz and Ray Bradley for organizing a stimulating and interesting conference from which this chapter emerged. A portion of this study was supported by the Office of Biological and Environmental Research, U.S. Department of Energy, as part of its Climate Change Prediction Program, and by the National Science Foundation. The National Center for Atmospheric Research is sponsored by the National Science Foundation.

8. REFERENCES

Balachandran, N.K., D. Rind, P. Lonergan, and D.R. Shindell. 1999. Effects of solar cycle variability on the lower stratosphere and the troposphere. *Journal of Geophysical Research* 104: 27321–27339.

Bonan, G.B. 1998. The land surface climatology of the NCAR land surface model (LSM 1.0) coupled to the NCAR Community Climate Model (CCM3). *Journal of Climate* 11: 1307–1326.

Cubasch, U., G.A. Meehl, G.J. Boer, R.J. Stouffer, M. Dix, A. Noda, C.A. Senior, S. Raper, and K.S. Yap. 2001. Projections of future climate change. In, Houghton, J.T., Y. Ding, D.J. Griggs, M. Noguer, P. van der Linden, X. Dai, K. Maskell, and C.I. Johnson (eds.). *Climate Change 2001: The Scientific Basis.* Contribution of Working Group I to the Third Assessment Report of the Intergovernmental Panel on Climate Change. Cambridge, UK: Cambridge University Press, pp. 525–582.

Cubasch, U., R. Voss, G.C. Hegerl, J. Waszkewitz, and T.J. Crowley. 1997. Simulation of the influence of solar radiation variations on the global climate with an ocean-atmosphere general circulation model. *Climate Dynamics* 13: 757–767.

Dai, A., T.M.L. Wigley, B.A. Boville, J.T. Kiehl, and L.E. Buja. 2001a. Climates of the 20th and 21st centuries simulated by the NCAR Climate System Model. *Journal of Climate* 14: 485–519 .

Dai, A., G.A. Meehl, W.M. Washington, and T.M.L. Wigley. 2001b. Ensemble simulations of 21st century climate changes: Business as usual vs. CO_2 stabilization. *Bulletin of the American Meteorological Society* 82: 2377–2388.

Dai, A., T.M.L. Wigley, G.A. Meehl, and W.M. Washington, 2001c. Effects of stabilizing atmospheric CO_2 on global climate in the next two centuries. *Geophysical Research Letters* 28: 4511–4514.

Delworth, T.L., and T.R. Knutson. 2000. Simulation of the early 20th century global warming. *Science* 287: 2246–2250.

Folland, C.K., T.N. Palmer, and D.E. Parker. 1986. Sahel rainfall and worldwide sea temperature. *Nature* 320: 602–607.

Folland, C.K., T.R. Karl, J.R. Christy, R.A. Clarke, G.V. Gruza, J. Jouzel, M.E. Mann, J. Oerlemans, M.J. Salinger, and S.-W. Wang. 2001. Observed climate variability and change. In, Houghton, J.T., Y. Ding, D.J. Griggs, M. Noguer, P. van der Linden, X. Dai, K. Maskell, and C.I. Johnson (eds.).*Climate Change 2001: The Scientific Basis.* Contribution of Working Group I to the Third Assessment Report of the Intergovernmental Panel on Climate Change. Cambridge, UK: Cambridge University Press, pp. 99–181.

Foukal, P., and L. Milano. 2001. A measurement of the quiet network contribution to solar irradiance variation. *Geophysical Research Letters* 28: 883–886.

Haigh, J.D. 1996. The impact of solar variability on climate. *Science* 272: 981–984.

Hoyt, D.V., and K.H. Schatten. 1993. A discussion of plausible solar irradiance variations, 1700–1992. *Journal of Geophysical Research* 98: 18895–18906.

Jones, P.D., M. New, D.E. Parker, S. Martin, and I.G. Rigor. 1999. Surface air temperature and its changes over the past 150 years. *Reviews in Geophysics* 37: 173–199.

Joussaume, S., and 35 co-authors. 1999. Monsoon changes for 6000 years ago: Results of 18 simulations from the Paleoclimate Modeling Intercomparison Project (PMIP). *Geophysical Research Letters* 26: 859–862.

Kiehl, J.T., J.J. Hack, G. Bonan, B. Boville, D. Williamson, and P. Rasch. 1998. The National Center for Atmospheric Research Community Climate Model (CCM3). *Journal of Climate* 11: 1131–1149.

Kutzbach, J., R. Gallimore, S. Harrison, P. Behling, R. Selin, and F. Laarif. 1998. Climate and biome simulations for the past 21,000 years. *Quaternary Science Reviews* 17: 473–506.

Labitzke, K., and H. van Loon. 1995. Connection between the troposphere and stratosphere on a decadal scale. *Tellus* 47A: 275–286.

Lean, J., J. Beer, and R. Bradley. 1995. Reconstruction of solar irradiance since 1610: Implications for climate change. *Geophysical Research Letters* 22: 3195–3198.

Lean, J., Y.-M. Wang, and N.R. Sheeley, Jr. 2002. The effect of increasing solar activity on the Sun's total and open magnetic flux during multiple cycles: Implications for solar forcing of climate. *Geophysical Research Letters* 29(24): 2224, doi: 10.1029/2002GL015880.

Meehl, G.A. 1997. The south Asian monsoon and the tropospheric biennial oscillation. *Journal of Climate* 10: 1921–1943.

Meehl, G.A., and J.M. Arblaster. 2002a. The tropospheric biennial oscillation and Asian-Australian monsoon rainfall. *Journal of Climate* 15: 722–744.

Meehl, G.A., and J.M. Arblaster. 2002b. GCM sensitivity experiments for the Indian monsoon and tropospheric biennial oscillation transition conditions. *Journal of Climate* 15: 923–944.

Meehl, G.A., W.D. Collins, B. Boville, J.T. Kiehl, T.M.L Wigley, and J.M. Arblaster. 2000. Response of the NCAR Climate System Model to increased CO_2 and the role of physical processes. *Journal of Climate* 13: 1879–1898.

Meehl, G.A., P. Gent, J.M. Arblaster, B. Otto-Bliesner, E. Brady, and A. Craig. 2001. Factors that affect amplitude of El Niño in global coupled climate models. *Climate Dynamics* 17: 515–526.

Meehl, G.A., W.M. Washington, T.M.L. Wigley, J.M. Arblaster, and A. Dai. 2003. Solar and greenhouse gas forcing and climate response in the 20th century. *Journal of Climate* 16: 426–444.

Parthasarathy, B., K. Rupa Kumar, and A.A. Munot. 1991. Evidence of secular variations in Indian monsoon rainfall-circulation relationships. *Journal of Climate* 4: 927–938.

Podgorny, I.A., W. Conant, V. Ramanathan, and S.K. Satheesh. 2000. Aerosol modulation of atmospheric and surface solar heating over the tropical Indian Ocean. *Tellus* 52B: 947–958.

Rind, D., and N.K. Balachandran. 1995. Modelling the effects of UV variability and the QBO on the troposphere-stratosphere system. Part II: The troposphere. *Journal of Climate* 8: 2080–2095.

Rind, D., J. Lean, and R. Healy. 1999. Simulated time-dependent climate response to solar radiative forcing since 1600. *Journal of Geophysical Research* 104: 1973–1990.

Shindell, D., D. Rind, N. Balachandran, J.U. Lean, and P. Lonergan. 1999. Solar cycle variability, ozone, and climate. *Science* 284: 305–308.

Shindell, D., G.A. Schmidt, R.L. Miller, and D. Rind. 2001. Northern hemisphere winter climate response to greenhouse gas, ozone, solar, and volcanic forcing. *Journal of Geophysical Research* 106: 7193–7210.

Stott, P.A., S.F.B. Tett, G.S. Jones, M.R. Allen, J.F.B. Mitchell, and G.J. Jenkins. 2000. External control of 20th century temperature by natural and anthropogenic forcings. *Science* 290: 2133–2137.

Stott, P.A., S.F.B. Tett, G.S. Jones, M.R. Allen, W.J. Ingram, and J.F.B. Mitchell. 2001. Attribution of twentieth century temperature change to natural and anthropogenic causes. *Climate Dynamics* 17: 1–21.

van Loon, H., and K. Labitzke. 1994. The 10–12-year atmospheric oscillation. *Meteorologische Zeitschrift* N.F. 3: 259–266.

van Loon, H., and D.J. Shea. 2000. The global 11-year solar signal in July–August. *Geophysical Research Letters* 27: 2965–2968.

Washington, W.M., J.W. Weatherly, G.A. Meehl, A.J. Semtner, Jr., T.W. Bettge, A.P. Craig, W.G. Strand, Jr., J.M. Arblaster, V.B. Wayland, R. James, and Y. Zhang. 2000. Parallel climate model (PCM) control and transient simulations. *Climate Dynamics* 16: 755–774.